The Green Six Sigma Handbook

This book is a hands-on, single-source reference of tools, techniques and processes integrating both Lean and Six Sigma. This comprehensive handbook provides up-to-date guidance on how to use these tools and processes in different settings, such as start-up companies and stalled projects, as well as established enterprises where the ongoing drive is to improve processes, profitability and long-term growth. It contains the "hard" Six Sigma approach as well as the flexible approach of FIT SIGMA, which is adaptable to manufacturing and service industries and also public sector organisations. You will also discover how climate change initiatives can be accelerated to sustainable outcomes by the holistic approach of Green Six Sigma.

The book is about what we can do now with leadership, training and teamwork in every sphere of our businesses. Lean, originally developed by Toyota, is a set of processes and tools aimed at minimising wastes. Six Sigma provides a set of data-driven techniques to minimise defects and improve processes. Integrating these two approaches provides a comprehensive and proven approach that can transform an organisation. To make change happen, we need both digital tools and analog approaches. We know that there has been a continuous push to generate newer approaches to operational excellence, such as Total Quality Management, Six Sigma, Lean Sigma, Lean Six Sigma and FIT SIGMA.

It is vital that we harness all our tools and resources to regenerate the economy after the Covid-19 pandemic and make climate change initiatives successful for the survival of our planet. Six Sigma and its hybrids (e.g., Lean Six Sigma) should also play a significant part. Over the last three decades, operational performance levels of both public sector and private sector organisations improved significantly and Lean Six Sigma has also acted as a powerful change agent. We urgently need an updated version of these tools and approaches.

The Green Six Sigma Handbook not only applies appropriate Lean and Six Sigma tools and approaches, fitness for the purpose, but it aims at sustainable changes. This goal of sustainability is a stable bridge between Lean Six Sigma and climate change initiatives. Hence, when the tools and approaches of Lean Six Sigma are focused and adapted primarily to climate change demands, we get Green Six Sigma.

The Green Six Sigma Handbook

A Complete Guide for Lean Six Sigma Practitioners and Managers

Ron Basu

A PRODUCTIVITY PRESS BOOK

First published 2023
by Routledge
605 Third Avenue, New York, NY 10158

and by Routledge
4 Park Square, Milton Park, Abingdon, Oxon, OX14 4RN

Routledge is an imprint of the Taylor & Francis Group, an informa business

ISBN: 978-1-03221-402-3 (hbk)
ISBN: 978-1-03221-401-6 (pbk)
ISBN: 978-1-00326-823-9 (ebk)

DOI: 10.4324/9781003268239

Typeset in Garamond
by Apex CoVantage, LLC

To my darling wife Moira—I am lucky
to be your husband

Contents

Foreword

Some Thoughts from the Younger Generation

Young voices often have not been heard loudly enough in the climate change debate, though ironically, we are the ones who will live long enough to feel the consequences of the decisions made now. Until the school striking movement led by Swedish activist Greta Thunberg, the opinions of students were not at the forefront of the debate. As young people in our teens and twenties, it is easy to feel discouraged and powerless by what we perceive is the complacency of many of the older generation. It can often be frustrating to witness the attitude of those who are in positions of power and have decision-making abilities that we lack. If you are CEO of an oil company, for example, you may make judgements based on what is best for your business and its profits. To our eyes, this is short-sighted and selfish, as by the time the quality of human life will be severely impaired by corporate actions, that generation will no longer be around to witness the impact of the choices they have made.

Young people often feel that previous generations have created the problem. Many have profited by the destruction of our earth, plundering its resources to mine oil and gas, farm palm oil, generate easy profits from fast fashion, or thoughtlessly cause devastation of the rainforests, specifically the Amazon, the 'green lungs' of the world on which we all depend. Our mother told us that, at age thirteen, she was taught in a geography lesson that an area the size of Wales was lost in the Brazilian rainforests every day. A generation later, this simple fact that resonated so strongly with her has an ironic potency given that the situation is now so much worse. It is absolutely imperative that every one of us, from schoolchild to business decision maker, take steps to do all we can to mitigate the risks of climate change. We must not resent the actions and effects of previous decades and older people, but work together constructively to find solutions and effect real change.

One of us, Georgina, was born in the summer of 2007, when the UK saw a heatwave followed by torrential summer rainfall. Within her lifetime, weather extremes and climate unpredictability have become normalized. As we write, a series of huge storms have battered Britain during the space of a single week with unprecedented wind speed leading to school closures, travel disruption and loss of life. The evidence that we should act now is overwhelming.

What can we do, and how can we reconcile our desire to enjoy life but live as sustainably and responsibly as possible? It is a question that as young women we have asked ourselves, and which we often understood intellectually but did not act upon. We have been just as guilty as others of our age of consumerism, in ordering online and loving the convenience of receiving our goods as we contribute to lots of cars and vans making these deliveries. We have indulged in fast fashion brands and bought toiletries all marketed at our demographic and have certainly enjoyed lots of beef burger fast-food meals. In the back of our minds of course we were aware that the deforestation that we loathe often occurs to manufacture the palm oil that is an ingredient of our beauty products, or that trees are lost to clear land to raise cattle for our beef consumption, but it is just so easy to make less sustainable choices when we all want to have fun and buy the products we covet.

So, as young people we are worried, angry, and even contradictory in our behaviour, but at least we are engaged. It is heartening to see that in the areas that we are interested in, fashion and beauty, companies are taking steps to make changes that will have positive impact. Many brands of toiletries produce refill packages and encourage reuse of their containers; jewellery firms are realising the value of recycling materials such as gold and sourcing stones ethically, and one of us, Harriet, has chosen the area of fast fashion to study for a school project and feels very strongly about the power of change that this industry can have by revising its working practices. Only one of us can drive at present, but our generation may see and embrace the widespread use of the electric car. Change must start with a mindset and follow through with positive and manageable actions that we can all embed into our daily lives. None of us are too small or unimportant to act.

Thank you for picking up this book. Our grandfather (Ron Basu) has devoted his working life to identifying how businesses may operate more efficiently, and his personal time to looking after and loving his children and grandchildren. In this volume, we see the two coalescing, as he truly believes that it is our generation that will benefit from any changes that we make now. We hope that you will learn how Green Six Sigma can make your business more sustainable and that you will put into practice his recommendations and ideas. We can all play our part, and we thank you for joining in his work to make the future better for everyone.

Catherine, Theresa, Harriet and Georgina Evans
(Aged 21, 19, 17 and 14)
Berkshire, February 2022

Preface

Background

When I wrote the book *FIT SIGMA* (2011) I included a chapter entitled 'FIT SIGMA in Green Thinking'. In that section I tried to share my concern about global warming as well as my optimism on green initiatives and the role that FIT SIGMA could play in this area. At this point, notwithstanding some consulting assignments, my involvement in the realm of climate change initiatives was purely reactive.

While enjoying a break sailing on a Princess Cruise liner a few years back I had the privilege of listening to a lecture by Mary Robinson, the former President of Ireland and also a UN High Commissioner for Human Rights. She spoke passionately/with compelling force for half an hour on how to avoid a climate change disaster. While listening to her words, the idea of writing a book to address these issues was conceived in my mind. Prior to hearing Mrs. Robinson's lecture, like so many television viewers I had also been influenced and affected by the *Blue Planet II* documentary on plastics pollution by David Attenborough and the powerful 2006 movie *An Inconvenient Truth* by Al Gore. Then on the eve of COP26 in Glasgow it seemed that nature was conspiring to show us that we could not ignore the issues on our own doorstep any longer, as there were a number of climate-related disasters in Western Europe, North America, China and South Asia in quick succession. I decided that it was now the time to write a book on climate change so that I could look into the eyes of my grandchildren and say that I had tried to do my bit.

I admit that I am not an environmental scientist, but I do have some expertise in the field of Six Sigma applications in achieving operations excellence within organisations. My research revealed that there were many publications on climate change out there, and also on Six Sigma and its hybrids (e.g. Lean Six Sigma). Crucially though, it seemed that there was a gap in the available literature—there was no bridge between the two. Thus the concept of 'Green Six Sigma', which is basically the adaptation of Lean Six Sigma for climate change initiatives, was developed.

I approached two publishers with my proposal to write a book on Green Six Sigma for climate change initiatives, hoping that I would get at least one positive response. I was pleasantly surprised when both publishers expressed interest. So the outcome was a book entitled *Green Six Sigma* and another book called *The Green Six Sigma Handbook*. Some degree of overlap of the contents between the two books is unavoidable. This handbook contains a greater level of detail, especially in the sections on tools, techniques and in the case studies of Green Six Sigma.

About This Book

There are twenty-four chapters in this volume. The first fifteen chapters are devoted to the concept of Green Six Sigma with its tools and techniques. The subsequent nine chapters deal with the causes of climate change and how we can mitigate and adapt to its consequences, assisted by Green Six Sigma. However, it must be understood that Green Six Sigma is not a silver bullet. It is a catalyst to accelerate climate change initiatives leading to sustainable processes and environmental standards.

Who Should Use This Book?

This book is aimed at a broad cross section of readership including the following:

- Functional managers, participants and practitioners in Six Sigma and operational excellence will find that this book will provide them with a comprehensive insight into the tools and techniques of sustainable improvement in a single package for climate change initiatives.
- Professional management and training consultants will also find that the comprehensive approach of tools and techniques offers an essential handbook for Six Sigma–related climate change assignments and seminars.

- Leaders of the global community with the responsibility for climate change initiatives will find this book informative to sponsor the training and applications of Green Six Sigma in both current and future projects regarding climate change.
- Senior executives, in both manufacturing and service industries, should discover that this book gives them a better understanding of basic tools and techniques and helps them to support climate change initiatives and sustain a strong competitive position.
- Universities, management schools, academies and research associations will find this book valuable in filling the visible gap in the literature covering the basics of operational excellence that is especially focused on climate change.

I anticipate that the readership will be global and will particularly cover North America, the UK, Continental Europe, Australia and the Asia-Pacific countries.

I have made an effort to furnish you with both simple and more complex concepts that are nonetheless, I hope, easy to understand. Finally, at the end of each chapter some simple guidelines named 'Green Tips' are included, which I hope will offer an invaluable and bite-sized practical resource.

Mahatma Gandhi once said, 'Be the change you wish to see in the world'. I hope that this book will in some way help you to bring about that shift in the field of climate change that we all so urgently need.

Ron Basu
Gerrards Cross, England
February 2022

Acknowledgements

Every effort has been made to credit the authors, publishers and websites for materials used in this book. I apologise if inadvertently any sources remain unacknowledged and if known I shall be pleased to credit them in the next edition.

My sincere thanks go to the staff of my publisher Taylor & Francis Group especially to Michael Sinocchi for getting this project off the ground.

Finally the project could not have been completed without the encouragement and help of my family, especially my wife Moira, sister-in-law Reena, son Robi and daughter Bonnie. Bonnie's contributions to edit draft chapters have been invaluable.

Ron Basu

About the Author

Ron Basu is director of RB Consultants and a Visiting Fellow at Henley Business School, England. He is also a visiting professor at SKEMA Business School, France. He specialises in operational excellence and supply chain management and has research interests in performance management and project management.

Previously he held senior management roles in blue-chip companies like GSK, Glaxo Wellcome and Unilever and led global initiatives and projects in Six Sigma, ERP/MRPII, supply chain re-engineering and total productive maintenance. Prior to this he worked as management consultant with A.T. Kearney.

He is the co-author of *Total Manufacturing Solutions*, *Quality Beyond Six Sigma*, *Total Operations Solutions* and *Total Supply Chain Management* and the author of the titles *Measuring e-Business Performance*, *Implementing Quality*, *Implementing Six Sigma and Lean*, *FIT SIGMA*, *Managing Project Supply Chains*, *Managing Quality in Projects*, *Managing Global Supply Chains*, *Managing Projects in Research and Development* and *Green Six Sigma*. He has authored a number of peer-reviewed papers in the operational excellence and project management fields.

After graduating in manufacturing engineering from UMIST, Manchester, he obtained an MSc in operational research from Strathclyde University, Glasgow. He also completed a PhD at Reading University. He is a Fellow of the Institution of Mechanical Engineers, the Institute of Business Consultancy and the Chartered Quality Institute. He is also the winner of the APM Project Management Award.

I

FOUNDATIONS OF SIX SIGMA AND LEAN

Chapter 1

Quality and Operational Excellence

We are what we repeatedly do. Excellence, then, is not an act but a habit.

—Aristotle

1.1 Introduction

The methodology of implementing a Green Six Sigma programme is rooted to the quality management and improvement programmes. Green Six Sigma programmes aim to achieve sustainable results. I shall explain Green Six Sigma in more detail later. These programmes can be varied. Such a programme is likely to have a different name or label, such as Total Quality Management (TQM), Six Sigma, Lean Six Sigma, kaizen, business process re-engineering (BPR) or operational excellence. Regardless of the methodology or name of the continuous improvement programmes, each organisation and programme team will certainly need to use a selection of tools and techniques in their implementation process. Most of these tools and techniques are simple to understand and can be used by a large population of the company. However, there are also some techniques that are more complex. These advanced techniques are used by specialists for specific problem-solving applications. It is vital that the tools and techniques are selected for the appropriate team and applied correctly to the appropriate process. Therefore,

DOI: 10.4324/9781003268239-2

the fundamental requirements for achieving repeatable and reliable results by these tools and techniques is a clear understanding, both of the tools and techniques themselves and the process by which they could be applied.

The objective of this chapter is to introduce to the reader the following areas:

- The tools and techniques
- The concept of quality and operational excellence

1.2 Tools and Techniques

In general, tools and techniques can be broadly defined as the practical methods and skills applied to specific activities to enable improvement. A specific tool has a defined role, and a technique may comprise the application of several such tools.

Dale and McQuater (1998) have suggested the following definition of tools and techniques.

1.2.1 Tools and Techniques

A single tool may be described as a device that has a clear role and defined application. It is often narrow in its focus and can be and is usually used on its own. Examples of tools are

- Cause and effect diagram
- Pareto analysis
- Relationship diagram
- Control chart
- Histogram
- Flow chart

A technique, on the other hand, has a wider application than a tool. There is also a need for a greater intellectual thought process and more skill, knowledge, understanding and training in order to use a technique effectively. A technique may even be viewed as a collection of tools. For example, statistical process control employs a variety of the tools, such as graphs, charts, histograms and capability studies, as well as other statistical methods, all of which are necessary for the effective deployment of a technique. The use of a technique may cause the necessity for a tool to be identified.

Examples of techniques are

- Statistical process control
- Benchmarking
- Quality function deployment
- Failure mode and effects analysis
- Design of experiments
- Self-assessment

Source: Dale and McQuater (1998).

1.3 What Is Quality?

If you were to ask quality experts to define 'quality', it is likely that you would receive many different answers, although you would elicit a set of common or comparable themes, such as 'fitness for purpose', 'right first time', 'what the customer wants', 'conformance to standards', 'value for money', 'right thing at the right time' and so on. A basic reason for differing perceptions of quality is arguably that each person has their own set of individual preferences.

A simple story from the Indian fables may illustrate this point. Four blind men went to visit an elephant, and each felt the creature to form an impression of it. On their way back, they discussed the experience. The first man said, 'The elephant is just like a swinging fan'. The second blind man replied, 'No, I disagree. I think that it is more like a pillar'. Then the third person protested, 'You're both wrong. I would describe it as being more like a huge, thick whip'. He added, 'I am absolutely sure, it's a long and very flexible object'. It is clear from their very different impressions and viewpoints that the three blind men were influenced by their varying attitudes and the way in which they touched the elephant in order to arrive at such contrary perceptions about the same animal. However, it was only by sharing their ideas that they realised that they had visualised one concept in a variety of ways.

There are many different definitions and dimensions of quality to be found in books and academic literature. We present three of these definitions selected from published literature and propose a three-dimensional definition of quality to reflect the appropriate application of tools and techniques for Green Six Sigma.

Table 1.1 Gravin's Product Quality Dimensions

Performance
Features
Reliability
Conformance
Durability
Serviceability
Aesthetics
Perceived quality

One of the most respected definitions of quality is given by the eight quality dimensions (see Table 1.1) developed by David Gravin of the Harvard Business School (1984).

Performance refers to the efficiency (e.g. return on investment) with which the product achieves its intended purpose.
Features are attributes that supplement the product's basic performance (e.g. tinted glass windows in a car).
Reliability refers to the capability of the product to perform consistently over its life cycle.
Conformance refers to meeting the specifications of the product, usually defined by numeric values.
Durability is the degree to which a product withstands stress without failure.
Serviceability is used to denote the ease of repair.
Aesthetics are sensory characteristics such as a look, sound, taste and smell.
Perceived quality is based upon customer opinion.

These dimensions of quality are not mutually exclusive, although they relate primarily to the quality of the product. Neither are they exhaustive. Service quality is perhaps even more difficult to define than product quality. A set of service quality dimensions (see Table 1.2) that is widely cited has been compiled by Parasuraman et al. (1988).

Tangibles are the physical appearance of the service facility and people.
Service reliability deals with the ability of the service provider to perform dependably.

Table 1.2 Parasuraman et al.'s Service Quality Dimensions

Tangibles
Service reliability
Responsiveness
Assurance
Empathy
Availability
Timeliness
Professionalism
Completeness
Pleasantness

Responsiveness is the willingness of the service provider to be prompt in delivering the service.

Assurance relates to the ability of the service provider to inspire trust and confidence.

Empathy refers to the ability of the service provider to demonstrate care and individual attention to the customer.

Availability is the ability to provide service at the right time and place.

Professionalism encompasses the impartial and ethical characteristics of the service provider.

Timeliness refers to the delivery of service within the agreed lead time.

Completeness addresses the delivery of the order in full.

Pleasantness simply means the good manners and politeness of the service provider.

Noriaki Kano (1996) demonstrates in the well-known Kano model of customer satisfaction that there are three attributes to quality (viz. basic needs, performance needs and excitement needs) and that to be competitive, products and services must flawlessly execute all three attributes of quality.

Our third authoritative definition of quality is taken from Ray Wild's *Operations Management* (2002), page 644 (see Table 1.3).

The list of quality dimensions by both Garvin and Parasuraman et al. are widely cited and respected. However, one problem with multiple dimensions is that of communication, and if allowed time, the reader could probably identify additional dimensions. It is not easy to devise a strategic plan on quality based on specific dimensions that could be interpreted differently by different departments. Wild's definition of design/process quality however provides a broad framework to develop a company-specific quality strategy.

Table 1.3 Wild's Definition of Quality

The quality of a product or service is the degree to which it satisfies customer requirements. It is influenced by the following: Design quality: the degree to which the *specification* of the product or service satisfies customers' requirements. Process quality: The degree to which the product or service, which is made available to the customer, *conforms* to specification.

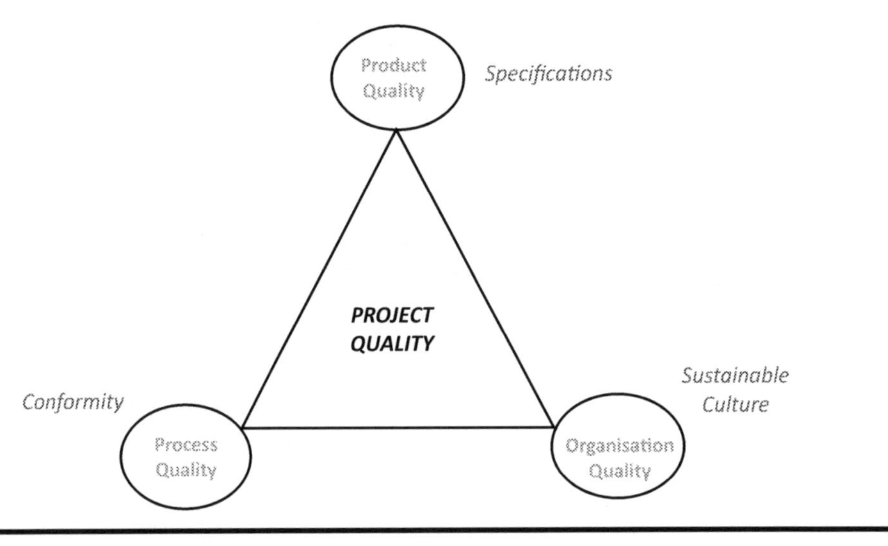

Figure 1.1 Three dimensions of quality.

Nonetheless, one important dimension of quality is not clearly visible in the previously mentioned models: the quality of the organisation. This is a fundamental cornerstone of the quality of a holistic process and an essential requirement of an approved quality assessment scheme such as EFQM (European Foundation of Quality Management).

Our three-dimensional model of quality is shown in diagrammatic form in Figure 1.1.

When an organisation develops and defines its quality strategy, it is important to share a common definition of quality and each department within a company can work towards a common objective. The product quality should contain defined attributes of both numeric specifications and perceived dimensions. Process quality relates to the appropriateness and effectiveness of the tools and techniques applied. The process quality, whether it relates to manufacturing or service operations, should also

Table 1.4 Basu's Organisation Quality Dimensions

Top Management Leadership and Commitment
Defined Roles and Responsibilities
Sales and Operations Planning
Single Set of Numbers
Stakeholder Management
Performance Management
Knowledge Management and Continuous Learning
Communication and Teamwork Culture
Self-Assessment

contain some defined criteria of acceptable service level so that the conformity of the output can be validated against these criteria. Perhaps the most important determinant of how we perceive sustainable quality is the functional and holistic role we fulfil within the organisation. It is only when an organisation begins to change its approach to a holistic culture emphasising a single set of numbers based on transparent measurement with senior management commitment that the 'organisation quality' germinates. I have compiled (see Table 1.4) a set of key organisation quality dimensions.

Top Management Leadership and Commitment means that organisational quality cannot exist without the total commitment of the competent top executive team.

Defined Roles and Responsibility means each member of the organisation has a written job description with a defined role and responsibility.

Sales and Operations Planning is a monthly senior management review process to align strategic objectives with operation tasks.

Single Set of Numbers provides the common business data for all functions in the company.

Stakeholder Management is the process of maintaining good relationships with the people who have the most impact on the operation.

Performance Management includes the selection, measurement, monitoring and application of key Performance Indicators.

Knowledge Management and Continuous Learning includes education, training and development of employees, sharing of best practice and communication media.

Communication and Teamwork Culture requires that teamwork should be practised in cross-functional teams with transparency to encourage a borderless organisation.

Self-Assessment enables a regular health check of all aspects of the organisation against a checklist or accepted assessment process such as EFQM.

It is evident all the dimensions of organisation quality are related to people. Hence the three dimensions of quality are also product quality, process quality and people quality.

1.4 Hierarchy of Quality

Our hierarchy of quality approximately follows the evolution of quality management from simple inspection to full quality management system as shown in Figure 1.2.

Quality by Inspection is an expensive method of achieving a basic level of quality. It requires the employment of people to check on the operation. Inspection and supervision do not add value to a product, they merely add to the cost. However the inspection of results with specified requirements are often necessary to ensure regulatory or approved standards.

Quality Control is the next stage above quality inspection. The control process is based on the statistical method that includes the phases of analysis, relation and generalisation. Activities relating to quality control include

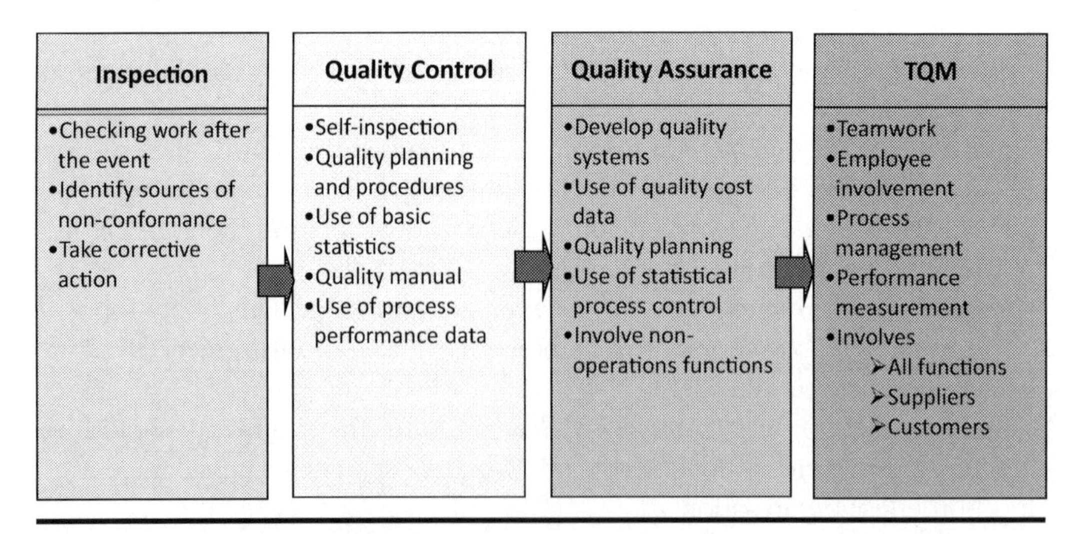

Figure 1.2 Hierarchy of quality.

- Monitoring Process Performance
- Acceptance Sampling
- Designing and Maintaining Control Charts

Quality Assurance (QA) relates to activities needed to provide adequate confidence that an entity will fulfil requirements for quality. The first two stages, inspection and control, are based on a detection approach and relate to 'after the event', while quality assurance is aimed at preventing mistakes. Quality assurance activities include

- Approved Supplier Scheme
- Operator Training
- Process Improvement

Total Quality Management (TQM) has been defined in ISO 8402:1995, as the 'Management approach of an organisation, centred on quality, based on the participation of all its members and aiming at long-term success through customer satisfaction, and benefits to all members of the organisation and society'. The holistic view of TQM supports the idea that quality is the responsibility of all employees and not just quality managers. TQM encompasses all three dimensions of quality as shown in Figure 1.1, with particular emphasis on organisational quality.

1.5 Cost of Quality

One frequently asked question about quality management is, 'Can you quantify the benefits?' The answer to this question is yes we can, albeit approximately. The benefits are quantified in terms of not having the right quality or the cost of poor quality. The concept of 'the cost of quality' is not new. In fact Juran (1951) first discussed the cost of quality analysis as far back as 1951. However, Feigenbaum (1956) should be credited with the definition of the cost of quality when he identified the four cost categories in 1956. These can be classified as Prevention Costs, Appraisal Costs, Internal Failure Costs and External Failure Costs. Both concept and categories have been followed basically in the same format ever since. Prevention Costs and Appraisal Costs are often defined using one of three terms: Cost of Control, Cost of Conformance or Cost of Good Quality. Regardless of the label used, this refers to the outlay of setting up and managing a quality management team with clearly defined processes. Similarly, Internal Failure Costs and External

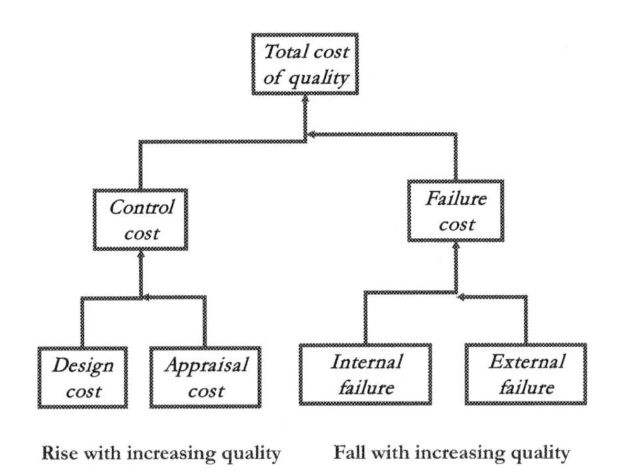

Figure 1.3 Cost of quality.

Failure Costs are also combined to be known as the Cost of Failure, the Cost of Non-conformance or the Cost of Poor Quality. These are the expenses of defects and reworks arising from poor quality management.

As shown in Figure 1.3, the cost of quality is derived from non-value-added activities or wastes in the process and is made up of costs associated with

- Prevention
- Appraisal
- External failure
- Internal failure

The emphasis will be on prevention rather than appraisal or detection, and thus the cost of supervision and inspection will go down. Prevention will go up because of training and action-orientated efforts. But the real benefits will be gained by a significant reduction in failures—both internal (e.g. scrap, rework, downtime) and external (handling of complaints, servicing cost, loss of goodwill). The total cost of quality will reduce over time as shown in Figure 1.4.

The concept of the law of diminished returns argues that there is a point at which investment in quality improvement will become uneconomical. However with the application of Total Quality Management (TQM) and Six Sigma processes, the traditional view outlined earlier becomes challenged. Increased quality is not achieved by more inspectors but instead by a culture of quality assurance underpinned by initial training. The ethic of continuous

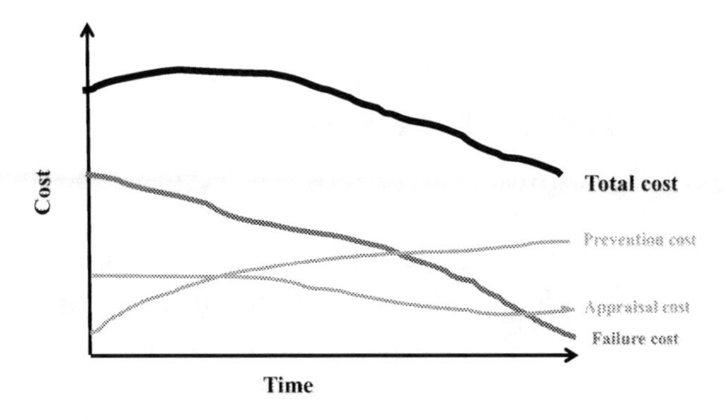

Figure 1.4 Total cost of quality.

improvement should aim at the appropriate or optimum level of quality and then sustain it. This is where Green Six Sigma with additional tools for sustainability can help.

1.6 Quality Movement

Total Quality Management (TQM) has its origins in Japan. In the 1960s Japan went through a quality revolution. Prior to this, 'Made in Japan' meant cheap or shoddy consumer goods. The approach used in Japan in the 1950s and 1960s to improve quality standards was to employ consultants or experts from America. The four best-known quality management experts (or 'Quality Gurus') are all American: Deming, Juran, Feigenbaum and Crosby. Because of the considerable influence of these four men in the development of TQM and then Six Sigma, their contributions are briefly explored later. This is followed by the adaptation of their contributions in Japan to enhance the quality movement.

1.7 W Edwards Deming

The most famous of these consultants was W Edwards Deming. Deming's philosophy was to establish the best current practices within an organisation, establish the best practice as standard procedure, and train the workers in the best way. In this manner, everyone would be using the same best way. Deming's approach was to involve everyone in the organisation and to win them over. He believed that quality was everyone's business. Deming said that to find the best way meant getting the facts, collecting data, setting standard procedures, measuring results and getting prompt and accurate feedback of results so as to eliminate variations to the standard. He saw this as a continuous cycle. He claimed that cultivating the know-how of employees was 98% of the quality challenge—as Gabor (2000) says, Deming has been criticised for hyperbole. No section on Deming is complete without reference to his famous 14 points of quality. Comments in italics are our notes and not direct quotes of Deming.

1. Create consistency of purpose toward improvement of product and service.
2. Adopt the new philosophy; *management has to learn its responsibilities and to take leadership. It is difficult for management to accept that 90% of problems lie with management and the process.*
3. Cease dependence on inspection to achieve quality *supervision and supervisors' wages do* ***not*** *add value, they are an extra cost, far better if staff take responsibility and supervise themselves. Deming also added that if quality is built into the design or process, then inspection will not be necessary.*
4. End the practice of awarding business on the basis of the price tag. *The cheaper the price is, the higher the number of failures. Move to dedicated suppliers and value reliability, delivery on time and quality.*
5. Improve constantly and forever the system of production and service. *This is an extension of the Japanese philosophy of kaizen whereby not a day should go by without some incremental improvement within the organisation.*
6. Institute training on the job. *Become a learning organisation with a willingness to share knowledge.*
7. Institute leadership. Everyone, *at all levels, especially supervisors should be team leaders not disciplinarians. Everyone should be encouraged to develop self-leadership. Quality is too important to be left to management.*
8. Drive out fear. *Encourage people to admit mistakes. The aim is to fix not to punish. However, it is expected that people will not go on making the same mistakes.*
9. Break down barriers between departments, *eliminate suspicion between departments. There need to be clear objectives with everyone striving to work for the common good.*

10. Eliminate slogans, exhortations and targets for the workforce. *There is no use asking for zero defects if the process or the product design is not perfect. Ten per cent across-the-board cost reduction demands are poor for morale if they are not possible.*
11. Eliminate work standards—quotas—on the factory floor. *For example, 100 pieces per hour with a bonus for 110 will result in 110 pieces, but not necessarily quality products. The focus will be on output numbers rather than quality. If the worker is encouraged to consider quality, 95 high-quality pieces per hour will be worth more than 110 if 15 (of the 110) are subsequently rejected or returned by the customer.*
12. (a) Remove barriers that rob the worker of the right to pride of workmanship. *Give them the right tools, right materials, right processes and comfortable working conditions and treat them with respect.*
    (b) Remove barriers that rob people in management and in engineering of their right to pride in craftsmanship. *This includes appraisal systems that reward on bottom-line results and keeping expense budgets down, and ignore customer satisfaction. If cost is the only driver, then training, maintenance, customer service etc. will suffer.*
13. Institute a program of education and self-improvement. *Encourage staff to seek higher educational qualifications. Become a knowledge-based organisation.*
14. Put everybody in the company to work to accomplish the transformation.
    Change of culture is difficult to achieve. Deming saw that everyone has to be involved in transforming the culture of an organisation.

Deming was not the only guru of quality used by the Japanese.

1.8 Joseph M Juran

Joseph M Juran was also associated with Japan's emergence as the benchmark for quality of products. Like Deming, Juran was an American statistician, and there are similarities between his work and that of Deming. Above all, both men highlight managerial responsibility for quality. Arguably Juran was the first guru to emphasise that quality was achieved by communication. The Juran trilogy for quality is planning, control and improvement (Juran, 1989). His approach includes an annual plan for quality improvement and cost reduction, and continuous education on quality. Juran's foundations are still valid and are embedded within Six Sigma and Lean Sigma and our FIT SIGMA philosophies. He argues, and few would disagree, that inspection at the end of the line, post-production, is too late to prevent errors. Juran says that quality monitoring needs to be performed during the production process to ensure that mistakes do not occur and that the system is operating effectively. Juran does this by examining the relationship between the process variables and the resultant product. Once these relationships have been determined by statistical experiment, the process variables can be monitored using statistical methods. Juran adds that the role of upper management is more than making policies; they have to show leadership through action—they have to walk the talk, not just give orders and set targets. Juran says quality is not free and that investment is needed (often substantial investment) in training, often including statistical analysis, at all levels of the organisation. Juran also believed in the use of quality circles. As he describes them, quality circles are small teams of staff with a common interest who are brought together to solve quality problems. Our constituents for a successful quality circle are discussed later in this chapter.

1.9 Armand V Feigenbaum

Feigenbaum is recognised for his work in raising quality awareness in the US. He was General Electric's worldwide chief of manufacturing operations for a decade until the late 1960s. The term Total Quality Management originated from his book *Total Quality Control* (1961). Feigenbaum states that total quality control has an organisation-wide impact that involves managerial and technical implementation of customer-orientated quality activities as a prime responsibility of general management and of the main line operations of marketing, engineering, production, industrial relations, finance and service as well as of the quality control function itself. He adds that a quality system is the agreed, company-wide operating work structure, documented in integrated technical and managerial procedures, for guiding the coordinated actions of the people, the machines and company-wide communication in the most practical ways with the focus on customer quality satisfaction.

Feigenbaum is one of the first writers to recognise that quality must be determined from the customer's perspective, and *not* the designer's (or the engineer's or the marketing department's concept of what quality is).

Feigenbaum also says that the best does not mean outright excellence, but means best for satisfying certain customer conditions. In other words, like FIT SIGMA, 'best' means sufficiently good to meet the circumstances. Feigenbaum, also like Deming and Juran, found that measurement is necessary. But, whereas Dening and Juran tended to measure production and outputs, Feigenbaum concentrated on measurement to evaluate if good service and product met the desired level of customer satisfaction.

According to Dale (1999), Feigenbaum's major contribution to quality was to recognise that the three major categories of cost are appraisal, prevention and cost of failure. According to Feigenbaum, the goal of quality improvement is to reduce the total cost of quality from the often quoted 25–30% of cost of sales (a huge per cent when you think about it) to as low a per cent as possible. Thus Feigenbaum takes a very financial approach to the cost of quality.

1.10 Philip B Crosby

Philip B Crosby, a guru of the late 1970s, was the populist who 'sold' the concept of Total Quality Management and 'zero defects' to the US. Although zero defects sounds very much like Six Sigma, in fact Crosby takes a much softer approach than does Deming, Juran, Feigenbaum or Six Sigma. His concept of zero defects is based on the assumption that it is always cheaper to do things right the first time, and quality is conformance to requirements. Note the wording 'conformance to requirements'. Thus any product that conforms to requirements, even where requirements are specified at less than perfection, would be deemed to be defect free.

Crosby developed the concept of non-conformance when recording the cost of quality. Non-conformance includes the cost of waste and scrap, downtime due to poor maintenance, putting things right, product recall, replacement, and at worst legal costs. All these can be measured, and according to Crosby the cost of non-conformance 'can be as much as 20 percent of manufacturing sales and 30 percent of operating costs in service industries'.

Crosby is famous for saying quality is free. And he wrote a book with that title (Crosby, 1979). He emphasized cultural and behavioural issues ahead of the statistical approach of Deming and Feigenbaum. Crosby was saying, if staff have the right attitude, know what the standards are and do things right the first time every time, the cost of conformance will be free. The flow on effect is that motivated workers will go further than just doing things right, they will detect problems in advance, they will be proactive in correcting situations and they will be quick to suggest improvements. Crosby concluded workers should not be blamed for errors, but rather that management should take the lead and the workers will follow. Crosby suggests that 85% of quality problems are within management's control. (Deming put this figure at 90%.)

1.11 What of the Japanese?

The most important of the Japanese writers on quality are Genichi Taguchi, Ishikawa, Shingo and Imai. And of course Toyota is widely cited as the epitome of lean production.

Of the Japanese approaches to quality, Taguchi methods have been the most widely adopted in America and Europe. Taguchi, an electrical engineer, used an experimental technique to assess the impact of many parameters on a single output. His method was developed during his work of rebuilding the Japanese telephone system in the 1970s. His approach to quality control is focused on 'off line' or loss of function (derived from telephone system failures).

The Taguchi approach is to

- Determine the existing quality level measured in the incidence of downtime—which he called 'off line'
- Improve the quality level by parameter and tolerance design
- Monitor the quality level by using statistical process control, to show upper- and lower-level variances

Taguchi promotes three stages in developing quality in the design of product or systems:

Determine the quality level, as expressed in his loss function concept
Improve the quality level in a cost-effective manner by parameter and tolerance design
Monitor the quality of performance by use of feedback and statistical control

1.12 Lean Enterprise

Ohno Taiichi, after visiting US car manufacturers in the sixties, returned to Japan and developed a new method of manufacturing, which became known as lean production. 'Lean is more than a system, it is a philosophy', began with Japanese automobile manufacturing in the 1960s, and was popularised by Womack, Jones and Roos in *The Machine that Changed the World* (1990). *The Machine that Changed the World* is essentially the story of the Toyota way of manufacturing automobiles. The characteristics of Lean, sometimes referred to as Toyotaism, are that materials flow 'like water' from the supplier through the production process onto the customer with little if any stock of raw materials or components in warehouses, no buffer stocks of materials and part finished goods between stages of the manufacturing process, and no output stock of finished goods. This 'just-in-time' (JIT) approach requires that materials arrive from dedicated suppliers on the factory floor at the right stage of production just when required, and when the production process is completed, the product is shipped directly to the customer. With no spare or safety stock in the system, there is no room for error. Scheduling of activities and resource has to be exact, communication with suppliers must be precise, suppliers have to be reliable and able to perform to exacting timetables, materials have to arrive on time and meet the specifications, machines have to be maintained so that there is no downtime, operators cannot make mistakes, there is no allowance for scrap or rework, and finally the finished product has to be delivered on time to customers. This is often implemented by circulating cards or kanbans between a workstation and the downstream buffer. The workstation must have a card before it can start an operation. It can pick raw materials out of its upstream (or input) buffer, perform the operation, attach the card to the finished part, and put it in to the downstream (or output) buffer. The card is circulated back to the upstream to signal the next upstream workstation to do the next cycle. The number of cards circulating determines the total buffer size. Kanban control ensures that parts are made only in response to a demand. With computer-controlled production, the kanban principle applies, but there is not a physical movement of cards, information is transferred electronically.

This JIT approach generally precludes large batch production, instead items are made in 'batches' of one. This means that operators have to be flexible, the system has to be flexible and 'single minute exchange of dies' (SMED) becomes the norm. A lean approach reduces the number of supervisors and quality inspectors. The operators are trained to know the production standards required and are authorised to take corrective action; in short they become their own inspectors/supervisors.

The original Toyota Production System (TPS) model of Lean Manufacturing, from which various hybrids were developed, comprised nine tools and approaches:

1. TPM (total productive maintenance)
2. 5 S's: These represent a set of Japanese words for excellent housekeeping (*Sein*—sort, *Seiton*—set in place, *Seiso*—shine, *Seiketso*—standardise, and *Sitsuke*—sustain)
3. JIT (just in time)
4. SMED (single minute exchange of dies)
5. Judoka or Zero Quality Control
6. Production work cells
7. Kanban (see earlier)
8. Poka-yoke
9. Mudas (wastes)

These terms, and others, are explained in various sections and also in the glossary at the end of the book.

1.13 Total Productive Maintenance

Total productive maintenance (TPM) includes more than maintenance, it addresses all aspects of manufacturing performance. The two primary goals of TPM are to develop optimum conditions for the factory through a self-help people/machine systems culture and to improve the overall quality of the workplace. It involves every employee in the factory. Since the mid-1980s, TPM has been promoted throughout the world by the Japan Institute of Plant Maintenance (JIPM). TPM is based on five key principles or 'pillars':

1. The improvement of manufacturing efficiency by the elimination of 'six big losses'
2. The establishment of a system of autonomous maintenance by operators working in small groups
3. An effective planned maintenance system by expert engineers
4. A training system to increase the skill and knowledge level of all permanent employees
5. A system of maintenance prevention where engineers work closely with suppliers to specify and design equipment that requires less maintenance

TPM has been applied both in Japan and outside Japan in three ways: as a stand-alone programme, as part of Lean manufacturing and as the manufacturing arm of TQM.

1.14 ISO 9000

In a discussion on the subject of quality, it would be wrong to ignore the effect that the International Organization for Standardization 9000 series (ISO 9000) has had on quality. The ISO 9000 series (accrediting criteria revised 2000/2001) and the more recent 14000 environmental series have been developed over a long period of time. The origins can be traced back to military requirements, for example the North Atlantic Treaty Organization (NATO) in the late 1940s developed specifications and methods of production to ensure compatibility between NATO forces in weapons and weapons systems. In Britain the predecessor to ISO 9000 was the British standard BS 5750 which was introduced in 1979 to set standard specifications for military suppliers.

ISO 9000 certification means that an organisation constantly meets rigorous standards, which are well documented, of management of quality of product and services. To retain certification the organisation is audited annually by an outside accredited body. ISO 9000 on the letterhead of an organisation demonstrates to themselves, to their customers and to other interested bodies that it has an effective quality assurance system in place.

TQM means more than just the basics as outlined in or ISO 9000, indeed ISO 9000 could be seen as running contrary to the philosophy of TQM.

1.15 Kaizen

The Japanese have a word for continuous improvement: 'kaizen'. The word is derived from a philosophy of gradual day-by-day betterment of life and spiritual enlightenment. Kaizen has been adopted by Japanese business to denote gradual unending improvement for the organisation. The philosophy is the doing of little things better to achieve a long-term objective. Kaizen is 'the single most important concept in Japanese management—the key to Japanese competitive successes' (Masaaki Imai, 1986).

Kaizen moves the organisation's focus away from the bottom line, and the fitful starts and stops that come from major changes, towards a continuous improvement of service. Japanese firms have, for many years, taken quality for granted. Kaizen is now so deeply ingrained that people do not even realize that they are thinking kaizen. The philosophy is that not one day should go by without some kind of improvement being made somewhere in the company. The far-reaching nature of kaizen can now be seen in Japanese government and social programs.

All this means trust. The managers have to stop being bosses and trust the staff; the staff must believe in the managers. This may require a major paradigm change for some people. The end goal is to gain a competitive edge by reducing costs and by improving the quality of the service. To determine the level of quality to aim for, it is first necessary to find out what the customer wants and to be very mindful of what the competition is doing.

The daily aim should be accepted as being 'kaizen'—that is, some improvement somewhere in the business.

1.16 Quality Circles

In the 1960s, Juran (1988, p. 111) said, 'The quality-circle movement is a tremendous one which no other country seems to be able to imitate. Through the development of this movement, Japan will be swept to world leadership in quality'. Certainly, Japan did make a rapid advance in quality standards from the sixties onwards, and quality circles were part of this advance. But quality circles were only one part of the Japanese quality revolution.

Quality circles have been tried in the US and Europe, often with poor results. In Japan, the quality circle traditionally meets in its own time rather than during normal working hours. Not only do circles concern themselves with quality improvement, but they also become a social group engaged in sporting and social activities. It is not expected in a European country that a quality circle would meet in the members' own time. Few workers are that committed to an organisation. However, there is no reason why, once the quality circle is up and running, management could not support and encourage social events for a circle, perhaps in recognition of an achievement.

1.17 Summary

If you cannot define it, you cannot measure it; if you cannot measure it, you cannot improve it. Green Six Sigma is a quality-focused methodology to achieve sustainable improvement of a process, project or organisation. In this chapter quality is defined. Quality has three dimensions, viz. product quality, process quality and people quality. The performance of quality can be measured by four elements, viz, prevention, appraisal, internal failure and external failure. Performance improvement should aim at the appropriate level of quality and then sustain it by Green Six Sigma.

GREEN TIPS

- Quality has three dimensions—product quality, process quality and people quality.
- External failure is less frequent but more significant.
- Four Americans (Deming, Juran, Feigenbaum and Crosby) started the quality movement.

Chapter 2

The Evolution of Six Sigma, Lean Six Sigma and Green Six Sigma

Quality is not something you install like a new carpet. You implant it. Quality is something you work at. It is a learning process.

—Edward Deming

2.1 Introduction

'Today, depending on whom you listen to, Six Sigma is either a revolution slashing trillions of dollars from corporate inefficiency, or it's the most maddening management fad yet devised to keep front-line workers too busy collecting data to do their jobs' (*USA Today*, 21 July 1998).

At the time of writing, it has been 23 years since this statement was made. During this time the 'Six Sigma revolution' has created a huge impact in the field of operational excellence, yet conflicting views are still prevalent.

Let us evaluate the arguments for both sides. On a positive note, the success of 'Six Sigma' in General Electric under the leadership of Jack Welch is undisputed. In the GE company report of 2000, their CEO was unstinting in his praise: 'Six Sigma has galvanized our company with an intensity the likes of which I have never seen in my 40 years of GE'. Even financial analysts and investment bankers compliment the success of Six Sigma at GE.

DOI: 10.4324/9781003268239-3

An analyst at Morgan Stanley Dean Witter recently estimated that GE's gross annual benefit from Six Sigma could reach 5% of sales and that share values might increase by between 10 and 15%.

However, the situation is more complex than such predictions would suggest. In spite of the demonstrated benefits of many improvement techniques such as Total Quality Management (TQM), business process re-engineering (BPR) and Six Sigma, most attempts by companies to use them have ended in failure (Easton and Jarrell, 1998). Sterman et al. (1997) conclude that firms have found it extremely difficult to sustain even initially successful process improvement initiatives. Yet more puzzling is the fact that successful improvement programmes have sometimes led to declining business performance causing lay-offs and low employee morale. Motorola, the originator of Six Sigma, announced in 1998 that its second-quarter profit was almost nonexistent and that consequently it was cutting 15,000 of its 150,000 jobs.

To counter heavyweight enthusiasts like Jack Welch (GE) and Larry Bossidy (AlliedSignal), it must be said in the interests of balance that there are sharp critics of Six Sigma. In fact, while Six Sigma may sound new, its critics say that it is really just statistical process control (SPC) in fresh clothing. Others dismiss it as another transitory management fad that will soon pass.

It is evident that like any good product, Six Sigma should also have a finite life cycle. In addition, business managers can be forgiven if they are often confused by the grey areas of distinction between quality initiatives such as TQM, Six Sigma and Lean Sigma.

Against this background, let us examine the evolution of total quality improvement processes (or in a broader sense operational excellence) from ad hoc upgrading, working up to TQM and then to Six Sigma and finally to Lean Sigma. Building on the success factors of these processes, the vital question is: how do we sustain the results? The author has named this sustainable process FIT SIGMA (see Basu, 2011).

So, what is FIT SIGMA? First, take the key ingredient of quality, then add accuracy in the order of 3.4 defects in 1,000,000. Now implement this across your business with an intensive education and training programme. The result is Six Sigma. Now let's look at Lean Enterprise, an updated version of classical industrial engineering. It focuses on delivered value from a customer's perspective and strives to eliminate all non-value-added activities ('waste') for each product or service along a value chain. The integration of the complementary approaches of Six Sigma and Lean Enterprise is known as Lean Sigma. FIT SIGMA is simply the next wave. If Lean Sigma provides agility and efficiency, then FIT SIGMA allows a sustainable fitness. In

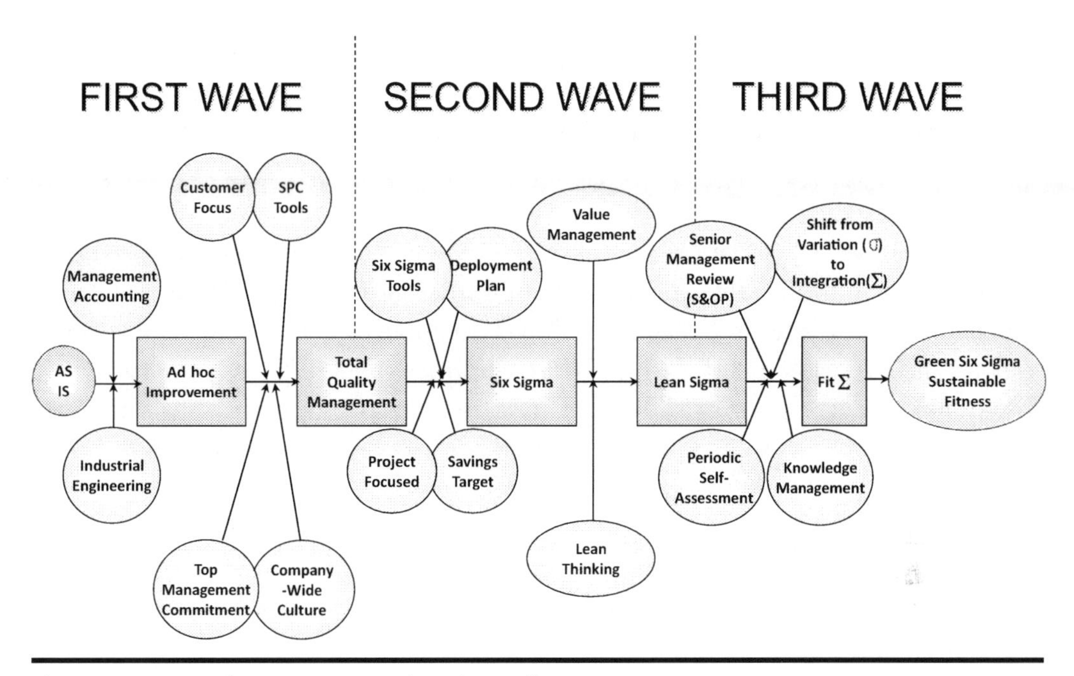

Figure 2.1 Road map to operational excellence.

addition, the control of variation from the mean (small Sigma 'σ') in the Six Sigma process is transformed to company-wide integration (capital Sigma 'Σ') in the FIT SIGMA process. Furthermore, the philosophy of FIT SIGMA should ensure that it is indeed fit for the organisation.

The road map to FIT SIGMA (see Figure 2.1) contains three waves, and it should be noted that the entry point of each organisation will vary:

First Wave : As Is to TQM
Second Wave : TQM to Lean Six Sigma
Third Wave : Lean Six Sigma to FIT SIGMA and Green Six Sigma

2.2 First Wave: As Is to TQM

The organised division of labour to improve operations may have started with Adam Smith in 1776. However, it is often the industrial engineering approach, which has roots in FW Taylor's 'Scientific Management', that is credited with the formal initiation of the first wave of operational excellence. This industrial engineering approach was sharpened by operational research and complemented by operational tools such as management accounting.

During the years following the Second World War, the 'first wave' saw through the rapid growth of industrialisation; but in the short term the focus seemed to be upon both increasing volume and reducing the cost. In general, improvement processes were 'ad hoc', factory centric and conducive to 'pockets of excellence'. Then in the 1970s the holistic approach of TQM initiated the 'second wave' of operational excellence. The traditional factors of quality control and quality assurance are aimed at achieving an agreed and consistent level of quality. However, TQM goes far beyond mere conformity to standard. TQM is a company-wide programme and requires a culture in which every member of the organisation believes that not a single day should go by within that organisation without in some way improving the quality of its goods and services.

2.3 Second Wave: TQM to Lean Six Sigma

Learning the basics from WE Deming and JM Juran, Japanese companies extended and customised the integrated approach and culture of TQM (Basu and Wright, 1997; Oakland, 2003). Arguably the economic growth and manufacturing dominance of Japanese industries in the 1980s can be attributed to the successful application of TQM in Japan. The three fundamental tenets of Juran's TQM process are first, upper management leadership of quality; second, continuous education on quality for all; and finally, an annual plan for quality improvement and cost reduction. These foundations are still valid today and embedded within the Six Sigma/Lean Sigma philosophies. Philip Crosby and other leading TQM consultants incorporated customer focus and Deming's SPC tools and propagated the TQM philosophy both to the US and the industrialised world. The Malcolm Baldridge Quality Award, ISO 9000 and the Deming Quality Award have enhanced the popularity of TQM throughout the world, while in Europe the EFQM (European Foundation of Quality Management) was formed. During the 1980s TQM seemed to be everywhere, and some of its definitions such as 'fitness for the purpose', 'quality is what the customer wants' and 'getting it right the first time' became so overused that they were almost clichés. Thus the impact of TQM began to diminish.

In order to complement the gaps to be found in TQM in specific areas of operational excellence, high-profile consultants marketed mostly Japanese practices in the form of a host of three-letter acronyms such as JIT (just in time), TPM (total productive maintenance), BPR and MRPII (Manufacturing Resource Planning). Total productive maintenance (TPM) has demonstrated successes outside Japan by focusing on increasing the capacity of individual processes. TQM was the buzzword of the 1980s, but it is viewed by many, especially in the US quality field, as an embarrassing failure—a quality concept that promised more than it could deliver. Philip Crosby pinpoints the cause of TQM 'failures' as 'TQM never did anything to define quality, which is conformance to standards'. Perhaps the pendulum swung too far towards the concept of quality as 'goodness' and the employee culture. It was against this background that the scene for Six Sigma appeared to establish itself.

Six Sigma began back in 1985 when Bill Smith, an engineer at Motorola, came up with the idea of inserting data based statistics into the softer philosophy of quality. In statistical terms, Sigma (σ) is a measure of variation from the mean; thus the greater the value of Sigma, the fewer the defects. Most companies produce results which are at best around four Sigma or more than 6,000 defects. By contrast at the Six Sigma level, the expectation is only 3.4 defects per million as companies move towards attaining this far higher level of performance.

Although invented at Motorola, Six Sigma has been experimented with by AlliedSignal and perfected at General Electric. Following the recent merger of these two companies, GE is truly the home of Six Sigma. During the last 5 years, Six Sigma has taken the quantum leap into operational excellence in many blue-chip companies including DuPont, Raytheon, Schneider Electric Ivensys, Marconi, Bombardier Shorts, Seagate Technology and GSK.

The key success factors differentiating Six Sigma from TQM are

- The emphasis on statistical science and measurement
- A rigorous and structured training deployment plan (Champion, Master Black Belt, Black Belt and Green Belt)
- A project-focused approach with a single set of problem-solving techniques such as DMAIC (define, measure, analyse, improve, control)
- Reinforcement of the Juran tenets (Top Management Leadership, Continuous Education and Annual Savings Plan)

Following their application in companies like GSK, Raytheon, Ivensys and Seagate, the Six Sigma programmes have moved towards the Lean Sigma philosophy, which integrates Six Sigma with the complementary approach of Lean Enterprise. Lean focuses the company's resources and its suppliers on the delivered value from the customer's perspective. Lean Enterprise begins with Lean production, the concept of waste reduction developed from industrial engineering principles and refined by Toyota. It expands upon these principles to engage all support partners and customers along the value stream. Common goals to both Six Sigma and Lean Sigma are the elimination of waste and improvement of process capability. The industrial engineering tools of Lean Enterprise complement the science of the statistical processes of Six Sigma. It is the integration of these tools in Lean Sigma that provides an operational excellence methodology capable of addressing the entire value delivery system.

2.4 Third Wave: Lean Six Sigma to FIT SIGMA and Green Six Sigma

Lean Six is the beginning of the 'third wave'. The predictable Six Sigma precisions combined with the speed and agility of Lean produces definitive solutions for better, faster and cheaper business processes. Through the systematic identification and eradication of non-value-added activities, optimum value flow is achieved, cycle times are reduced and defects are eliminated.

The dramatic bottom-line results and extensive training deployment of Six Sigma and Lean Sigma must be sustained with additional features for securing the longer-term competitive advantage of a company. The process to do just that is FIT SIGMA. The best practices of Six Sigma, Lean Sigma as well as other proven operational excellence best practices underpin the basic building blocks of FIT SIGMA.

Four additional features are embedded in the Lean Sigma philosophy to create FIT SIGMA. These are

- A formal senior management review process at regular intervals, similar to the sales and operational planning process
- Periodic self-assessment with a structured checklist which is formalised by a certification or award, similar to the EFQM award but with more emphasis on self-assessment
- A continuous learning and knowledge management programme
- The extension of the programme across the whole business with the shifting of the theme of the variation control (σ) of Six Sigma to the integration of a seamless organisation (Σ)

Green Six Sigma is an adaptation of Six Sigma, Lean Six Sigma and FIT SIGMA to the specific applications for climate change initiatives. A new feature of Green Six Sigma is the extension of DMAIC to DMAICS (define, measure, analyse, improve, control, sustain).

2.5 More about Six Sigma

Six Sigma is an approach that takes a whole system attitude to the improvement of quality and customer service so as to enhance the bottom line. The Six Sigma concept matured between 1985 and 1986 and grew out of various quality initiatives at Motorola. Like most such quality initiatives since the days of Deming in the 1960s and in particular the concept of TQM, Six Sigma requires a total culture throughout an organisation. This means that everyone at all levels should possess a passion for continuous improvement with the ultimate aim of achieving virtual perfection. The difference with Six Sigma is the setting of a performance level that equates to 3.4 defects per 1 million opportunities. To ascertain whether Six Sigma has been achieved requires a common language throughout the organisation (at all levels and within each function) and standardised, uniform measurement techniques of quality. The overall Six Sigma philosophy has a goal of total customer satisfaction.

A survey (Basu and Wright, 2003) was conducted with the following leading companies in the UK who had adopted the Six Sigma approach to quality:

Motorola
AlliedSignal (Honeywell)
General Electric
Raytheon
DuPont Teijin
Bombardier Shorts
Seagate Technology
Foxboro (Invensys)
Norando
Ericson

The results indicated that the main driver leading to the application of Six Sigma within a company is cost savings rather than customer satisfaction.

In coming to this conclusion, the firms benefited from informal networking with members of these companies as well as leading consulting groups such as Air Academy Associates, Rath and Strong, PricewaterhouseCoopers, Iomega and Cambridge Management Consulting. The surveyed companies reported between them a long list of intangible and indirect benefits. However, these plus points did not seem to be supported by any employee or customer surveys.

Nonetheless, very real results from the adoption of Six Sigma continue to be noted. For example in 1997 Citibank undertook a Six Sigma initiative and after just 3 years it was reported that defects had reduced by ten times (see Erwin and Douglas, 2000, for details). Likewise, General Electric states that the initial $300 million invested in 1997 in Six Sigma will deliver between $400 million and $500 million savings with additional incremental margins of $100 million to $200 million. Wipro Corporation in India says that from a start in 1999, after just 2 years defects were reduced to such an extent as to realise a gain of eight times over their initial investment in Six Sigma.

The application of operational excellence concepts is now extended to non-manufacturing processes. 'Firms such as Motorola and General Electric . . . successfully implemented Six Sigma. Motorola saved $15 billion in an 11 year period. General Electric saved $2 billion in 1999 alone . . .. Although Six Sigma initiatives have focused primarily on improving the performance of manufacturing processes, the concepts are widely applied in non-manufacturing, administrative and service functions' (Weinstein et al., 2008).

2.6 What Is Six Sigma?

So just what is the enigma of Six Sigma? Sigma is a classical Greek letter (σ) which is used in mathematical and statistical models to signify the standard deviation from the mean. This might sound like statistical mumbo jumbo but in reality is a very simple concept. The mean (more correctly referred to as the arithmetic mean) is what most of us would call the average. For example, if a player in cricket batted ten times and the total of the player's ten scores is 650 then the average is 65 (even though the player might have 'scored' nil on one occasion and 250 off another ball). This is because mathematically, each turn at bat contributes to the average. In statistical terms the arithmetic mean of the total score of 650 is 65, arrived at by dividing the total number of runs by the number of bats (650/10).

The next basic concept in statistics is frequency distribution. An often quoted example in statistical textbooks is the tossing of ten coins 100 times. The result of each throw of the ten coins could range from ten heads and no tails, to ten tails and no heads, or any combination in between (i.e. one head and nine tails, two heads and eight tails and so on). We would expect that if the coins are evenly balanced, we are more likely to have a probability of five heads and five tails than we are to get ten heads and no tails.

Table 2.1 shows the results of tossing ten coins 100 times. This can be shown as a histogram as in Figure 2.2, and also as a distribution curve as shown in Figure 2.3.

The curve shown in Figure 2.3 is an example of a normal distribution curve. The curve is bell shaped (i.e. it is symmetrical from the midpoint). Of course not all distributions will provide this outline, but under normal circumstances and given a large enough population, in our example

Table 2.1 Results of Tossing Ten Coins 100 Times

Number of heads	0	1	2	3	4	5	6	7	8	9	10
Frequency	1	2	5	12	18	23	16	10	9	3	1

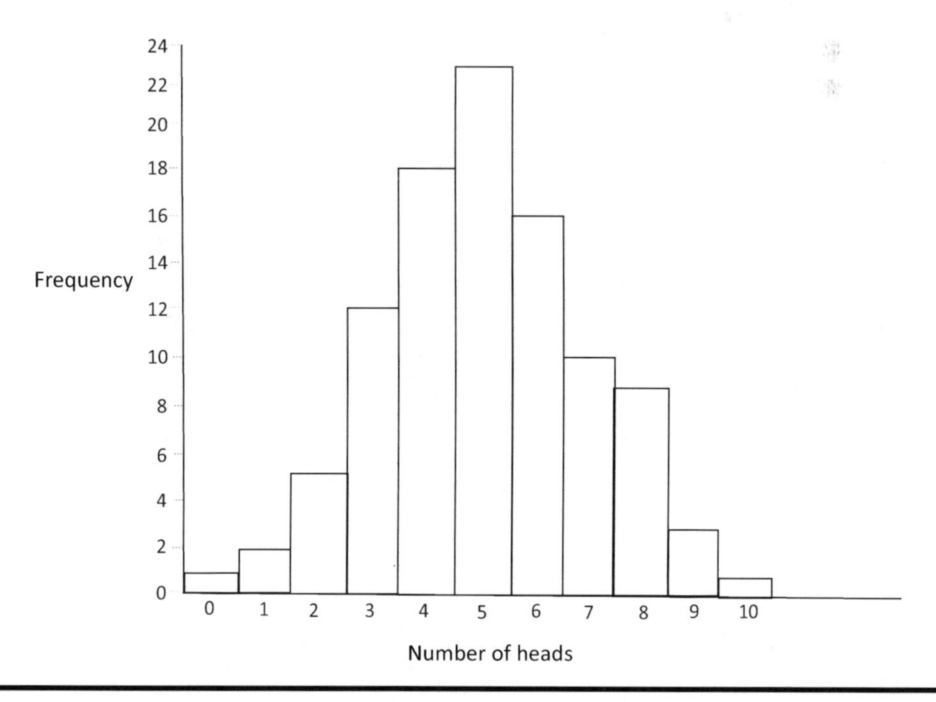

Figure 2.2 Histogram of tossing ten coins.

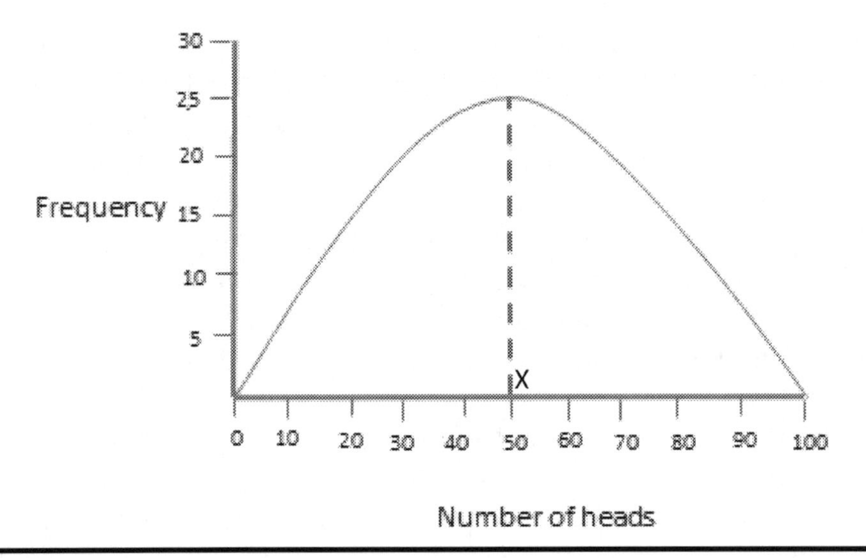

Figure 2.3 Normal distribution curve.

concerning 100 throws of ten coins, it is very likely that the distribution curve will be similar to that shown in Figure 2.2. The midpoint is shown on our curve as '*x*'. In statistical language, *x* represents the measure of central dispersion, but in everyday English the term *midpoint* indicates the same thing and is good enough for us.

If we assume a normal distribution curve as shown in Figure 2.3, one standard deviation from both sides of the midpoint (midpoint plus or minus one sigma) will include 68.27% of the total, and two standard deviations (two sigma) from both sides of the midpoint will include 95.45% of the total. Thus three standard deviations (three sigma) will cover 99.73%. If we extend out to six standard deviations (six sigma) from each side of the midpoint we cover 99.99966% of the total.

Now let us look at the quality program known as Six Sigma, our subject. Regarding this we can observe that for a process, the higher the Sigma, the more the outputs of the process (be they products or services) and the closer they will be to always meeting the customer requirements. In other words, the higher the Sigma is, the fewer the defects. It should be noted that the higher multiple Sigma does not increase the 'variation', it only increases the area covered under the distribution curve. For example:

With One Sigma, 68.27% of products or services will meet customer requirements, and there will be 317,300 defects per million opportunities.

Whereas with Three Sigma, 99.73% of products or services will meet customer requirements, and there will be 2,700 defects per million opportunities.

But, with Six Sigma, 99.99966% of products or services will meet customer requirements, and there will be 3.4 defects per million opportunities.

Thus, in effect, Six Sigma is a theoretical statistical measurement allowing the assessment of the quality of products and services to a position where there exist practically zero defects for any product or process in an organisation.

The key success factors differentiating Six Sigma from TQM are as follows:

- Emphasis on statistical science and measurement
- Structured training plans at different levels (Champion, Master Black Belt, Black Belt and Green Belt)
- Project-focused approach with a single set of problem-solving techniques such as DMAIC
- Reinforcement of Juran's tenets such as top management leadership, continuous education and annual savings plan
- Effects quantified in tangible savings (as opposed to TQM where it was often said that we cannot measure the benefits, but if we did not have TQM who knows what losses might have occurred?)

This last point concerning the quantification of tangible savings is a major selling point for Six Sigma.

It is usually possible to measure the cost of poor quality (COPQ) with the sigma level at which the process consistently performs. The COPQ of the performance level at six Sigma will be less than 1% of cost of sales, whereas at three Sigma (three Sigma is regarded by many organisations as a very acceptable level of process quality), the corresponding COPQ will range from 25 to 30% of cost of sales. These figures are explained/justified later in this book.

2.7 The Structured Approach of Six Sigma

Following the rigorous application of Six Sigma in many organisations, including Motorola, AlliedSignal, General Electric, Bombardier, ABB,

American Express, Wipro, GSK and others, a proven structured approach has emerged for product and process improvement. This structured and hierarchical process is as follows:

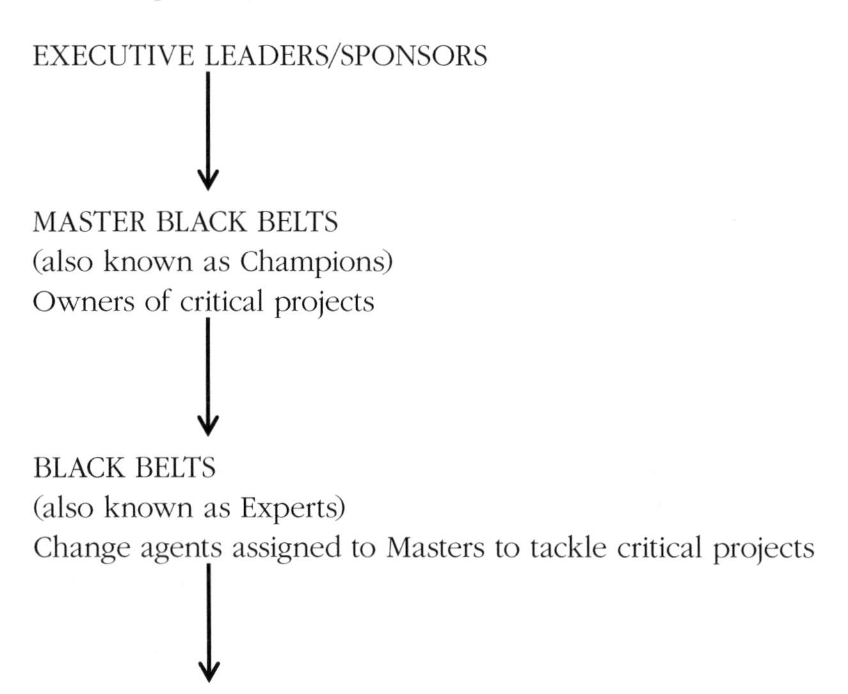

GREEN BELTS
(also known as Agent Advocates)
Grassroots support to implement changes

The top management must have a total commitment to the implementation of Six Sigma and accomplish the following tasks:

1. Establish a Six Sigma leadership team.
2. Develop and roll out a deployment plan for the training of Master Black Belts, Black Belts and Green Belts.
3. Assign Master Black Belts to identify and own critical projects related to key business issues.
4. Provide support for Black Belts to make breakthrough improvements in critical projects.
5. Encourage Green Belts to identify and implement 'just do it' projects.
6. Set aggressive Six Sigma Targets. This is done by the Six Sigma leadership team.

7. Continuously evaluate the Six Sigma implementation and deployment programmes and make changes if necessary.

Six Sigma Deployment

A critical piece of successful Six Sigma experience is the Six Sigma Deployment Plan. A typical plan includes four parts.

1. Business alignment planning.
2. First wave of Black Belt training.
3. Second wave (and subsequent waves as required) of Black and Green Belt training.
4. Infrastructure development to deliver results and sustain culture.

Business alignment planning assures that Six Sigma projects align with business strategy and drive results. Preliminary project selection criteria start with projects having potentially substantial savings (US$1 million or more). During the leadership education programme which usually lasts from 5 days, Master Black Belts are selected and Masters are assigned to key business issues and projects. Intensive Master Black Belt instruction takes 3 to 4 weeks. Concurrently while the Masters are being trained, the selection process for those to be qualified as Black Belts is carried out.

Master Black Belts (MBBs) have a thorough understanding of the improvement process, DMAIC, Six Sigma associated statistics and change management.

An MBB is competent to lead and manage significant end-to-end projects. They are also capable of coaching Green Belts.

Black Belt training is a combination of formal classroom education and onsite project work. Teaching is spread over 4 to 5 months and includes 4 or 5 weeks of classroom work with the balance being onsite project experience.

Green Belts (GBs) have an awareness of Six Sigma principles and in particular apply the DMAIC cycle to work with and support Black Belts in end-to-end projects. Green Belt training is usually distributed over 6 or perhaps 7 weeks including 5 days of formal classroom work.

External consultants and experienced trainers are usually needed to manage and train the first deployment wave. For the second and third waves of the deployment plan, in-house MBB take charge of both the training and the management of specific projects.

2.8 Certification of Black Belts and Master Black Belts

The certification authority is the Six Sigma Academy. They have set standards and guidelines as follows:

Black Belts are awarded a certificate after completing
Three weeks of formal classroom Black Belt training
One week of change management and project management training
Completion, as a team leader, of an end-to-end $1 million Black Belt project
Demonstrated Six Sigma commitment evidenced by mentoring Green Belts and the preparation of Green Belt training material

Master Black Belts are awarded a Master certificate after

Black Belt certification
Completion of five end-to-end Black Belt projects as team leader
Delivery of three Green Belt training sessions
Demonstrated commitment to the Six Sigma philosophy through mentoring Black Belt and/or Green Belt projects
Designing Black Belt training materials

Table 2.2 shows a sample schedule of Black Belt training.

Table 2.2 Sample of Black Belt Training

Week 1	*Week 2*	*Week 3*	*Week 4*
Define and measure Variation Data collection Basic statistics Cost of poor quality Define and measure tools	Analyse Advanced statistics Analyse and tools	Improve and control Generating improvement Change management Improve and control tools	Advanced statistical techniques Design of experiments Process capability Quality function deployment Multivariance analysis Design for Six Sigma Measurement system analysis

2.9 What Is Lean Six Sigma?

Lean Six Sigma (also known as Lean Sigma) is the combination of two world-class approaches (viz. Six Sigma and Lean Manufacturing) to organisational performance improvement. You have been introduced to Six Sigma in the preceding section. Let us now explain Lean manufacturing and then show how these two approaches are combined in Lean Six Sigma.

The term 'lean manufacturing' was first employed by Womack and Jones in 1998 as a description of the way in which manufacturing was carried out at Toyota. Toyota called their manufacturing process the Toyota Production System (TPS). Ohno (1988), who is said to be the founder of the TPS system, noted that the basis of this TPS method is the elimination of waste. In order to achieve this, two pillars are used:

- Just in time
- Automation with a human touch

'Just in time' in production means that the right parts reach the assembly line at the time they are needed and only in the amount required. The basic tool for achieving JIT is the kanban system. 'Automation with a human touch' goes back to Sakichi Toyoda, the founder of Toyota, who invented the automatic loom. This loom would automatically stop if a thread broke, thereby preventing any work being carried out after the defect had occurred. According to Ohno (1988) an important part of the TPS system is the ability to prevent mistakes from happening again. This is achieved by solving the cause of the problem rather than the actual mistake. Examples of tools for problem-solving and finding root causes of difficulties are Genshi Genbutso which means going to the source to directly observe the problem; cause and effect diagrams (also known as fish-bone diagrams); and the '5 Whys', or asking questions five times.

Ohno (1988) identified seven types of waste or 'mudas':

1. Waste of overproduction
2. Waste of time on hand (waiting)
3. Waste in transportation
4. Waste of processing itself (processing too much)
5. Waste of stock on hand (inventory)
6. Waste of movement
7. Waste of making defective products

According to Ohno the biggest source of waste is overproduction as it leads to many of the other types of squandering and misuse.

Liker (2004) has written several books about Toyota, and he describes 14 management principles garnered from the company. Liker writes in his book that Lean is not just about implementing tools such as 5S and JIT. It also concerns applying a complete system of Lean Thinking and culture. This view is supported by Womack and Jones (1998) in a concept for Lean Thinking described as follows: 'Lean Thinking is lean because it provides a way to do more and more with less and less'. Their five principles of Lean Thinking are value, value stream, flow, pull and perfection.

However, the application of Lean Thinking has moved with time and the experience of organisations in both the manufacturing and service sectors. Basu and Wright (2017) have extended Lean Thinking to supply chain management. The competition for gaining and retaining customers and market share is between supply chains rather than other functions of companies. A supply chain therefore has to be lean with four interrelated key characteristics or objectives:

1. Elimination of waste
2. Smooth operation flow
3. High level of efficiency
4. Quality assurance

2.9.1 Elimination of Waste

The lean methodology as laid out by Womack and Jones (1998) is sharply focused on the identification and elimination of 'mudas' or waste. Indeed, their first two principles (i.e. value and value stream) are centered on the elimination of waste. Their motto has been 'banish waste and create wealth in your organisation'. It starts with value stream mapping to identify value and then pinpoints waste with the process mapping of valued processes, followed by systematically eliminating them. This emphasis on waste abolition has probably made 'lean' synonymous with the absence of waste. Waste reduction is often a good place to start in the overall effort to create a lean supply chain because it can often be achieved with little or no capital investment.

2.9.2 Smooth Operational Flow

The well-publicised JIT approach is a key driver of a lean supply chain, and as we indicated earlier, it requires that materials and products flow 'like water' from the supplier through the production process onto the customer. The capacity bottlenecks are eliminated, the process times of workstations are balanced, and there are few buffer inventories between operations. Smooth operation flow requires the applications of appropriate approaches. Three of the most frequently applied methods are

- Cellular manufacturing
- Kanban pull system
- Theory of constraints

In the cellular manufacturing concept, the traditional batch production area is transformed into flow line layouts so that ideally a single piece flows through the line at any time. In practice, an optimum batch size is calculated starting with the most critical work centers and the largest inventory carrying costs. Action is taken for improvement both at these work centers and in methods that have greatest impact on the throughput, customer satisfaction, operating cost and inventory carrying charges. Second, kanban (literally meaning 'card') is a way of pulling parts and products through the manufacturing or logistics sequence as needed. It is therefore sometimes referred to as the 'pull system'. Finally, the 'theory of constraints' (TOC) is a management philosophy developed by Goldratt (1999). It enables the managers of a system to achieve more of the goal that system is designed to produce. However, the concept or the objective is not new. Nonetheless, in service operations where it is often difficult to quantify the capacity constraint, TOC could be very useful.

2.9.3 High Level of Efficiency

The more popular concepts of lean operations tend to be the principles of mudas, flow and the pull system. However, a preliminary analysis of all these methods, as we described earlier, highlights the fact that all assume sufficient machine availability exists as a prerequisite. In our experience, for many companies attempting a lean transformation, this assumption is just not true. Machine availability depends on maximising the machine uptime by eliminating the root causes of downtime. The ratio of uptime and planned operation time is the efficiency of the operation. Therefore, in order to make lean concepts work, it is vital that the precondition of running the operations at a high level of efficiency should be met.

2.9.4 Quality Assurance

Womack and Jones (1998) propose 'perfection' as the fifth Lean principle. According to this tenet, a lean manufacturer sets the targets for perfection in an incremental (kaizen) path. The idea of TQM also is to systematically and continuously remove the root causes of poor quality from the production processes so that the organisation as a whole and its products are kept moving towards perfection. This relentless pursuit of the faultless has to be a key attitude of an organisation that is 'going for lean'.

2.10 More on Lean Six Sigma

If the objective of Six Sigma is the reduction of variation, then Lean Six Sigma aims to accomplish the mission of the organisation better, faster and cheaper. To put it another way, Lean Six Sigma combines the focus on efficiency by Lean and the emphasis on quality by Six Sigma. There is a trade-off between quality (better), faster (delivery time) and cheaper (cost). The key is to leverage Lean Six Sigma to enable the organisation to get better, faster and cheaper—all at the same time. In general, Six Sigma is used to reduce defects and errors, thus making products or processes better. However, Six Sigma also reduces waste and streamlines processes like Lean. In fact, although Lean is primarily used to remove waste, in doing so it also reduces rework. When you reduce rework, you also reduce defects. Therefore, it is more sensible to combine the two methodologies of Lean and Six Sigma to make one approach of Lean Six Sigma. For example, consider a problem of lead time which varies between 20 and 12 weeks with an average of 16 weeks. With a Six Sigma approach it now fluctuates between 18 and 17 weeks with a mean value of 16 weeks. When Lean Six Sigma is applied, the lead time is reduced to 10 weeks, varying between 11 and 9 weeks.

There appear to be two approaches of Lean Six Sigma (De Carlo, 2007). One is to follow the DMAIC methodology of Six Sigma, to focus on the business objective of value creation and to apply additional Lean tools such as

Table 2.3 DMAIC and SCORE

Six Sigma (DMAIC)	*Lean Kaizen (SCORE)*
Define	Select
Measure	Clarify
Analyse	Organise
Improve	Run
Control	Evaluate

value stream mapping and rapid changeover. The second approach is centered on Kaizen Event (SCORE [select, clarify, organise, run, evaluate]) rather than DMAIC. It is argued that the DMAIC process often takes longer when root causes are not known. On the other hand, you may employ a Kaizen Event when the root causes are known (e.g. for cycle time reduction). This may follow a war room–like environment for a short period (also called the Kaizen Blitz). However, a drawback of this process is that it may not bring the rigour of data-driven measurement and analysis of DMAIC. As shown in Table 2.3, although there are similarities between DMAIC and SCORE, the former is more related to project management, and the latter is more linked to industrial engineering.

2.11 Why FIT SIGMA?

We live in a competitive world. The pace of change is increasing, and businesses and national economies are continuously being disrupted. In recent times the biggest external factors have been the collapse of banking and the finance market causing job losses and a sharp decline of gross domestic products, followed by the Covid-19 pandemic. We also experienced the impact of the Internet and e-business. The spectacular rise in 1999 and fall in 2000 of so many so-called dot.com companies showed that without substance businesses are not sustainable. When the huge bubble bursts, it is the innocent consumers who will feel the effect. Added to this, we need to factor in large-scale mergers, and acquisitions are continuously taking place in all sectors of manufacturing and service industries.

Globalisation and the economic growth particularly in China are reshaping the economic power balance and consequent business strategies. We can conclude that change is here to stay, and it often comes quickly from unexpected quarters. The challenge for all businesses is to find the benefits of change with an appropriate value-adding change programme.

The failed dot.com companies of the last two decades and the more current problems of the banking industry have demonstrated that their early apparent success was not sustainable. There are similar stories of unsustainable improvements in traditional businesses in the 'old economy' (before e-business). In spite of the demonstrated benefits of many improvement techniques such as TQM, BPR and Six Sigma/Lean Sigma, many attempts to implement and uphold improvements have fizzled out, not with a bang but with a whimper. What is more puzzling is that some companies (such as Motorola and Ford) that successfully implemented a quality initiative (such as Six Sigma) have subsequently experienced overall drops in business performance resulting in layoffs and lower employee morale. Therefore, we need a sustainable continuous improvement programme.

The well-publicised results of Six Sigma were first exhibited by Motorola, General Electric and AlliedSignal and then by Ford and Dow Chemical. This created an image of Six Sigma as a process primarily for large manufacturing multinationals who can sanction a huge budget for consultants to train hundreds of Black Belts. In addition to the apparent misconceptions of high start-up cost and the process being only suitable for manufacturing, an aversion to sophisticated statistical techniques also discouraged the service sector and small and medium-sized enterprises from applying Six Sigma. Therefore, there is a need for a 'fit for purpose' improvement programme suitable for all types or sizes of organisations.

As described in the following chapters, the holistic and appropriate methodology of FIT SIGMA aims to address these gaps by focusing on three fundamentals:

- Fitness for purpose
- Fitness for integration
- Fitness for sustainability

2.12 What Is Green Six Sigma?

Green Six Sigma is the updated version of FIT SIGMA tailored to climate change initiatives. As explained further in the following chapters, Green Six Sigma includes three additional tenets over FIT SIGMA:

- More emphasis on Lean and moving towards the circular economy
- Addition of sustainability of the environment to the sustainability of the process and performance
- Extension of DMAIC to DMAICS

A circular economy is aimed at eliminating waste and the continual use of resources. The approach is to reuse resources, repair defects, refurbish facilities, rebuild products to original specifications, and recycle wastes to create a closed-loop system. The outcome is minimising the use of resource input, wastes, pollution and carbon emissions.

The measuring and monitoring of carbon emissions, carbon reduction and carbon offsetting is a major component of the sustainability of the environment. A carbon offset is a reduction in emissions of greenhouse gases made in order to compensate for emissions made elsewhere. The offsetting is done by planting trees or buying carbon offsets from emission trading companies. Carbon reduction is done by direct actions such as energy economy, changing the energy sources to renewables, reducing travel and reducing wastes.

2.13 Summary

Quality management has evolved over the years through various stages, from inspection to control to assurance to TQM. The new waves of Six Sigma, Lean processes and FIT SIGMA are embedded in the holistic programmes of operational excellence. Our belief is that quality is not a new or separate discipline but rather that it pervades all management actions. Our philosophy is that quality is too important to be left to the managers, and that in fact, quality is everybody's concern, not only in the organisation but also in the interest of customers and suppliers as well as any other stakeholder.

Quality has two main aspects. It can be measured from the customer's perspective—customer satisfaction—and it can be viewed from the standpoint of the efficient use of resources. These two seemingly separate objectives are in fact inseparable when quality is considered. An organisation that wishes to compete in the global market must be efficient and provide a high level of customer satisfaction. No organisation will be able to afford to provide a world-class service unless its use of resources is efficient and non-value-adding activities have been minimised, and no organisation can afford not to be world class.

With this logic, the application of Six Sigma or Lean Sigma programmers should address both customer service and resource utilization in a cost-effective way. Quality has three dimensions—product quality, process quality and organisation quality (Basu, 2012). The results should also be sustainable and underpinned by people related to organisation quality. Results should be focused towards the sustainability of the environment and net zero carbon emissions. As explained in the following chapters, this is where lies the value of Green Six Sigma.

GREEN TIPS

- TQM is the company-wide programmer in improving products, processes, services and culture.
- Six Sigma is the data-driven TQM to achieve zero defects underpinned by assessed training deployment, structured project cycle and project-driven improvements.
- Lean Six Sigma is a combined approach of Six Sigma and Lean Thinking to systematically eliminate wastes and reduce variation in processes.
- Green Six Sigma is the adaptation of Lean Six Sigma for climate change initiatives with an additional project cycle of 'sustain'.

Chapter 3

More of Green Six Sigma

The golden rule is that there is no golden rule.

—George Bernard Shaw

3.1 Introduction

The success of Six Sigma and Lean Six Sigma cannot be faulted. The rigorous Six Sigma process combined with the speed and agility of Lean Six Sigma has produced definitive solutions for better, faster and cheaper business processes. Through the systematic identification and eradication of non-value-added activities, an optimum value flow is achieved, cycle times are reduced and defects are eliminated. However, business managers understand the grey areas of distinction between different quality initiatives and justifiably are expressing concerns, including raising the question 'how do we sustain these results?' Thus we need FIT SIGMA. The next question is 'how do we sustain the environment?' Thus we need Green Six Sigma. It follows that as Green Six Sigma is based on FIT SIGMA, any reference to FIT SIGMA will also apply to Green Six Sigma.

A survey by Basu (2001)[1] has shown that there are considerable barriers to achieving and sustaining results in quality initiatives. These are illustrated in Figure 3.1.

The biggest obstacle appears to be the packaged approach of the programme, causing a paucity of customised local solutions. Furthermore, due to the 'top-down' directive, middle managers are often not 'on board'. The

DOI: 10.4324/9781003268239-4

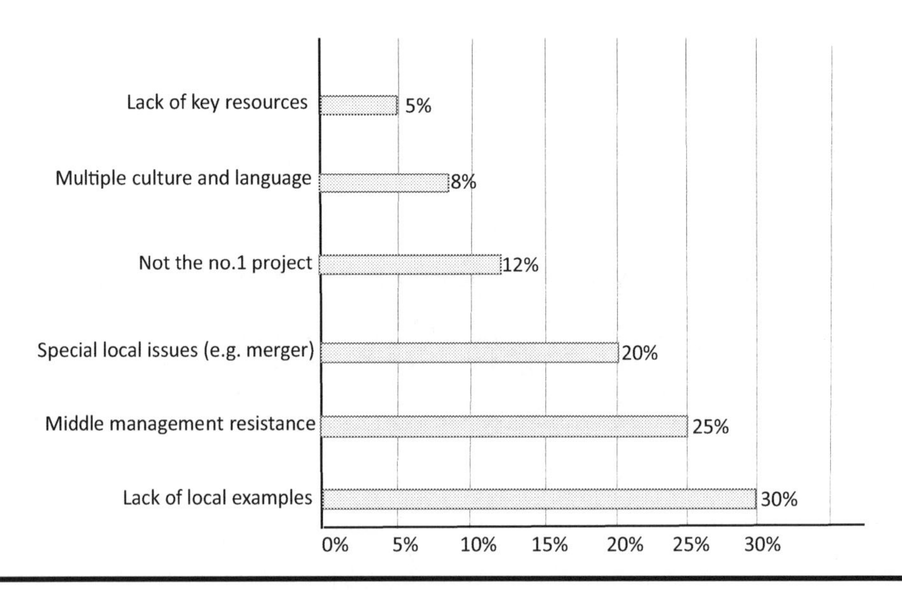

Figure 3.1 Main barriers of quality initiatives.

initiative is not owned by employees. We can identify additional and complementary areas of concern including the following:

- 'Some star performers of Six Sigma have shown poor business results' (e.g. site closures by Motorola)
- 'Incomplete initiatives' (e.g. Marconi abandoned Six Sigma during the economic downturn of 2001)
- 'Change of management and loss of sponsors' (e.g. the decline of Six Sigma at AlliedSignal after the departure of Larry Bossidy)
- 'External push by high powered consultants' (e.g. the dominance of the Air Academy consortium in the GSK programme)
- 'Excellent early results not sustained' (e.g. Raytheon relaunched Lean Sigma after a drop in performance)
- 'High start-up costs impede small and medium-sized enterprises' (the initial training start-up cost for Six Sigma is reported to be over $1 million)
- 'Still regarded as tools for manufacturing' (in spite of the success of Six Sigma in GE Capital)

The dramatic bottom-line results and extensive training deployment of Six Sigma and Lean Sigma must be sustained with additional features in order to secure the long-term competitive advantage of a company. If Lean Sigma

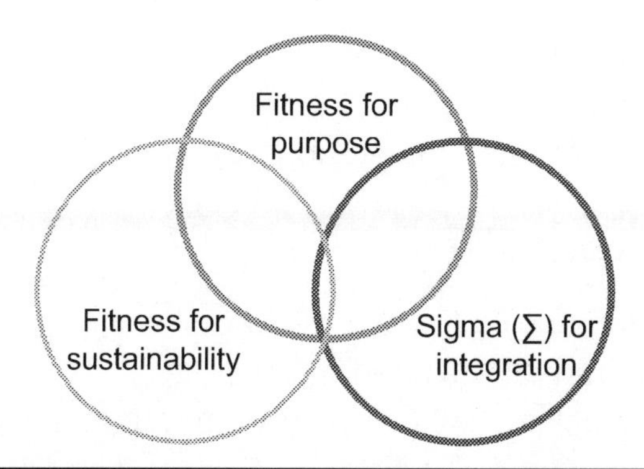

Figure 3.2 FIT Σ fundamentals.

provides agility and efficiency, then measures must be in place to develop a sustainable fitness. The process to do just that is FIT SIGMA.[2] In addition, the control of variation from the mean in the Six Sigma process (σ) is transformed to company-wide integration in the FIT SIGMA process (Σ). FIT SIGMA is therefore synonymous with 'FIT Σ'. Furthermore, the philosophy of FIT Σ should ensure that it is indeed fit for all organisations—whether large or small, manufacturing or service. In this book both FIT SIGMA and FIT Σ should be treated as identical terms.

FIT Σ is a solution for sustainable excellence in all operations. It is a quality process beyond Six Sigma. The fundamentals of FIT Σ are underpinned by the three cornerstones as shown in Figure 3.2.

Taken individually, the main components of these cornerstones are not new, but they do constitute proven processes. However, the combination of these components is both novel and unique to create the total process called FIT Σ. These elements are as follows:

FITNESS FOR THE PURPOSE

- Initial assessment
- All functions
- Any size of organisation
- Aligning with business strategy

SIGMA (Σ) FOR IMPROVEMENT AND INTEGRATION

- Appropriate Six Sigma tools (see Chapter 4)
- Appropriate DMAIC (define, measure, analyse, improve, control) methodology (presented in this chapter)

- Learning deployment
- Project plan and delivery
- Shift from variation (σ) to integration (Σ)

FITNESS FOR SUSTAINABILITY

- Performance management
- Self-assessment and certification
- Senior management review (sales and operations planning [S&OP])
- Knowledge management

3.2 Fitness for Purpose

Joseph Juran coined the phrase 'fitness for the purpose' relating to the basic requirements for quality. In the context of FIT Σ, 'fitness for the purpose' has wider implications. Here, 'fitness' means that the FIT Σ methodology is tailored to 'fit' all types of operations (whether manufacturing, service or transport) as well as all sizes of organisations (whether a multi-billion-dollar global operation or a small local enterprise). The customisation of the improvement programme to identify the right fit appropriate to the type and size of operation is determined by a formalised initial assessment process.

Treacy and Wiersema (1993) described three value disciplines in a business strategy:

- Operational excellence
- Customer intimacy
- Product leadership

They suggested that in a business context, organisations must select and excel at one of these value disciplines as a core operating model, while remaining adept at the other two. Regardless of an organisation's specific core competence or the sector in which it operates (e.g. manufacturing, retail, service or technology), the value discipline shapes the choices made by managers on a day-to-day basis. Applying the same logic in FIT SIGMA, each organisation must align the objectives and scope of the programme with the value disciplines of the business strategy.

It is advantageous, although not essential, to apply the initial assessment process based on a set of questions that could be applied later for the periodic self-assessment or certification process. This certification process, which is described in more detail later, may be adapted either from international quality awards (such as the European Foundation of Quality Management [EFQM] or America's Malcolm Baldridge Quality Award) or a holistic, published checklist such as Basu and Wright's Total Solutions 200 Questions (Basu and Wright, 1997), or even the company's own inventory. Regardless of the scheme, it is essential that the checklist is customised and that the key players of the company believe in it. As part of the Six Sigma or Lean Sigma programme, a 'baseline analysis' is carried out to identify areas of improvement, but this is performed at a later stage after committing the company to a rigorous deployment programme. The initial assessment is similar to that of a 'baseline analysis', but it is carried out right at the onset before the start of the deployment plan.

We believe that there are substantial benefits to be gained from designing your own assessment process rather than following a consulting firm's standards. However, it is imperative that the company is aware of the requirements and criteria of the assessment. The initial evaluation requires a good understanding of one's own processes, while an objective training to international standards is necessary for the certification. In both cases, the assessor is a trained expert.

Many organisations have dived into a Six Sigma programme without going through the earlier crucial stages of identifying the real requirements. Arguably, a significant number of companies that initiated a Six Sigma programme did so because they felt threatened in terms of their very survival, or they became victims of management fads. The 'GE factor' was too strong to ignore. But we should not interpret this 'GE factor' as a licence to blindly copy General Electric, but rather an exhortation to 'Get Effective' by applying the FIT Σ methodology. The starting point is an initial assessment as illustrated in Figure 3.3.

The initial assessment process can benefit from the application of a rating scale to each criterion or question on a ladder of say 1 to 5, with 1 being marked as 'poor' and 5 ranked as 'excellent'. A spider diagram can be constructed from the scores of each criterion to highlight the gaps.

The 'fitness for the purpose' philosophy also applies to the type of business, whether it is manufacturing operations in a factory or a service procedure in an office. The success of GE Capital after the application of Six Sigma is well documented. However, Jack Welch has often been quoted as the success factor rather than the application of the principles of Six Sigma, and after all, GE Capital is a huge operation of billions of dollars.

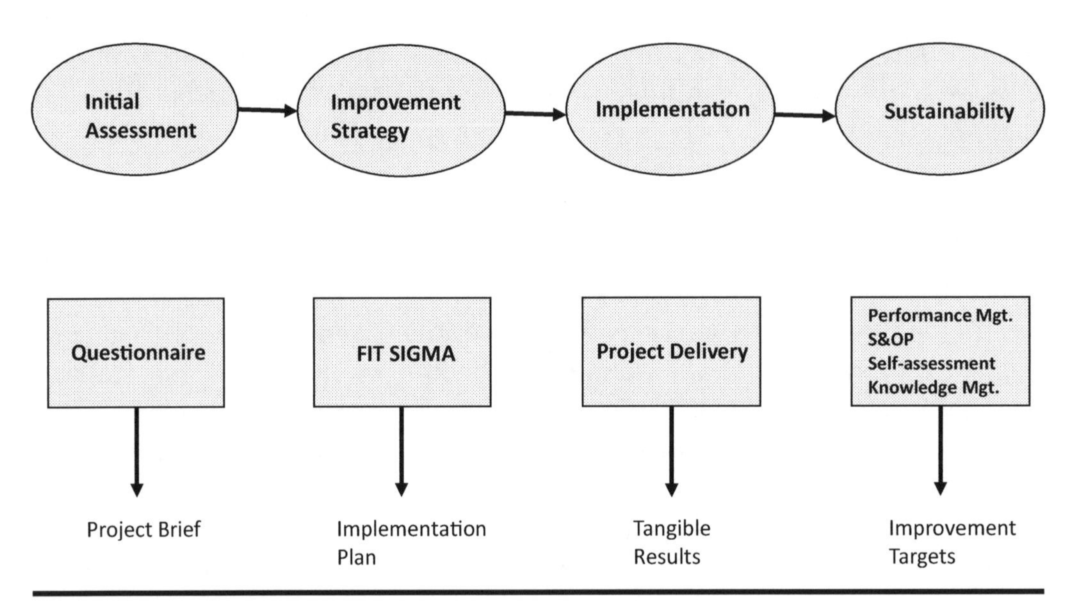

Figure 3.3 Fitness for purpose.

3.3 Sigma (Σ) for Improvement and Integration

The key cornerstone of FIT Σ methodology is 'Sigma (Σ) for improvement and integration'. The 'improvement' aspect is essentially based upon the proven tools and processes of Six Sigma/Lean Sigma initiatives. While the contents of Six Sigma or Lean Sigma approaches vary according to a company, a consultant or an author, the common features with FIT Σ are

- Appropriate Six Sigma tools
- Appropriate DMAIC methodology
- Learning deployment
- Project plan and delivery
- Shift from variation (σ) to integration (Σ)

3.3.1 The Appropriate Green Six Sigma Tools

When subjected to rigorous and more detailed analysis, it is true that Six Sigma or Lean Sigma tools are not original, and we should acknowledge their antecedents. For example, the focus on variation is historically known as the control chart devised by Deming and Schewhart. A design of experiments (DOE) can be linked to Taguchi's methods. The proactive use of Pareto and 'fish-bone' diagrams in Six Sigma may seem laudable, but these

Table 3.1 Appropriate Six Sigma Tools for Green Six Sigma

Basic Tools:
■ Pareto diagram
■ Flow process chart
■ Upper control limit (UCL)/lower control limit (LCL) control chart
■ Cause and effect diagram
■ Input-process-output (IPO) diagram
■ Brainstorming
■ Scatter diagram
■ Histogram
■ The Seven Wastes
■ The Five S's
Advanced Tools:
– Failure mode and effect analysis (FMEA)
– Design of experiments (DOE)
– *Design for Six Sigma (DFSS)*

are really 'old hat', being the clothing of Total Quality Management (TQM) and tools originally developed by Pareto and Ishikawa. The flow process chart of Lean Sigma is also a classic industrial engineering tool. Therefore, we do not propose to introduce any new so-called FIT Σ tools, but rather to refer to them as 'Six Sigma tools'. Thomas Edison once said, 'Your idea has to be original only in its adaptation to the problem you are currently working on'. The adaptation of the existing apparatus constitutes the 'appropriate Six Sigma' tools. Table 3.1 presents these tools.

Some of the tools outlined in Table 3.1 are described in previous chapters, and further details can be found in Appendix 1. We recommend that the learning deployment of FIT Σ should ensure both the understanding and application of these 'appropriate tools'.

3.3.2 Appropriate DMAIC Methodology

- The basic framework of DMAIC cannot be doubted, and it is also a core requirement of FIT SIGMA. However, the time and cost of using DMAIC should be weighed against the complexity and potential benefits of each project. Depending on the requirements of an individual undertaking, three levels of methodology are suggested:

- DMAIC Full
- DMAIC Lite
- Kaizen Event

DMAIC Full is the pure-play data-driven Six Sigma framework in five stages of define, measure, analyse, improve and control. For Green Six Sigma, DMAIC Full becomes DMAICS Full.

DMAIC Lite is the modified framework of FIT SIGMA for small and medium-sized enterprises (SMEs) comprising three stages (define, measure and analyse and improve and control).

A Kaizen Event is a structured continuous improvement process constituting a small group of people to improve a specific aspect of the business process in a rapid and focused manner. This method is applicable to all types and sizes of organisations and is usually employed when the root causes of the problem are known.

3.3.3 Learning Deployment

In order to use tools 'appropriate for Six Sigma' to achieve longer-term benefits, it is essential that an extensive learning deployment programme is dedicated to the education and training of employees at all levels. This learning deployment plan should be formulated after the 'initial assessment', and details will vary according to the 'quality level' and size of the organisation. It is recommended that the proven paths of previous Six Sigma and Lean Sigma programmes should be treated with respect, and a deployment plan can then be built around the outline as shown in Table 3.2.

Through a rigorous training deployment programme, Six Sigma has contributed to the creation of a people infrastructure within an organisation to enable the roll-out of a comprehensive programme. The issue is not who should be trained but rather who should do the training? The original source of Six Sigma education was the Six Sigma Academy, founded in 1994 in Scottsdale, Arizona. It is run by former Motorola experts Mikel Harry and Richard Schroeder. Their fees have been reported to start at $1 million per corporate client. Although there are many capable consultancy firms offering instruction, such tutoring costs are still running on an average threshold of $40,000 per 'Black Belt'.[3] Thus the initial start-up and

Table 3.2 Learning Deployment Programmes

Programme	*Target Audience*	*Duration*	*Approximate Number*
Leadership education	Senior management	2 days	3–5% of employees
Expert training (Black Belt)	Senior and middle management	4–6 weeks (in waves over 6 months)	1% of employees
Advocate training (Green Belt)	Supervisors and functional staff	1 week	10% of employees
Appreciation and cultural education	All employees	2 × half days	All

training expenses have prevented many SMEs from embracing a Six Sigma programme.

It is essential that high-quality input is provided to the training programme. This is usually available from specialist external advisers. At the same time a 'turnkey' consultancy support is not only expensive but also contains the risk that 'when consultants leave, expertise leaves as well'. In the FIT Σ learning deployment programme, we recommend two options:

Option 1:

- Retain consultants for, say, 3 months
- Run part of the leadership education and expert training programmes
- Develop with the assistance of consultants a firm's own trainers and in-house experts to complete the remaining waves of leadership and expert training
- Coach advocates (Green Belts) by own experts
- Cultural education by line managers

Option 2:

- Deploy consultants for one top-level leadership education workshop
- Prepare a small team of experts (two to five people) as trainers and develop a deployment plan with the assistance of consultants
- Roll out the deployment plan with own experts
- Ensure that consultants are available if required
- Train advocates (Green Belts) by own experts
- Cultural education by line managers

At General Electric, known to be the 'cathedral' of Six Sigma, the Six Sigma programme has been supported globally by a corporate team (known as CLOE—Centre of Learning and Operational Excellence), based at Stanford, Connecticut. Regardless of the type or size of operations, the development of a firm's own training capability is the foundation of sustainable performance.

The learning deployment to educate and develop your own experts or Black Belts provides a successful balance between a well-measured job structuring by a central team of industrial engineers and self-managed work teams. Over the years the principles of industrial engineering and Taylorism became corrupted to the extent where time and motion study people found the best method and then imposed that best method upon the worker. Such external control impeded teamwork and created the tedium of repetitive operations. This was followed by the quality circles (mainly in Japan) and self-managed teams (mainly in Scandinavia). The failure of the experimental Volvo factory at Uddevalle in the early 1990s was a wake-up call to realise that planners and team leaders should be trained in analytical tools. The experience of both GM-Toyota joint ventures in California and the Ford-VW factory in Portugal demonstrated that group performance can be improved by training the teams in industrial engineering principles. The collaboration of expert (or Black Belt) training and the team comprising advocates (or Green Belts) provides a balance of empowerment, motivation and measured efficiency.

3.3.4 Project Plan and Delivery

The success of any project is underpinned by management commitment, organisation, resources and formal reviews. A FIT Σ programme is no exception to the basic rules of project management. The process logic of a FIT Σ programme is shown in Figure 3.4, where the positions of a project plan and delivery have been highlighted.

The structure of a project organisation varies according to the operations and culture, but it must comprise some essential requirements:

- A sponsor or 'torchbearer' should be at the highest level of the organisation
- Project team leaders should be multifunctional and 'Black Belt' trained
- Two-way communication—both top-down and bottom-up
- RACI (responsibility, accountability, consulting and information) roles should be defined clearly

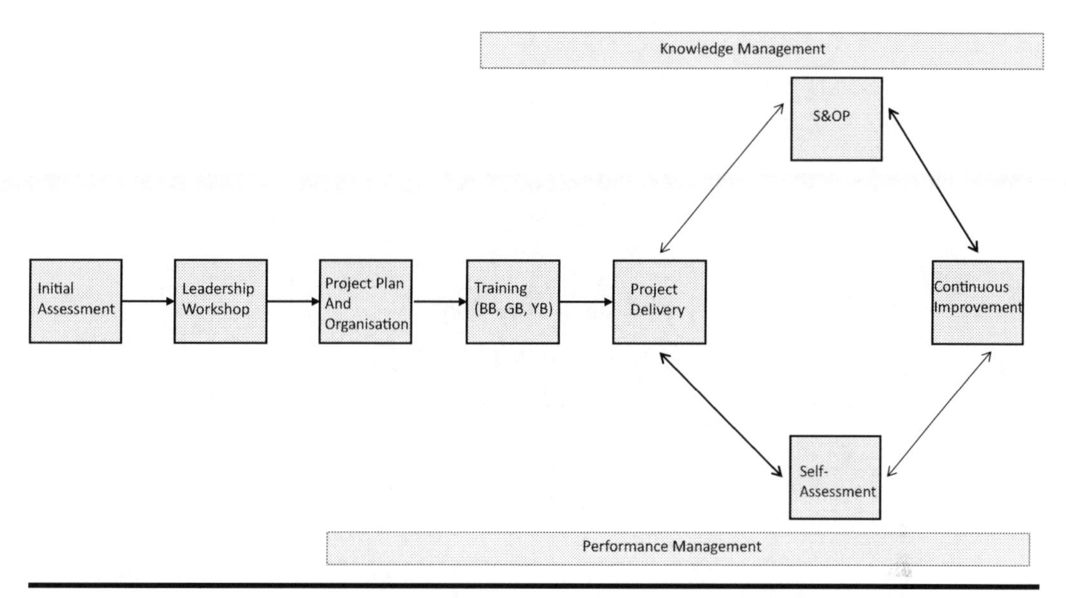

Figure 3.4 FIT SIGMA process logic.

Figure 3.5 shows an example of a typical project organisation for a FIT Σ programme.

It is recommended that at an early stage, following the initial assessment, a project brief or project charter is prepared to define clearly the following:

- Project organisation
- Time plan
- Learning deployment
- Project selection criteria
- Key deliverables and benefits

The project selection criteria cover two broad categories of ventures within the FIT Σ programme:

1. Large projects (managed by Black Belts)
2. Small 'just do it' assignments

Project selection can rely on both the 'top-down' and 'bottom-up' approaches. The top-down method usually relates to a large Black Belt project and considers a company's major business issues and performance objectives. Teams identify processes, CTQ (critical to quality) characteristics, process base lining and opportunities for improvement. This tactic has the

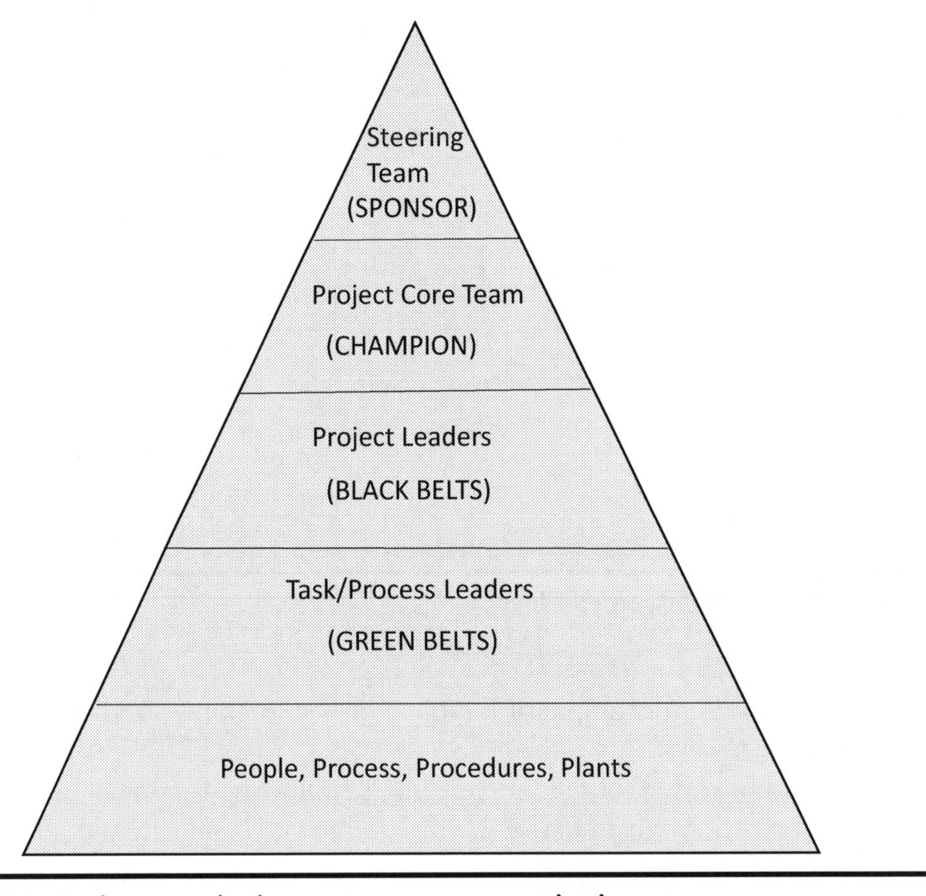

Figure 3.5 Project organisation—a two-way communication process.

advantage of aligning FIT Σ projects with strategic, corporate objectives. The 'bottom-up' approach can apply to both large and just do it schemes.

A rule of thumb for large Six Sigma projects was to attain savings of $1 million per year per Black Belt, deriving from between four and six projects each year. A just do it task usually does not incur a significant investment and varies from producing $5,000 to $100,000 savings annually.

Adapted from Oakland (2003), following are five key sources for identifying savings projects:

1. DPMO (defects per million opportunities)—The number of defects per unit divided by the number of opportunities for defects multiplied by 1,000,000. This number can be converted into a Sigma value.
2. COGS (cost of goods sold)—The variance is both fixed and variable costs.

3. COPQ (cost of poor quality)—This comprises internal failure costs, external failure costs, appraisal costs, prevention costs and lost opportunity costs.
4. Capacity—Overall Equipment Effectiveness—This is the number of good units a process is able to produce in a given period of time.
5. Cycle time—This is the length of time it takes to produce a unit of product or service.

It is important that, at the early stage of the programme, both large and just do it projects are straightforward and manageable. These types of ventures are often referred to as 'low-hanging fruits'—where improvements can be achieved easily by 'basic tools' (e.g. fish-bone diagram, flow process charts, histograms etc.). A simple selection process for projects is to choose the projects with a high potential for savings and which offer ease of implementation.

However, FIT Σ is far from completing savings projects. Over time, undertakings are selected and implemented to strengthen the company's knowledge base, stabilise processes and procedures and expand the cross-functional boundaries.

3.3.5 Shift from Variation (σ) to Integration (Σ)

The FIT Σ process fully accepts the importance of variation reduction. The risk of an improvement process based upon average values alone has been incontrovertibly proven. Likewise, there is an abundance of real-life examples where added values of lower variation or 'span' are well established. It is essential to concentrate on the variation control of subsystems and individual processes. However, the Six Sigma theme of variation control has often caused the focusing of an improvement plan in a relatively narrow sector or department.

According to web reports, 'Bob Glavin, former CEO of Motorola, has stated that the lack of an initial Six Sigma initiative in non-manufacturing areas was a mistake that cost Motorola $5 billion over a four year period' (Finkelstein, 2006).

The success of Six Sigma within General Electric was further enhanced by moving from a 'quality focus' to a 'business focus' and extending the initiative to its financial services area (e.g. GE Capital).

It is indicative, though not conclusive, that maximum benefit will be obtained from Six Sigma by integrating it with other proven continuous improvement initiatives and extending the programme to encompass the total business. When that happens, then Six Sigma embraces the FIT Σ philosophy of integration. The shift becomes complete from a small σ (standard deviation) to a capital Σ (summation or integration). In order to ensure that FIT SIGMA spans across all related functions, it is recommended that a value stream mapping is carried out for each key product. This will also help to identify bottlenecks and obvious areas of waste. The road map to sustainable success for the companies engaged in the pure-play Six Sigma programme constitutes great progress towards a company-wide integration of solutions in the FIT Σ process.

3.4 Fitness for Sustainability

Sustainability, in a general sense, is the capacity to maintain a certain process or state indefinitely. In the context of preserving the environment, it relates to uniting 'the needs of the present without compromising the ability of future generations to meet their own needs' (United Nations, 1987). The sustainability of project outcomes, in the context of FIT SIGMA, is not 'environmental'; rather, it is the longer-lasting stability of the deliverables and processes as illustrated in Figure 3.6.

The sustainability of FIT SIGMA project outcomes is comparable to the stability of a process. Stability involves attaining consistent and, ultimately, higher process yields defined by a metric or a set point. This is achieved through the application of an improvement methodology and continuous review. Such stability is ensured by minimising the variation of the set point. The approach of the PDCA cycle (Deming, 1986) enables both temporary and longer-term corrections. The short-term action is aimed at fixing the problem, while the permanent corrective steps consist of investigating and eliminating the root causes. This therefore targets the sustainability of the improved process. Thus, a sustainable project is expected to deliver long-lasting outcomes of acceptable quality performance criteria (Mengel, 2008).

This explanation of sustainable outcome is in agreement with the following definition of sustained success in British standards: 'The sustained success of an organisation is the result of its ability to achieve and maintain its objectives in the long-term. The achievement of sustained success for any organisation is a complex and demanding challenge in an ever-changing environment' (BSI, 2009).

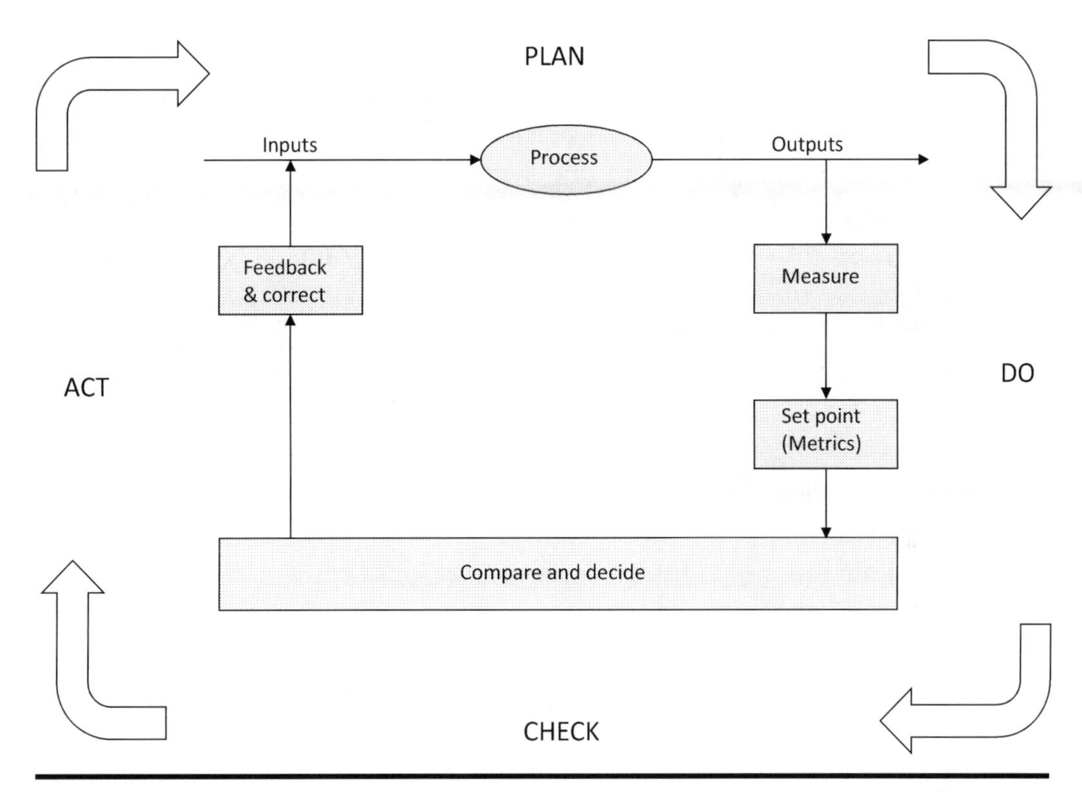

Figure 3.6 Process towards a longer-lasting project outcome in project management (plan-do-check-act [PDCA] cycle) (developed from Deming, 1986).

Having steered your way through the challenges of Six Sigma and Lean Sigma programmes over recent years, you are quite possibly very proud of their achievements in terms of results and the leanness of operations. You certainly deserve to congratulate yourself on attaining your goal, but it must be said that this is just the beginning. Realistically, you have only just embarked upon your path of success.

In a Lean programme, the reduction of overheads and elimination of non-value-added activities are all excellent accomplishments in themselves. However, no Lean practitioner can afford to be complacent. In layman's terms, they can be likened to someone who is keen to lose weight and so 'crash diets' without incorporating an appropriate fitness programme to back up their new regime. Sure, such a dramatic reduction in calorific intake will produce speedy results, but this will be in the short term only. The target weight, once attained, will be very difficult to uphold without making more far-reaching, holistic changes to nutrition, exercise and lifestyle. Experts always tell us that the only way to achieve and maintain a goal weight long-term is to completely review every aspect of our lives. We all know that crash diets have an immediate but unsustainable effect—in the end, we are sure to break them, however noble our intentions.

This analogy is one we can all understand and bear in mind when considering how to achieve long-term success and maintain goals that are important to us. Thus, in a FIT Σ programme the sustainability of performance is instilled right through the process and not just after the implementation of the deployment plan.

Fitness for the sustainability of performance is underpinned by four key processes:

- Performance management
- Senior management review
- Self-assessment and certification
- Knowledge management

3.4.1 Performance Management

The fact that the success of Six Sigma is highly focused on measurements, both statistical and savings, makes performance management a logical and essential component of the programme. In the context of FIT Σ, we address some relevant issues including what we measure, when we measure and how we measure.

There is little doubt, even in the present environment of advanced information technology, that a company's performance is governed by quarterly or annual financial reports. These accounts create an immediate impact on the share value of the company. As the financial reports are linked, an accounting model was developed a long time ago for site-centric activities. The majority of performance measures are still rooted to this traditional accounting practice. The senior managers of a company are usually driven to improve the share values which affect their personal share options in the firm. Thus, we find the traditional accounting model is still being used, even by information-age companies.

In addition to reporting basic financial measures (e.g. sales value, net profit, equity, working capital and return on investment), other conventional measures have been extended to assess customer service (market effectiveness) and resource utilisation (operations efficiency). Wild (2002) argues that the three aspects of customer service—specifications, cost and timing—can be measured against set points or targets. Given many resources as input to a process, resource utilisation can be measured as 'the ratio of useful output to input'. Resource utilisation is cost driven, while the objective of customer service is 'value added' to the business.

The models of financial accounting, customer service and resource utilisation may also be applicable to some areas of FIT Σ, but these may not incorporate the key aspect of the programme.

Kaplan and Norton (2004) argued that 'a valuation of intangible assets and company capabilities would be especially helpful since, for information age companies, these assets are more critical to success than traditional physical and tangible assets'. They created a new model called the 'Balanced Scorecard' as illustrated in Figure 3.7.

The Balanced Scorecard retains traditional financial measures, customer services and resource utilisation (internal business process) and includes additional measures for learning (people) and growth (innovation). This approach complements measures of past performance with drivers for future development. The Balanced Scorecard can be applied to a stable business process following good progress with the FIT Σ programme.

Performance management in FIT Σ should also be 'fit for the purpose', and the appropriate metrics should depend on the stages of the programme. There are three key stages of a FIT Σ initiative in the context of measuring its performance as shown in Figure 3.8.

As discussed earlier, larger projects in a FIT Σ programme are selected based upon an organisation's strategic goals and requirements. The viability of the project is then established dependent upon certain quantifiable criteria including return on investment (ROI). At the project evaluation stage of a FIT Σ initiative, similar criteria should prevail. Although attempts must be made to show an 'order of magnitude' of ROI data, the emphasis should be focused more on strategic goals and requirements.

The measurement process at the project implementation stage is basically the monitoring of the key factors considered during the project evaluation phase. The following six elements are suggested.

The first factor is to determine the project's value to the business which can be reflected in the company's overall financial performance. This aspect can be applied by monitoring the savings on a monthly basis.

The second characteristic is the resources required. If resources are outsourced, then this cost is measured and monitored. The timescale of the project is also included in this factor.

The third element is the metrics that many be required to monitor the performance of specific large projects. Examples of this factor

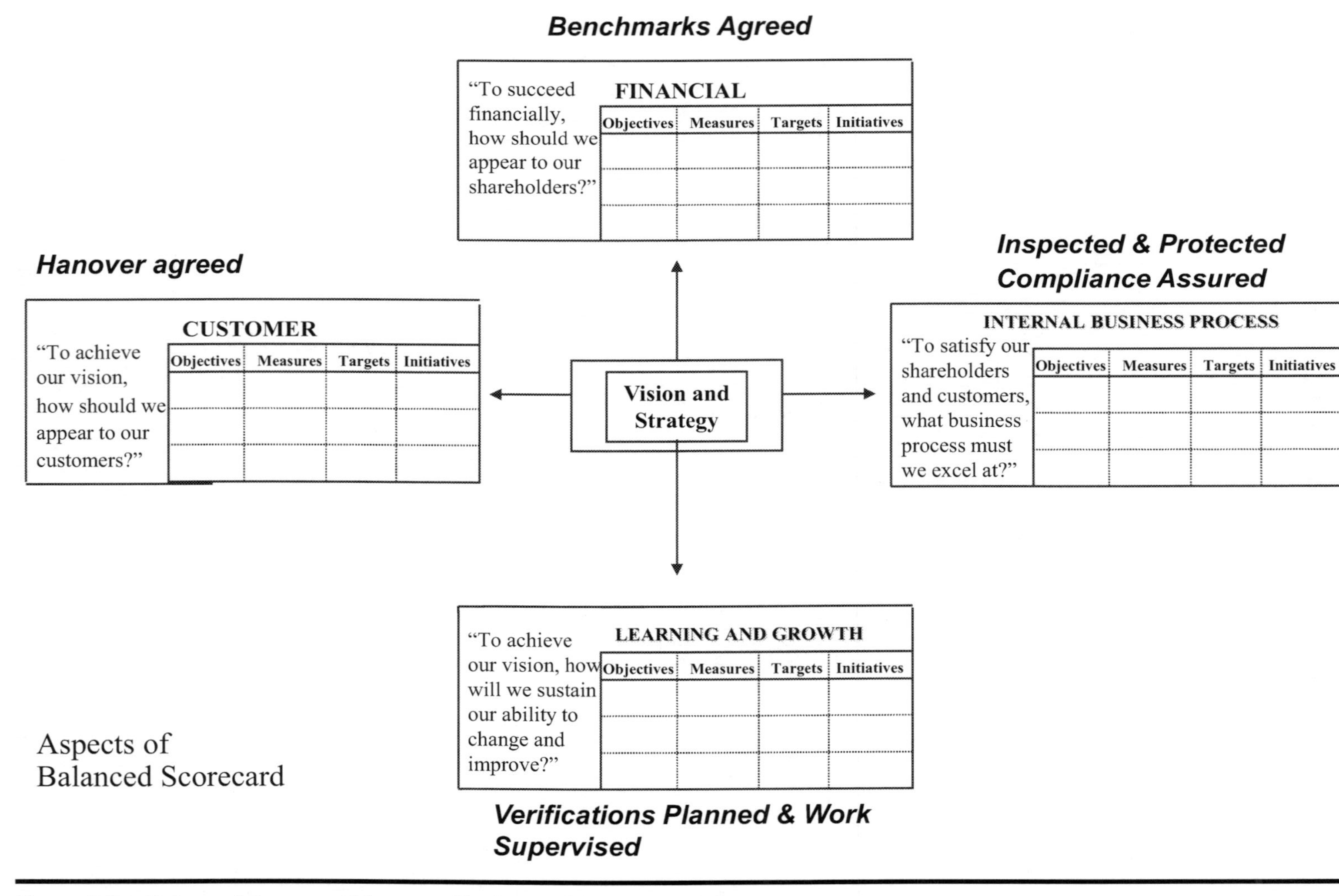

Figure 3.7 **Balanced Scorecard (*source*: Kaplan and Norton, 1996).**

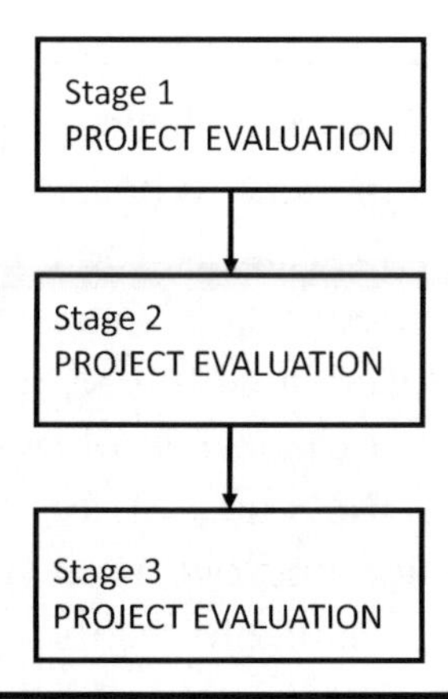

Figure 3.8 Stages of performance management.

are defects per million opportunities (DPMO) and rolled throughput yield (RTY).

Fourth, it is necessary to monitor the impact of the project on the external market: whether there is an eroding customer service or sales revenue as a result of key company resources deployed in the project.

The fifth phase is to ensure that the FIT Σ initiative continues to align with the overall mission and strategy of the business.

Finally, selective key performance indicators (KPIs) must be established for the next stage of the stable business process.

The transition period of a FIT Σ initiative from the project juncture to a stable business operation is often difficult to pinpoint. The main reasons for contribution to this blurred situation are due to the relatively short project duration and the need for continuous modifications in a dynamic technological environment. Therefore, five of the six factors (excluding the second dynamic) should be monitored for a stable business operation. However, additional emphasis should be given by focusing on the sixth step, the KPIs, and gradually all factors can then be incorporated in selective KPIs. A customised Balanced Scorecard should be appropriate for a stable business process.

3.4.2 Sales and Operations Planning (S&OP)

S&OP is also known as senior management review. A recurring challenge exists for companies that have invested significant time and resources in implementing proven improvement plans such as Six Sigma. Put simply, this is how to ensure their sustainable performance beyond the duration of a one-off corporate exercise. The annual review of the change programme during the budget planning is ineffective because 12 months is a long time in a competitive marketplace. We can use the analogy of travelling in a car to understand this issue. In order to steer the benefits of the programme and the business objectives to a sustainable future, the senior managers who are in the driving seats must have a clear view of both the front screen and the rear-view mirrors. In addition, they must look at them as frequently as possible to decide on their direction and optimum speed.

In recent years the pace of change in technology and marketplace dynamics have been so rapid that the traditional methodology of monitoring actual performance against predetermined budgets set at the beginning of the year may no longer be valid. It is fundamental that businesses are managed based on current conditions and up-to-date assumptions; there is also a vital need to establish an effective communication link, both horizontally across functional divisions and vertically across the management hierarchy, to share common data and decision processes. One such solution to these continuous review requirements is S&OP.

S&OP has become an established company-wide business planning process in the Oliver Wight MRP II methodology (Ling and Goddard, 1988). The diagram in Figure 3.9 shows the five steps in the process that will usually be present. The course of action can be adapted to specific organisational requirements.

New Product Planning (Step 1): Many companies follow parallel projects related to the new products in research and development (R&D), marketing and operations. The purpose of this review process in Step 1 is to

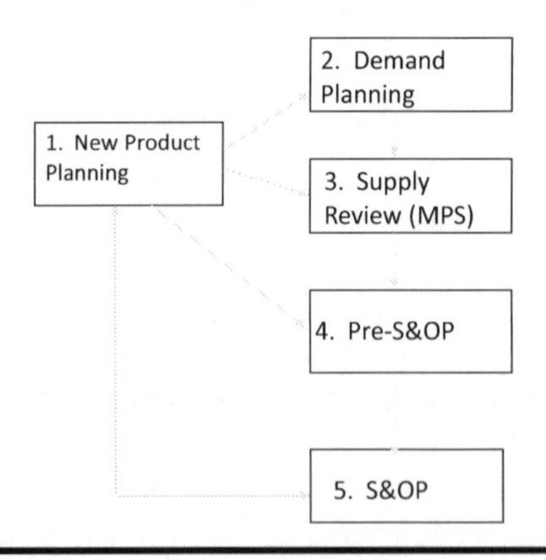

Figure 3.9 Five steps of S&OP process.

examine the different objectives of various departments at the beginning of the month and resolve new product-related assumptions and concerns. The issues raised will impact upon the demand plan and the supply chain at a later stage of the process.

Demand Planning (Step 2): Demand planning is more of a consensus art than a forecasting science. Demand may change from month to month depending on market intelligence, customer confidence, exchange rates, promotions, product availability and many other internal and external factors. This review at the end of the first week of the month, between marketing, sales, IT and logistics departments, establishes agreement and accountability for the latest demand plan identifying changes and issues arising.

Supply Review (Step 3): In the current climate of increasing outsourcing and supply partnership, the capacity of supply is highly variable, and there is a need to ensure the availability and optimisation of supply every month. This review, usually on the second week of the month, between logistics, purchasing and production departments, establishes the production and procurement plans and raises capacity, inventory and scheduling issues.

Reconciliation Review (Step 4): Problems would have been identified in previous reviews of new products, demand and supply. The reconciliation step goes beyond the balancing of numbers to assess the business advantage and risk for each area of conflict. This review looks at issues from the business point of view rather than departmental objectives. This is also known as the pre-S&OP review, and its aim is to minimise issues for the final S&OP stage.

Senior Management Review (Step 5): Senior managers or board members, with an MD or CEO in Chair, will approve the plan that will provide clear visibility for a single set of members driving the total business forward. The agenda includes the review of KPIs, business trends of operational and financial performance, issues arising from previous reviews and corporate initiatives. This is a powerful forum to adjust business direction and priorities. This is also known as the S&OP review.

At each process step the reviews must address a planning horizon of 18 to 36 months in order to make a decision for both operational and strategic objectives. There may be a perceived view that S&OP is a process of aggregate/volume planning for a supply chain. However, it is also a top-level forum to provide a link between business plan and strategy. The continuous improvement and sustainability of company performance by a FIT Σ programme can only be ensured in the longer term by a well-structured S&OP or senior management review process. The results and issues related to FIT Σ should be a regular item in the S&OP agenda.

3.4.3 Self-Assessment and Certification

In order to maintain a wave of interest in the quality programme and also to market the competitive advantage that quality undoubtedly affords, many companies channelled their efforts in two respects. They directed their energies to the pursuit of an approved accreditation such as ISO 9000, or to a prize such as the Malcolm Baldridge Award (in the US) or derivatives of the Baldridge Award (in other countries). The process of certification and awards has had a chequered history. After a peak during the early 1990s, the Baldridge Awards gradually lost their impact in the US and companies such as GE or Johnson and Johnson started developing their own customised quality assessment process. Encouraged by customer demand for the ISO stamp of approval, there was a rush for ISO 9000 certification in the 1990s. However, many firms became disillusioned by the auditors' focus on ensuring compliance with mainly current procedures without necessarily improving standards. A number of consultancy companies attempted to introduce their own awards to progress an improvement programme (e.g. Class 'A' by Oliver Wight).

It is essential to incorporate a self-assessment process in a FIT Σ programme in order to sustain a performance and improvement culture. Put simply, there are two choices: either select an external certification or develop your own checklist based on proven processes. Table 3.3 highlights the relative pros and cons of these two options.

Table 3.3 Self-Assessment Options

Option	*Pros*	*Cons*
Standard accreditation	– Proven process – Known to customers and suppliers – Trained auditors and consultants available – External networking	– Too generic to fit business – Invasion of auditors and consultants – More expensive – Not improvement driven
Customised self-assessment	– Process ownership – Customised to business needs – Improvement orientated – Common company culture – In-house knowledge based – Enables self-assessment	– Lack of external benchmark – Time to develop and pilot

There are several examples where a company achieved an external award based on a set of criteria but without improving business performance. There are also cases where, after a burst of initial publicity, the performance level and pursuit for excellence were not maintained. If the process is not underpinned by self-assessment, then the award will gradually lose its shine, just like lack of maintenance of an expensive new car. We therefore recommend that a FIT Σ programme should adopt a self-assessment process developed from proven procedures. Two such courses of action are described next: EFQM and Total Solutions.

3.4.3.1 European Foundation of Quality Management

The EFQM accolade is derived from America's Malcolm Baldridge National Quality Award. There are similar tributes available in other countries, such as the Canadian Excellence Awards and the Australian Quality Award (Figure 3.10).

The EFQM Global Award was established in 1991. It is supported by the European Union (EU), and countries within the EU have their own support unit (e.g. British Quality Foundation in the UK). As shown in Table 3.4, the ERQM model provides a set of checklist questionnaires under nine categories, each providing a maximum number of points.

The first five categories (Leadership to Process) can be seen as 'enablers', while the remaining four categories are 'performance' related.

Table 3.4 Nine categories of EFQM checklist questionnaire

Category Number	*Category*	*Maximum Number of Points*
1	Leadership	100 points
2	People Management	90 points
3	Policy and Strategy	80 points
4	Resources	90 points
5	Processes	140 points
6	People Satisfaction	90 points
7	Customer Satisfaction	200 points
8	Impact on Society	60 points
9	Business Results	150 points
TOTAL:		1,000 points

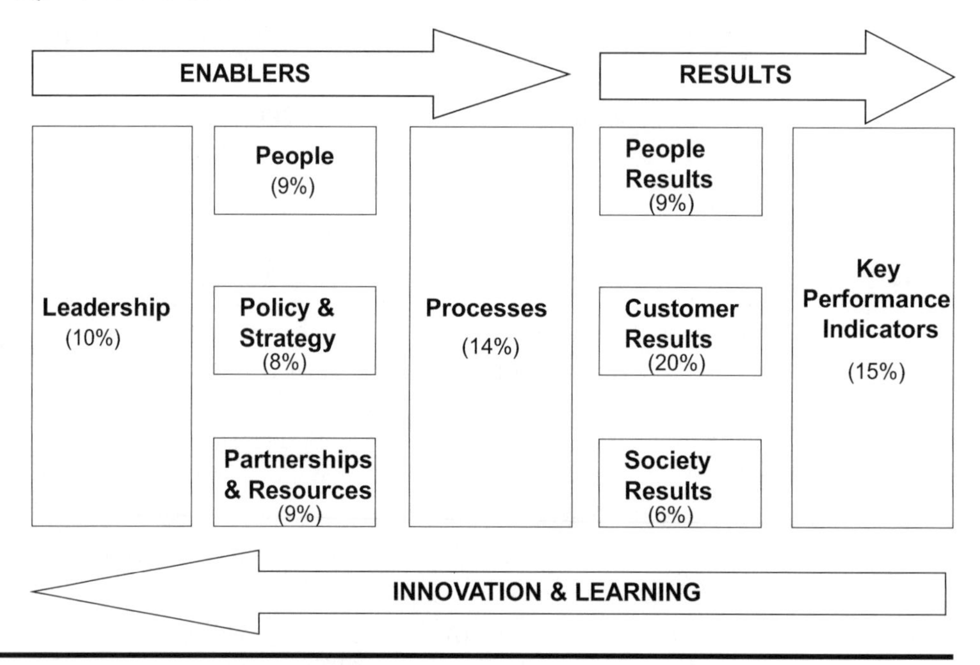

Figure 3.10 Balanced Scorecard hierarchy.

3.4.3.2 'Total Solutions'

This holistic approach of self-analysis covering all aspects of the business has been described in detail in 'Total Manufacturing Solutions' (Basu and Wright, 1997). As shown in Figure 3.11, 'Total Solutions' enables self-assessment against 20 defined areas ('foundation stones') to identify areas of improvement for achieving the full potential of the business.

The business is built from the foundation stones up and consists of the 'six pillars' of total solutions. There are 200 questions in the checklist, with ten questions for each foundation stone. These pillars are as follows:

- Marketing and innovation
- Supply chain management
- Environment and safety
- Facilities
- Procedures
- People

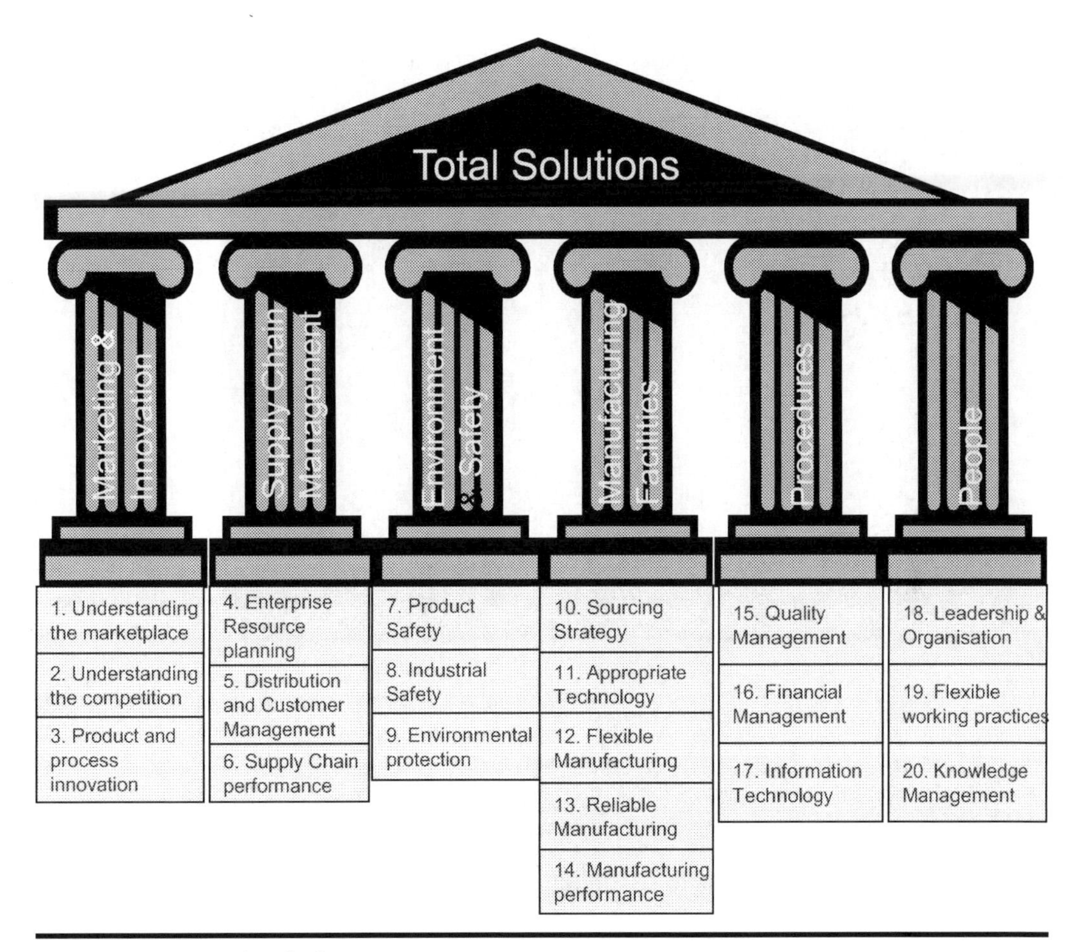

Figure 3.11 Total Solutions.

Although the checklist is aimed at manufacturing operations, it can be adapted easily to service operations. A 'spider diagram' can be constructed from scores of each foundation stone to highlight the current performance profile and gaps (see Figure 3.12).

The recommended methodology of the self-assessment in a FIT Σ programme consists of the following features:

1. Establish the policy of external certification or customised self-assessment in line with the company culture and business characteristics.
2. Develop or confirm the checklist of assessment.
3. Train internal assessors in the common company assessment process (one assessor for every 500 employees as a rough guide). These assessors should also carry out normal line or functional duties.

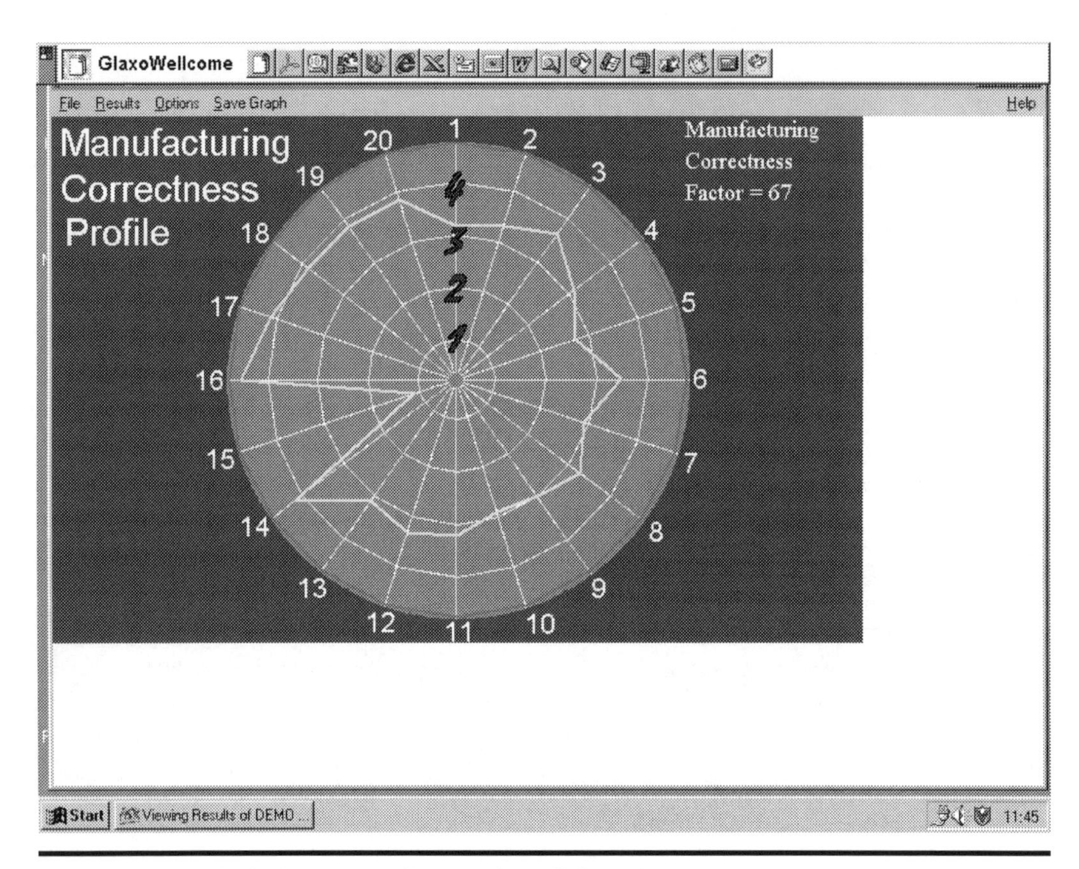

Figure 3.12 A holistic approach spanning all functions.

4. Train experts (Black Belts) and department managers in the self-assessment checklist and process.
5. Carry out quarterly self-assessment by departmental managers.
6. Ensure six-monthly (at the initial stage) and annual (at the later phase) assessments by internal assessors.
7. Analyse gaps and implement measures to minimise these shortcomings.
8. Consider corporate awards, depending on the performance attained, by the CEO.
9. Review the checklist with the change of business every 2 years.
10. Consider external accreditation if it adds value to the business.

This methodology is applicable to all types of business, both manufacturing and service, and all sizes of operations whether large, medium or small. A larger organisation is likely to possess its own resources to develop and maintain the process. A smaller organisation may require the assistance of external consultants to develop the procedure.

3.4.4 Knowledge Management

Almost 400 years ago, Francis Bacon stated that 'knowledge is power'. Peter Drucker wrote in the *Atlanta Monthly*, 11 September 1995, 'Knowledge has become the key economic resource and the dominant, if not the only, source of comparative advantage' (mentioned in Ruggles, 1998).

Webster's Dictionary defines knowledge as 'familiarity or understanding gained through experience or study'. In the context of a FIT Σ initiative, an essential sustainable driver of performance management is the sharing of knowledge and best practice. Although not explicit in his 14-points philosophy, Deming advocated the principle 'to find the best practices and train the worker in that best way'.

The key tenets of knowledge management in FIT Σ methodology are as follows:

1. Systematically capture knowledge from proven 'good practices'.
2. Select examples of 'best practices' based upon added value to the business.
3. Do not differentiate between the sources, regardless of level of technology or economic power.
4. Inculcate knowledge sharing between all units.

The essential ingredient for benefiting from knowledge management is the establishment of a 'learning organisation' culture. The key element here is the faith of the participants in the process. Unless members at all levels of a company involved in sharing knowledge believe that their business can benefit from it, then the exercise has little value. If a company thinks that they already know the best way or that the 'best practice' is actually not appropriate to their circumstances, then sustainable improvement just will not happen. The development of a 'learning organisation 'culture does not, of course, happen overnight—it takes time, and it requires the appropriate infrastructure to be in place. The experience suggests that time and money spent in knowledge management are also invested in the most valuable resource of competitive advantage—people. The support structure for such a knowledge-sharing process should include

- A champion to act as a focal point to coordinate the process
- A regular best practice forum to learn from each other and to allow networking
- Internal and external benchmarking to assess targets and gaps
- Continuous communication through websites, newsletters, videos and 'visual factors'

Sustainability in Green Six Sigma has the additional feature of the sustainability of environment standards and to achieve that, additional tools are required. These tools include material flow analysis and carbon footprint monitoring. These are described in Chapter 13.

3.5 Summary

FIT SIGMA is a natural extension of the third wave of the quality movement offering a historically proven process to improve and sustain the performance of all businesses, both manufacturing and services, whether large or small. Green Six Sigma is an adaptation of FIT SIGMA specifically for climate change initiatives with DMAICS in place of DMAIC. Green Six Sigma includes additional tools in the Sustain phase to ensure the sustainability of the environment.

Green Six Sigma is not a statistic. It is both a management philosophy and an improvement process. The underlying belief is that of a total business-focused approach underpinned by continuous reviews and a knowledge-based culture to sustain a high level of performance. In order to implement the Green Six Sigma philosophy, a systematic approach is recommended. The process is not a set of new or unknown tools; in fact, these tools and cultures have been proven to yield excellent results in earlier waves.

The differentiation of Green Six Sigma is the process of combining and retaining success factors. Its strength is that the process is not a rigid programme in search of problems but an adaptable solution for a specific climate change initiative or business.

Small wonder then that Green Six Sigma can be seen to offer new and exciting possibilities in the field of operational excellence and climate change initiatives. There is no magic formula in a new name or brand; what counts are the underlying total business philosophy process and culture of Green Six Sigma.

A unique selling proposition of Green Six Sigma is that it will form a bridge between the two camps of Lean and Six Sigma and the topic of climate change towards sustainability. It will thus act as a catalyst for the implementation of climate change initiatives.

GREEN TIPS

- FIT SIGMA is a derivative of Lean Six Sigma aimed at all types of organisations with three fundamentals: fitness for purpose, sigma for improvement and integration and fitness for sustainability.
- Green Six Sigma is the adaptation of Lean Six Sigma and FIT SIGMA for climate change initiatives.

Notes

1 Unpublished GSK survey, January 2001.
2 FIT SIGMA is a registered trademark of Performance Excellence Ltd.
3 Unpublished GSK survey, January 2001.

Chapter 4

Managing Green Six Sigma Projects with DMAICS

4.1 Introduction

During the early part of the 1980s, Total Quality Management (TQM), especially in the US and Europe, concentrated on the application of quality management principles and statistical process control (SPC) tools within all aspects of the organisation. The focus was upon their integration with key business processes. There were considerable successes in communication and culture, but despite this something was missing—the rigour of project management principles. The successes of Six Sigma and Lean Six Sigma programmes are primarily measured by the outcome of projects managed by Black Belts and Green Belts.

DMAIC (define, measure, analyse, improve, control) was introduced by Motorola as the life-cycle discipline for Six Sigma projects in the late 1980s. Since then, DMAIC has become the essential component of all Six Sigma initiatives and improvement programmes. While Deming's PDCA (plan-do-check-act) cycle has been extensively used in the development and deployment of quality policies, DMAIC has added the rigour of project life cycle to the implementation and closeout of Six Sigma projects.

In order to emphasise the sustainability of processes and also the environment standards during and after the control phase in DMAIC, an additional phase (sustain) is added in Green Six Sigma. Thus, DMAIC has become DMAICS in Green Six Sigma.

DOI: 10.4324/9781003268239-5

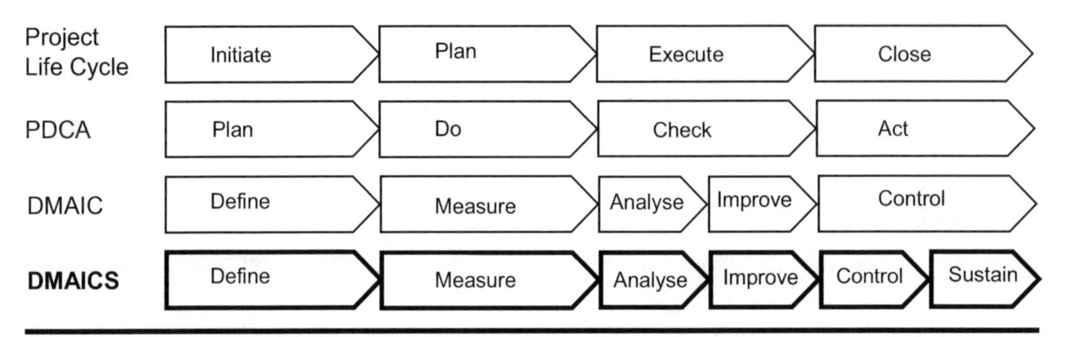

Figure 4.1 DMAICS life cycle.

Figure 4.1 shows the relationship between DMAIC and DMAICS with PDCA and a typical project life cycle.

4.2 Definition

DMAICS refers to a data-driven life-cycle approach to Green Six Sigma projects for improving processes and is an essential part of a company's Green Six Sigma programme. The simplified definitions of each phase are

Define by identifying, prioritising and selecting the right project.
Measure key process characteristics, the scope of parameters and their performances.
Analyse by identifying key causes and process determinants.
Improve by changing the process and optimising performance.
Control by consolidating the gain.
Sustain by sustaining the process and environmental standards.

4.3 Define Phase

The starting point of the Define stage is the selection of a project. The project is usually chosen by project leaders (Black Belts or Green Belts) from the environment and processes they are familiar with. It is also a good practice to apply some objective criteria, such as savings, cost, the probability of success and completion time. The probability of success often identifies so-called low-hanging fruits.

During the Define phase a project team is formed and then the project team drafts a project charter. A project charter includes definition and scope of the problem, project objectives and benefits. Often at this stage a formal goal or target for the improvement is set. It does not include the causes of the problem. If you want to remember anything about the Define phase, you should remember this question, 'what is the problem'. If you don't know what the problem is, you cannot fix it.

When we talk about improving performance, we actually talk about improving processes. So we need tools to understand processes that we are trying to improve. All our processes depend on customers who are willing to pay for our output. We should look at what is important to customers and how that 'voice' can enable us to define a Green Six Sigma project.

The following tools for the Define phase are described in Chapter 8:

- Project charter
- Value stream mapping
- SIPOC (suppliers, inputs, process, outputs, customers) diagram
- Flow diagram
- CTQ (critical to quality) tree

Before moving on to the Measure phase, the team refines their project focus and ensures they are aligned with the goals of organisational leadership.

There is a formal checkpoint to sign off the project before moving on from the Define phase to the Measure phase.

4.4 Measure Phase

After the Define phase comes the Measure phase. It is important because the Green Six Sigma approach is data driven. It emphasises the importance of working with hard evidence rather than opinion. The goal of the Define phase is to identify what is wrong with the process, and the goal of the Measure phase is to figure out how bad the process is. It involves validating the problem by using data and measuring what is happening.

The Measure phase is critical throughout the life of the project because it provides key indicators of process health and pointers to where process issues are happening. These pointers are further examined in the Analyse phase to identify opportunities for improvement.

The following tools for the Measure phase are described in Chapter 9:

- Run chart
- Histogram
- Cause and effect diagram

- Pareto chart
- Control charts

In the Measure phase a detailed flow diagram is often used to identify 'fail points'. Likewise, a cause and effect diagram is used in the Analyse phase for the root cause analysis.

There is also a formal checkpoint to sign off the project before moving on from the Measure phase to the Analyse phase.

4.5 Analyse Phase

The primary objective of the Analyse phase is to pinpoint the root causes of the problem. This phase can be seen as an opportunity to develop hypotheses to find out the root causes of the problem. Such hypotheses are validated by analysis, and the root causes of the problem are identified. The Analyse phase is vital because without proper analysis, teams can implement solutions that do not resolve the issue.

The following tools for the Analyse phase are described in Chapter 10:

Regression analysis
Force field analysis
SWOT (strengths, weaknesses, opportunities, threats) analysis
PESTLE (political, economic, social, technical, legal, environmental) Analysis
Failure mode and effect analysis (FMEA)
Five Whys
Interrelationship diagram
Capability analysis

The detailed process chart with fail points created in the Measure phase will help to identify potential causes of fail points (e.g. delays, defects, excess stock, rework etc.). A capability analysis and control chart can help to identify outliers and trends. The Five Whys and a cause and effect diagram (or fish-bone diagram) identify root causes. A force field analysis and a cause and effect diagram also enable stakeholders' buy-in.

There is also a formal checkpoint to sign off the project before moving on from the Analyse phase to the Improve phase.

4.6 Improve Phase

In the Improve phase ideas are developed to remove the root causes of problems identified in the Analyse phase. Solutions are tested, and the solutions that seem to work are implemented and results are measured. A structured improvement effort can lead to innovative changes that improve the baseline measure of the process.

The following tools for Improvement are described in Chapter 11:

- Single minute exchange of dies (SMED)
- Five S's
- Mistake proofing
- Brainstorming
- Overall equipment effectiveness (OEE)

Brainstorming is a method to generate ideas and solutions to solve problems with the team. The tools from Lean Thinking (e.g. SMED, 5S and mistake proofing) provide solutions to eliminate wastes. OEE is applied to improve the effectiveness of manufacturing plants. The potential solutions are usually evaluated by simple indicators, such as throughput rate, lead time or defects. For more complex analysis, simulations or design of experiments (DOE) are applied.

There is also a formal checkpoint to sign off the project before moving on from the Improve phase to the Control phase.

4.7 Control Phase

In the Control phase the improved process is continuously monitored to check that the improved level of performance is consolidated. To establish sufficient process control, the support of the project sponsor or project champion is important.

The following tools for Control are described in Chapter 12:

- Gantt chart
- Activity network diagram
- Radar chart
- PDCA cycle

The Gant chart and activity network diagram will ensure the effectiveness of the project control. A radar chart will monitor the profile of performance level. It is useful to use mistake proofing again which means making it unlikely to make a mistake. In this phase, it is also recommended to update the FMEA because the old process has changed.

There is also a formal checkpoint to sign off the project before moving on from the Control phase to the Sustain phase.

4.8 Sustain Phase

In the Control phase, measures have been taken to consolidate and sustain the improved performance. The performance of a process is not sustained by control measures alone. You need enablers to change and instil the culture of continuous improvement. Furthermore, especially for climate change initiatives, you need to ensure environmental standards. This is where you need the Green Six Sigma tools and processes at the Sustain stage.

The following tools for the Sustain phase are described in Chapter 13:

- Balanced Scorecard
- EFQM (European Foundation of Quality Management)
- S&OP (Sales & Operations Planning)
- Material flow account
- Carbon footprint tool

Balanced Scorecard, EFQM and S&OP are enablers to sustain the improved process and continuous improvement culture. The sustainability of environmental standards is ensured by the circular economy aided by material flow account and the monitoring of greenhouse gas emissions by the carbon footprint tool.

It is important to note that the DMAICS cycle starts again searching and defining new problems for further improvement. As shown in Figure 4.2, DMAICS is a continuous improvement cycle, and it literally does not stop.

4.9 Application

The tools of Green Six Sigma and operational excellence, as described in this book, are applied most often within the framework of DMAICS. As such, DMAICS is an integral part of a Green Six Sigma initiative.

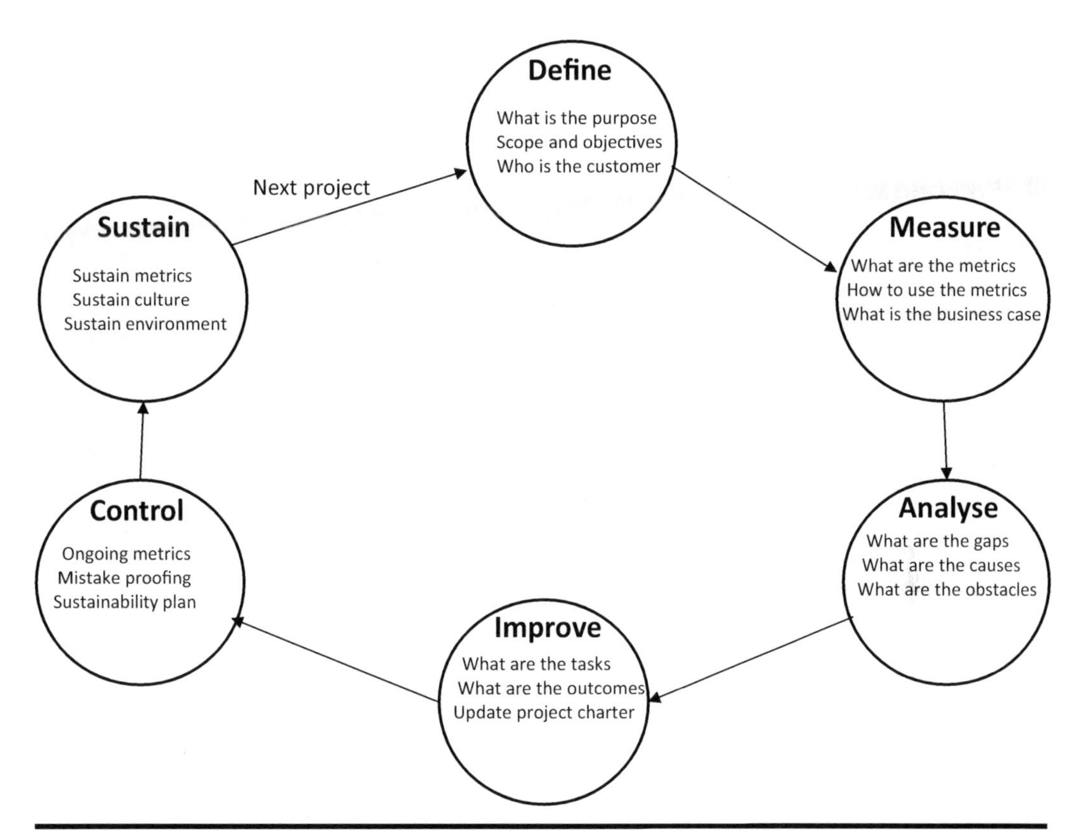

Figure 4.2 DMAICS cycle as a 'gated process'.

DMAICS is also used to create a 'gated process' for project control. For example, the criteria for a particular phase are defined, the project is reviewed, and if the criteria are met then the next phase starts (see Figure 4.2). As indicated earlier this is the cycle of continuous improvement.

As a summary of the application of the DMAICS technique, if you cannot define your process, you cannot measure it. That means if you cannot express the data, you are not able to utilise DMAICS in your development actions. Therefore, you cannot improve and sustain the quality.

4.9.1 Basic Steps

The basic steps of DMAICS are described in Sections 4.3 through 4.7.

4.9.2 Worked-Out Example

Although DMAICS is a data-driven and quantitative technique, it is also a methodology that may not produce unique output from a given set of input data. We have therefore chosen an actual case (from Dow Chemical) to illustrate a worked-out example.

4.9.3 Define

The Dow Chemical Company began using Six Sigma in 1998 to improve the operations of its subsidiary the FilmTec Corp. of Minneapolis, Minnesota, a manufacturer of membranes that was having difficulty in meeting customer demand.

Membrane quality is determined by two criteria: (a) flux or the amount of water the membrane lets through during a given period and (b) how much impurity is removed from the water.

4.9.4 Measure

Membrane elements were tested prior to shipping to ensure the prescribed quality standards. The specifications were assured but the speed with which their customers were serviced suffered. The shortfall of customer service was costing FilmTec approximately $500,000 a year.

4.9.5 Analyse

Six Sigma tools and methodologies were used at FilmTec to reduce product and process variation. Statistical analysis was used to identify variables that affected membrane flux most significantly.

One of the variables identified for improvement was the concentration of a chemical component used in the manufacturing process. The problem stemmed from the inconsistencies in concentration caused by the interruptions in feeding the chemical in the manufacturing process. The feeding was done from a movable container that was replaced every day. When the empty container was replaced by a full one, the chemical did not always reach the process area. Empty containers were often not noticed.

4.9.6 Improve

To reduce the variation, an inexpensive reservoir was added to feed the chemical while containers were exchanged. Additionally, a level transmitter with an alarm was installed to alert operators to containers that were nearly empty.

The improvements have been significant. For one of FilmTec's water membranes, the standard deviation of the product out of spec was reduced from 14.5% to only 2.2%. This resulted in several benefits to FilmTec and its customers including the bonus that membranes were made available to customers faster than before.

4.9.7 Control

To sustain the gains from the project, the Six Sigma team made additional changes. Prior to DMAICS, numbers used to track trends were displayed in tabular form. However, the tables were difficult to read and had little impact on monitoring. Now measurements are displayed on Excel charts that illustrate graphically and with visual immediacy the trends in flux on the finished membrane. The results and learnings from this project were incorporated into a report and posted on the knowledge management website of Dow Chemical.

4.9.8 Sustain

Dow Chemical did not follow the Sustain phase as such. However, in order to emphasise the sustainability of gains during the Control phase they added the process of Leverage (L) and thus extended DMAIC to become DMAICL. In view of the challenges of climate change especially for a chemical company, Dow Chemical should also consider the tools of the Sustain phase.

4.9.9 Benefits

The success of DMAICS is so embedded in the success of Six Sigma initiatives that DMAICS is often regarded as being synonymous with Six Sigma. The training programmes (e.g. Black Belt and Green Belt) are also centred around DMAICS. The grouping of the tools in this book is also roughly structured along DMAICS. The main benefits of DMAICS include the following:

1. It provides a systematic approach which is common to everyone involved in the Green Six Sigma programme.
2. DMAICS represents the life cycle of each Six Sigma project—thus imparting a disciple and rigour of project management to achieve its objectives.

3. The results-orientated approach of DMAICS ensures the tracking of benefits and savings.
4. Unlike other advanced quantitative techniques of Six Sigma (viz. SPC, quality function deployment [QFD]), DMAICS is easy to follow for all members of the project team.
5. DMAICS includes the addition of a Sustain phase especially to ensure the sustainability of environmental standards.

4.9.10 Pitfalls

DMAIC is not immune to pitfalls, however. These include the following:

1. The critics of Six Sigma often argue that there is nothing new in DMAIC. Similar results, they point out, could be achieved by Deming's PDCA cycle or the basic steps of classical industrial engineering.

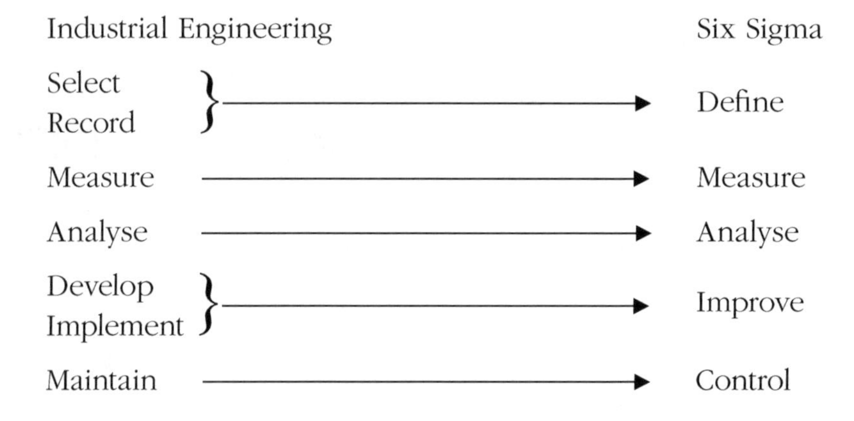

2. It is conceivable that some less complex projects (or the so-called low-hanging fruits) may be subjected to the rigour of DMAIC more than it is necessary and thus delay the realisation of results.

4.9.11 Training Requirements

The training process of DMAICS can be carried out at three levels:

1. Awareness of DMAICS principles in a 1-day workshop.
2. Detailed training of DMAIC (for Green Belts) in 1 week.
3. Training and application of DMAICS as part of Black Belt training, over 5 or 6 weeks. This would include hands-on training of DOE, SPC, FMEA and QFD.

4.10 DMAICS and DMADV

DMADV (define, measure, analyse, design, verify) is especially useful when developing a new product, service or process or implementing new strategies. It is driven by data, it identifies success at each phase and it is thorough in analysis. DMADV is a modified version of design for Six Sigma (DFSS).

The DMADV methodology is appropriate when a new product or process needs to be developed at a company or when an existing product or process needs to meet a Six Sigma level of performance standard.

The first three phases of DMADV (define, measure and analyse) are similar to DMAICS but with more emphasis on customers and markets. The appropriate tools of Green Six Sigma are used at each phase.

4.10.1 Define

The goals of the Define phase are to identify the purpose of the product, process or service, to identify and then set realistic and measurable targets as seen from the perspectives of the customers and market demand.

A clear definition of the project is established during this phase, and every strategy and target must be aligned with the corporate strategy and the expectations of customers.

4.10.2 Measure

In the Measure phase the product or process specifications are CTQ. Steps in this phase include defining market requirements, identifying the critical design parameters, designing performance metrics that will evaluate the design components more important to the quality, reassessing risks and assessing the product and process capability. It is important here to determine which metrics are critical to the customer and to translate the customer requirements into clear project goals.

4.10.3 Analyse

In the Analyse phase actions taken include developing design alternatives, identifying the optimal combination of requirements to achieve value within constraints, developing conceptual designs, evaluating and selecting the best components, then developing the best possible design. It is during this stage a business case is prepared containing an estimate of the total life-cycle cost of the design. After thoroughly evaluating the different design alternatives, the best design option is chosen.

4.10.4 Design

This is an important stage of DMADV, and actions include both a conceptual and detailed design for the selected alternative. The elements of the conceptual design are prioritized and from there a detailed design is developed. Once this step is complete, a more detailed model will be prototyped in order to test it in the laboratory environment to make necessary modifications.

4.10.5 Verify

In the final Verify phase, the team validates that the design is acceptable to all stakeholders. Several pilot and production runs will be necessary to ensure that the quality is the highest possible. Here, expectations will be confirmed, deployment will be expanded and all lessons learned will be documented. The Verify phase also includes a production plan for the product or service to a routine operation and to ensure that the final outcome is sustainable.

4.11 Summary

DMAICS is an integral part of Green Six Sigma. It is systematic and fact based and provides a rigorous framework of results-orientated project management. The methodology may appear to be linear and explicitly defined, but it should be noted that the best results from DMAICS are achieved when the process is flexible and continuous, thus eliminating unproductive steps. An iterative approach may be necessary as well, especially when the team members are new to the tools and techniques. DMAICS has some similarities with the PDCA cycle, and in some ways DMAICS is more rigorous because it has a sign-off checkpoint after each phase.

DMADV is a modified version of DFSS to develop a new product or a process more focused to customers.

GREEN TIPS

- DMAIC is an essential project life cycle for Six Sigma
- DMAICS is the application of DMAIC for climate change initiatives with an additional phase Sustain.
- DMAIC improves an existing process, DMAICS improves and sustains a process and DMADV creates a new process.

Chapter 5

The Scope of Green Six Sigma Tools and Techniques

> It must be considered that there is nothing more difficult to carry out, no more doubtful to success, no more dangerous to handle than to initiate a new order of things.
>
> **—Machiavelli (1513)**

5.1 Introduction

Tools and techniques are essential ingredients of a process. These are instrumental to the success of a quality programme. Similarly many companies have used tools and techniques without giving sufficient thought and have then experienced barriers to progress (McQuater et al., 1995). Tools and techniques can be a double-edged sword—they are effective in the right hands and can be dangerous in the wrong ones. Tools and techniques applicable to different stages of the project should be identified. Tools and techniques should also be classified as core and optional. For example Green Belts may use core tools only and projects managed by Black Belts may require the use of both core and optional tools. Tools and techniques have been described with their appropriate applications in Part II and Part III.

In this chapter, we aim to address the general guidelines and scope for tools and techniques under the following headings:

DOI: 10.4324/9781003268239-6

- The Drivers for Tools and Techniques
- The Problems of Using Tools and Techniques
- The Critical Success Factors

Tools and techniques are not a panacea for quality problems, but rather can be seen as a means of solving them.

5.2 The Drivers for Tools and Techniques

It is clear that an organised approach of continuous improvement will require the use of a selection of tools and techniques for any effective problem-solving process. There are a number of good reasons for this including

- They help to initiate the process
- They pinpoint the problem
- They offer a basis for systematic analysis leading to a solution
- Employees using them feel involved
- They enhance teamwork through problem-solving
- They provide an effective medium of communication at all levels
- They form a single set of methodology
- They facilitate a mindset of quality culture

With the continuous growth of outsourcing and collaborative partnership between suppliers and customers, some tools and techniques offer a common platform for service-level agreements. A customer may insist upon the use of a specific technique as part of an agreement with its supplier. For example, automotive component suppliers have developed a learning experience to apply failure mode and effects analysis (FMEA) to satisfy the customers that the technique is applied in an effective way. For this manner, both the supplier and customer share the improvement programme and enhance the mutual competitive advantage.

In DMAICS (define, measure, analyse, improve, control and sustain) life cycle some tools, with various degrees of details, are used in more than one phase. For example, process chart is used in both Measure and Analyse phases or mistake proofing is used in both Improve and Control phases. Most of the tools are used for a specific project. However, some tools in the Sustain stage (e.g. Balanced Scorecard and S&OP) are also applicable to the whole organisation to sustain the continuous improvement culture.

The following three key factors should be considered carefully when selecting tools and techniques for a quality initiative:

1. Rigour in Purpose—The tools or technique selected must be meeting the main purpose or reason for its application. No single tool is more important in isolation but could be most significant for a specific application. The approach must not be 'a solution in search of problems'.
2. Rigour in Training—It is imperative that all users of a tool or technique are trained to a level of competence so that they feel comfortable to apply it effectively. It is like giving someone the best golf club and expecting that person to win automatically a grand tournament. It just is not possible to become a good player without proper training and practice—otherwise, every one of us could be a Tiger Woods or a Jack Nicklaus.
3. Rigour in Application—After the appropriate selection of a tool or technique followed by adequate training, its success will be determined by the results of its application. The key criteria are as follows: has it solved the problem or has it improved the process? There are instances when a company created a high expectation by selling the virtues of one specific technique. A single tool or technique on its own will produce results in a limited area. It is the cumulative effect of a number of appropriate tools and techniques that would create sustainable benefits for the whole organisation: Dale et al. (1993) developed a table which would assist in identifying the specific application of tools and techniques (see Table 5.1).

5.3 The Problems of Using Tools and Techniques

There are many examples where difficulties were encountered during the application of complex techniques such as the design of experiments (DOE), failure mode and effects analysis (FMEA) and quality function deployment (QFD). However there are also more frequent instances when the use and application of all types of tools and techniques were not addressed to the specific requirements of companies and their people using them. There are common problems in the use of all tools and techniques. Dale and McQuater (1998) identified five areas of difficulties, viz, resources, management commitment, detection-based mentality, knowledge and understanding and resistance to change.

Table 5.1 Application of Tools and Techniques

Application	*Tools and Techniques*
1. Checking	Checklists, control plans
2. Data collection/presentation	Check sheets, bar charts, tally charts, histograms, graphs
3. Setting priorities/planning	Pareto analysis, arrow diagram, quality costs
4. Structuring ideas	Affinity diagrams, systematic diagrams, brainstorming
5. Performance/capability measurement/assessment	Statistical process control, departmental purpose analysis
6. Understanding/analysing problems/process	Flow chart, cause and effect diagrams, process decision programme chart (PDPC)
7. Identifying relationship	Scatter diagrams/regression/correlation/ matrix diagrams
8. Identifying control parameters	Design of experiment
9. Monitoring and maintaining control	Mistake proofing, FMEA, matrix data analysis
10. Interface between customer needs and product features	Quality function deployment

Our research (Basu and Wright, 2003) has pinpointed clearly four key factors leading to problems for the effective application of tools and techniques:

1. Inadequate training
2. Management commitment of resources
3. Employee mindset
4. Poor application of tools and techniques

Let us address these areas in a little more detail.

5.3.1 Inadequate Training

The proper training of employees for the use of tools and techniques depends on three key factors:

- Technical facts transfer
- Organisational culture and education
- Qualified trainers

Most of the tools and techniques are conceptually simple but detail rich. It is important to separate the technical aspect of a tool or technique and transfer the facts in the context of the known environment of the organisation. It is very helpful to demystify a concept or definition by a worked-out example based upon familiar data from the company.

The users of the tools are usually derived from a different level of the organisation and therefore the learning process must be geared to the specific capability and culture of the organisation.

The initial training is usually carried out by specialists or external consultants. The employees are likely to find the training more sustainable when it is carried out by their line managers or colleagues who have been trained as trainers. Furthermore, if the training depends on external consultants then the learning ceases with the departure of these consultants.

5.3.2 Management Commitment of Resources

Members of senior management are often unsure what they may expect from tools and techniques. There is also the existence of rivalry between different departments, and middle managers fail to see the benefits of a change beyond their own turf. Consequently, adequate resources are not always made available to support the training and improvement activities arising from the application of tools and techniques.

In order to have any success with the use of tools and techniques, it is essential to ensure total commitment and leadership from the top management. This form of support is more significant for basic tools for company-wide applications (e.g. cause and effect diagram, Pareto analysis, flow diagram). For more advanced and specialist tools and techniques (e.g. DOE, FMEA, QFD), departmental involvement is even more important.

5.3.3 Employee Mindset

In congruence with the commitment of senior management, another important hurdle encountered while implementing tools and techniques is the mindset of all the employees of a company. As part of a company-wide quality programme, a group of people are brought together who may have never worked together before to achieve a novel task. In spite of the full support of top management, failure to convince the project stakeholders may cause your project to fail. In the case of tools and techniques, the main stakeholders are employees. Managing people is a complex area encompassing such considerations as human motivation, attitude and culture. There is an abundance of published works available for the reader who is interested in pursuing this area in greater depth. For our specific purposes regarding the problems of using tools and techniques, three models or approaches are most relevant.

Tuckman (2001) outlines that members of the team have to develop a set of common values or norms before they can work together effectively as a group. As shown in Figure 5.1, project teams typically go through five stages, which can be dubbed: forming, storming, norming, performing and mourning.

At forming, the team members come together with a sense of high anticipation and motivation. As the team begins to work together, differences occur during the 'storming' stage, and the team's level of performance drops. Gradually the team develops a sense of group identity and a set of values or norms. The team members start to work together, and their motivation and effectiveness begin to increase and thus group performance reaches a plateau. Finally towards the end of the project where the future of the team is uncertain, the mourning stage sets in.

The second model is based upon McElroy and Mills (2000). The two attributes of employees, commitment and knowledge, are mapped onto a chart,

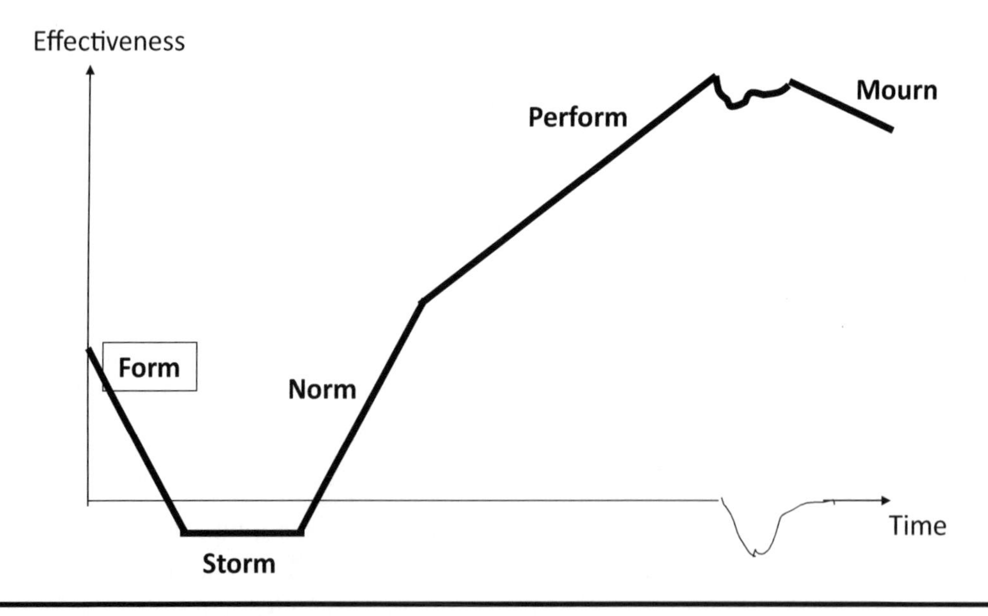

Figure 5.1 The Tuckman model of team formation.

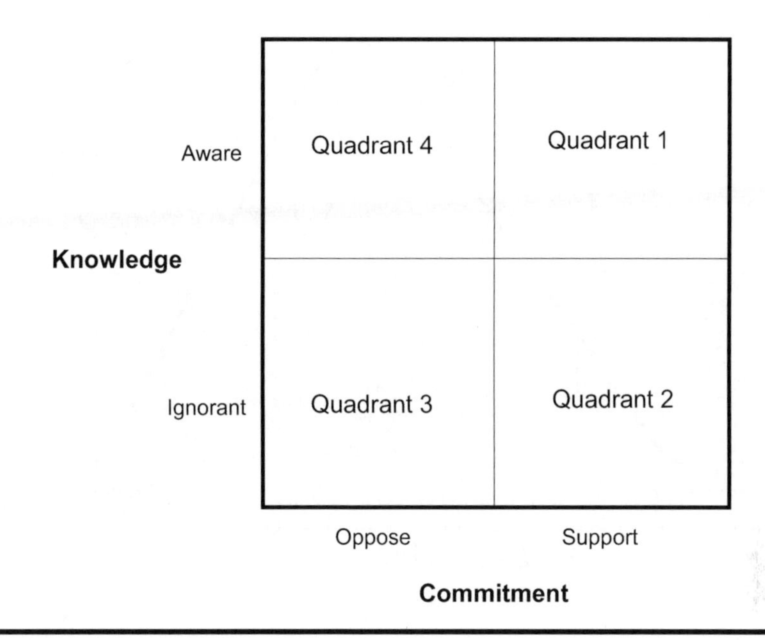

Figure 5.2 Knowledge-commitment matrix.

as shown in Figure 5.2, to group the knowledge base across employees into four quadrants. To populate this chart certain assumptions are made regarding the knowledge base and attitudes of employees. These assumptions could be tested by a voluntary survey.

Quadrant 1: Support/Aware: These supporters are the key players for tools and techniques.
Quadrant 2: Support/Ignorant: This support is vulnerable and should be reinforced by training.
Quadrant 3: Oppose/Ignorant: This is a key target area to achieve commitment by a combination of culture and facts transfer.
Quadrant 4: Oppose/Aware: This will be the most difficult group of employees to convert. The only way to move them to support the project could be to negotiate a training role for them.

In the third model, Wallace (1990) suggests that the distribution of three groups of employees (naysayers, the silent majority and enthusiasts) can be transformed by a properly administered education programme for employees. This change is illustrated qualitatively in Figure 5.3 and Figure 5.4.

Even after education, there will be a small group of people (the 'naysayers') who do not believe that tools and techniques will actually work.

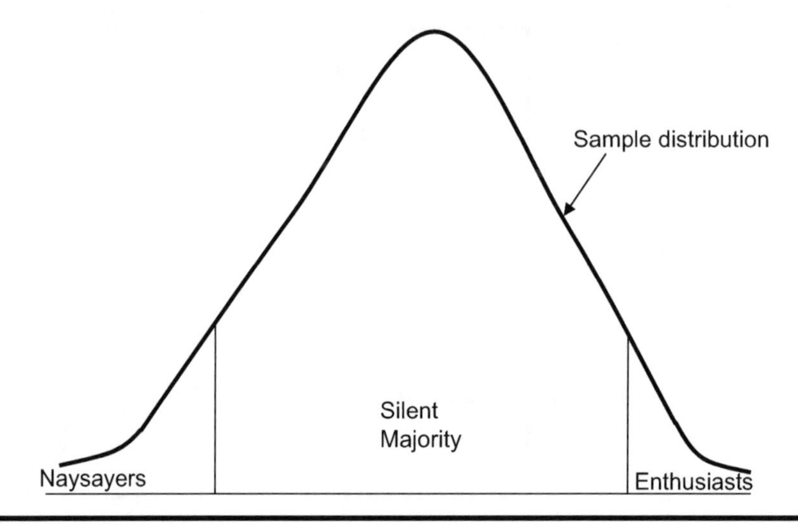

Figure 5.3 Commitment before education.

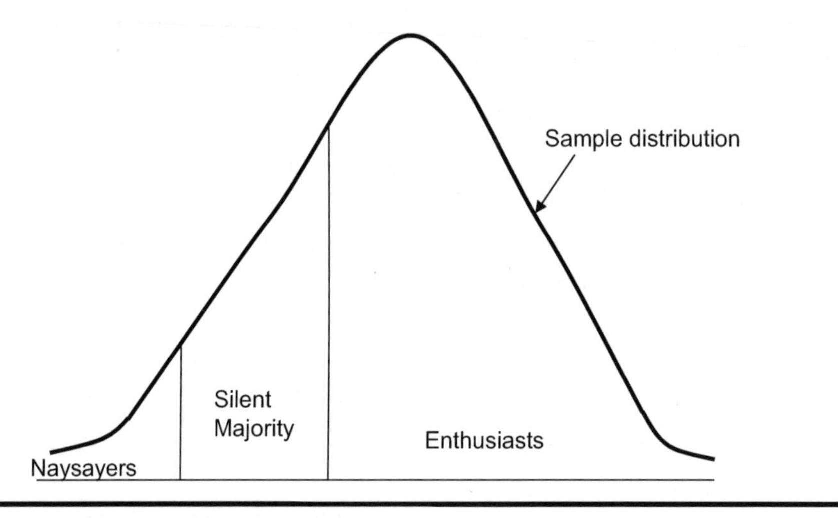

Figure 5.4 Commitment after education.

5.3.4 Poor Application of Tools and Techniques

The inappropriate application of tools and techniques introduces the additional problems of using them to achieve and sustain results. This is often a more difficult area to correct when the user has applied a tool which is specific for a departmental target. For example, a packing line supervisor is striving to maximise the overall equipment effectiveness (OEE) of a particular soap wrapping machine, ACMA 791. The supervisor's aim will be to maximise output and minimise the loss time of the machine. As a consequence, more than the scheduled quantity is likely to be produced. This creates an obstacle to achieving the scheduling performance. The appropriate application of tools and techniques aims to improve and sustain the business performance as a whole rather than conflicting departmental efficiency. An effective way to overcome these 'turf management' issues is to ensure that training includes an understanding of the holistic approach of using tools and techniques for the business rather than the department.

The typical signs of the poor application of tools and techniques are

- Department managers are using tools and techniques as a routine to publicise departmental performance rather than as a basis to identify deficiencies for improvement.
- There is dependence on computer software rather than the rigour of data collection, correct input and interpretation of results.
- Tools and techniques are used to find excuses for not making changes.
- Knowledge is restricted to specialists and not shared by relevant employees.

5.4 The Critical Success Factors

The factors influencing the success of tools and techniques are naturally contributed by the 'drivers for tools and techniques'. However, in spite of the apparent paradox, the 'problems' of using tools and techniques also contribute to the success by offering restraining forces. The understanding and anticipation of problems at an early stage ensures the avoidance of pitfalls during a later aspect of the application. The restraining forces provide the basis for both quality assurance and quality control to the drivers of tools and techniques leading to critical success factors as shown in Figure 5.5.

Force field analysis, created by Lewin (1951), is also a useful tool for making decisions by analysing the forces for and against a change. The analysis of the driving forces and restraining forces also helps to communicate the reasoning of differing points of view to reach a consensus. Force field analysis is described in Chapter 8.

There are a number of resultant success factors in the effective use of tools and techniques, and these are mostly in common with success factors for a company-wide change programme. The critical success factors are

1. Top management commitment
2. Availability of resources

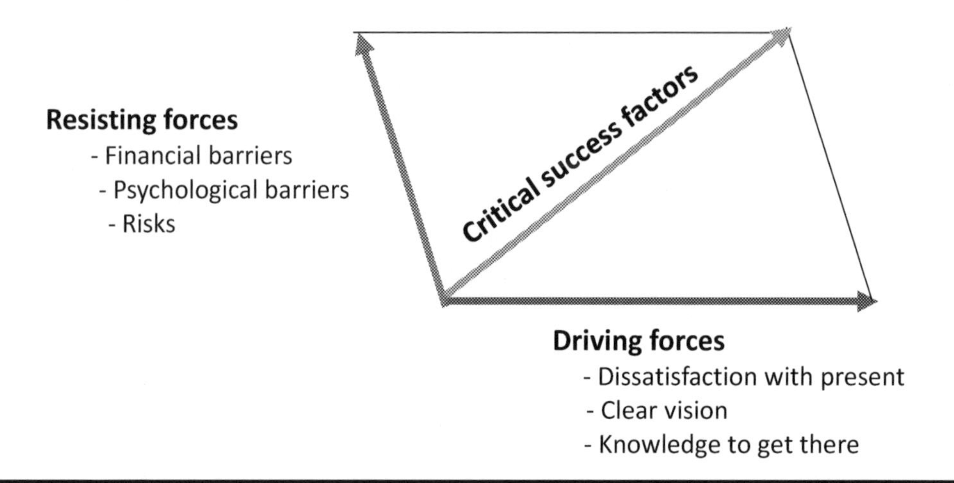

Figure 5.5 Forces of change management.

3. Well-designed education and training programme
4. Rigorous project management approach

5.4.1 Top Management Commitment

Various case studies and research reports have concluded over and over again that without the total commitment of senior management, a company-wide project or change programme can never succeed. The essential requirement is that the top management must be convinced that there is a business case for the use and application of tools and techniques. The secondary requirement is that they can be embedded in the corporate programmes. Critical to the successful introduction of tools and techniques is a champion, a senior manager whose position, enthusiasm and knowledge can be used to good effect.

5.4.2 Availability of Resources

It is paramount that, once a decision has been made by the management to implement tools and techniques as part of a company-wide quality programme, the management ensures that all relevant resources for the initiative are available and used effectively. Resources in this context relate to people, facilities, systems, cost and money. The availability of one's own employees is most critical, because the demands of the 'day job' to meet customer-dominated business requirements often introduce a conflict between the project and the operation. The 'people resource' is required at two levels: they are necessary as project team members or trainers as well as other employees fulfilling the role of users. A good practice of resource utilisation is to train a small group of carefully selected people in tools and techniques and then use them to transfer their knowledge to other employees of the company. This approach has the advantage of both reducing the cost of consultants and external trainers and customising the training to the needs and culture of the company.

5.4.3 Well-Designed Education and Training Programme

Education is fundamental to making things happen. Education comprises both the 'facts transfer' related to tools and techniques and the 'behaviour change' necessary for employees to do their jobs differently. Education lies in teaching employees how to use the tools and getting them to accept that they can work with these tools as a team.

Training is the process of making smaller groups of people become proficient in the more complex techniques, such as QFD, FMEA and DOE.

Many Total Quality Management (TQM) education and training programmes have failed because of their emphasis on cultural issues and were nicknamed the 'country club approach'. The rigorous training deployment plan of Six Sigma is a good model for a well-designed programme. Employees are educated and trained following defined modules and then assessed to gain 'Green Belt' or 'Black Belt' status. Green Belts are educated in tools while Black Belts are additionally trained in the application of techniques.

5.4.4 Rigorous Project Management Approach

The success of the tools and techniques is further assured when these are used as part of a well-managed project or programme. The rigours of project management in the context of this discussion must include

- A well-defined project brief
- Project appraisal before any major commitment
- Project organisation
- Stakeholder management
- Regular monitoring of time, cost and benefits
- Continuous feedback, learning and sharing of good practice

Common failures are the overrun in time and budget as well as incomplete achievement of the original objectives of the projects. Projects generally have five basic elements: scope, time, cost, quality and risk. With the use and application of tools and techniques, we take a whole systems approach to each of the six basic elements and the related issues listed earlier.

5.5 Summary

In this chapter the scope and importance of appropriate tools and techniques in Green Six Sigma projects have been analysed. A review of both the positive drivers and the problems conducive to restraining forces for the use and application of tools and techniques is included. It can be concluded that success factors are derived from the resultant vector of the mutually contradicting driving forces and restraining forces. The critical success factors are top management commitment, availability of resources, well-designed education and training programmes, stakeholder management and a rigorous project management approach.

Although tools and techniques are designed to address problems in specific stages of the project life cycle, there are some tools which are used in more than one stage of a project.

It is important to note that the effectiveness of a tool or technique depends on the rigour in purpose, rigour in training and rigour in application.

GREEN TIPS

- Tools and techniques are essential requirements of Green Six Sigma projects.
- Choose them for their purpose and apply them after a training programme.

Chapter 6

The Digital Revolution and Climate Change

Whether you like it or not we are now in a digitally interconnected world.

—Barak Obama

6.1 Introduction

Nicholas Negroponte (1995) predicted over three decades ago: 'Like a force of nature the digital age cannot be denied or stopped. It has four very powerful qualities that will result in ultimate triumph: decentralizing, globalizing, harmonizing and empowering'. The digital revolution is definitely here to stay, and we can safely say that each generation will become more digitally competent than the preceding one. It is reinventing business models, reshaping economic sectors as well as changing societal infrastructures. Big data and artificial intelligence (AI) along with several other technological developments are now fundamentally altering the ways in which economies work and how we live our lives. Perhaps surprisingly, this movement to digital technologies is also envisaged to play a significant role in the planet's ecosystem along with climate change.

The focus of information and communication technology within organisations has shifted dramatically over the last 40 years, moving from improving the efficiency of business processes within companies to enhancing

DOI: 10.4324/9781003268239-7

the effectiveness of the value chain reaching suppliers, customers and consumers. During the 1960s and 1970s, businesses focused on the use of mainframes to process large quantities of data. In the 1980s and early 1990s organisations concentrated instead on using personal desktop computers to improve individual efficiencies.

The last decade with the revolution of the widespread use of the Internet has seen the use of technologies to create electronic communication networks within and between organisations and individuals. The implementation of enterprise resource planning (ERP), websites, e-commerce and email systems during the past 15 years have allowed individuals within organisations to communicate together and share data. Information technology (IT) has now grown into information and communication technology (ICT). In this chapter we consider the following broad areas:

- Information technology and systems
- E-business
- Big data and AI
- Digital tools for Green Six Sigma
- Digital applications in climate change

6.2 Information Technology and Systems

IT is rapidly changing and becoming more powerful. It is a continuing source of competitive advantages for manufacturers and the supply chain if used correctly. By 2000 the personal computer (PC) on the desk of an average operations manager even then offered more computing power than the average US $100 million a year manufacturing plant had provided ten years earlier in 1990. The beauty of it was that this amazing IT revolution was available to everyone. However, it is how a company puts it to work that determines the extent of their competitiveness in the global market.

The rapid growth of IT has created both problems and challenges. Many senior managers of companies lack any detailed understanding of the complexity of technology. They either follow the current fashion (e.g. 'no one was ever fired for choosing IBM') or they are discouraged by the cost of technology, or by a lack of evidence that savings can be made in a new field. When executives read about all the clever things that seemingly low-cost computer technology can do, understandably they feel frustrated when the systems experts caution them, 'But it will take 3 years to develop the software'.

Most senior managers also feel lost in a blizzard of buzzwords and are conscious of well-publicised failures. A notorious example can be found in the UK National Health Service's electronic care records project of 2002–2011. This scheme was discontinued after US $2.3 billion had been spent, and was described by members of parliament as one of the worst and most expensive contracting fiascos ever encountered. Further examples of mismanagement of the potential of technology can be seen in the US Air Force Enterprise Resource Planning project 2005–2012, which was cancelled at a cost of US$1.1 billion; and in Denmark's 2007–2012 POLSAG plan concerning police case file management. This scheme was cancelled after an enormous DKK500 million expenditure. The list of similar and extremely costly failures is, unfortunately, endless.

Yet another issue is the implementation of systems for the benefit of the users. When a company looks for an IT solution to a problem without re-engineering the process, instead seeking to refine the existing database or concentrating upon training the end users, the application is doomed to fail. Real disasters can be very expensive. For example the US$60 million Master Trust accounting system for Bank of America had to be scrapped because it could not fulfil the simple brief of keeping accurate accounts.

Figure 6.1 shows a framework of IT strategy comprising three levels of hardware strategy, software strategy and implementation strategy.

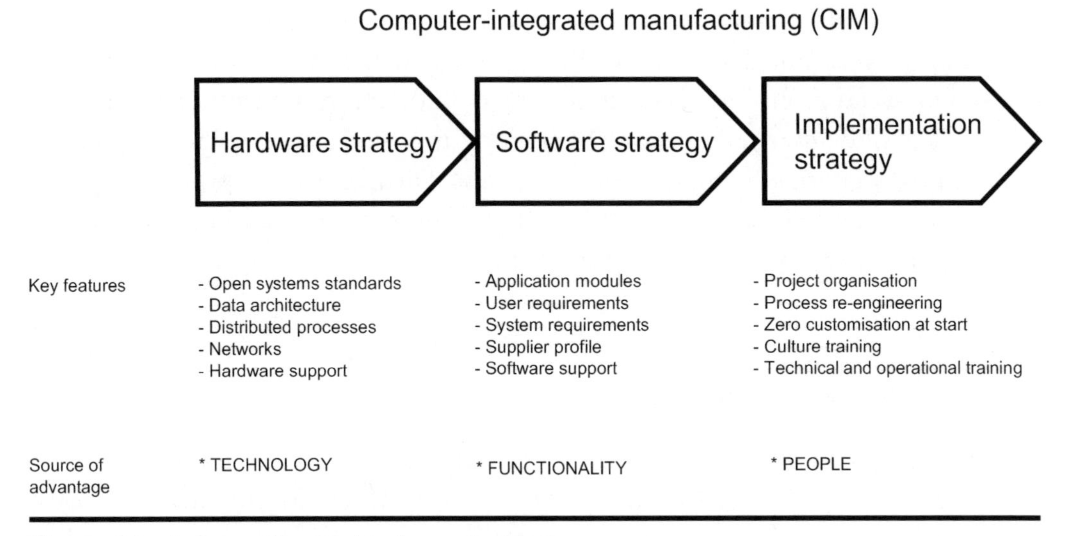

Figure 6.1 Information technology strategy.

6.2.1 IT Hardware Strategy

Hardware requires auditing with a refresh cycle of ideally 3 years but no more than 5 years. Chief information officers (CIOs) have the ongoing challenge of possessing sufficient IT capability to respond to new business needs and balancing the cost of capital spending with operating efficiency, while at the same time they must be mindful of risk. The rapid growth in data storage has placed pressure on CIOs and has tempted them to retain old storage hardware. However, keeping older and not always reliable unsupported hardware can be costly and risky. Using hardware until it breaks down or the supporting application is no longer available might save capital expenditure in the short term but could prove disastrous over a longer period. Across a supply chain there will be significant differences in the way organisations run their operations and in the applications that are used. New hardware is quicker, consumes less power, requires less maintenance, is more reliable and reduces labour (help desks, fixing and maintenance). A well-managed refresh audit will ensure the correct trade-offs between capital expenditure and lowering maintenance and power costs, as well as identifying actual and expected needs.

The hardware strategy should also include the capability of local hardware support both by suppliers and the company's own staff. The support capability may influence the selection of hardware. A sensible strategy is to go with the market leaders who are setting the de facto standards.

ERP systems are supply chain IT systems that exchange information across all functions of an organisation or enterprise and can be extended across the supply chain to gain integration and the sharing of information. There are several modules of an ERP system which can be installed and are either stand alone or which function by interaction with other modules. Some of the key components are Finance, Purchasing, Master Production Scheduling, Materials Management, Sales and Distribution, Supplier Management and Human Resources. ERP systems clearly hold major advantages over 'legacy systems' in terms of functionality, scope and flexibility of applications.

6.2.2 IT Software Strategy

At the early stage of IT, applications software was limited to financial and commercial areas. Now a company is faced with a bewildering array of software ranging from design/process engineering, to manufacturing, to supply chain, to administration. Versions of specific software and systems technology will continue to change. Therefore, it is vital that a manufacturing company formulates a software strategy by careful planning.

The first step is to identify the areas of application depending on the size of activities and priorities of the company. Figure 6.2 shows a framework of application software in five key areas, namely financial administration, supply chain management, factory administration, and 'client' workstation. The traditional computing modules of accounts and payroll are in the realm of financial management. The biggest area of application lies in supply chain management starting from sales forecasting and ranging to customer service and electronic data interchange (EDI). At the factory shop floor there are two application areas, namely factory administration—comprising management information systems—and factory automation—encompassing design, process engineering and automation of equipment.

The software for client workstations is PC based. During the late 1980s many manufacturing companies searched for one turnkey package and

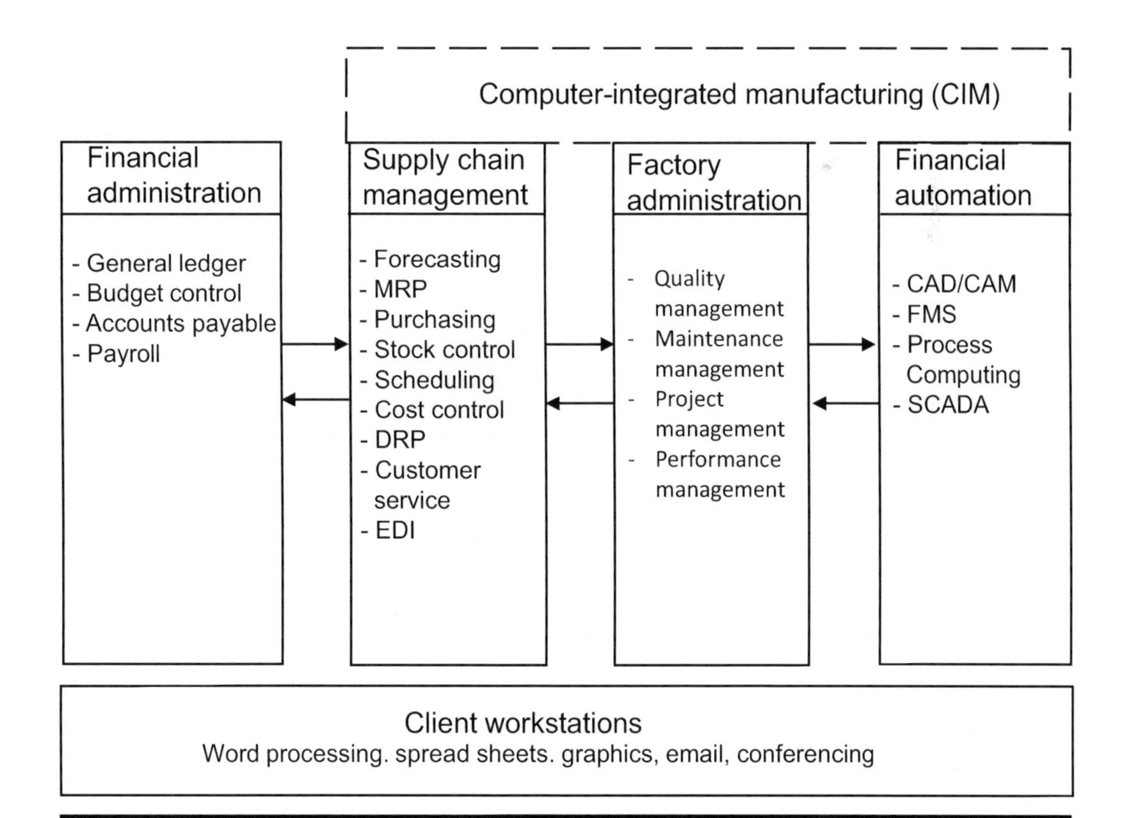

Figure 6.2 Application software modules.

invested in what is known as computer-integrated manufacturing (CIM) although with limited success. If a company follows an 'open systems' policy for a hardware and relational database, then different proprietary software packages stand a better chance of being interfaced and database information can be shared in a client-server environment. Probably the most significant advantage lies in the enterprise-wise view of a business that ERP systems allow.

The software policy should include standard packages for the company in specific areas of application. The selection of software should conform to the key criteria of user requirements, systems requirements, supplier profile and software support. The earlier examples of applications software were relatively inflexible, and the approach was to 'systematise the customer' rather than 'customise the system'. Many disillusioned customers attempted to build their own software and burnt their fingers in the process. In the present climate, software tools have become flexible, the IT technology is advancing rapidly and competitive expert support is provided by specialist software houses; thus it is prudent to buy appropriate software rather than to develop your own (see Figure 6.3). The software should conform to open systems requirements, and the supplier should be both reputable and locally available for support. The company should also build up its own IT support staff, especially a 'user support' service.

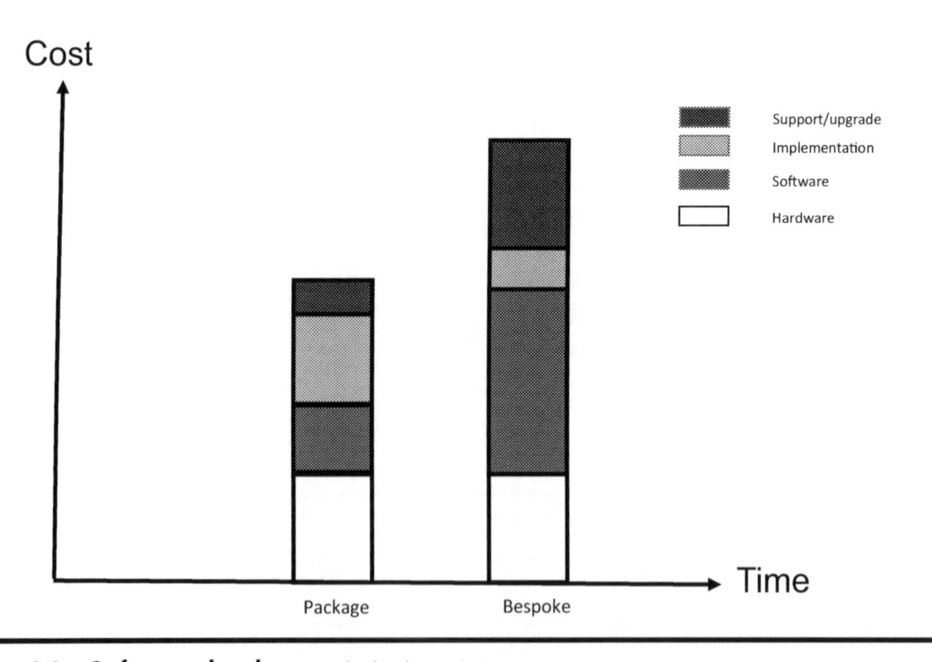

Figure 6.3 Software development strategy.

There is a major conflict in developing a software strategy between a 'best of breed' approach and a 'single integrator' methodology. In a 'best of breed' methodology the optimum functionality solution of each individual function is chosen, and companies attempt to integrate different types of systems. Although the business is likely to obtain the best solutions in each area, the problems of integration often offset the advantages of such top-quality solutions. ERP providers offer flexible modules that allow a single integrated framework of different functions of a supply chain stage. A single integrator approach also offers the advantage of technical support and a maintenance contract with a single-source supplier.

6.2.3 Market-Making Applications

There are broadly two types of market-making applications. The first allows businesses to buy or sell by online auctions or bidding. Buyers place an open order to purchase an item and the sellers have the opportunity to bid. The second mechanism is the exchange or two-way auction platform containing a high-speed bid/ask bartering process.

6.2.4 Enterprise Resource Planning Applications

Internet technology has certainly enhanced the collaborative business culture by enabling the transparency of online information and transactions. Company-centric enterprise application vendors including SAP and Oracle (J D Edwards Enterprise One) have extended into supply chain supervision and customer relationship management (CRM).

6.2.5 Customer Relationships Management Solutions

It is fair to state that most businesses regard the retention of customers as an important goal and therefore the criteria of CRM are not new or unfamiliar to most enterprises. However, the collaborative power of the Internet-based network has enhanced the need for customer intimacy and personalisation. A number of software solutions have been developed to provide some powerful holistic functionalities including

- Using a customer database for knowing and understanding customer characteristics
- Managing the relationship with key business partners (e.g. customers)

- Providing value-added services to retain customer loyalty
- Providing transparency and real-time acceptability of information for both customers and suppliers
- Optimising cross-selling opportunities

6.2.6 Supply Chain Management Solutions

There are now few companies that do not recognise that the Internet has had a profound effect on supply chain performance. Applications that fall into this category are essentially decision support software packages for optimising multiple levels of demands and supply in the global supply chain.

The collaborative planning, forecasting and replenishment (CPFR) for key stakeholders of the total supply chain has emerged. In CPFR, data and process model standards are developed for collaboration between suppliers and an enterprise with prescribed methods for planning (agreement between the trading partners to conduct business in a certain way); forecasting (agreed-to methods, technology and timing for sales, promotions and order forecasting); and replenishment (order generation and order fulfilment). These solutions take into account the constraints of transportation, supply capacity and inventory requirements. The ultimate objective is order fulfilment within the set time and at a cost acceptable to customers.

The early leading vendors in the market, for example i2 (now part of IBM) and Manugistics (acquired by JDA Software in 2006), have been surpassed by SAP and Oracle's J D Edwards Enterprise One.

6.2.7 Implementation Strategy

The success of an IT strategy depends on well-managed implementation as much as it does on the selection of the appropriate hardware and software.

Similar to a company-wide programme such as Total Quality Management (TQM), the implementation must have top management commitment. This should be reflected in setting up a project team comprising members from all users (marketing, logistics, manufacturing, accounts) and business systems. The project manager is usually chosen from the main user group. For example if the application software is for supply chain management (SCM) then the project manager should ideally have a logistics background.

The project team should receive both technical training and operational coaching (functionality of the software). The project manager then prepares a clearly stated action plan with target dates and resources for key activities.

This plan must include review points and steering by the members of the board.

It is essential that the existing procedures and processes are thoroughly and systematically reviewed. There are various tools for analysing the flow and requirements of these existing systems. Statistical process control (SPC) techniques are widely used. Nowadays some companies are applying computer-aided software engineering (CASE) tools to analyse the structure, database and flows of the existing processes and compare them with the proposed software for implementation. With the success of the business process re-engineering (BPR) approach of Hammer and Champy (1993), some companies are utilising an IT application as a catalyst and applying the principles of BPR to re-engineer the total business processes of the company. The approach chosen should depend on the depth and breadth of the application systems, but there is no doubt that the existing procedures must be reviewed and refined when implementing a new method.

One important rule is that the user should not try to customise the system at the outset. Often, having acquired experience on the new system the user may find that in reality, the need for and nature of customisation could be very different. However, it is necessary that a 'prototype' is tested for any new system using the company's own data.

After the preparation and teaching of the project team, the training programmes should be extended to all potential users of the system. The training features should contain both cultural education to establish acceptance by everyone concerned and operational instruction to understand the functionality and operations of the new system. Training documents must be designed specifically for the users' needs. The next stage is the data input and 'dry run' of the new system in parallel with the existing procedure before the system goes live. There are numerous benefits to forming users' groups for exchanging experience, with members drawn both from within and from outside the company.

6.3 E-business

One might gain the impression from today's press that all business problems can be solved by e-business while, at the same time, the media do tend to blame all business failures and any economic downturn on e-business as well. Given the volume of news items on the subject, it may appear that defining 'e-business' is merely to state the obvious—or is it?

The distinctions between e-commerce, e-marketplace and e-business are poorly interpreted. For example, the most popular perception of e-business is that it is best exemplified by online shopping. However, it should be noted that in 2015 in the UK only 25% of retail sales were made online, rising to only 26.2% of sales in 2020, even taking into account the effect of Covid-19 lockdowns. (*source*: www.statista.com/statistics/315520/online-share-of-retail-sales-in-the-uk/)

Let's take a moment to clarify some aspects. E-commerce is the transactional electronic exchange that takes place involving the buying and selling of goods and services.

The 'e-marketplace' is the online intermediary for electronic transactions between buyers, sellers and brokers. This is also referred to as the digital marketplace, portals or hubs.

Early opportunities were observed in the enabling infrastructures and Internet-based networks (Internet, intranet and extranet), which replaced existing telephone, fax and EDI networks. The early success of e-procurement vendors (e.g., Commerce One, Ariba, Info Bank) was well received. However, the old suppliers suffered many problems including that of authorisation with no conformity of systems between business partners. It was rather like having different telephone systems for each of the people to whom you speak. This has been transformed by trading portals that interconnect the contents of different suppliers thus making them usable by all buyers.

A report by Basu (2003) indicated that the complex web and infrastructure of e-business applications have been simplified as shown in Figure 6.4 which illustrates the 'building blocks'.

There are five key types of e-business application systems that enable businesses to trade and conduct electronic transactions or communications:

- E-commerce solutions for both the 'sell-side' and 'buy-side' applications
- Market-making applications that enable multiple buyers and sellers to collaborate and trade
- CRM solutions to facilitate improved business partnerships with customers
- ERP solutions for site-based planning and execution of operations
- SCM solutions for optimising the demand and supply in the total supply network including for the suppliers

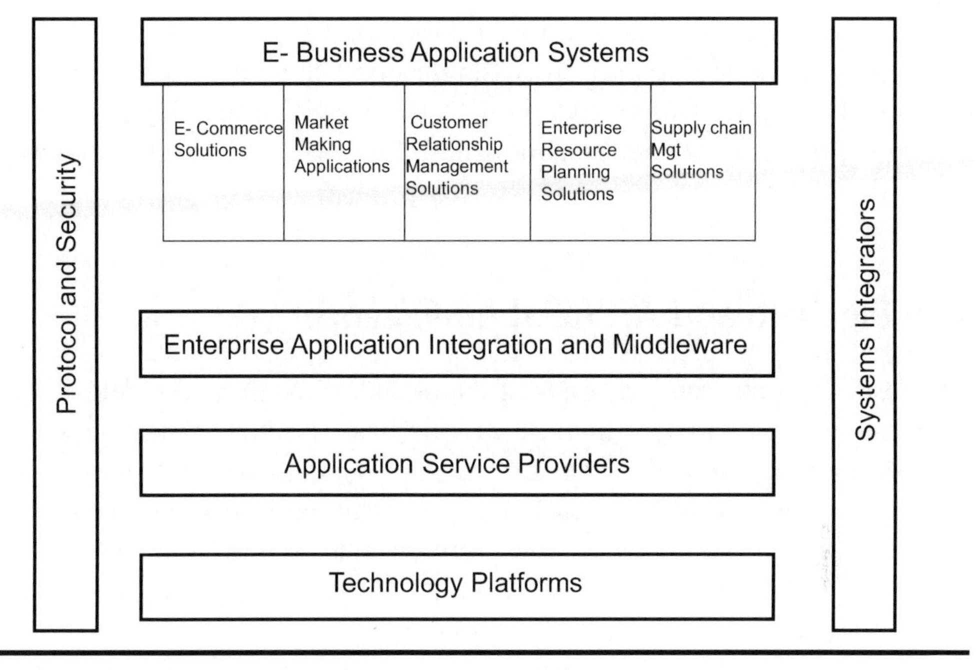

Figure 6.4 E-business building blocks.

6.3.1 E-commerce Solutions

The buy-side applications of e-commerce, initially targeted at larger buyers, enable companies to levy across new or existing vendors. Solutions are increasingly aiming at integrating ERP systems with the organisation's own suppliers and customers. The new application developers are utilising the opportunities created by the lack of integration of ERP systems with other Internet systems and outside companies.

Initially, buy-side application vendors (including Commerce One and Ariba) were driven by pure-play solutions for the purchase of MRO (maintenance, repairs and operations) or indirect goods. The huge potential of e-procurement offered up by 'pure companies' has been recognised and seized by established ERP vendors such as SAP and Oracle, and software vendors like Netscape and Datastream.

The buy-side vendors, whether pure-play or not, are focusing on packaged buy-side application suites and looking to move into the direct procurement area. This requires a greater degree of understanding of business processes in specific industries and rigorous validation of the data processing.

The sell-side application vendors are looking to provide services content management and transaction processing. Hence, there are some subcategories of software within this group. These include cataloguing, profiling, configurations and payment technologies. As a result, this sector is highly fragmented.

6.4 Big Data and Artificial Intelligence

The term 'big data' refers to piles of data that are meaningless unless you have the power to analyse them. Everyone has a limited vertical view of data sitting in its own silo. The advantage of big data is that it lets you get a 360-degree view of information in both vertical and horizontal planes and right across your organisation. Data from discrete silos is brought together by a single data lake in web services (such as AWS or Amazon Web Services), now commonly known as cloud computing.

A typical flow of big data analytics is shown in Figure 6.5 where unprocessed information from multiple sources is extracted into a daily batch of raw evidence. This data is then cleansed and brought into a data lake using cloud computing. The data is then transformed and queried by a set of web services tools (e.g Apache) and then personalised and visualised by using Tableau.

One of the key benefits of cloud computing is the opportunity to replace up-front capital infrastructure expenses with low variable costs that scale

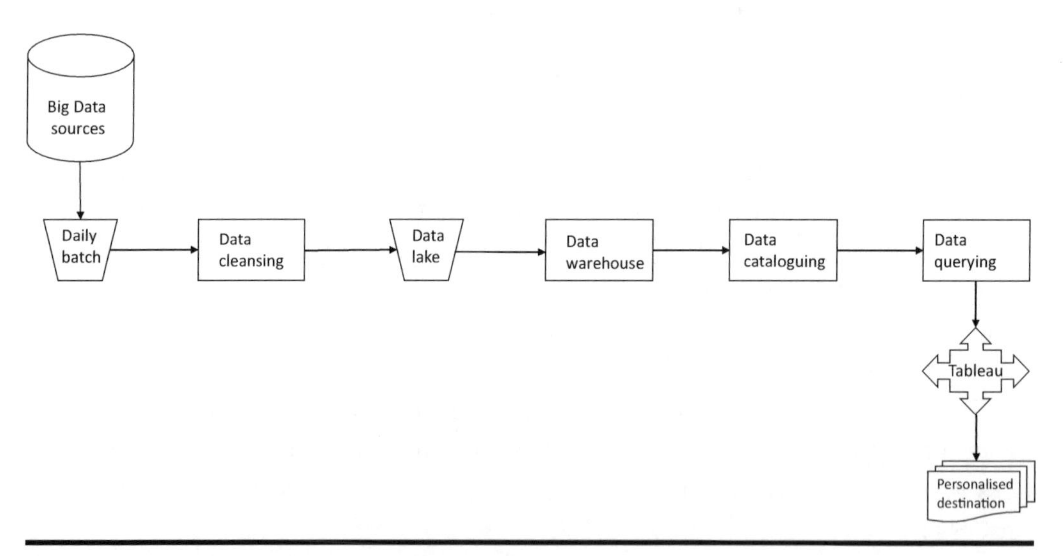

Figure 6.5 A typical flow of big data analytics.

with your business. Companies no longer need to plan for and procure servers and other IT infrastructure weeks or months in advance. As part of cloud computing, small and medium-sized enterprises (SMEs) can afford to invest in licensed and expensive ERP systems (also called 'on-demand software') on a so-called SaaS (Software as a Service) platform.

AI is the simulation of human intelligence processes by computer systems. AI research has been divided into subfields such as particular goals (e.g. 'robotics') and the use of specialised tools ('machine learning'). Specific applications of AI include expert systems, natural language processing, speech recognition and machine vision.

AI programming concentrates upon three cognitive skills: learning, reasoning and self-correction. Learning processes of AI programming focus on acquiring data and creating rules or algorithms to provide computing devices with step-by-step instructions for how to complete a specific task. Machine learning (ML), a fundamental concept of AI research, is the study of computer algorithms that improve automatically through experience. Reasoning processes of programming focus on choosing the right algorithm to reach a desired outcome. Finally, self-correction processes of programming are designed to continually fine-tune algorithms and ensure they provide the most accurate results possible.

AI is relevant to any intellectual task and advanced robotics. Modern AI applications are too numerous to list here; however some high-profile examples include autonomous vehicles (such as drones and self-driving cars), medical diagnosis, creating art (e.g. poetry), proving mathematical theorems, playing games (such as chess), search engines (e.g. Google Search), online assistants (such as Siri), image recognition in photographs, spam filtering, the prediction of judicial decisions and the forecasting of demands, to name a few.

Since the introduction of iPhone by Apple Inc. in June 2007, the iPhone iOS has dominated the mobile market, but Google Android has now demonstrably overtaken iPhone in terms of market share. In addition to the social media revolution, mobile technology and its 'apps' have encroached upon and indeed come to dominate our daily lives. Mobile and wireless devices are also enabling organisations to conduct business more effectively. Mobile applications can be used to support e-commerce with customers and suppliers, and to conduct e-business within and across organisational boundaries. There are two broad choices involved in deploying a system to mobile users, such as creating custom native apps targeted at major mobile platforms or developing a Web application optimised for mobile access.

The Internet of things (IoT) describes the network of physical objects connecting and exchanging data with other devices and systems over the Internet. IoT technology is generally related to smart homes including security devices and home appliances that can be controlled by smartphones. The IoT can also be used in industries and healthcare systems. There are however some concerns regarding IoT, especially in the areas of security and privacy.

A significant disruption in digital technology has been created by blockchain technology which is popularly used as a generic term that most people associate with cryptocurrency (e.g. Bitcoin). However, arguably it is more useful as distributed ledger technology, also known as 'trustware', because it replaces interpersonal trust with technological verification. This new form of technological trust can reduce the capacity of operators to behave opportunistically, enhance input verification and ensure transparency and traceability during transaction time.

6.5 Digital Tools for Green Six Sigma

Like pure-play Six Sigma, Green Six Sigma is also a data-driven process. Both the simpler tools (e.g. histogram, Pareto chart, and standard deviation) and more complex instruments and techniques (e.g. ANOVA report, process capability analysis, and design of experiment) require statistical analyses. There are many software systems available for such statistical evaluations, the most commonly used of which are Minitab and SPSS.

Minitab (Minitab.com) is a statistics package developed at Pennsylvania State University and distributed by Minitab LLC. This powerful statistical analysis package is a favourite with Six Sigma practitioners and offers three areas of functions—Discover, Predict and Achieve. Minitab can empower all parts of an organisation to predict better outcomes, design enhanced products and improve processes. It can access modern data analysis and explore data even further with advanced analytics and open-source integration. Visualisations can help present the findings through scatterplots, bubble plots, histograms, parallel plots, time series plots and more.

The many areas of Minitab output include the following:

Graphics

- scatterplots, boxplots, histograms, time series plots

Basic Statistics

- Descriptive statistics
- Correlation and covariance
- Normality test
- Linear regression
- Nonlinear regression
- Partial least squares
- Orthogonal regression
- Poisson regression
- Plots: residual, factorial, contour, surface, etc.

Analysis of Variance

- ANOVA
- MANOVA
- Test for equal variances
- Analysis of means

Quality Tools

- Run chart
- Pareto chart
- Cause and effect diagram
- Process capability
- Variability chart

Design of Experiments

- Definitive screening designs
- General factorial designs
- Response surface designs
- Taguchi designs
- User-specified designs
- Response prediction and optimisation

Another type of software package is SPSS (Pallant, 2010) which is used for interactive, or batched, statistical analysis. It has been produced by SPSS Inc. for some years and was acquired by IBM in 2009. The initial Version 1 appeared in 1968 and SPSS has now reached Version 26. The package has run on MS Windows since Version 16. Like Minitab, SPSS is a powerful data analysis package that can handle complex statistical processes. It is assumed that the user is versed in the fundamentals of statistics; in fact, instruction in statistics and the use of Minitab and SPSS constitute a vital part of Black Belt training.

SPSS tools and techniques for statistical data analyses include the following:

Descriptive statistics

- Histogram and assessing normality
- Checking outliers
- Bar chart and line chart
- Scatter diagram
- Box plot

Statistical techniques

- Regression analysis
- Correlation analysis
- Factor analysis
- T-tests
- ANOVA
- MANOVA
- Analysis of covariance

6.6 Quality 4.0

Quality 4.0 is a term introduced by ASQ (American Society of Quality) that references the future of quality and operational excellence within the context of the digital disruption especially in the 21st century. Quality 4.0 tools (including ERP, big data analytics, cloud computing, AI, machine learning and blockchain) are described here. A transitory period of disruption caused by Quality 4.0 is expected. During this period many of the traditional activities (e.g. data collection, inspection and corrective systems) of quality professionals will be digitised.

Quality 4.0 will also affect the implementation of Green Six Sigma in a positive way. This gives us more real-time information and digital tools and systems (in addition to Minitab and SPSS) to deliver Green Six Sigma solutions at various stages of DMAICS (define, measure, analyse, improve, control and sustain).

6.7 Digital Technology Applications in Climate Change

Various approaches and tools are currently being tried and advocated regarding how best to tackle climate change and sustainable development and the creation of ecological value. To this, digital sustainability activities can be added that focus on digital innovations to create scalable socio-ecological value. The technologies most commonly used in such digital sustainability tasks include distributed ledger technologies (blockchain), AI and machine learning, big data analytics, mobile technology and applications, sensors and other IoT devices, and additional tools like satellites and drones.

George et al. (2020) explore the opportunities and challenges of digital technology in climate change initiatives and define them as 'digital sustainability activities'. The authors give credit to entrepreneurial organisations that have adopted innovative approaches to ecological challenges and support them with examples including the following:

- Poseidon, a Malta-based foundation, is tokenising carbon credits by blockchain tokens from conservation programs in the Andean rainforest. It changes or tokenises those credits into 'carbon by the gram'.
- Ecosia is a search engine that uses 80% of its advert revenue to plant trees to fight global warming. Ecosia enables users to contribute to tree planting by simply installing Ecosia as their default search engine.
- Efforce is a blockchain-based energy-saving trading platform seeking to revolutionise the market for Energy Performance Contracting (EPC) to achieve infrastructure upgrades that reduce energy costs. An energy service company (ESCO) proposes improvements to an industrial facility, which are then funded by a finance partner.
- At the supply stage, Olam, an agribusiness multinational based in Singapore, is working to digitise the origination process for crops like cocoa across its global network. By equipping small-scale farmers with mobile phones armed with a digital sales platform, Olam cuts out price-setting middlemen and provides higher prices to farmers.

Many proven software products are already available in the market to measure and monitor carbon footprint of business operations (e.g. *evizi, Emitwise, GaBi* and more). These digital tools are providing practical help in climate change initiatives.

6.8 Summary

When looking at IT, a more general approach has been taken. This section is equally applicable to all functions of the organisation involved in climate change initiatives. The key issue in any new IT system is knowing what you want, going with a system that has local support, and initially making do with off-the-shelf software.

Topics such as uninterrupted power supply, disaster recovery, the need to backup files and so on have not been discussed. All these issues are 'nuts and bolts' and should be second nature to your IT manager. This section was not written for the professional IT executive, but rather to give the average manager an understanding of the strategy of IT implementation. The opportunities and challenges emerging from e-business technologies, big data analytics, AI, mobile technology, IoT and blockchain technology were also addressed.

During the last ten years we have experienced the growth of e-business applications and enabling infrastructures that have rapidly increased productivity by streamlining existing business processes. The time has come to take a fresh look at Internet technology, cloud computing and AI. We need to see such digital technology as a powerful enabling tool that can be used in almost any business and become part of almost any strategy. The key question now is not whether to implement digital technology but how to deploy it.

The proactive and ambitious international and national plans and support for meeting the challenge of the Paris Climate Agreement cannot be doubted, along with their laudable ambition to achieve net zero carbon emissions by 2050. However, the challenge is to make it happen at the sources of emissions including power plants, factories, infrastructures, transport systems and buildings. Both sustainable digital activities and Green Six Sigma tools and techniques underpinned by its holistic approach can be vital factors in such implementation programmes.

GREEN TIPS

- As part of the green supply chain large organisations are benefitting from comprehensive information systems such as ERP and CRM and Internet-based e-business.
- SMEs are taking advantages from the services of SaaS platform and cloud computing.
- Keep an eye on the dynamic contributions from disruptive technology such as AI, big data analytics and blockchain.
- Complex statistical analysis of advanced of Green Six Sigma techniques have the digital support of Minitab and SPSS.
- Use proven software to measure and monitor carbon footprint.

Chapter 7

Green Six Sigma in Manufacturing, Services, Projects and SMEs

7.1 Introduction

The first wave of Six Sigma, following the grand groundwork carried out by Motorola, included AlliedSignal, Texas Instruments, Ratheon and Polaroid (to name a few). GE entered the arena in the mid-1990s and in turn was followed by many powerful corporations including Sony, Dow, Bombardier and GSK. The ability to leverage the experience of successful Six Sigma players proved highly attractive both as a competitive issue and also to improve profit margins. However, the early applications were focused on large manufacturing as reflected by these examples. In the second wave of Six Sigma came Lean Six Sigma and we saw successful applications of Lean Six Sigma in service industries, including large financial services such as American Express and Lloyds Bank. Lean Six Sigma also had successful applications in major projects, especially the projects manged by Bechtel. As discussed in preceding chapters Green Six Sigma is an adaptation of Lean Six Sigma for sustainable climate change initiatives. Therefore, it can be concluded that Six Sigma and Lean Six Sigma applications are also valid for Green Six Sigma. If the earlier Six Sigma or Lean Six Sigma projects in manufacturing, services and projects are managed again with the Green Six Sigma approach, then an additional phase (viz. Sustain) will be added to ensure sustainable environment standards. Green Six Sigma can also be applied to small

DOI: 10.4324/9781003268239-8

and medium-sized enterprises (SMEs) with its approach of the 'fitness for purpose'.

The following case examples provide insights for organisations in all sectors (e.g. manufacturing, services, projects and SMEs) that can achieve success in their business performance through the use of Green Six Sigma programmes.

7.2 Green Six Sigma in Large Manufacturing Operations

In the Green Six Sigma approach to large manufacturing operations, it is important to establish the level of process accuracy to satisfy the 'fitness for purpose' criteria. We talk of Six Sigma–level implementations having an accuracy of 99.99966%. Do we need to be so accurate? Isn't it enough to be accurate to, say, 99% of the time? According to the American Society for Quality being 99% accurate means that

- There is no electric power for nearly 7 hours every month.
- Every hour, at least 20,000 letters are lost in the mail.
- Over 200,000 errors are made in medical prescriptions on an annual basis.

Now, doesn't that additional 0.99966% (six sigma) sound worthwhile? Well, the answer to the question is 'that depends on the application'. For example for a surgeon's operation or aircraft manufacture, the accuracy of six sigma is justified, while for the manufacture of bin bags an accuracy of one sigma is adequate.

Many large organisations' manufacturing precision (e.g. aircraft systems) and life-saving (e.g. pharmaceutical) products should allocate adequate resources to aiming towards achieving six sigma accuracy. This approach conforms to the Green Six Sigma tenet of 'fitness for purpose'.

The focus on a data-driven approach of Green Six Sigma implies that the process information is available. If this is not the case, it is possible to waste a lot of resources in obtaining process data. The key components of this Green Six Sigma tenet for large manufacturing organisations (e.g. appropriate DMAICS [define, measure, analyse, improve, control and sustain] methodology, learning deployment, project plans etc.) are similar to those of a comprehensive Six Sigma or Lean Sigma programme. However, the Green Six Sigma approach recognises that even for a large organisation it may overcomplicate the investigation if advanced techniques are used on simple problems. The following two case studies illustrate this point of appropriate application. Although General Electric and Seagate did not adopt the name Green Six Sigma for their programmes, their applications of Six Sigma principles do conform to the principles of Green Six Sigma. The case examples are a few years old, but the application principles and learning points are still valid.

CASE EXAMPLE 7.1 GENERAL ELECTRIC (GE)

With over 4,000 Black Belts and 10,000 Green Belts across its businesses, and Six Sigma savings of $2 billion in 1999 alone, GE is both a comprehensive Six Sigma organisation and can be seen as a benchmark for Six Sigma programmes.

GE has been at the top of the list of Fortune 500's most admired companies for the last several years, and without doubt their Six Sigma programme has played a key role in their continued success. In 2001 GE's turnover was over $125.8 billion, they employed 310,000 people worldwide and their market value was $401 billion. With earnings growing at 10% per annum GE also had the enviable record of pleasing Wall Street and financial analysts year after year. GE's products and business categories span a wide spectrum of automotive, construction, healthcare, retail, transport, utilities, telecommunications and finance.

A pervious CEO of GE, Jack Welch, is reported to have become attracted to the systematic and statistical method of Six Sigma in the mid-1990s. He was ultimately convinced of the power of Six Sigma after a presentation by AlliedSignal's former CEO, Larry Bossidy, to a group of GE employees. Bossidy, a former Vice Chairman of GE, had witnessed excellent returns from AlliedSignals' experience with Six Sigma.

In 1995 GE retained the Six Sigma Academy, an organisation started by two early pioneers of the process, both ex-Motorola employees, Mikel Harry and Richard Schroeder. It was pointed out that the gap between Three Sigma and Six Sigma was costing GE between $7 and $10 billion annually in scrap, rework, transactional errors and lost productivity. GE identified four specific reasons for implementing Six Sigma:

1. Cost reduction
2. Customer satisfaction improvement

3. Wall Street recognition
4. Corporate synergies

Although Motorola pioneered the Six Sigma programme in the 1980s to improve manufacturing quality and to eliminate waste in production, GE broke the mould of Motorola's original process by applying Six Sigma standards to its service-oriented businesses—GE Capital Services and GE Medical Systems. The Six Sigma programme was launched in 1995 with 200 separate projects supported by a massive training effort. In the following 2 years a further 9,000 projects were successfully undertaken, and the reported savings were $600 million. The training investment for the first 5 years of the programme was close to $1 billion. Total savings in 1999 amounted to $2 billion.

Following are some specific examples from business units:

GE Medical Systems: In the introductory year there were 200 successful projects.
GE Capital: Invested $6 million over 4 years to train just on 5% of the workforce who were involved full time on quality projects, 28,000 of which were successfully completed.
GE Aircraft Engines: The time taken to overhaul engines reduced by an average of 65 days.
GE Plastics: In just one project, a European polycarbonate unit increased capacity by 30% in 8 months.

At one level, the lofty aim of emulating GE may be considered as being beyond the reach of many companies. It is cash rich, and its business generates over $6.6 billion per month ($80 billion sales for 2020). It makes real things like turbines and refrigerators, and people buy their products with real money. However, on closer examination there are some strong learning points from the GE Six Sigma programme that can benefit any company embarking on a quality programme. These learning points include leadership support, defined objectives, application of appropriate tools, alignment with career paths and Six Sigma for services (e.g. GE Capital).

In addition, GE could benefit from the Sustain phase of Green Six Sigma to ensure the sustainability of processes, continuous improvement culture and environment standards.

Source: Best Practices LLC (2002).

CASE EXAMPLE 7.2 SEAGATE TECHNOLOGY INTERNATIONAL

Seagate's position as the world's largest manufacturer of disc drives, magnetic discs and read-write heads, and as a leader in Storage Area Network solutions puts it at the heart of today's "information-centric" world. Seagate Technology is a global company employing nearly 50,000 people, with R&D and product sites in the Silicon Valley, California; Pittsburgh, Pennsylvania; Longmont, Colorado; Bloomington and Shakopee, Minnesota; Oklahoma City, Oklahoma; Springtown, Northern Ireland; and Singapore. Manufacturing and customer service sites are located in California, Colorado, Minnesota, Oklahoma, Northern Ireland, China, Indonesia, Malaysia, Mexico, Singapore and Thailand.

Seagate is the world's leading provider of storage technology for Internet, business and consumer applications. Seagate's market leadership is based on delivering award-winning products, customer support and reliability to meet the world's growing demand for information storage. So why implement Six Sigma? The market leadership of the company is continuously challenged in a highly competitive and dynamic environment as indicated by the following measures:

- Volume products remain in production only 6 to 9 months
- Technology content doubles every 12 months
- Worldwide shipments of hard disc drives increase 10–20% per year
- Cost per unit of storage drops 1% every year

In 1998 Seagate's senior executive team was concerned that business performance was not on par with expectations and capabilities. The quality group was charged with recommending a new model or system with which to run the business. The Six Sigma methodology was selected and launched in 1998 to bring common tools, processes, language and statistical methodologies to Seagate as a means to design and develop robust products and processes.

Six Sigma was one of the three key activities seen as essential for Seagate's continuing prosperity. The other two were

- Supply Chain—how to respond to demand changes in a timely manner, execute to order commitments and provide flexibility to customers

- Core Teams—how to manage product development life cycle from research to volume manufacture.

For example, Seagate's Lean Manufacturing activities are a key part of Seagate's Supply Chain improvements and are increasingly tightly bound with Six Sigma. Thus, their Six Sigma was a Lean Six Sigma application. Lean manufacturing's value stream mapping approach and Six Sigma's analytical strength fit together extremely well to define, solve and then prevent problems.

The Six Sigma Academy was employed to guide the implementation and provide the initial waves of training for executives, champions and Black Belts throughout the latter period of 1998 and also 1999. Black Belt candidates were trained in the US and Singapore. All sites were assigned hands-on Champions, members of senior staff familiar with the operational requirements of the sites and who were trained on project selection and support. Seagate developed and customised training materials so as to be self-sufficient in educating up to the Master Black Belt level. Training centres of excellence exist in the US, Europe and Asia-Pacific areas. Green Belt coaching was required for all Seagate's professional and technical staff.

Seagate achieved savings of nearly $700 million. All savings were validated independently and audited by the finance team, using very strict criteria. It is also certain that the so-called soft savings that Seagate has achieved, but not counted, far exceeded this value.

In April 2002 a diverse group of representatives gathered for a 2-day, intensive planning session to help define Seagate's future strategy for Six Sigma.

The 14 representatives formed a collective vision for Six Sigma's future over the next 3 years and included specific elements that help to build a solid basis for future growth. Future plans included an ongoing assessment of the programme to ensure that Six Sigma was delivering what it was intended to deliver. It is evident that Seagate's future plan is in line with the sustainability of continuous improvement culture of Green Six Sigma. In addition Seagate would benefit further from application of the Sustain phase of Green Six Sigma especially in ensuring environment standards.

Source: Rob Hardeman, Seagate Northern Ireland.

7.3 Green Six Sigma in Service Operations

It is generally agreed that 75% of the workforce in the UK is engaged in service industries. This high proportion is not unique to the UK; indeed it is representative of employment statistics for developed nations throughout the world. In fact the US Census Bureau shows that over 80% of the workforce in the US are employed in service industries. There are many service businesses such as healthcare, banking, food services and retailing. In addition, there exist numerous service functions in both manufacturing and service organisations, for example marketing, finance, information technology, procurement, legal and human resources.

Continual advances in technology mean that manufacturing is considerably less labour intensive than during previous times. Automation, robotics, advanced information technology, new materials and improved work methods all have led to the decimation of manual labour.

Organisations can no longer regard themselves as being purely in manufacturing and hope to survive. The market first and foremost now demands quality of product and service. Market expectations of the level of quality are driven by perceptions of what technology is promising and by perceptions of what the competition is offering.

A service organisation is when two or more people are engaged in a systematic effort to provide services to a customer, the objective being to serve a customer. For any service to be provided there has to be a customer. Without the existence of such a customer, and interaction between the customer and the service organisation, the objective of providing service cannot happen. The degree of intensity of interaction between the customer and people of a service organisation varies and depends on the type of service offered. For example a specialist medical consultant will have a high degree of 'face-to-face' interaction with the customer, and likewise so will a hairdresser.

Businesses in the service sector have often questioned the tangible value that a Six Sigma programme delivers in services where there are many intangible and transactional processes. Historically, the first organisations to adopt Six Sigma were mainly in the manufacturing sector. This was due to the fact that the core Six Sigma methodology revolved around the reduction of defects in a process. As with aircraft engines, this might be a fault in a component that could typically lead to a catastrophic failure. Hence a robust quantitative methodology of a very high confidence level such as Six Sigma lends itself well to the prevention of problems in this type of scenario.

Services, by their nature, are very often bound by time in terms of the processes that are run and lead to the delivery of an outcome that then benefits a customer. The confidence level in services is likely to be lower than an aircraft engine in this scenario although the customer expects a consistent service level. This is where Green Six Sigma comes in as a methodology that looks at how waste (in terms of time) may be taken out of a process and allows that process to become more efficient and, in turn, build a sustainable service to the customer.

So if the process characteristics and management tools of service organisations overlap with those of manufacturing is it not logical to assume that the Green Six Sigma approach for both sectors should be the same? The broad answer is yes. The 'fitness for purpose' methodology of Green Six Sigma can accommodate the variability between service and manufacturing as much as the variability with the manufacturing sectors. However, there are areas where some differentiation may be applicable, in particular, in 'service level' and 'culture'.

The following case examples illustrate how a Lean Six Sigma approach conforms to Green Six Sigma in service organisations. In the first example of an insurance firm, the objective was to reduce waste and increase efficiency with a confidence level of 90%. The start-up cost was not significant as the firm initially did not deploy an extensive training programme for Black Belts and Green Belts.

CASE EXAMPLE 7.3 LEAN SIX SIGMA IN AN ACCOUNTANCY FIRM

Rea & Associates Inc., is an Ohio-based accountancy firm with 11 locations and 250 employees. The firm started a Lean Sigma programme in 2007 by recruiting a Six Sigma Master Black Belt. Initially, with about a year of start-up costs, including tuition for the additional Black Belts and Master Black Belts, the programme was slowly rolled out as a Lean Six Sigma tax programme. The approach of Lean Sigma conforms to FIT SIGMA in this application.

Much of the waste in tax processes was viewed as being small in nature and therefore was overlooked. Nearly every step of a process had some form of waste. It was important to identify this, however small, and to quantify it to show the entire impact to the procedure. 'Minor' wastes

Table 7.1 Lean Six Sigma Results

Lean Metrics	*From*	*To*
Waiting Time for Review	Two-week average	Less than 1 week
Waiting Time to Clear Review Notes	One-week average	Less than 2 days

added up to 30% of efficiency with a confidence level of 90%. Table 7.1 shows the results in Autumn 2008.

The same principles discussed for the tax preparation function also applied to audit, payroll, bookkeeping and pension administration, to name a few. These ethics applied if the desire to improve efficiency and quality remained competitive. This approach also worked well when applied to internal administrative tasks, such as billing. Regardless of the internal service being performed, this method was also used to analyse and improve the process.

This example illustrates the Green Six Sigma principle of 'fitness for purpose'. Furthermore, the firm should also benefit from the application of the Sustain phase of Green Six Sigma.

Source: Journal of Accountancy, *January 2010.*

'Fitness for purpose' in Green Six Sigma methodology also conforms to the application of Six Sigma and Lean Sigma in large service organisations such as GE Capital, American Express and the Bank of America. In this scenario the organisations can afford a large budget for a comprehensive deployment of the training programme for Black Belts and Green Belts.

The following case example in a large health service organisation illustrates how a DMAIC (define, measure, analyse, improve, control) Full approach is applied to integrate all departments and improve processes.

CASE EXAMPLE 7.4. LEAN SIX SIGMA IN A COMMUNITY HOSPITAL

Floyd Medical Centre, a stand-alone safety net community hospital, is located in northwest Georgia. Floyd engaged the Health Management Corporation (HMC) to benchmark the organisation in the early 2000s and began its journey into Lean Six Sigma by investigating what measurement and improvement methodologies other organisations were using. Floyd's initiative began in the summer of 2006 with a management retreat with Chip Caldwell and Associates, who were engaged to assist the initiative and remain for a year to conduct training. After the retreat, leaders were educated about the cost of poor quality and the concept of driving out waste to improve quality, cost, time and performance, and the change model of the 100-day workout was introduced. These 'workouts' are comparable to Green Six Sigma training modules for Black Belts and Green Belts.

The first 100-day workout was a waste walk. Future 100-day workouts utilised the SIPOC (suppliers, inputs, process, outputs, customers) tool, the DMAIC process and other Six Sigma tools. At the 100-day mark, a summation of all the changes and savings was reported and the next workout was kicked off. This exercise was used to train and certify the first group of Green Belts and Black Belts. Four employees continued on to obtain Black Belt certification in August 2007, and 12 employees were trained as Green Belts.

In this first workout 288 changes were instituted and 270 were validated, for a savings of $2,458,200. In storage costs alone, $63,500 was saved. 'The Greatest Financial Impact' award was given for the recommendation to switch Zofran to a generic drug. This yielded a savings of $223,800.

The second workout dealt with waste and quality staffing. Results of this study were reductions in overtime, decreased utilisation of agency staff and a reduced number of positions through attrition. In total, 145 changes were instituted, 131 were validated, and a savings of $1,543,600 was achieved.

The third workout concerned Waste and SIPOC. Leaders identified flow and throughput issues that could be improved. This exercise made leaders think about the customers of each step in a process. This workout resulted in savings of $635,895. The next three workouts also produced excellent outcomes, more tools were implemented and this also led to a refined reporting process. However, after the sixth workout, the management felt that a change was needed to sustain the momentum and the level of performance.

At the end of the sixth 'workout' or workshop, the Lean Sigma programme at Floyd Medical Centre in Georgia began to lose momentum.

Leaders were becoming weary. While the first wave had been very successful with significant savings in each workout, the management decided to change focus. Some years earlier, Floyd had adopted a Balanced Scorecard we call the 'value compass' comprising five aspects (e.g. financial, people, strategy, customer satisfaction and internal quality). This value compass became the focal point of subsequent workouts.

For the seventh workout, called the Fall Value Compass Workout, seven teams were chartered to enhance the value compass in seven areas:

1. Emergency Care Centre Throughput (Strategy)
2. Orthopedic and Neurology Service Line (Strategy)
3. Patient Satisfaction (Customer Satisfaction)
4. Employee Choice (People)
5. Hygiene and Antibiotic Stewardship (Internal Quality)
6. Revenue Cycle (Financial)
7. Time Management (People)

These value compass items focused on projects in the seventh workout and also applied DMAIC as well as contributing savings for $668,000. The focus on a Balanced Scorecard–based performance management conforms to the FIT SIGMA tenet of sustaining performance. Floyd Medical Centre is committed to 'workouts' based on the value compass and also aiming towards the following conditions of sustainability:

- Requiring all departments to have control charts.
- Having Black Belts serve as full-time Belts and then rotating them to other leadership positions.
- Requiring all directors to obtain Green Belt certification, and making that a prerequisite for anyone who wants to advance into a director position.
- Having more nurses trained as Green Belts.
- Having one or two Master Black Belts within the organisation.
- Establishing a network of similarly minded hospitals throughout the country that want to share their Lean Six Sigma experiences.

There are several important lessons learnt from this case example.

First, the initiative must be driven from the top down. Second, knowledge and training in the methodology are critical at all levels of the organisation. Third, it is necessary to identify a Black Belt to work on projects full time. Belt duties were often carried out after regular work hours, and it was found increasingly difficult to have Belts available to meet project needs and grow the initiatives. Therefore, one Black Belt was identified to focus full time on such ideas. Fourth, it is essential to limit the number of changes and efforts in a particular area at any one time. You are likely to set them up for failure with too many efforts running in tandem. Finally there is need to have processes in place to sustain the enthusiasm of the team and the performance level. These learning points conform to the tenets of Green Six Sigma.

Source: Frontiers of Health Services Management, *Volume 26, No. 1.*

7.4 Green Six Sigma in Project Management

Interest in Six Sigma is growing rapidly within the professional project management community, and the most common question coming from that group is something like 'How does Six Sigma relate to the Project Management Body of Knowledge (PMBoK)?' PMBoK (2006) concludes that Six Sigma and PMBoK do have connections, similarities and distinctions, and it is clear that Six Sigma complements and extends professional project management but does not replace it. Both disciplines make important contributions to successful business outcomes. As described earlier the core methodology of Green Six Sigma, i.e. DMAICS, is closely linked to the methodology, rigour and stages of life cycle of project management.

DMAICS has added the rigour of project life cycle to the implementation and close-out of Green Six Sigma schemes. Figure 7.1 shows the relationship between DMAICS and a typical project life cycle.

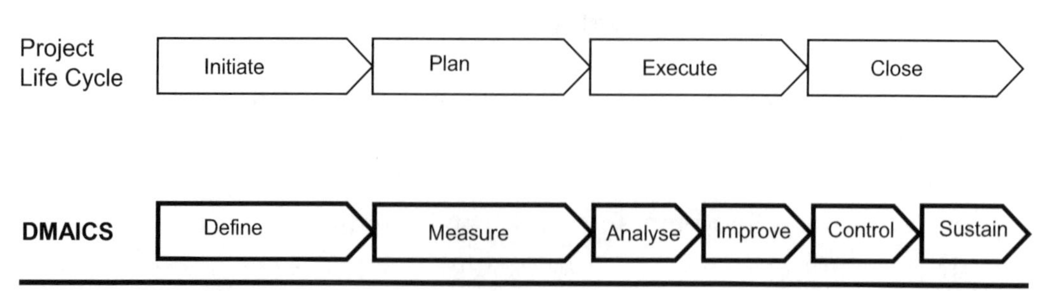

Figure 7.1 DMAICS cycle and project life cycle.

Project organisations are showing a positive interest in Six Sigma, and courses and conferences are on offer for project members. Bechtel was one of the early users of Six Sigma in delivering their multinational assignments as the following case example illustrates.

CASE EXAMPLE 7.5 SIX SIGMA AT BECHTEL

Founded in 1898, Bechtel is one of the world's premier engineering, construction and project management companies. Nearly 40,000 employees are teamed with customers, partners and suppliers on a wide range of schemes in nearly 46 countries.

Bechtel has completed more than 22,000 projects in 140 countries, including the Hoover Dam, the Channel Tunnel, Hong Kong International Airport, the reconstruction of Kuwait's oil fields after the Gulf War and the Jubail industrial city.

Bechtel was the first major engineering and construction company to adopt Six Sigma, a data-driven approach to improving efficiency and quality. Although it was originally developed for manufacturing companies, the company was confident that Six Sigma would work in professional services organisations such as Bechtel. Six Sigma has improved every aspect of Bechtel's business, from construction projects to regional offices, saving time and money for both customers and the company.

Six Sigma uses a rigorous set of statistical and analytic tools to produce dramatic improvements in their work processes, see Basu and Wright (2003). Bechtel launched Six Sigma in 2000, when the company was experiencing unprecedented growth—and facing corresponding process challenges. The company has now implemented Six Sigma in its key offices and business units around the world. About half of its employees have had Six Sigma training, and most of its major projects employ its methods from start to finish.

The investment of Bechtel in Six Sigma reached the break-even point in less than 3 years, and the overall savings have added substantially to the bottom line, while also benefiting customers. Some examples include the following:

- On a big rail modernisation project in the UK, a Bechtel team used Six Sigma to minimise costly train delays caused by project work and reduced the 'break-in' period for renovated high-speed tracks.

- At a US Department of Defense site in Maryland, Six Sigma helped achieve significant cost savings by streamlining the analysis of neutralised mustard gas at a project to eliminate chemical weapons.
- To speed up the location of new cellular sites in big cities, Bechtel developed a way to let planners use computers to view video surveys of streets and buildings, making it easier to pick the best spots.
- In a mountainous region of Chile, Six Sigma led to more efficient use of equipment in a massive mine expansion, with significant cost savings.
- In the 'High Speed 1 Project', a prominent flagship project for high-speed 'Eurostar' trains between London and Brussels and Paris, Bechtel (in partnership with Arup, Halcrow and Systra) applied Six Sigma and achieved a saving of US $40 million in over 500 improvement schemes.

'Six Sigma is the most important initiative for change we have ever undertaken. We're happy to report that it's becoming "the way we work."'

Source: https://www.bechtel.com (2006).

The application of Green Six Sigma in a major project is justifiable when the duration of a project is longer (e.g. 2 to 5 years) and a budget for Black Belt and Green Belt training is cost effective. The undertaking of a major project in this situation may be considered as an enterprise. In 1996, the London and Continental Railway (LCR) was awarded the concession to build the Channel Tunnel Rail Link, which was later known as the High Speed 1 (HS1) project. This US $5.8 billion project lasted for 7 years and justifiably adopted Lean Six Sigma. The Lean Six Sigma programme trained 23 Black Belts and around 250 Green Belts and Yellow Belts. Over 500 improvement projects were completed, which led to a cost savings of more than US $40 million. As Green Six Sigma is an adaptation of Lean Six Sigma with additional features of fitness for purpose and sustainability, the success of Lean Six Sigma in HS1 can also be attributed to Green Six Sigma.

It is evident that there are significant potentials and benefits to be gained by adapting and adopting operational excellence concepts such as Green Six Sigma in major projects. Because of the perception of the one-off unique nature of a project and the repetitive nature of operations, the mindsets of many project managers may have a prejudice against or lack of confidence to adopt operational excellence concepts. Leading project contractors such as Bechtel, however, are also training their managers with operational excellence skills (such as those leaders tutored as 'Black Belts'). In a major project or programme with a time span of 2 years or more, it could be argued that the activities, processes and objectives in the organisation could be closely related to those of operations management. This could be conducive to the application of operational excellence concepts.

7.5 Green Six Sigma in Small and Medium-Sized Enterprises

SMEs are companies whose headcount or turnover falls below certain limits. As shown in Table 7.1, an enterprise with a turnover of less than €50 million and employing fewer than 250 people is classified as an SME.

SMEs are underpinning the economy of a country, accounting for 50% of the annual turnover in the private sector in the UK (Berr, 2008). One of the more familiar challenges in business today is how to implement Six Sigma in SMEs. This trial is compounded when larger companies are beginning to demand Six Sigma applications to their supply base comprising SMEs as a condition of conducting future business. However, when the SMEs solicit advice and proposals from Six Sigma consulting companies, they learn that the traditional Six Sigma or Lean Six Sigma implementation approach can require a disproportionate investment and the dedication of their best full-time resources.

There are examples of many successful applications of the Six Sigma methodology in SMEs in the UK (Davies, 2005), Germany (Wessel and Butcher, 2004), Denmark (Barfod, 2007), India (Kumar et al., 2006) and Indonesia (Amar and Davies, 2007). The findings indicate that the results

Table 7.1 SME Definition

Enterprise Category	*Headcount*	*Annual Turnover*	*Balance Sheet*
Medium	<250	<€50 million	<€43 million
Small	<50	<€10 million	<€10 million
Micro	<10	<€2 million	<€2 million

Source: European Commission (2003).

are quicker and more visible in smaller companies than in larger corporations. However, the DMAIC or DMAICS approach had to be modified to be applicable and valuable in the SME environment. This modified approach is DMAIC Lite (Basu, 2011, p. 50) where the boundaries between 'Measure' and 'Analyse' and also between 'Improve' and 'Control' are more flexible. The following case example to retain customers in a call center illustrates this point.

CASE EXAMPLE 7.6 GREEN SIX SIGMA TO SOLVE A CALL CENTER PROBLEM

Six Sigma methodologies can be tailored to be as simple or as complex as you want them to be, dependent on the project the business has in mind. This approach then conforms to Green Six Sigma.

A customer relations project in a call center demonstrates how problems that are perceived to be at the 'end' of a process are actually shown to be issues that have arisen earlier on.

The company was experiencing problems within the customer call center. It was inundated with calls, most of them problems. Significant money had been spent on extra staff, new training, looking at limiting the time taken per customer call etc. The call center was deemed crucial as the business had a 50% customer turnover annually and had made no real effort to focus on and improve customer retention. The business then decided to a undertake a Six Sigma project to look at the problem, which began by identifying whom the call center was actually trying to serve—the company or the customer.

Initial analysis captured how their customers defined 'value', and one of the key areas critical to quality (CTQ) was 'customer focus'. The key attributes of 'customer focus' were responsiveness, consistency, problem resolution, ease of dealing with the call center and an understanding of customer needs.

These key elements all scored poorly when applied to the company. The business then used this data to map out the 'order to delivery' process and identify areas where improvements could be made. Interestingly, while many of the problems were within the call center, some difficulties occurred earlier in the sales process. The root cause analysis indicated that

- Salespeople were being paid commission on the number of new customers leading to incomplete applications and generating more calls from customers to the call centre.

- Customers were not being trained in how to use the product properly at the time of the sale, leading to more calls to the call center.

The solution had three parts:

1. The business introduced an electronic order form that could not be submitted until all field processes had been completed.
2. Sales team job descriptions were changed to make sure that customer training was emphasised.
3. Call center personnel were offered training on how to retain customers.

Outcome—customer turnover reduced from 50 to 25%

As you can see from this example, all that was needed to resolve the call center issue was an examination of the process that the customer was subject to and to fix any areas that were deemed as problematic to the customer. This did not involve any extra cost or any staff redundancies, and customer retention was greatly improved.

This example shows that Six Sigma can be used effectively in the SME marketplace, and it can be made simple and relevant to many sectors and functions. This is Green Six Sigma in action.

In the current economic downturn, enterprises all over the world, including SMEs, are forced to do more with less and to trim costs while growing profits. It is likely that the reduction of waste by Lean processes is more attractive to SMEs than a commitment to a Six Sigma programme. In this scenario the appropriateness of the three tenets of Green Six Sigma, especially 'fitness for purpose', becomes more relevant for SMEs.

7.6 Summary

The evidence-based discussions in this chapter have established that the data-driven holistic approach of both Lean Six Sigma and Green Six Sigma can be effectively applied to improve the operations of large manufacturing organisations, service operations and major projects. Green Six Sigma has the additional contributions to sustain the processes, continuous improvement culture and environmental standards. The 'fitness for purpose' tenet of Green Six Sigma also enables the adaptable approach of Green Six Sigma to SMEs.

Therefore, it can be concluded that Green Six Sigma is an effective approach to all organisations regardless of their sector or size to improve sustainable performance and culture and the environmental standards of climate change.

With the dynamic growth of emerging new economies (e.g. China, India, Eastern Europe and Latin America) and the globalisation of manufacturing businesses, in order to remain competitive and eco-friendly, manufacturing companies need to implement Green Six Sigma which achieves sustainable changes. As the service sector accounts for about 75% of the economy this is an area of significant improvement by Green Six Sigma. There are also significant benefits to be gained by adapting Green Six Sigma concepts in major projects. The methodology applied in SMEs should be DMAICS Lite.

GREEN TIPS

- Deploy DMAICS Full methodology for larger manufacturing operations and major projects.
- Consider Green Six Sigma for projects lasting over 2 years.
- Deploy DMAICS Full methodology or DMAICS Lite methodology in service operations depending on process sigma targets.
- Deploy DMAICS Lite methodology for SMEs.

TOOLS FOR GREEN SIX SIGMA

Chapter 8

Tools for Definition

The greatest challenge to any thinker is starting the problem in a way that will allow a solution.

—Bertrand Russell

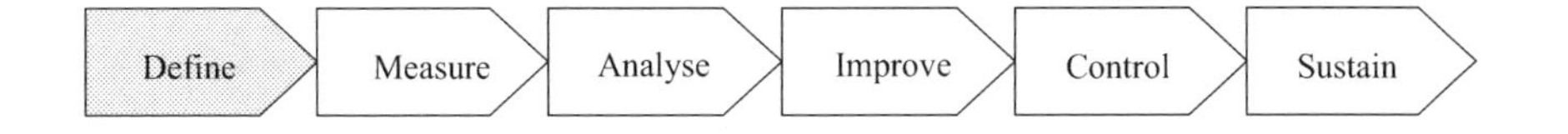

8.1 Introduction

A single tool is a device which has a clear role and defined application. However, numerous tools jostle for space in the training literature. Although most of these tools are basically simple, their inappropriate selection and application often ended in failure. Thus it is important to present them in a structured way.

DMAICS (define, measure, analyse, improve, control and sustain; see Chapter 4 for details) provides the rigour of a proven project management life cycle in Green Six Sigma programmes. DMAICS will probably be recognised and understood quite easily. Hence the next chapters describing Green Six Sigma tools have been structured according to the DMAICS model as follows:

Chapter 8:	Tools for Definition	(Define)
Chapter 9:	Tools for Measurement	(Measure)
Chapter 10:	Tools for Analysis	(Analyse)
Chapter 11:	Tools for Improvement	(Improve)
Chapter 12:	Tools for Control	(Control)
Chapter 13:	Tools for Sustainability	(Sustain)

DOI: 10.4324/9781003268239-10

Although we present the tools under specific stages of DMAIC, it is to be noted that some tools can be used in more than one stage. For example, the flow process chart can be deployed at all stages, but we included it in Chapter 8 where it is likely to be most frequently used.

Each of the tools will be presented according to the following format:

- Definition
- Application
- Basic Steps
- Worked-Out Examples
- Training Requirements
- Final Thoughts

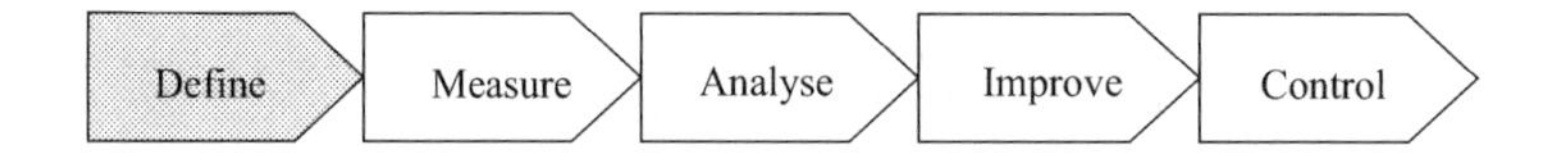

8.2 Description of Tools for Definition

The project definition is arguably the most important part of the project life cycle, because it establishes the basis for the other project management subprocesses. In the context of a Six Sigma or related operational excellence programme, if we follow DMAIC as the basis of the life cycle then the 'Define' stage sets the baseline which allows the subsequent stages to follow a structured methodology and expected quality standards.

The tools at this 'Define' stage are primarily for data collection which influences the management of the project start and terms of reference. The important tools for data collection include

D1: Input-process-output (IPO) diagram
D2: SIPOC (suppliers, inputs, process, outputs, customers) diagram
D3: Flow diagram
D4: Critical to quality (CTQ) Tree
D5: Project charter

8.2.1 D1: IPO Diagram

8.2.1.1 Definition

An IPO diagram, also known as a general process diagram, provides a visual representation of a process by defining a process and demonstrating the relationships between input and output elements.

The input and output variables are known as 'factors' and 'responses', respectively.

8.2.1.2 Application

Whether we are performing a service or manufacturing a product or completing a task, an IPO diagram is very useful to define a process as an activity that transforms inputs in order to generate corresponding outputs.

An IPO diagram is very often the starting point of a Six Sigma or similar improvement project. This high-level mapping of the processes is then followed by flow diagrams, process mapping and design of experiments (DOE) to understand fully the process and related subprocesses.

It is necessary to develop an IPO diagram to determine the factors and responses before carrying out a design of experiment exercise (see Chapter 13).

8.2.1.3 Basic Steps

1. In building an IPO diagram we first choose a process.
2. Next we define the outputs or responses. They are also called 'quality characteristics' or CTQs or *y*-variables. They are usually defined from a customer perspective.
3. We then define the input factors which will be required to make the process valuable to the customer.
4. Draw the IPO diagram with incoming arrows for inputs and outgoing arrows for outputs. (See Figure 8.1.)

8.2.1.4 Worked-Out Example

Figure 8.1 shows an example of an IPO diagram of a tablet manufacturing process in a pharmaceutical company.

8.2.1.5 Training Requirements

An IPO diagram is a simple tool in principle and a 1-hour training session (including exercises) should be adequate for a project team member to develop a practical diagram.

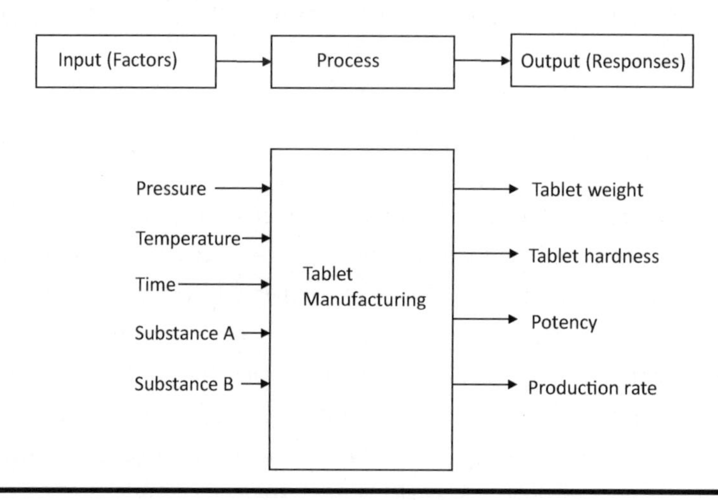

Figure 8.1 An IPO diagram.

8.2.1.6 Final Thoughts

An IPO diagram is a simple tool to define a process and focus on its key variables. This is closely linked with a cause and effect diagram and DOE. We recommend the extensive use of IPO diagrams

8.2.2 D2: SIPOC Diagram

8.2.2.1 Definition

SIPOC is a high-level map of a process to view how a company goes about satisfying a particular customer requirement in the overall supply chain. SIPOC stands for

Supplier: The person or company that provides the input to the process (e.g. raw materials, labour, machinery, information etc). The supplier may be both external and internal to the company.

Input: The materials, labour, machinery, information etc. required for the process.

Process: The internal steps necessary to transform the input to output.

Output: The product (both goods and services) being delivered to the customer.

Customer: The receiver of the product. The customer could be the next step of the process or a person or organisation.

8.2.2.2 Application

A SIPOC diagram is usually applied during the data collection of a project or at the 'Define' stage of DMAIC in a Six Sigma programme. However, its impact is utilised throughout the project life cycle.

SIPOC not only shows the inter-relationships of the elements in a supply chain, but also CTQ indicators such as 'delivered in 7 days'.

8.2.2.3 Basic Steps

1. Select the process for the SIPOC diagram and identify CTQ parameters.
2. Determine the input requirements.
3. Identify the suppliers for each of the input elements.
4. Define the output and validate CTQ parameters.
5. Identify the customers.
6. Draw the SIPOC process diagram.
7. Retain the diagram for the rest of the improvement project.

8.2.2.4 Worked-Out Example

Figure 8.2 shows a SIPOC diagram for a company that leases equipment (adapted from George, 2002, p. 185).

8.2.2.5 Training Requirements

The training of SIPOC should be combined with that for the CTQ tree and the IPO diagram as they complement one another. The combined training is expected to be carried out for both Black Belts and Green Belts in half a day.

8.2.2.6 Final Thoughts

SIPOC is a useful tool to identify the customer, stakeholders and CTQs of a process before the development of a project charter. However, it is not an essential tool, and a process can be analysed at a high level by an IPO or a flow diagram.

Suppliers	Inputs	Process	Outputs	Customers
List the suppliers (internal or external) of any inputs to the process	List the transformed resource inputs to the process (materials, information, etc)	Describe the process and/or list the key process steps	List the outputs of this process (goods and/or services)	Identify the customers (internal or external) of these process outputs
Sales department	Order specification	Customer order received	Lease agreement	Equipment lessor
Credit agency	Credit report	Customer credit review		
Engineering department	Equipment specification	Preparation of lease documents		
Transforming Resources		Signing off lease documents		
List the transforming resource inputs (staff, facilities, equipment, etc.) that are needed for the process				
IT system Finance department staff		Despatch of lease documents		

Figure 8.2 SIPOC process diagram.

8.2.3 D3: Flow Diagram

8.2.3.1 Definition

A flow diagram (also called a flow chart) is a visual representation of all major steps in a process. It helps a team to understand a process better by identifying the actual flow or sequence of events in a process that any product or service follows.

There are variations of a flow diagram depending on the details required in an application. We have included two other forms of diagrams in this family. These are the flow process chart (see Chapter 9) and process mapping (see Chapter 10).

The type of flow diagram described in this section is a top-level mapping of the general process flow and uses four standard symbols:

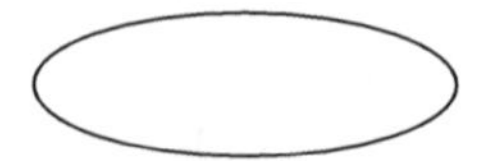

An oval is used to show the start and the end of a process.

A box or a rectangle is used to show a task or activity performed in the process.

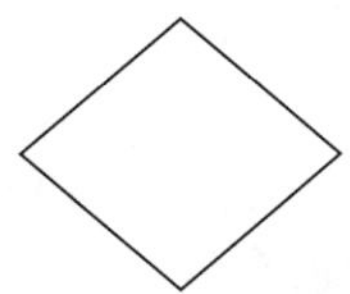

A diamond shows those points in the process where a decision is required.

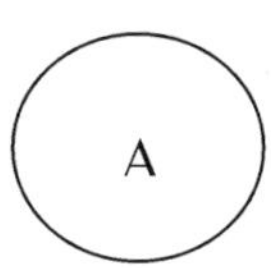

A circle with either a letter or a number identifies a break and connects to another page or part of the diagram.

8.2.3.2 Application

A flow diagram can be applied to any type of process, to anything from the development of a product to the steps in making a sale or servicing a product.

It allows a team to come to a consensus regarding the steps of the process and to identify critical and problem areas for improvement.

In addition, it serves as an excellent training aid to understand the complete process.

8.2.3.3 Basic Steps

1. Select the process and determine the scope or boundaries of the process:
   a. Clearly define where the process understudy starts and ends.
   b. Agree on the level of detail to be shown on the flow diagram.
2. Brainstorm a list of major activities and determine the steps in the process.
3. Arrange the steps in the order in which they are carried out. Unless you are developing a new process, sequence what actually is and not what should be.

4. Draw the flow diagram using the appropriate symbols.
5. There are a number of good practices when charting a process including
   a. Use Post-it notes so that you can move them around in a large process.
   b. For a large-scale process, start by charting only the major steps or activities.
   c. Come back to develop further details for major steps if necessary.
   d. Consider process mapping (see Chapter 6) to apply to a larger process.

8.2.3.4 Worked-Out Example

Figure 8.3 shows an example of a flow diagram to illustrate a purchasing process.

8.2.3.5 Training Requirements

The basic principles of a flow diagram are very simple and can be acquired in a self-teaching process by the user. However, good experience is required

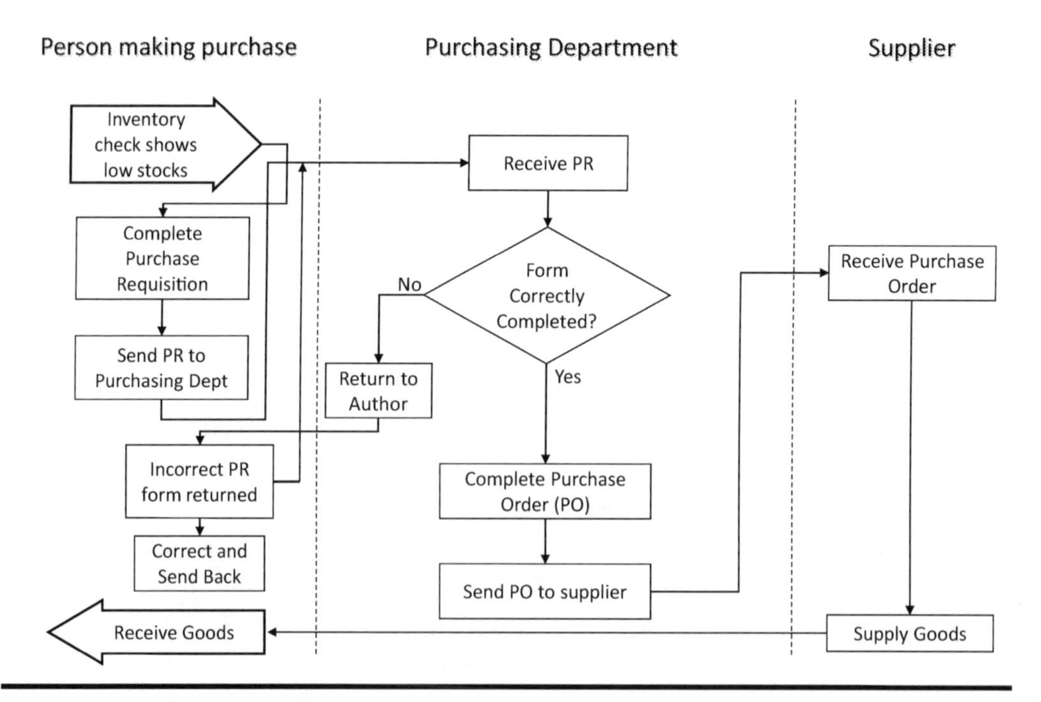

Figure 8.3 A flow diagram.

to develop a flow diagram of a complex process. This experience can only be gained by working as a team in improvement projects.

8.2.3.6 Final Thoughts

We recommend the use of a flow diagram as a tool to map the major steps of a process. This should be useful at the start of a project to understand the process and agree on the steps of the process. For a detailed assessment or analysis of the process, a flow process chart or a process mapping tool will be more appropriate.

8.2.4 D4: Critical to Quality Tree

8.2.4.1 Definition

CTQ is a term widely used within the field of Six Sigma activities to describe the key output characteristics of a process. An example may be an element of a design or an attribute of a service that is critical in the eyes of the customer.

A CTQ tree helps the team to derive the more specific behavioural requirements of the customer from their general needs.

8.2.4.2 Application

A CTQ tree is a useful tool during the data collection stage (Define) of an improvement project. Once the project team has established who their customers are, the team should then move towards determining the customer needs and requirements. The need of a customer is the output of a process. Requirements are the characteristics to determine whether the customer is happy with the output delivered. These constitute what is 'critical to quality', and a CTQ tree helps to identify these CTQs in a systematic way.

8.2.4.3 Basic Steps

1. Identify the customer.
2. Identify customers' general needs in Level 1.
3. Identify the first set of requirements for that need in Level 2.
4. Drill down to Level 3 if necessary to identify the specific behavioural requirements of the customer.

5. Validate the requirements with the customer. The process of validation could be one-to-one interviews, surveys or focus groups depending on the CTQ.

8.2.4.4 Worked-Out Example

Figure 8.4 shows a worked-out example of a CTQ tree for room service in a hotel (adapted from Eckes, 2001, p. 55).

8.2.4.5 Training Requirements

A CTQ tree is a simple tool in principle, and a 1-hour training session (including an exercise) should be sufficient for a team member to get involved in developing a practical CTQ tree. This is appropriate for both Black Belts and Green Belts.

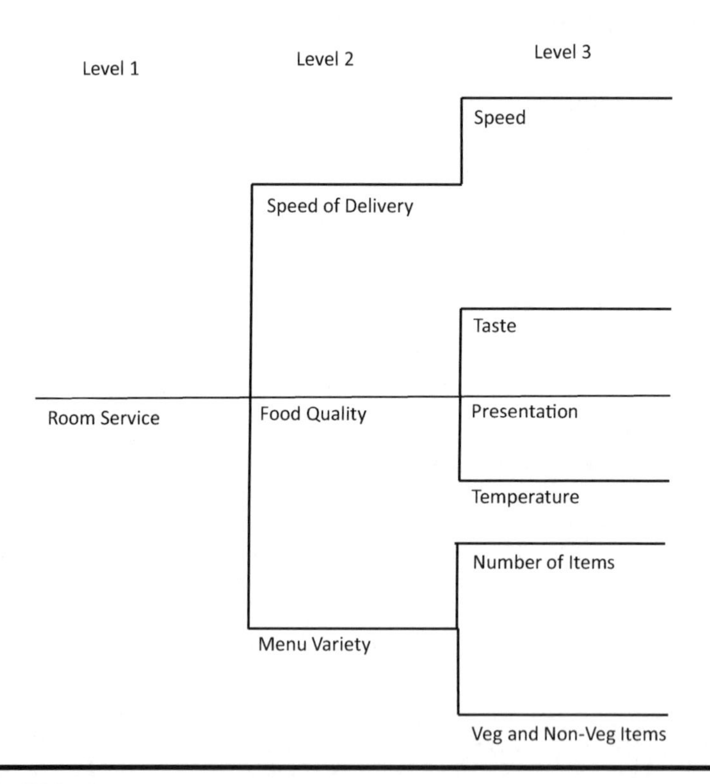

Figure 8.4 A CTQ tree.

8.2.4.6 Final Thoughts

A CTQ tree is a simple but powerful tool for capturing the details of customer requirements, and we recommend its use at a very early stage of a Six Sigma project.

8.2.5 D5: Project Charter

8.2.5.1 Definition

A project charter is a working document for defining the terms of reference of each Six Sigma project. The charter can make a successful project by specifying necessary resources and boundaries that will in turn ensure success.

The necessary elements of a project charter include the following:

Project Title: It is important to use a descriptive title that will allow others to quickly identify the project.
Project Type: Whether it is for quality improvement, increasing revenue or reducing cost.
Project Description: A clear description of the problem, the opportunity and the goal.
Project Purpose: Why you are carrying out this project.
Project Scope: Project dimensions, what is included and not included.
Project Objectives: Target performance improvement in measurable terms.
Project Team: Sponsor, team leader and team members. It is important to identify Black Belts and Green Belts within the team.
Customers and CTQs: Both internal and external customers and CTQs specific to each customer.
Cost Benefits: A draft business case of savings expected and the cost required to complete the project.
Timing: Anticipated project start date and end date. Target completion dates of each phase (e.g. DMAIC) of the project are useful.

8.2.5.2 Application

A project charter is the formalised starting point in the Six Sigma methodology. It takes place in the Define stage of DMAIC.

We recommend that larger (e.g. over $1 million savings) projects of a Six Sigma programme which is usually led by a Black Belt should have a project charter. A project charter can also be useful in a multidiscipline, medium-sized project.

A project charter should be used as a working document that is updated as the project evolves. The version control of the charter is therefore very important.

8.2.5.3 Basic Steps

1. Select the project by taking into account the following criteria:
   a. Not capital intensive
   b. Achievable in 6 months
   c. High probability of success
   d. Good fit with Six Sigma techniques
   e. Clearly linked to real business need
   f. Historical and current data accessible
2. Identify the customers and their specific CTQ requirements by a SIPOC diagram.
3. Estimate order of magnitude figures for the costs and savings for the project. (These estimates would be updated when more accurate data become available as the project advances.)
4. Obtain top management support and sponsorship.
5. Select the project team with clear leadership and ownership for delivery.
6. Develop the project charter following a defined template (see Figure 8.5).
7. Obtain the approval of the sponsor.
8. Review and update the charter with the necessary version control as the project progresses.

8.2.5.4 Worked-Out Example

Figure 8.5 shows an example of a project charter for a 'safety performance' project of DuPont Teijin Films Ltd., UK.

8.2.5.5 Training Requirements

The training of a project charter is included in the training programmes for both Black Belt and Green Belt attainment. The basic principles of a project

Project:	Safety Performance Predictor Model				
Date:	3/28/21				
Project Type:	Quality X Revenue Cost Reduction				
Problem Statement:	Plant needs early warning signals so programs can be put in place to prevent injuries/incidents				
Goal Statement:	Predict injuries/incidents prior to occurring Performance Level: 1.47 DPMO Time Frame: 6 months				
Project Scope and Approach:	Study safety data to determine what factors influence injuries and incidents. Use the most significant factors and develop a predictor model to serve as early warning signals for a potential deterioration in our safety performance.				
Team Members:	Terry Leadership John Debbie Leader: Vickie Sponsor: Terry Mentor: Tim				
Customers:	Leadership Terry Network Leaders				
CTQs:	Early warning signals of downward shifts in safety climate Combined data source for all site safety data To be able to identify problem areas before injuries occur				
Defect Definition:	Injuries and Incidents				
Opportunities per Unit:	Exposure Hours				
Approximate DPMO:	1.47	Z st: 6.19			
Goal DPMO:	1.25	Z st goal: 6.19			
Stake:	Confidence Interval:	Upper:	Lower		
Capital:					
Timing:	Define	Measure	Analyse	Improve	Control
Target Completion:	04/30/00	06/30/00	07/31/00	08/31/00	09/30/00
Data Issues:	Safety Climate Indicator measurements system needs to be validated				

Figure 8.5 Green Six Sigma project charter.

charter are relatively simple and can be explained to team members in less than an hour. The development of a chapter with appropriate data may need a couple of days.

8.2.5.6 Final Thoughts

A project charter is an essential working document of larger Six Sigma projects. The success of a project may often be determined by the project charter at the Definition stage. However, for smaller projects and those of a shorter duration (e.g. less than 3 months), a project charter is not essential and may even create unnecessary bureaucracy.

8.3 Summary

It is important to clearly define our problem to help focus our efforts on solving it. Einstein also emphasised, 'If I were given one hour to save the planet, I would spend 59 minutes defining the problem and one minute solving it'.

The starting process of a project is characterised by the pressure to 'kick off' the task as soon as possible, but there is a danger of too quick a start leading to a poor foundation. The simple data collection tools as described in this chapter should help the project team to identify the key requirements of defining the project on the right basis. The use of these tools does not require any specific expertise or lengthy training, but it does add significant value to develop the project charter. The project charter clearly sets the scope and key deliverables of the project.

GREEN TIPS

- Remember that if you cannot define a problem, you cannot improve it.
- Use the tools described in this chapter to define the problem and set the scope and deliverables in a project charter.

Chapter 9

Tools for Measurement

When you can measure what you are speaking of, and express it in numbers, you know that on which you are discoursing. But if you cannot measure it and express it in numbers, your knowledge is of a meagre and unsatisfactory kind.

—Lord Kelvin

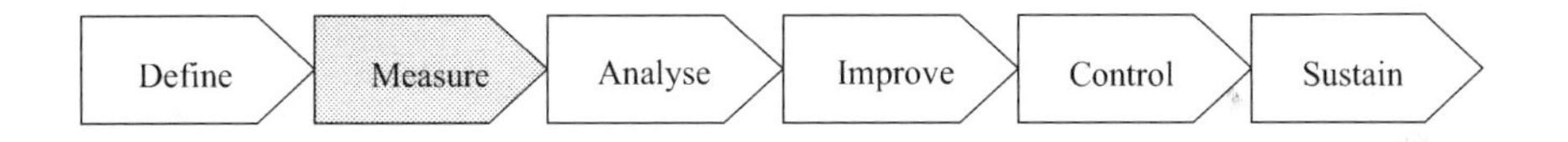

9.1 Introduction

It is generally accepted that when we start a journey, the most essential thing to know is where we are going. We agree that a road map showing where we want to go is important, and we have covered this aspect in the section on project charters. Even if you have a reliable road map, you will get lost if you do not know where you are at the time of reading it. Therefore, it is vital to know the current 'as is' situation of your process (i.e. where you are now). The tools for measurement are aimed at doing just that.

The measurement stage means turning the ideas and objectives of the project charter into a structured appraisal process. Resources will be expanded by validating ideas and identifying further opportunities. It is during the measurement stage that the business case for the principal options of a quality improvement project is developed. Another key objective of this stage is to measure the current performance of the core business process

DOI: 10.4324/9781003268239-11

involved in the project. The main deliverables of the 'Measure' stage are as follows:

1. Data for the project—data collected using a data collection plan to map the current performance of the process
2. Improvement goal for the project—specific goal or standard for improving the process performance
3. Process capability of the project—a quantitative assessment of how well the current process meets the performance standard of the project

9.2 Description of Tools for Measurement

The important tools for measurement should include

M1. Check sheets
M2. Histograms
M3. Run charts
M4. Scatter diagrams
M5. Cause and effect diagrams
M6. Pareto charts
M7. Control charts
M8. Flow process charts
M9. Process capability measurement

The techniques for the Measure stage include statistical process control (SPC; see Chapter 14) and benchmarking (see Chapter 15).

9.2.1 M1: Check Sheets

9.2.1.1 Definition

The check sheet is a simple and convenient recording method of collecting and determining the occurrence of events. These sheets or forms allow a team to systematically record and compile data from observations so that trends can be shown clearly.

9.2.1.2 Application

The check sheets are very easy to apply and are used to record nonconforming data and events including

- The breakdown of machinery
- Non-value-added activities in a process
- Mistake or defects recording in a process or a problem

The forms are prepared in advance of recording the data by the operatives being affected by the problem. It makes patterns in the data clearer, based on facts, from a simple process that can be applied to any performance area. The supplement attribute quality control charts and histograms and in a complex process provide a form of 'mistake proofing'.

9.2.1.3 Basic Steps

1. Agree on the type of data to be recorded. The data could relate to the number of defects and type of defects and apply to equipment, the operator, process, department, shift and so on.
2. Decide which characteristics and items are to be checked.
3. Determine the type of check sheet to use, e.g. tabular form, defect position or tally chart.
4. Design the form to allow the data to be recorded in a flexible and meaningful way.
5. Decide who will collect data, over which period and from what sources.
6. Record the data on check sheets and analyse data.

9.2.1.4 Worked-Out Examples

The design of check sheets is highly flexible, but they can be grouped into three main categories: tabular form, defect position and tally charts (Table 9.1).

Figure 9.1 shows an example of a check sheet where defects positions are marked on the drawing of a product.

9.2.1.5 Training Needs

The basic principles of a check sheet are so simple that it does not require any classroom training. A member of a project usually gains experience in a check sheet by on-the-job training.

Table 9.1 Example Check Sheet in Tabular Form

Check Items	*Week Number*							
	Day 1	*Day 2*	*Day 3*	*Day 4*	*Day 5*	*Day 6*	*Day 7*	*etc.*
Incorrect brand specification								
Incorrect print density								
Ink smudging								
Mis-registration								

Mistakes	*Check Sheet for: Typing February 03*				
	Week 1	*Week 2*	*Week 3*	*Week 4*	*Total*
Centring	11	1111	11	~~1111~~	13
Spelling	~~1111~~ 1	1111	11	111	15
Punctuation	~~1111~~ ~~1111~~ 11	~~1111~~ ~~1111~~ ~~1111~~ 1	111	1111	35
Missed paragraph	11	1	1	111	7
Wrong page number	1	1	11	1	5
Total	23	26	10	16	65

Figure 9.1 Example of a check sheet as a tally chart.

9.2.1.6 Final Thoughts

Check sheets are very simple to apply and effective as a tool for data collection and team building. This is a case of low investment and high return and therefore cannot be missed.

9.2.2 M2: Histograms

9.2.2.1 Definition

A histogram is a graphical representation of recorded values in a data set according to frequency of occurrence. It is a bar chart of numerical variables giving a graphical representation of how the data are distributed.

9.2.2.2 Applications

The histogram is used extensively in both statistical analysis and data presentation. A histogram displays the distribution of data and thus reveals the amount of variation within a process. There are a number of theoretical models for various shapes of distribution of which the most common one is the normal or Gaussian distribution.

There are several advantages of applying histograms in continuous improvement projects including

- It displays large amounts of data that are difficult to interpret in tabular form.
- It illustrates quickly the underlying distribution data revealing the central tendency and variability of a data set.

9.2.2.3 Basic Steps

1. Collect at least 50 to 125 data points for establishing a representative pattern.
2. Subtract the smallest individual value from the largest in the data set.
3. Divide this range by 5, 7, 9 or 11 depending on the number of data points. As a rough guide take the square root of the total number of data points and round it to the nearest integer. For example for 50 data points divide the range by 7 and 125 data points by 11.
4. The resultant value determines the interval of the sample. It should be rounded up for convenience.
5. Calculate the number of data points in each group or class.
6. Plot the histogram with the intervals in the x-axis and the frequency of occurrence on the y-axis.
7. Clearly label the histogram.
8. Interpret the histogram related to centring, variation and shape (distribution).

9.2.2.4 Worked-Out Example

The following example is taken from Schmidt et al. (1999, pp. 135–137).

Table 9.2 shows the data points of miles per gallon data of a carpool. (NB. Data should presented in Table 9.2 as 18, 16, 30, 92 etc.)

Table 9.2 Data Points of Miles per Gallon of Carpool

18	16	30	29	28	21	17	41	8	17
32	26	16	24	27	17	17	33	19	18
31	27	23	38	33	14	13	26	11	28
21	19	25	22	17	12	21	21	25	26
23	20	22	19	21	14	45	15	24	34

Table 9.3 Individual Data in Each Class

Class Number	*Class Range*	*Values*	*Frequency of Occurrence*
1	6–12	8,11	2
2	12–18	16,16,17,17,14,12,14,17,17, 13,15,17	12
3	18–24	18,21,23,19,20,23,22,22,19,21, 21,21,21,19,18	15
4	24–30	26,27,25,29,24,28,27,26,25,24, 28,26	12
5	30–36	32,31,30,33,33,34	6
6	36–42	38,41	2
7	42–48	45	1

For a 50-points data set we would use $\sqrt{50} = 7.06 \rightarrow 7$ classes. Since our smallest data point is 8 and the largest is 45, the interval should be (45 – 8)/7 = 5.3. We round it up to make it 6. In Table 9.3 we place individual data in each class as shown.

Next we construct the histogram as shown in Figure 9.2. We see the most values are between 12 to 30 mpg with 18 to 24 mpg representing the median. The data are also skewed to the right by some high mileage values.

9.2.2.5 Training Needs

Although a histogram can lead to complex statistical analysis, its basic principles are relatively simple. Most members of the project team are likely to

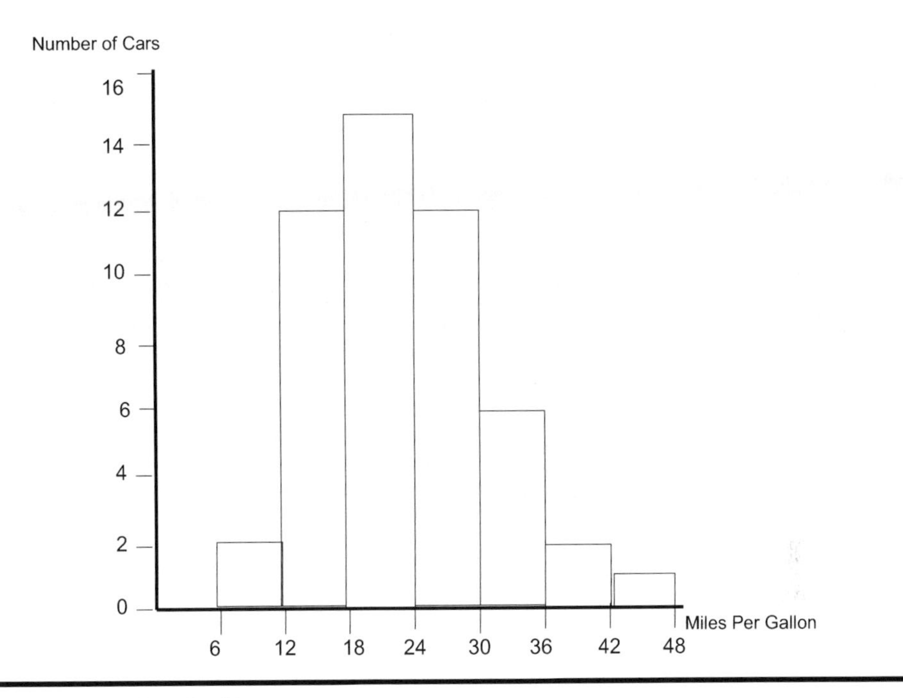

Figure 9.2 Example of a histogram.

have some familiarity with and knowledge of histograms. However, a 1-hour session of classroom training should be valuable to establish the methodology, especially related to the range, class and intervals.

9.2.2.6 *Final Thoughts*

A histogram gives a pictorial presentation of how the data are distributed. It is easy to understand and use, and it should be regarded as an essential tool of continuous improvement.

9.2.3 *M3: Run Charts*

9.2.3.1 *Definition*

A run chart is a graphical tool to allow a team to study observed data for trends over a specified period of time. It is basically a simple line graph of x and y axes.

9.2.3.2 *Application*

A run chart has a wide range of applications to detect trends, variation or cycles. It allows a team to compare performances of a process before and after the implementation of the solution. The application areas include sales analysis, forecasting, performance reporting and seasonality analysis.

9.2.3.3 Basic Steps

1. Select the parameter and time period for measurement.
2. Collect data (generally 10 to 20 data points) to identify meaningful trends:
   a. *x*-axis for time or sequence cycle (horizontal)
   b. *y*-axis for the variable parameter that you are measuring (vertical)
3. Plot the data in a line graph along the *x* and *y* axes.
4. Interpret the chart. If there are no obvious trends, then calculate the average value of the data points and draw a horizontal line at the average value.

9.2.3.4 Worked-Out Example

The following table shows the operational efficiency (%) of a packaging machine.

x	Jan	Feb	Mar	Apr	May	Jun	Jul	Aug	Sept	Oct	Nov	Dec
y	55	60	45	40	65	60	65	30	60	65	60	55

The run chart for the above data is shown in Figure 9.3.

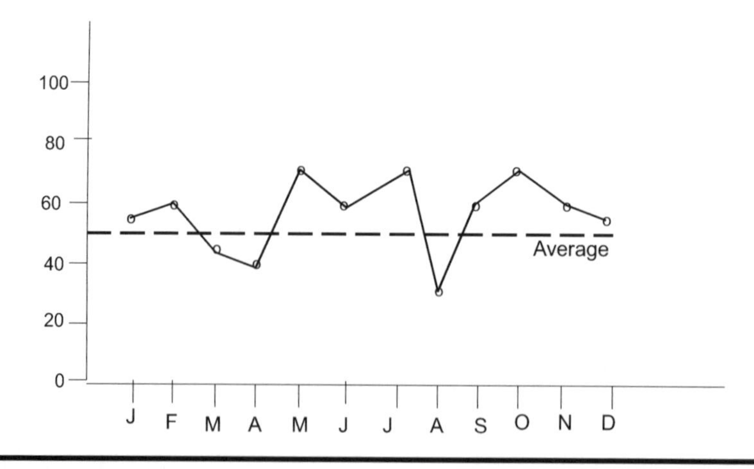

Figure 9.3 Example of a run chart.

9.2.3.5 Training Requirement

The understanding and application of a run chart is so simple that it is not essential to conduct a classroom training specifically for it. The team members should be able to use a run chart after a preliminary briefing, in a practical application. The training process involving a run chart is usually included in a training programme for continuous improvement tools.

9.2.3.6 Final Thoughts

Due to the fact that it could be seen as encompassing simplicity to a fault, a danger in using a run chart is the inclination to interpret every variation as significant. For any statistically significant result, control charts or regression analyses should be more appropriate. We recommend the use of a run chart for detecting a visual trend only and identifying areas of further analysis.

9.2.4 M4: Scatter Diagrams

9.2.4.1 Definition

A scatter diagram is a plot of points to study and identify the possible relationship between two variables, characteristics or factors. The knowledge provided by a scatter diagram can be enhanced more accurately by regression analysis.

9.2.4.2 Application

A scatter diagram is used, as an initial step before regression analysis, to show in simple terms if the variables are associated (a linear pattern) or unrelated (non-linear random pattern). Analysis should investigate the scatter of the plotted points and if some linear or non-linear relationship exists between two variables. In this the scatter diagram is very useful for diagnosis and problem-solving.

A scatter diagram is often used in a follow-up to a cause and effect diagram to detect if there are more than two variables between causes and the effect.

9.2.4.3 Basic Steps

1. Collect paired samples of data (at least ten) of two variables that may be related.

2. Draw the x and y axes of the diagram.
3. Plot the data points on the diagram.
4. If some values are repeated, circle those points depending on the number of times they are repeated.
5. Interpret the results. The scatter diagram does not predict cause and effect relationships. It can, however, indicate a possible positive or negative correlation or no correlation between two variables.

9.2.4.4 *Worked-Out Example*

The following example is taken from Schmidt et al. (1999, pp. 154–156).

Table 9.4 shows a set of paired data points of two variables.

We plot these data with weight as the x-axis (cause) and mpg as the y-axis (effect), as an example of a scatter diagram, as shown in Figure 9.4.

It appears that a negative correlation exists between two variables, because when one variable (weight) gets bigger, the other variable (mpg) becomes smaller.

9.2.4.5 *Training Requirements*

It is important that training of regression analysis is conducted together with that of scatter diagrams. Although the topic of scatter diagrams may not

Table 9.4 Data Points of Two Variables

Observations	*Weight (lbs)*	*Miles per Gallon (mpg)*
1	3000	18
2	2800	21
3	2100	32
4	2900	17
5	2400	31
6	3300	14
7	2700	21
8	3500	12
9	2500	23
10	3200	14

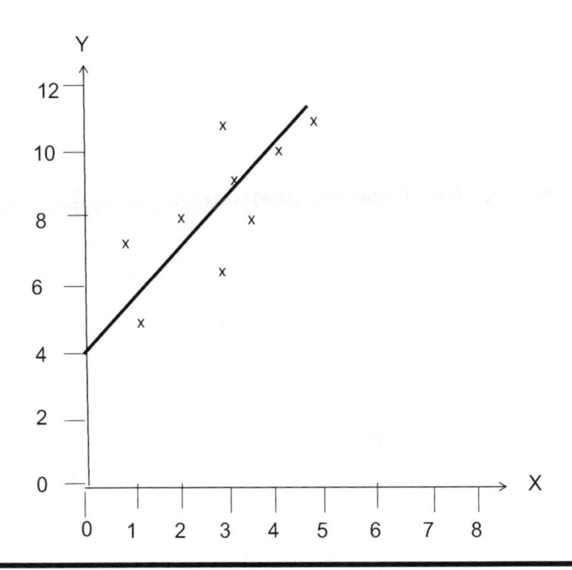

Figure 9.4 Example of a scatter diagram.

require a classroom session on its own, their link with regression analysis should be understood clearly.

It is also useful to have a few minutes' training to assess whether the two variables have possible positive correlation, no correlation or a possible negative correlation.

9.2.4.6 Final Thoughts

Scatter diagrams are easy to understand and apply. We recommend scatter diagrams as the first step to study the relationship of two variables before embarking on more complex regression analysis.

9.2.5 M5: Cause and Effect Diagrams

9.2.5.1 Definition

The cause and effect diagram is a graphical representation of potential causes for a given effect.

Since it was first used by Ishikawa, this type of illustration is also known as an Ishikawa diagram. In addition, it is often referred to as a 'fish-bone' diagram due to its skeletal appearance.

The purpose of the diagram is to assist in brainstorming and enabling a team to identify and graphically display, in increasing detail, the root causes of a problem.

9.2.5.2 Application

The cause and effect diagram is arguably the most commonly used of all quality improvement tools. The 'effect' is a specific problem and is considered to constitute the head of the diagram. The potential causes and sub-causes of the problem form the bone structure of the skeletal fish.

They are typically used during the measurement and analysis phases of the project. Their wide application area covers Six Sigma teams, Total Quality Management (TQM) teams or continuous improvement teams as part of brainstorming exercises to identify the root causes of a problem and offer solutions. It focuses the team on causes, rather than symptoms.

There are a number of variants in the application of a cause and effect diagram. The two most common types are the 6M diagram and CEDAC. Let us examine each of these terms a little more closely.

In a 6M diagram, the main bone structure or branches typically consist of the self-explanatory 6Ms:

- Machine
- Manpower
- Material
- Method
- Measurement
- Mother Nature (Environment)

A CEDAC (cause and effect diagram assisted by cards) works slightly differently. A blank, highly visible fish-bone chart is displayed in a meeting room. Every member of the team must post potential causes and solutions on a card (or Post-it notes) considering each of the categories.

A CEDAC also consists of two major formats:

- Dispersion analysis type
- Process classification type

First, the dispersion analysis type is used usually after 6M diagrams or CEDAC have been completed. The major causes identified are then treated as separate branches, and their sub-causes are identified.

However, the process classification type uses the major steps of the process in place of the major cause categories. This form is usually used when the problem encountered cannot be isolated to a single department. Each stage of the process is then analysed by using a 6M or CEDAC approach.

A typical sequence of cause and effect diagrams for a complex problem could be

Process classification type

↓

6M or CEDAC

↓

Dispersion analysis type

9.2.5.3 Basic Steps

1. Select the most appropriate cause and effect format. If the problem can be isolated to a single section or department choose either a 6M (small team) or a CEDAC (large team) approach.
2. Define with clarity and write the key effect of the problem in a box to the right-hand side of the diagram.
3. Draw a horizontal line from the left-hand side of the box. Draw main branches (fish bones) of the diagram after agreeing on the main categories (e.g. 6M) of causes.
4. Brainstorm for each category the potential sub-causes affecting the category.
5. List the sub-causes of each category in a flip chart. In a CEDAC approach these would be a collection of Post-it notes for each category.
6. Rank the sub-causes in order of importance by a group consensus (or multi-voting) and select up to six top sub-causes for each category.
7. Construct the diagram by posting the top sub-causes in each category. These are the 'root causes'.
8. Decide upon further dispersion analysis or gather additional data needed to confirm the root causes.
9. Develop solutions and improvement plans.

9.2.5.4 Worked-Out Examples

The following example is taken from Basu and Wright (2003, pp. 29–30).

Consider the situation where customers of a large international travel agency sometimes find that when they arrive at their destination, the hotel has no knowledge of their booking.

In this case, to get started you might begin with 'Hotel Not Booked' as the effect and the 6Ms (machine, manpower, method, material, measurement

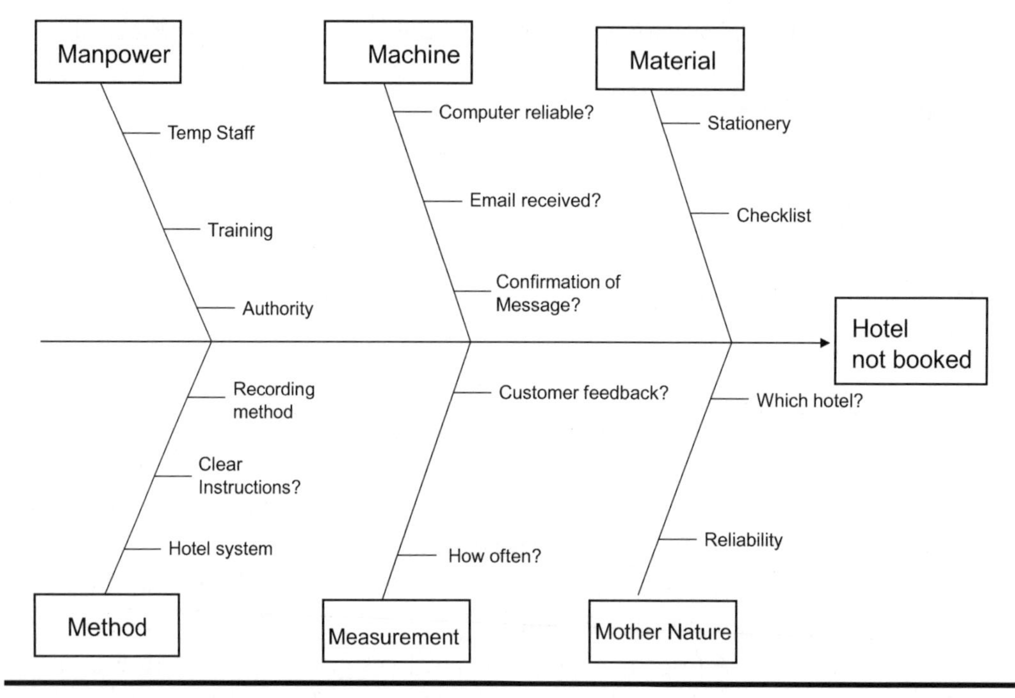

Figure 9.5 Cause and effect diagram.

and Mother Nature) as the causes. When the sub-causes are further investigated, the diagram may look like that shown in Figure 9.5.

The diagram points out that many sub-causes including training, email systems and customer feedback appear to be worthwhile areas to follow up.

9.2.5.5 Training Requirements

As cause and effect diagrams are essential to a quality improvement programme, it is important that members of the team receive at least 1 hour of hands-on training. An effective method of training is by participating in a brainstorming exercise where a cause and effect diagram is developed. It is more important to identify the root causes than to assign the cause in a specific category. For instance, in the example given in Figure 9.5, 'clear instructions' could also be grouped under Manpower or 'checklist' could be assigned to the category of Measurement.

9.2.5.6 Final Thoughts

Cause and effect diagrams are invaluable in brainstorming the root causes for a given problem. We strongly recommend their application on as wide a basis as possible.

9.2.6 M6: Pareto Charts

9.2.6.1 Definition

A Pareto chart is a special form of bar chart that rank orders the bars from highest to lowest in order to prioritise problems of any nature.

It is known as 'Pareto' after a 19th-century Italian economist Wilfredo Pareto who observed that 80% of the effects are cause by 20% of the causes: 'the 80/20 rule'.

9.2.6.2 Applications

Pareto charts are applied to analyse the priorities of problems of all types, e.g. sales, production, stock, defects, sickness, accident occurrences, etc. The improvement efforts are directed to priority areas that will have the greatest impact.

There are usually two variants in the application of Pareto charts. The first type is the standard chart where bar charts are presented in descending order. The second type is also known as the 'ABC Analysis' where

- Cumulative percentage (%) values of causes are plotted along the x-axis.
- Cumulative percentage (%) values of effects are plotted along the y-axis.
- 80% of the effects with corresponding percentage (%) of causes are grouped as 'A' category.
- 80–96% of the effects and corresponding causes are 'B' items.
- The remaining values are 'C' category.

9.2.6.3 Basic Steps

The following steps apply for the preparation of a Pareto chart:

1. Identify the general problem (e.g. IC Board Defects) and its causes (e.g. soldering, etching, moulding, cracking, other).
2. Select a standard unit of measurement (e.g. frequency of defects or money loss) for a chosen time period.
3. Collect data for each of the causes in terms of the chosen unit of measurement.
4. Plot the Pareto chart with causes along the x-axis and the unit of measurement along the y-axis. The causes are charted in descending order of values from left to right.
5. Analyse the graph and decide on the priority for improvement.

The following steps apply for the preparation of an ABC analysis:

1. Decide on the causes (e.g. number of customers) and effect (e.g. sales) of the problem areas that you want to prioritise.
2. Select a standard unit of measuring the effect (e.g. dollars for sales values).
3. Collect data for each cause (e.g. customer) and the corresponding effect (e.g. sales in dollars).
4. Rank the cause according to the value of the effects (e.g. customers in descending order of dollar sales).
5. Calculate the cumulative percentage (%) values of the causes and effects.
6. Plot the cumulative percentage (%) values of the causes (e.g. number of customers) along the *x*-axis and the cumulative percentage (%) values of effects (e.g. sales in dollars) along the *y*-axis.
7. Identify A, B and categories (e.g. A for 80% of sales, B for 80–96% of sales and C for the remainder).
8. Analyse the graph and decide on the priority for improvement.

9.2.6.4 Worked-Out Examples

The following example of a Pareto chart is taken from Schmidt et al. (1999, p. 144).

The major defects identified during the manufacture of integrated circuit boards for a given period were as shown in Table 9.5.

We plot the causes along the *x*-axis and frequencies along the *y*-axis as shown in Figure 9.6.

Table 9.5 Major defects identified during the manufacture of Integrated Circuit Boards

Causes	*Frequency*	*Percentage (%) Frequency*
Soldering	60	40
Etching	40	27
Moulding	30	20
Cracking	15	10
Other	5	3
Total	150	100

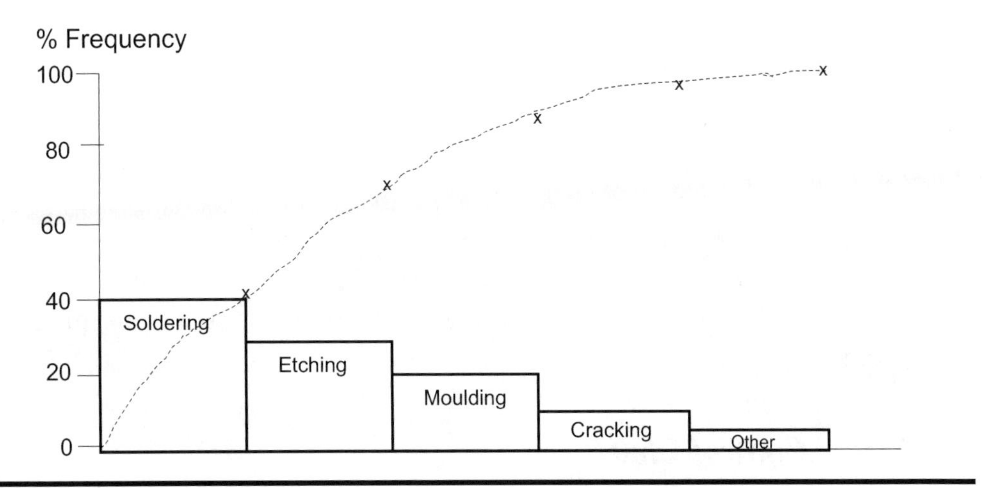

Figure 9.6 Pareto chart.

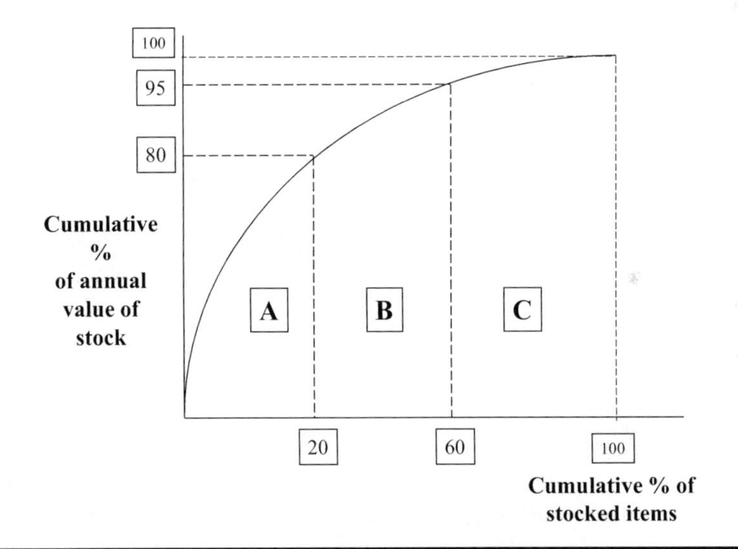

Figure 9.7 ABC analysis.

An example of ABC analysis for a set of inventory items is shown in Figure 9.7.

9.2.6.5 *Training Needs*

The basic principles of a Pareto chart and ABC analysis are easy to follow. The training for this tool of prioritising is usually included in the classroom training sessions related to the tools for measurement. A member of a team can be adept in the application of Pareto charts after, say, 1 hour's practice on a few practical problems.

9.2.6.6 Final Thoughts

Pareto charts are extremely useful in optimising efforts and 'going for gold'. They have the twin advantages of being both easy to understand and to apply. Thus we highly recommend the use of Pareto charts for all quality improvement projects.

9.2.7 M7: Control Charts

9.2.7.1 Definition

A control chart consists of a graph with time on the horizontal axis and an individual measurement (such as mean, or range) on the vertical axis.

A control chart is a basic graphical tool of SPC (see Chapter 14) for determining whether a process is stable and also for distinguishing usual (or common) variability from unusual (special assignable) causes. Three control limits are drawn: the central line (CL), the lower control limit (LCL) and the upper control limit (UCL). The points above the UCL or below the LCL indicate a special cause. If no signals occur the process is assumed to be under control, i.e. only common causes of variation are present.

9.2.7.2 Application

Control charts can be used for examining a historical set of data and also for current data. The current control based on current data underpins a feedback control loop in the process.

There are many good reasons why control charts have been applied successfully in both quality control and improvement initiatives since Walter Shewhart first introduced the concept at the Bell Laboratories in the early 1930s.

First, control charts establish what is to be controlled and force their resolution. Second, they focus attention on the process rather than on the product. For example, a poor product can result from an operator error, but a poor production process is not capable of meeting standards on a consistent basis. The third factor is that they comprise a set of prescribed techniques that can be applied by people with appropriate training in a specified manner.

For many probability distributions, most of the probability is within three standard deviations of the mean. So μ and σ are respectively the stable process mean and standard deviation, then

$$CL = \mu$$
$$UCL = \mu + 3\sigma$$
$$LCL = \mu - 3\sigma$$

When the mean (μs) and standard deviation (σs) values are calculated from a sample of n then

$$\mu s = \mu$$
$$\sigma s = \sigma/\sqrt{n}$$

There are two types of control charts. A variable chart is used to measure individual measurable characteristics, whereas an attributes chart is used for go/no-go types of inspection.

The x-bar chart (also called the mean chart), the s-chart (also called the standard deviation chart) and the R-Chart (also called the range chart) are used to monitor continuous measurement or variable data.

The stable control charts are used to determine process capability, that is whether a process is capable of meeting established customer requirements or specifications.

9.2.7.3 Basic Steps

1. Choose the quality characteristic to be charted. A Pareto analysis is useful to identify a characteristic that is currently experiencing a high number of non-conformities.
2. Establish the type of control chart to ascertain whether it is a variable chart or an attribute chart.
3. Choose the sub-group or sample size. For variable charts, samples of 4 to 5 are sufficient, whereas for attribute charts samples of 50 to 100 are often used.
4. Decide on a system of collecting data. The automatic recording of data by a calibrated instrument is preferable to manually recorded data.
5. Calculate the mean and standard deviation and then calculate the control limits.
6. Plot the data and control limits on a control chart and interpret results.

9.2.7.4 Worked-Out Example

The following example illustrates the construction of variable control charts based on the data of packing cartons of a morning shift. The data in Table 9.6 are taken from Ledolter and Burnhill (1999).

Hence,

$$CL = \mu_s = 25.118$$
$$\sigma_s = \sigma / \sqrt{n} = 0.159 / \sqrt{10} = 0.159 / 3.163 = 0.05$$
$$UCL = \mu_s + 3\ \mu_s = 25.118 + 3 \times 0.05 = 25.27$$
$$LCL = \mu_s - 3\ \sigma_s = 25.118 - 3 \times 0.05 = 24.97$$

The control limits and the data are plotted as shown in Figure 9.8.

Similar charts are drawn for the standard deviation and the range of the data. The control limits are constructed in such a manner that we expect approximately 99.7% of all data to fall between them. In Figure 9.8 all data are within the control limits, indicating that the process is stable.

9.2.7.5 Training Requirement

The construction and interpretation of control charts require a good understanding of SPC. The concepts of variable data versus attribute data,

Table 9.6 Data of Packing Cartons

Reading Number	*Measurements*				*Average*	*Standard Deviation*	*Range*
1	25.1	25.5	25.0	25.1	25.175	0.222	0.50
2	24.8	25.2	25.1	24.9	25.000	0.183	0.40
3	25.1	25.2	25.2	25.2	25.175	0.050	0.10
4	25.1	25.4	24.8	25.0	25.075	0.250	0.60
5	25.2	24.7	24.9	25.3	25.025	0.275	0.60
6	25.2	25.2	25.0	25.1	25.125	0.096	0.20
7	25.2	25.2	25.2	25.3	25.225	0.050	0.10
8	25.2	25.1	25.3	25.0	25.150	0.129	0.30
9	24.9	25.1	25.2	24.8	25.000	0.183	0.40
10	25.1	25.1	25.3	25.4	25.225	0.150	0.30
Average					25.118	0.159	0.35

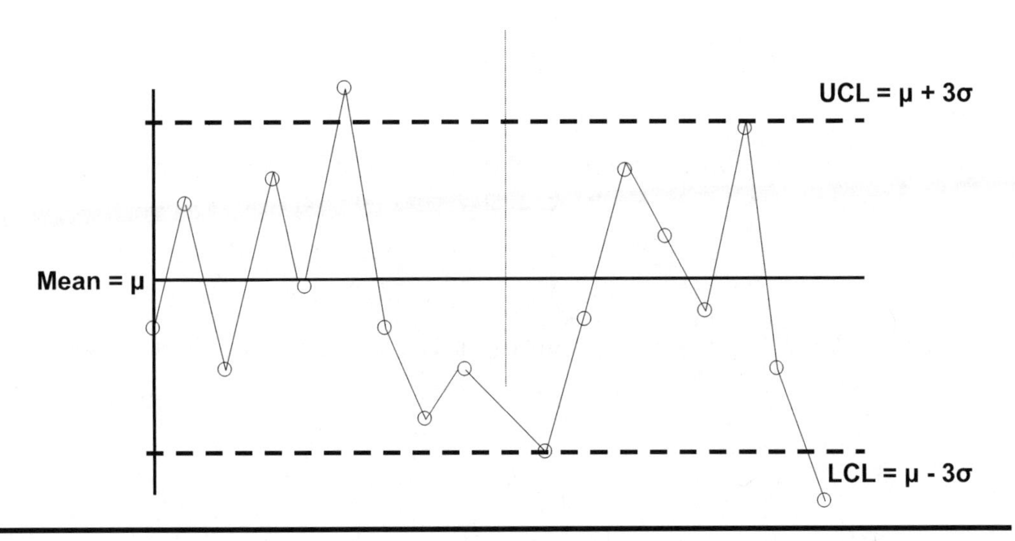

Figure 9.8 Control chart.

common causes versus special causes and control limits are essential to the effective application of control charts. Hence a few hours of classroom training are recommended before the members start the application of this tool. Care should be taken not to confuse a control chart with a run chart.

9.2.7.6 Final Thoughts

Control charts are useful to identify data and their causes outside the control limits. However, nothing will change just because you charted it. You need to do something and eliminate the causes.

9.2.8 M8: Flow Process Charts

9.2.8.1 Definition

A flow process chart is a symbolic representation of a physical process linked to correspond to the sequence of operation.

It was originally introduced as a classical industrial engineering tool with five symbols (collectively known as activities) as follows:

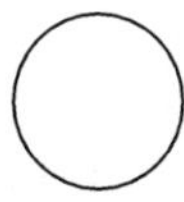

Operation: An operation consists of an activity that changes or transforms an input.

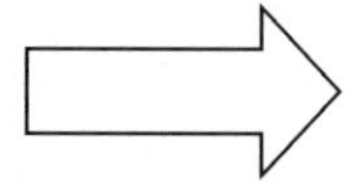

Transport: A transport consists of the physical movement of an input.

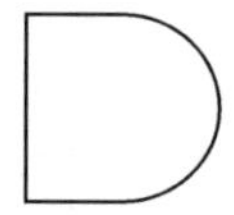

Delay: A delay is caused when an input is waiting for the next activity.

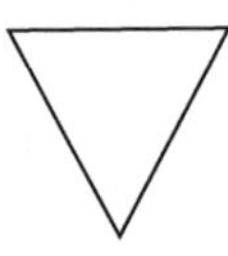

Storage: A storage is created when an input is somewhere so that a decision is required to move it.

Inspection: An inspection is caused through a check on an input for possible conformance.

9.2.8.2 Application

Flow process charts have been used extensively in manufacturing and supply chain operations to identify

- The hierarchical structure of operations
- The sequence of activities
- Non-value-added activities

Over the years, many specialised forms of flow charts have evolved to analysed the hierarchical structure and sequence of activities. Two such specialised derivatives are flow diagrams (for data processing) and process mapping (for process sequencing). However, the classical flow process charts are still being applied to identify non-value-added activities. Their applications have also been extended to service and transaction activities. A more recent application of flow process charts has been in the analysis of Lean processes (Basu and Wright, 2003) where seven forms of wastes or 'mudas' (non-value-added activities) have been defined:

- Excess production
- Waiting
- Transportation
- Motion
- Process
- Inventory
- Defects

Traditionally, there are two formats for flow process charts: a pre-printed format and a descriptive arrangement. In the pre-printed format, each activity is recorded in the form, the corresponding symbols are marked and then joined in sequence by a line. This is the classical industrial engineering format where the number of activities for the present and proposed methods are compared.

In the descriptive set-up, each activity is charted in sequence and represented by process symbols. This chart provides a systematic start to analyse the process and does not require any extra stationery.

9.2.8.3 Basic Steps

1. Select the process for analysis. Start with a high-level block diagram to gain a broad process view.
2. Identify the scope of the flow process chart by defining physical and functional boundaries.
3. Develop the sequence of process activities. Capture activity details as they really are, rather than what could be, according to written procedures.
4. Develop the flow process chart by assigning the appropriate symbols for each activity.
5. Identify non-value-added activities and opportunities for improvement.
6. Develop the flow process chart for the improved process and quantify that improvement.

9.2.8.4 Worked-Out Example

The following example is taken from Kolarik (1995, p. 212).

Consider the development of a flow process chart for a cattle feed manufacturing process. Figure 9.9 shows a high-level 'block diagram' of the process.

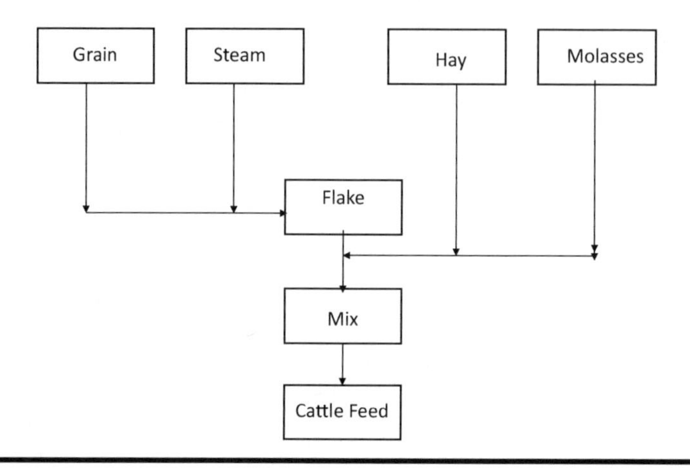

Figure 9.9 Block diagram.

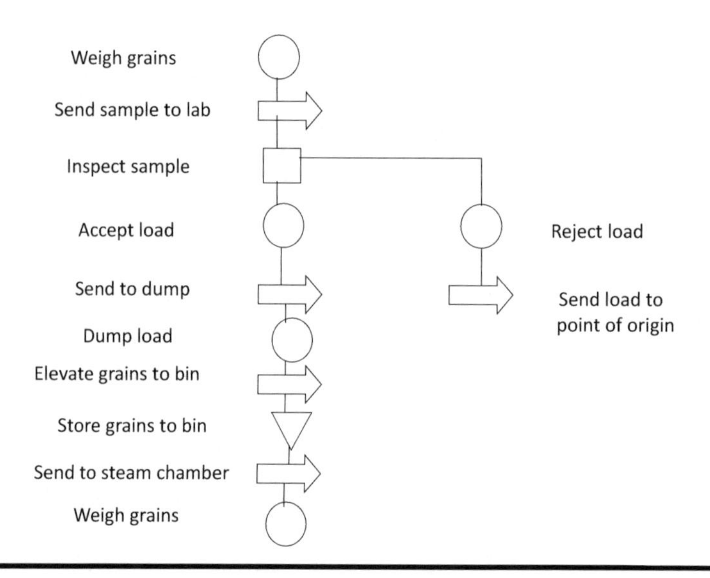

Figure 9.10 Flow process chart.

We scope the flow process chart for 'Grain' related activities only for the purposes of this example. The details of this process are shown in the flow process chart in Figure 9.10.

The chart in Figure 9.10 identifies the non-value-added activities of transport, inspection and storage and points towards a direct loading of grains to the steam chamber. This would lead to the feasibility of the quality assurance of grains before loading and an automated weight control system so that grains can be blown directly to the steam chamber.

9.2.8.5 Training Requirements

The basic principles of the flow process chart are fundamental and can be applied consistently. This makes the training process relatively simple and thus team members can use flow process charts effectively after 1 hour's practice on a practical problem. This is a manual process and does not require any computer software like process mapping.

9.2.8.6 Final Thoughts

With the advent of computerised process mapping facilities, the use of flow process charts is not fashionable any more. However, it is still an effective manual tool to identify non-value-added activities. We recommend its application in a simple process or as part of a more complex course of action.

9.2.9 M9: Process Capability Measurement

9.2.9.1 Definition

Process capability is the statistically measured inherent reproducibility of the output (product) turned out by a process.

A commonly used measure of process capability is given by the capability index (C_p):

$$C_p = \frac{USL - LSL}{6\sigma}$$

where USL and LSL are upper and lower specification limits and σ is the standard deviation.

The C_p index measures the potential or inherent capability of process. The C_{px} index measures the realised process capability relative to the actual operation and is defined as

$$C_{pk} = \text{minimum}\left(\frac{\mu - LSL}{3\sigma}, \frac{USL - \mu}{3\sigma}\right)$$

If $C_{pk} > 1$, we declare that the process is capable; if $C_{pk} < 1$, then we declare that the process is incapable.

C_{pk} is a more practical measure of capability than C_p.

9.2.9.2 Application

A customer requires that specifications are given in terms of a target value, a lower specification limit (LSL) and upper specification limit (USL). They are the 'tolerances' of the specification. The process capability index determines the reliability of these tolerances to be delivered by the process used by the supplier.

The process capability information serves many purposes including

- Selection of competing processes or equipment that best satisfies the specification
- Predicting the extent of variability that processes could exhibit to establish realistic specification limits
- Testing the theories of cause and effect during quality improvement programmes

Motorola introduced the concept of Six Sigma as a statistical way of measuring total customer satisfaction. Given that the process is a Six Sigma process, we know that USL – LSL = 6σ + 6σ = 12σ. Hence,

$$C_p = \frac{USL - LSL}{6\sigma} = \frac{12\sigma}{6\sigma} = 2.0$$

Another metric, defects per million opportunities (DPMO) is also used to assess the performance of a process. DPMO is represented by the proportion of area outside the specification limit multiplied by 1 million. For example if 3.5% of the area is outside specification limits, then

$$\text{DPMO} = 1{,}000{,}000 \times \frac{3.5}{100} = 35{,}000$$

The three metrics DPMO, C_p and C_{pk} all give numerical values that indicate how well a process is doing with respect to these specification limits.

9.2.9.3 Basic Steps

1. Using a stable control chart, determine the process grand average (χ), the average (R) and process standard deviation (s).
2. Determine the USL and the LSL based upon customer requirements.

3. For a stable process, assume that the values of the sample mean and standard deviation are the same as the corresponding values of the population, that is, $\sigma = s$ and $\mu = \chi$.
4. Calculate the process capability indices C_p and C_{pk} by using the formulae:

$$C_p = \frac{USL - LSL}{6\sigma}$$

$$C_{pk} = \text{minimum}\left(\frac{\mu - LSL}{3\sigma}, \frac{USL - \mu}{3\sigma}\right)$$

Figure 9.11 illustrates three states of the potential process capability for $C_p < 1$, $C_p = 1$ and $C_p > 1$.

9.2.9.4 *Worked-Out Example*

Consider the graph in Figure 9.12.

From the graph we get the following:

$\mu = \chi = 16$
$\sigma = s = 2$
$USL = 20$
$LSL = 10$

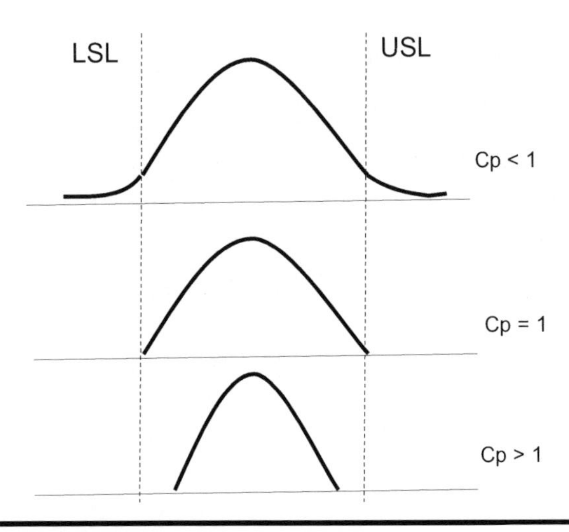

Figure 9.11 Process capability.

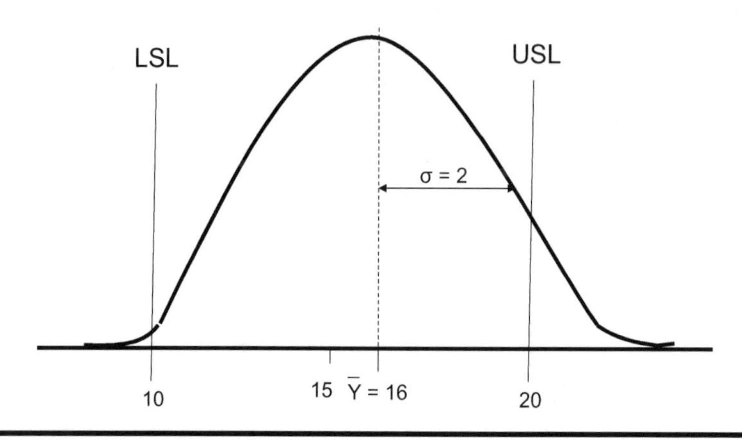

Figure 9.12 A distribution graph.

Therefore, we calculate

$$C_p = \frac{USL - LSL}{6\sigma} = \frac{20-10}{6\times 2} = 10/12 = 0.83$$

$$C_{pk} = \text{minimum}\left(\frac{\mu - LSL}{3\sigma}, \frac{USL-\mu}{3\sigma}\right)$$

$$C_{pk} = \text{minimum}\left(\frac{16-10}{6}, \frac{20-16}{6}\right)$$

$$= \text{minimum}\left(1, \frac{2}{3}\right) = 0.67$$

9.2.9.5 Training Requirements

The application of process capability requires a good understanding of SPC and the relationship between the sample data and the population data. It is important to recognise that USL and LSL are based on customer requirements and are different from the corresponding values of the control chart.

A few hours of classroom training for process capability along with control charts and SPC is recommended for specialist members of the team. In a Six Sigma programme, Black Belts go through a formal training schedule in process capability.

9.2.9.6 Final Thoughts

Process capability is a useful tool in determining whether a process can deliver the customer specifications. However, we recommend that its use and interpretation should be handled by Black Belts or an adequately trained member.

9.3 Summary

This chapter includes the basic tools for the Measure phase of DMAICS (define, measure, analyse, improve, control and sustain) in Green Six Sigma. Depending on the complexity of the project more advanced techniques (e.g. SPC) may be used during the Measure phase. There are also some tools described in this chapter (e.g. cause and effect diagram and flow process chart) that may be used again in the Analyse phase of the project.

It is advisable to approach the Measure phase in four stages. First, create a process map to understand the process and identify non-value-added activities. The next stage is to collect empirical data regarding how the process is functioning. The third stage is to validate the data. Finally, it is important to quantify the process performance. Appropriate tools are included in this chapter for each stage of the Measurement phase.

GREEN TIPS

- Consider four stages of the Measure phase: mapping, data collection, validation and process performance.
- Tools for measurement can be used again in subsequent phases of DMAICS.

Chapter 10

Tools for Analysis

All intelligent thoughts have already been thought; what is necessary is only to try to think them again.

—Johann Wolfgang Goethe

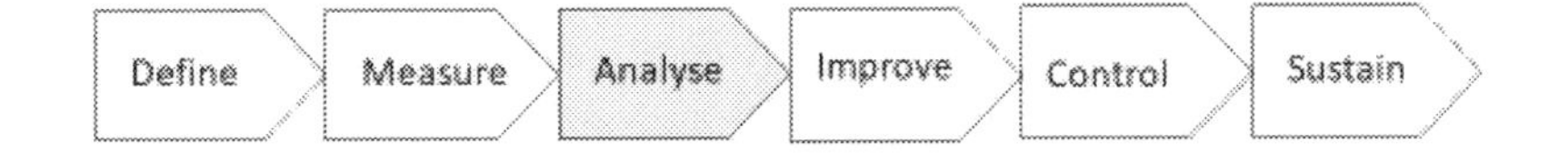

10.1 Introduction

Once the project is understood and defined in the Define stage, and then the baseline performance has been documented and validated at the Measure stage to ascertain that there is a real opportunity, it is now time to perform an in-depth analysis of the process. At this Analysis stage, tools and techniques are applied to identify and validate the root causes of problems. The objective is to identify all possible sources of variation in the process and distinguish between special and common causes of variation. Having got to the root causes of the problem, the business cause of the Measure stage can be updated with more accurate data. The data collected in the Measure stage are examined to generate a prioritised list of sources of variation.

The key deliverables of the Analysis phase are

1. A prioritised list of variables: A prioritised list of important sources of variation (particularly special causes) that affect the process output

DOI: 10.4324/9781003268239-12

2. Quantified financial opportunity: The financial benefit expected from the completion of the project

10.2 Description of Tools for Analysis

The important tools for analysis should include the following:

A1: Process mapping
A2: Regression analysis
A3: Resource utilisation and customer service (RU/CS) analysis
A4: SWOT (strengths, weaknesses, opportunities, threats) analysis
A5: PESTLE (political, economic, social, technical, legal, environmental) analysis
A6: 5 Whys
A7: Interrelationship diagram
A8: Overall equipment effectiveness (OEE)
A9: Tree diagram

During the Analyse stage, some of the tools from the Define and Measure stages are revisited, in particular:

M4: Scatter diagram
M5: Cause and effect diagram
M6: Pareto chart
M7: Control charts

There is also an overlap with some tools of the Improve stage and the Analyse stage, such as

I7: Brainstorming.

The Analyse stage also depends heavily on advanced techniques including SPC, failure mode and effect analysis (FMEA) and design of experiments (DOE).

10.2.1 A1: Process Mapping

10.2.1.1 Definition

Process mapping is a tool to represent a process by a diagram containing a series of linked tasks or activities which produce an output.

It is a further development of a flow diagram by using computer software so that the user can link quickly the activities and drill down to gain a more detailed picture.

10.2.1.2 Application

With the advent of well-supported software (e.g. 'Control' by Enigma Ltd.), process mapping is becoming a way of life for analysing a process or an organisation.

A process map does not use symbols like a flow process chart or a flow diagram. Only boxes and arrows are used, and different colours are often applied to identify types of activities (e.g. non-value-added or value added). There are several benefits of applying process mapping including the following. Process mapping means that the team

- Can clarify what is happening within an organisation
- Can simulate what should be happening
- Can show a process at various levels of detail
- Can allocate ownership of each activity and promote teamwork
- Can reflect the end-to-end process and its visibility
- Can add resources, costs, volumes and duration to build up sophisticated cost models
- Can identify how the performance of this process can be measured

10.2.1.3 Basic Steps

1. Decide on the organisation, function or process for analysis.
2. Agree on higher-level functions and their relationships.
3. Agree on input, process and output for each activity.
4. Construct a process diagram for selected functions.
5. Validate the process diagram with stakeholders.
6. Add resources, costs, volumes and duration if required.
7. Apply a 'what if' analysis and simulation to achieve sustainable process improvement.

10.2.1.4 Worked-Out Example

The following example is taken from the demonstration package of the 'Control' software.

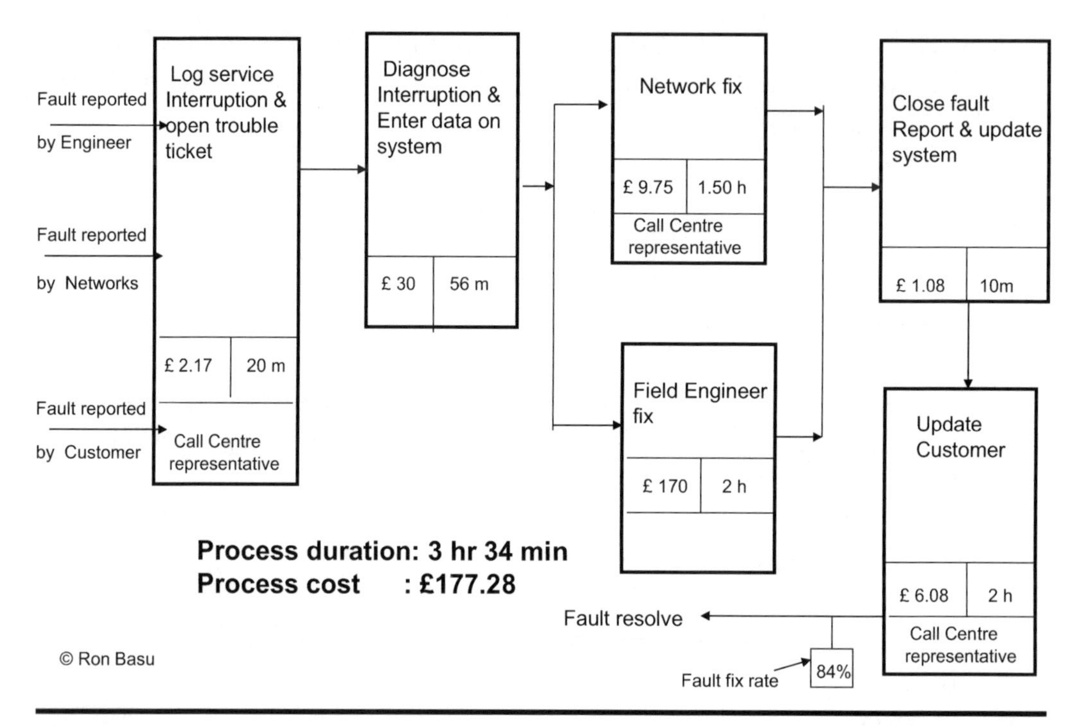

Figure 10.1 Process mapping.

(courtesy: Enigma Ltd., Oxford)

Consider a case example of dealing with faults in a computer network. A high-level process map is shown in Figure 10.1.

10.2.1.5 Training Requirements

Although the basic principles of process mapping are simple and logical, it is important that users receive at least 1 day of hands-on training in the chosen software for process mapping. It is also useful to gain a good understanding of the 'what if' simulation processes.

10.2.1.6 Final Thoughts

Process mapping has become a very useful computer-aided tool for process improvement. However, it should be used for process mapping's sake. Process maps do not—in isolation—change individual behaviour.

10.2.2 A2: Regression Analysis

10.2.2.1 Definition

Regression analysis is a tool to establish the 'best-fit' linear relationship between two variables.

The knowledge provided by the scatter ciagram is enhanced with the use of regression.

10.2.2.2 Application

The topic of regression analysis is usually studied at school in algebra lessons where different methods of 'curve fitting' are considered. Two common methods are

- Method of intercept and slope
- Method of least square

In a practical business environment, the team members normally resort to drawing an approximate straight line by employing their visual judgement. Sometimes they use the 'method of intercept and slope'. Both of these practices are the estimated 'best-fit' relationship between two variables. The reliability of such estimates depends on the degree of correlation that exists between the variables.

Regression analysis is used not only to establish the equation of a line but also to provide the basis for the prediction of a variable for a given value of a process parameter. The scatter diagram on the other hand does not predict cause and effect relationships. Given a significant co-relation between the two variables, regression analysis is very useful tool enabling one to extend and predict the relationship between these variables.

10.2.2.3 Basic Steps

We have considered the 'method of intercept and slope' for developing the basic steps as follows (*courtesy*: Moroney, 1973, p. 284):

1. Consider the equation of $y = mx + c$, where m is the slope, c is the intercept and x and y are the two variables.

2. In the equation of $y = mx + c$, substitute each of the pairs of values for x and y and then add the resulting equations.
3. Form a second similar set of equations, by multiplying through each of the equations of Step 2 by its co-efficient of m. Add this set of equations.
4. Steps 2 and 3 will each have produced an equation in m and c. Solve these simultaneous equations for m and c.
5. Plot the straight-line graph for $y = mx + c$ for the calculated values of m and c.

10.2.2.4 Worked-Out Example

The following example is taken from Moroney (1973, pp. 278–285).

Consider an investigation is made into the relationship between two quantities y and x and the following values were observed:

y	5	8	9	10
x	1	2	3	4

The values are plotted as shown in Figure 10.2. Now we follow the basic steps to calculate m and c in the equation $y = mx + c$.

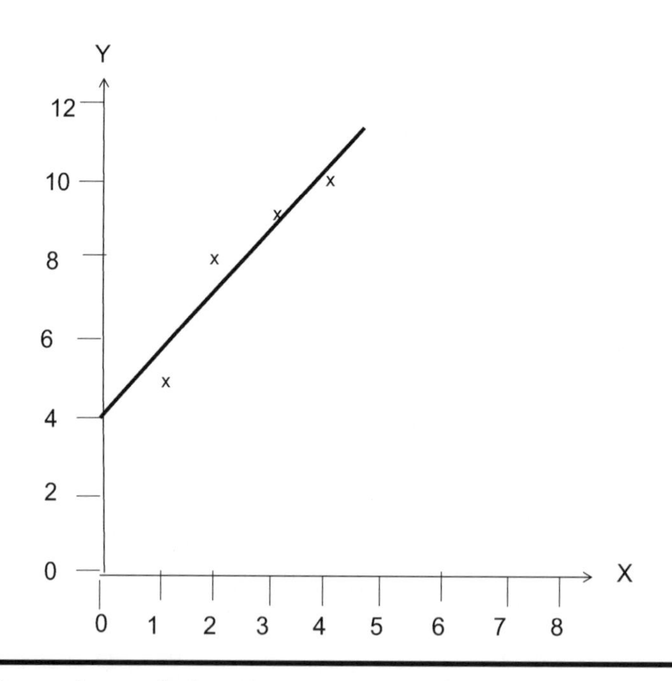

Figure 10.2 Regression analysis.

Substituting the observed values of x and y, the resulting equations are

$$5 = m + c$$
$$8 = 2m + c$$
$$9 = 3m + c$$
$$10 = 4m + c$$
$$32 = 10m + 4c \qquad (1)$$

Multiplying each of the equations by its co-efficient of m, the resulting equations are

$$5 = m + c$$
$$16 = 4m + 2c$$
$$27 = 9m + 3c$$
$$40 = 16m + 4c$$
$$88 = 30m + 10c \qquad (2)$$

We then solve simultaneously equations (1) and (2) for m and c.

We find that

$m = 1.6$
$c = 4$
Hence $y = 1.6x + 4$.

We calculate two pairs of values for x and y to draw a straight line as shown in Figure 10.2.

y	4	8.8
x	0	3

10.2.2.5 Training Requirements

The training of regression analysis is like brushing up on school algebra. A few classroom exercises for approximately 1 hour should be adequate for team members to prepare themselves for the practical application of regression analysis.

10.2.2.6 Final Thoughts

Regression analysis is a powerful tool to extend data from a scatter diagram. We recommend the application of regression analysis preferably by a team member who has competence and an interest in mathematics.

10.2.3 A3: RU/CS Analysis

10.2.3.1 Definition

The resource utilisation and customer service (RU/CS) analysis is a simple tool to establish the relative importance of the key parameters of both resource utilisation and customer service and to identify their conflicts.

Wild (2002) suggests the starting point of the RU/CS analysis with the operations objectives chart as shown in Figure 10.3.

The relative importance of the key parameters for RU (i.e. Machines, Materials and Labour) and CS (i.e. Specification, Cost and Time) can be given a rating of 1, 2 or 3 (3 being the most important).

10.2.3.2 Application

In any business or operation, a manager has to find a balance between two conflicting objectives. Customer service is of course the primary objective of the operation. For simplicity, three key parameters of customer service are considered. These are Specifications, Cost (or Price) and Timing. The customer expects the goods or service to be delivered according to acceptable standards, to be of an affordable price and that they arrive on time. However, the relative importance of Specification, Cost and Time could change depending on the market condition, competition and the desirability of demand. The second objective of the operation manager is to utilise resources to meet customer service requirements. Given infinite resources, any system can provide adequate customer service, but many companies have gone out of business in spite of possessing satisfied customers. Therefore, it is essential to provide an efficient use of resources. The RU/CS analysis aims to point out the way forward to a balanced approach of 'effective' customer service and 'efficient' resource utilisation.

An organisation in a normal condition will not aim to maximise all three parameters. Likewise, few organisations will aim to maximise the utilisation

	Resource Utilisation			*Customer Service*		
	Machines	*Materials*	*Labour*	*Specification*	*Cost*	*Time*
Operation						

Figure 10.3 RU/CS conflicts in operations objective chart.

of all resources. Hence there is some room for adjustment, and the operations manager must attempt to balance the parameters of these two basic objectives—resource utilisation and customer service. The RU/CS analysis is applied to examine the relative importance of the parameters to lead to a balanced solution of objectives.

10.2.3.3 Basic Steps

1. Identify the key parameters of resource utilisation. An operation may have several types of resources as input (e.g. machine, facilities, labour, information etc.). Choose three important resources.
2. For customer service parameters, select Specification, Cost and Time. As discussed in Chapter 1, there are other dimensions of quality as perceived by customers; for the sake of simplicity only Specifications has been chosen as the key parameter for the quality of service.
3. Draw two matrices for RU and CS showing the six parameters.
4. Allocate a rating of 1, 2 or 3 (3 being most important) to the parameters of both RU and CS. The ratings are influenced by internal processes for RU and external customer requirements for CS.
5. Rate separately what is actually achieved for each aspect of RU and CS.
6. Compare the two sets of figures (from Steps 4 and 5) and identify the shortfalls or misalignments.
7. Review the criticality of shortfall in CS and examine which resources are inhibiting customer service performance.
8. Draw a combined RU/CS matrix, with the allocated ratings outlined in Step 4, and identify their conflicts. It is important to note that the high importance of Specification and Time will require a lower resource utilisation. The high importance of cost, on the other hand, will demand a lower price, and this will require a higher resource utilisation. The tables in Figure 10.4 can be used as a ready-reckoner to identify conflicts.
9. Having identified the conflicts, the next step is to examine the relative importance of each parameter in order to minimise conflicts.

10.2.3.4 Worked-Out Example

Consider a mail order company where customers are expecting good value for money and do not mind receiving goods from catalogues within a

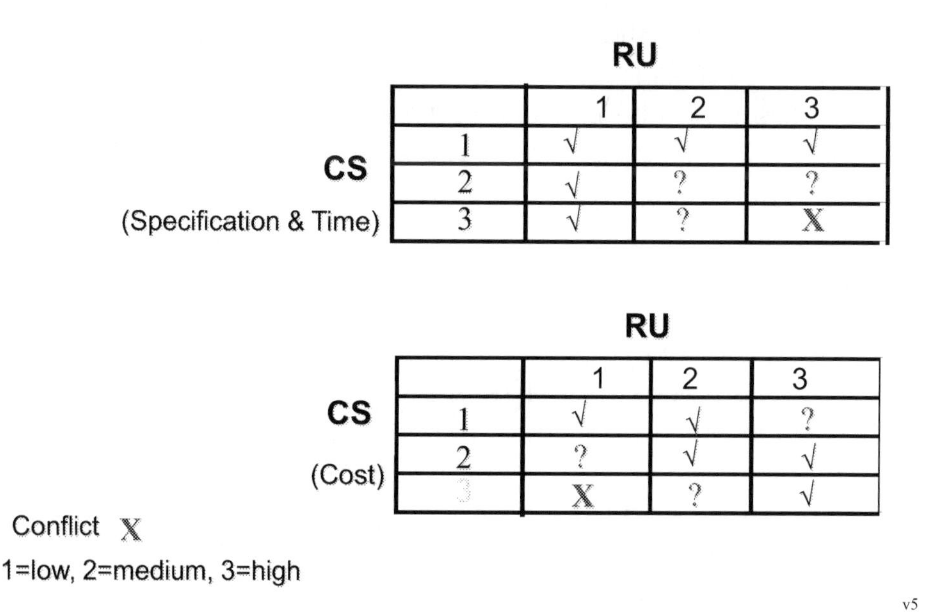

Figure 10.4 The balance of objectives for mail order company.

reasonable delivery time. The operation manager has focused on the utilisation of own resources to minimise operational costs.

Figure 10.5 shows the ratings of objectives, the actual performance and highlights the misalignment. It is evident that further examination is required for Timing and Material.

As shown in Figure 10.6, there is a conflict between Cost and Materials, and further attention or a change of policy is required to resolve this conflict.

10.2.3.5 Training Requirement

The training workshop for the RU/CS analysis is likely to be of half a day's duration and it could be combined with the training for other tools. The understanding of rating and alignment could be simple, but the identification of conflicts is likely to require a few hands-on exercises.

10.2.3.6 Final Thoughts

The RU/CS analysis is a simple but powerful tool to establish quickly the conflicts between resource utilisation and customer service and to reflect upon how to go about resolving them. This balance will vary between different operations or organisations.

	Machinery/Space	People	Materials
Utilisation Objectives	3	3	1
Actual Utilisation	3	3	2
Alignment	✓	✓	✗

	Specification	Cost	Timing
Customer Service Objectives	1	3	2
Actual Level of Service	2	3	1
Alignment	✗	✓	✗

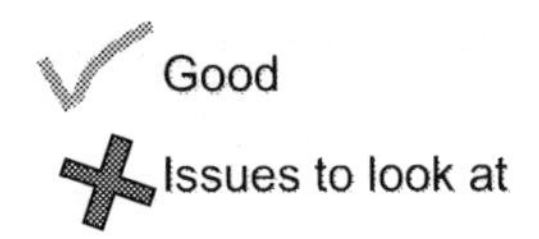

1=low, 2=medium, 3=high

v5

Figure 10.5 The balance of objectives: mail order company.

3 High Relative Importance

1 Low Relative Importance

Machinery/Space	People	Materials
3	3	1

Specifications	Cost	Time
1	3	2

	Machinery/Space	People	Materials
Specifications			
Cost			✗
Time			

Figure 10.6 RU/CS conflicts in a mail order company.

10.2.4 A4: SWOT Analysis

10.2.4.1 Definition

A SWOT (strengths, weaknesses, opportunities and threats) analysis is a tool for analysing an organisation's competitive position in relation to its competitors.

In the context of a quality improvement programme, a SWOT analysis refers to a summary of the gaps and positive features of a process following the analytical stage.

10.2.4.2 Application

Based on a SWOT framework, the team members can focus on alternative strategies for improvement. For example, an S/O (strengths/opportunities) strategy may be considered to consolidate the strengths and open further leverage from the process. Similarly an S/T (strengths/threats) strategy may be considered to maximise the strength of the process and minimise risks.

Thus a SWOT analysis can help the team to identify a wide range of alternative strategies for the next stage.

10.2.4.3 Basic Steps

1. Create two categories (internal factors and external factors) and then further sub-divide into positive aspects (strengths and opportunities) and negative aspects (weaknesses and threats).
2. Ensure that the internal factors may be viewed as a strength or weakness of a process depending on their impact on the outcome or the process output.
3. Similarly assess the external factors bearing in mind that threats to one process could be an opportunity for another process.
4. Summarise the key features and findings derived from previous analyses in each of the SWOT categories.
5. Develop the improvement strategy for the next stage as pointers from the SWOT analysis.

10.2.4.4 Worked-Out Example

The following example is taken from Kotabe and Helsen (2000, p. 277).

Table 10.1 shows the framework of a SWOT analysis.

Table 10.1 Framework of a SWOT Analysis

External Factors	*Internal Factors*	*Strengths*	*Weaknesses*
		Brand Name, Human Resources, Technology, Advertising	*Price, Lack of Financial Resources, Long Product Development Cycle*
Opportunities	Growth market, de-regulation, stable exchange rate, investment grant	S/O Strategy Maximise strengths and maximise opportunities	W/O Strategy Minimise weaknesses and maximise opportunities
Threats	New entrants, change in consumer preference, local requirements	S/T Strategy Maximise strengths and minimise threats	W/T Strategy Minimise weaknesses and minimise threats

10.2.4.5 Training Requirements

The training of the SWOT analysis should be conducted in stages. The first part is in a classroom where the basic principles can be explained based on a known marketing product. The second part of the training is delivered to the team member when the results following the analysis of process are summarised.

10.2.4.6 Final Thoughts

Although a SWOT analysis is primarily a marketing tool, it can be applied effectively in order to summarise the key features of a process after the analytical stage.

10.2.5 A5: PESTLE Analysis

10.2.5.1 Definition

The PESTLE (political, economic, social, technical, legal and environmental) analysis is an analytical tool for assessing the impact of external contexts on a project or a major operation and also the impact of a project on its external contexts. There are several possible contexts including

- Political
- Economic

- Social
- Technical
- Environmental

This is remembered easily by the acronym 'PESTLE' or 'Le Pest' in French. It is thus also known as the PEST analysis.

10.2.5.2 Application

Very few major changes, whether they are caused by a major operation or a project, are unaffected by the external surrounds.

Political: A project is affected by the policies of international, national or local government. It is also influenced by company policies and those of stakeholders, managers, employees and trade unions.

Economic: The project is affected by national and international economic issues, inflation, interest rates and exchange rates.

Social: The change is influenced by social issues, the local culture, the lives of employees, communications and language.

Technological: The success of implementation is affected by the technology of the industry and the technical capability of the parent company.

Legal: The project is affected by the legal aspects of planning, registration and working practices.

Environmental: The impact of the change on environmental emission, noise, health and safety is assessed.

10.2.5.3 Basic Steps

A PESTLE analysis is carried out in four stages:

1. Develop a good understanding of the deliverables of the operation and the project. At this stage, the relevant policy and guidelines of both the local company and the parent organisation are reviewed.
2. List the relevant factors affecting the various aspects of the project related to PESTLE. It is important that the appropriate expertise of the organisation is drawn into the team for this analysis.
3. Validate the factors in Step 2 with the stakeholders and functional leaders of the company.
4. Review progress and decide on the next steps by asking two questions:
   How did we do?
   Where do we go next?

5. For more information on the PESTLE analysis, see Turner and Simister (2000, pp. 165–215).

10.2.5.4 Worked-Out Example

Consider the policy renewal management process of an insurance company based in Finland. The company implemented an online renewal process within Finland and wanted to expand the process in the European Union (EU).

A PESTLE analysis was carried out as shown in Table 10.2.

The PESTLE analysis shows that the expansion of online service, in general, has direct influence upon and opens up opportunities in the EU.

10.2.5.5 Training Requirements

The principles of PESTLE analysis are best learned by the process of the group working together during the project life cycle. The programme of

Table 10.2 PESTLE Analysis

Contexts	*Key Factors*	*Impact on Company (0–10)*
POLITICAL	Within EU, countries are moving towards a more common political structure	6
ECONOMIC	Slowing economy of Finland; gross domestic product forecast to grow by 3.9% in 2003 Dynamic changes in client business environment in Finland and Europe	8
SOCIAL	Difference in buying habits in Finland versus EU	7
TECHNOLOGICAL	Accelerating pace of change in information and communication technology in Finland Online opportunity in EU with little extra cost New cyber-related risk in client business	9
LEGAL	New constraints or requirements initiated by regulatory bodies (e.g. Insurance Supervisory Authority in EU)	9
ENVIRONMENTAL	No significant impact	1

Black Belt training normally includes a session on PESTLE analysis and team leaders should receive a broad understanding of this tool. The analysis is of a strategic nature and thus may not involve all members of the project team.

10.2.5.6 Final Thoughts

A PESTLE analysis is most appropriate for the total programme rather than being used for individual operations. We recommend that this tool should be applied to all Six Sigma and operational excellence initiatives.

10.2.6 A6: The Five Whys

10.2.6.1 Definition

The Five Whys is a systematic technique of asking five questions successively. The aim is to probe the causes of a problem and thus hopefully get to the heart of the problem.

10.2.6.2 Application

The Five Whys is a technique that is widely used to analyse problems in both manufacturing and service operations. It is a variation on the classic work study approach of 'critical examination', involving six questions: Why, What, Where, When, Who and How?

The objective is to eliminate the root cause rather than patch up the effects.

10.2.6.3 Basic Steps

1. Select the problem for analysis.
2. Ask five 'close' questions, one after another, starting with why.
3. Do not defend the answer or point the finger of blame at others.
4. Determine the root cause of the problem.

10.2.6.4 Worked-Out Example

The following example is taken from Stamatis (1999, p. 183).

Consider a problem: 'Deliveries are not completed by 4 p.m.'.

Question 1: Why does it happen?

Answer: The routing of trucks is not optimised.

Question 2: Why is it not optimised?

Answer: Goods are loaded based on their size rather than the location of the delivery.

Question 3: Why are they loaded by size?

Answer: The computer defines the dispatch based upon the principle of 'large items first'.

Question 4: Why are large items given preference?

Answer: Large items are delivered first.

Question 5: But why?!

Answer: Current prioritisation policy puts large items first on the delivery schedule.

10.2.6.5 Training Requirements

As can be seen from the question and answer model, the principle of this analytical tool is very straightforward. Thus the application of the Five Whys does not require any rigorous classroom training. The members of a problem-solving team can easily understand and apply this simple tool after just one such group exercise.

10.2.6.6 Final Thoughts

The Five Whys is an uncomplicated but very effective tool that can be used to identify the root causes of a problem. We recommend that, taking advantage of the fact that it is such a quick and unfussy approach, it can be applied on a far wider basis than at present.

10.2.7 A7: Interrelationship Diagram

10.2.7.1 Definition

An ID is an analytical tool to identify, systematically analyse and classify the cause and effect relationships among all critical issues of a process. The key drivers or outcomes are identified leading to an effective solution.

10.2.7.2 Application

An ID is often applied to enable the further examination of causes and effects after these are recorded in a fish-bone diagram. ID encourages team members to think in multiple directions rather than merely in a linear sense.

This simple tool enables the team to set priorities to root causes even when credible data do not exist.

10.2.7.3 Basic Steps

1. Assemble the team and agree on the issue or problem for investigation.
2. Lay out all of the ideas or issues that have been brought from other tools (such as a cause and effect diagram) or brainstormed.
3. Look for the cause and effect relationships between all issues and assign the 'relationship strength' as
   3—Significant
   2—Medium
   1—Weak
4. Draw the final ID in a matrix format and insert the 'relationship strength' given by members.
5. Total the relationship strength in each row to identify the strongest effect of an issue on the greatest number of issues.

10.2.7.4 Worked-Out Example

The following example is taken from Brassard and Ritter (1994, p. 81).

Consider five key issues to improve customer service.

- Logistics support
- Customer satisfaction
- Education and training
- Personal incentives
- Leadership

The ID is plotted in a matrix (see Figure 10.7) with appropriate 'relationship strengths'.

From the analysis in the 'Total' column in Figure 10.7, it is evident that customer satisfaction and leadership are the two most critical issues.

	Logistics Support	Customer Satisfaction	Training	Personal Incentives	Leadership	**Total**
Logistics Support		○	□	△	□	8
Customer Satisfaction	○		□	○	□	10
Training	□	□		□	○	9
Personal Incentives	△	○	□		○	9
Leadership	□	□	○	○		10

Relationship Strength

○	Significant	3
□	Medium	2
△	Weak	1

Figure 10.7 Interrelationship diagram.

10.2.7.5 Training Requirements

The team can be adept in the application of ID after less than 1 hour's training in a classroom or a practical environment. The principles are simple, but it is important that a consensus is reached in attributing the relationship strength numbers.

10.2.7.6 Final Thoughts

ID is not an essential tool to analyse the cause and effect relationship of issues, but because of its simplicity we recommend that it is used selectively to set priorities to causes or issues.

10.2.8 A8: Overall Equipment Effectiveness

10.2.8.1 Definition

The OEE is an index of measuring the delivered performance of a plant or equipment based on good output.

The method of monitoring OEE is devised in such a way that it would highlight the losses and deficiencies incurred during the operation of the plant and identify the opportunities for improvement.

There are many ways to calculate OEE (see Shirose, 1992; Hartman, 1991). In this section we describe the methodology of OEE that was developed and applied by the author at both Unilever[1] and GlaxoWellcome.[2]

OEE is defined by the following formula:

OEE % = Actual Good Output × 100
Specified Output
where Specified Output = Specified Speed × Operation Time.

10.2.8.2 Application

The application of OEE has been extensive, especially when driven by the total productive maintenance (TPM) programmes, to critical plant and equipment. It can be applied to a single equipment, a packing line, a production plant or processes. In order to appreciate the usefulness of OEE it is important to understand equipment time analysis as shown in Figure 10.8 and described next.

Total Time defines the maximum time within a reporting period, such as 52 weeks a year, 24 hours a day, 8,760 hours in a year.

Available Time is the time during which the machine or equipment could be operated within the limits of national or local statutes, regulation or convention.

Operation Time is the time during which the machine or equipment is planned to run for production purposes. The operational time is normally the shift hours.

Production Time is the maximum time during which the machine or equipment could be expected to be operated productively after adjusting the operation time for routine stoppages such as changeover and meal breaks.

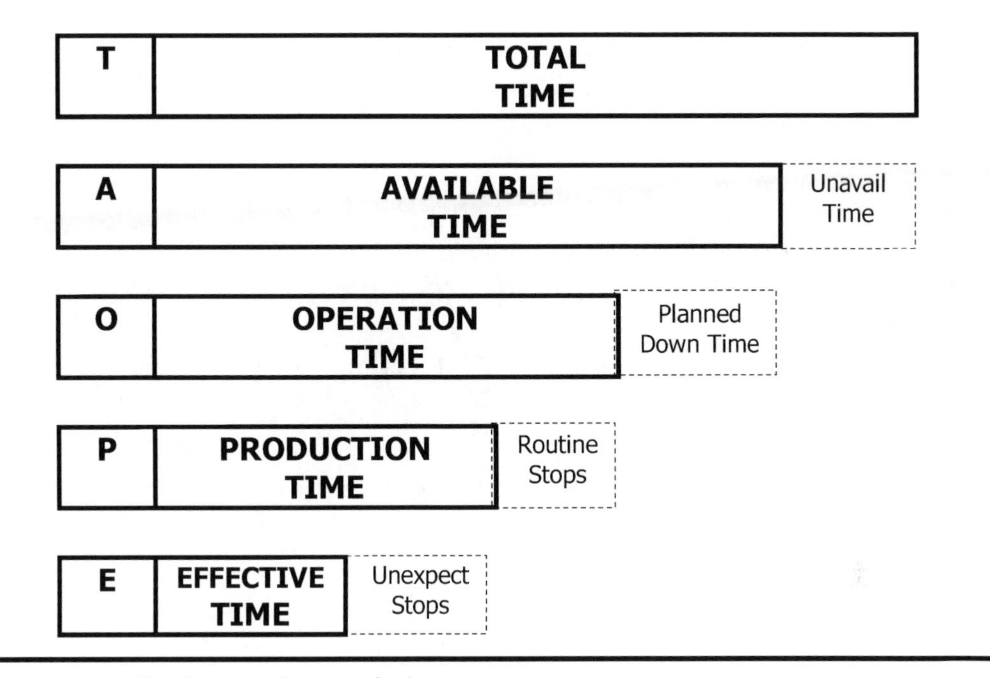

Figure 10.8 Equipment time analysis.

Effective Time is the time needed to produce a 'good output delivered' if the machine or equipment is working at its Specified Speed for a defined period. It includes no allowances for interruptions or any other time losses.

It is important to note that Effective Time is not recorded, it is calculated from the Specified Speed as

Effective Time = Good Output/Specified Speed

where Specified Speed is the optimum speed of a machine or equipment for a particular product without any allowances for loss of efficiency. It is expressed as quantity per unit such as tons per hour, bottles per minute, cases per hour or litres per minute.

In addition to OEE, two other indices are commonly used:

$$\text{Production Efficiency (\%)} = \frac{\text{Effective Time (E)}}{\text{Production Time (P)}} \times 100$$

$$\text{Operational Utilisation (\%)} = \frac{\text{Operation Time (O)}}{\text{Total Time (T)}} \times 100$$

A properly designed and administered OEE scheme offers a broad range of benefits and a comprehensive manufacturing performance system. Some of its key benefits are

- It provides information for shortening lead time and changeover time and a foundation for single minute exchange of dies (SMED).
- It provides essential and reliable data for capacity planning and scheduling.
- It identifies the 'six big losses' of TPM leading to a sustainable improvement of plant reliability.
- It provides information for improving asset utilisation and thus reduced capital and depreciation costs in the longer term.

10.2.8.3 Basic Steps

1. Select the machines, equipment or a production line where the OEE scheme could be applied. The selection criteria will depend on the criticality of the equipment in the context of the business. It is useful to start with a single production line as a trial or pilot.
2. Establish the specified speed of the production line governed by the control or bottleneck operation. As shown in the following example (Figure 10.9) of a soap packaging line, the specified speed is 150 tablets per minute (i.e. this constitutes the speed of the wrapper, which is the slowest piece of equipment).

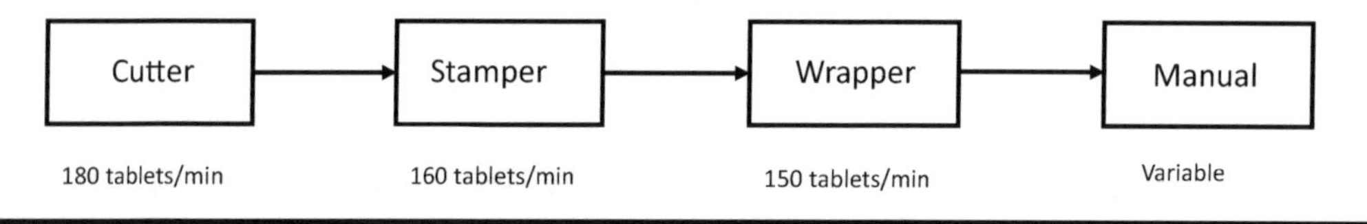

Figure 10.9 Soap production line.

3. Set up a data recording system so that the output data and various stoppages and losses can be recorded.
4. Compile the data every day and validate the results. At this stage, detailed calculations are not necessary.
5. Monitor the results, comprising OEE and key indices, major losses as a percentage of the Operation Time and the trends of indices. The reporting is normally on a weekly basis for the department and on a monthly basis for senior management.

6. Use the results for continuous improvement, planning and strategic changes.

10.2.8.4 Worked-Out Example

Consider the production data of a toilet soap packing line where the control station governing the specified speed is an ACMA 711 wrapping machine:

Week Number:	31
Operation Time:	128 hours
Specified Speed:	150 tablets per minute
Good Output:	4,232 cases
Routine Stoppages:	11 hours 30 minutes
Unexpected Stoppages:	27 hours 15 minutes

Given that each case contains 144 tablets,

Good Output = 4,232 × 144 = 609,408 tablets

$$\text{Effective } Time = \frac{GoodOutput}{SpecifiedSpeed} = \frac{609,408}{150 \times 60} = 67.71 \text{ hours}$$

Production Time = Operation Time – Routine Stoppages
= 128 – 11.5 = 116.5 hours
Total Time = 7 × 24 = 168 hours

$$\text{OEE} = \text{Effective Time/Operation Time} = \frac{67.71}{128} = 0.53 = 53\%$$

$$\text{Production Efficiency} = \frac{EffectiveTime}{ProductionTime} = \frac{67.71}{116.5} = 58\%$$

$$\text{Operation Utilisation} = \frac{OperationTime}{TotalTime} = \frac{128}{168} = 76\%$$

It is important to note that the Effective Time was calculated and not derived from the recorded stoppages. There will be an amount of unrecorded time (also known as Time Adjustment) as, in the example, given by

Unrecorded Time = (Production Time – Unexpected Stoppages) – Effective Time

= (116.5 − 27.25) − 67.71
= 21.54 hours

10.2.8.5 Training Requirements

The success of an OEE scheme depends heavily on the rigour of continuous training. It is important that each operator, supervisor and manager of a production department receives a half-day training programme covering the definitions, purpose and application of the OEE scheme. The training is continuous because of the turnover of staff. Senior management should also benefit from a 1-hour awareness session.

10.2.8.6 Final Thoughts

The principles of OEE are conceptually simple but detail rich. The main strength of this tool is that it highlights the areas of deficiency for improvement and the key results cannot be manipulated. The specified time is calculated from tangible 'good output', Operation Time is well-established shift hours and Total Time is absolute.

10.2.9 A9: Tree Diagram

10.2.9.1 Definition

A tree diagram in analysing a problem or process reveals the level of complexity in the achievement of any goal. It breaks a broad goal graphically into increasing levels of detail actions or the components of the goal.

10.2.9.2 Application

A tree diagram encourages team members to express their ideas in graphical forms to the overall goal of the process. It is a useful tool to analyse the complexity of a problem and is useful during a brainstorming session.

A tree diagram is an excellent communication tool to graphically represent the team members' ideas to analyse the complexity of a problem and breaking down a broad goal.

There are variations of the tree diagram. For example the process decision programme chart (Brassard et al., 2002) is based on the tree diagram.

Another form of the tree diagram is used in mathematics to show the combinations of two or more events. The probability of each event is written alongside the lines and the outcome of each branch is labelled at the end.

10.2.9.3 Basic Steps

1. Choose the goal of the tree diagram from the problem or process under review.
2. Assemble the team (usually four to six members) and brief the members.
3. Identify major sub-goals by brainstorming the major task area.
4. Generate tree headings from the major sub-goals.
5. You may use Post-it notes to encourage the level of details from team members.
6. Review the completed tree diagram for its completeness and the logical flow from sub-goals to the goal.
7. Chose the most effective measures for each level of the tree diagram and build them into a plan to achieve the goal.

10.2.9.4 Worked-Out Example

The following example is taken from Basu and Wright (2017), page 373.

Consider a case example of increasing return on investment (ROI) in a manufacturing company. The sustainable enhancement of ROI is one of the primary goals of all businesses, especially of investment intensive enterprises.

In this example as shown in the tree diagram in Figure 10.10, the primary goal of 'increasing ROI' is gradually broken down into four levels of sub-goals. It is important to focus on the total objective of ROI so that the inter-relationship between different elements and their relative weight can be visualised to develop an improvement plan.

10.2.9.5 Training Requirements

Although the basic principles of the tree diagram are simple and logical, it is important that users receive at least 1 day of hands-on training. The understanding of a tree diagram becomes clearer during a team meeting or brainstorming session.

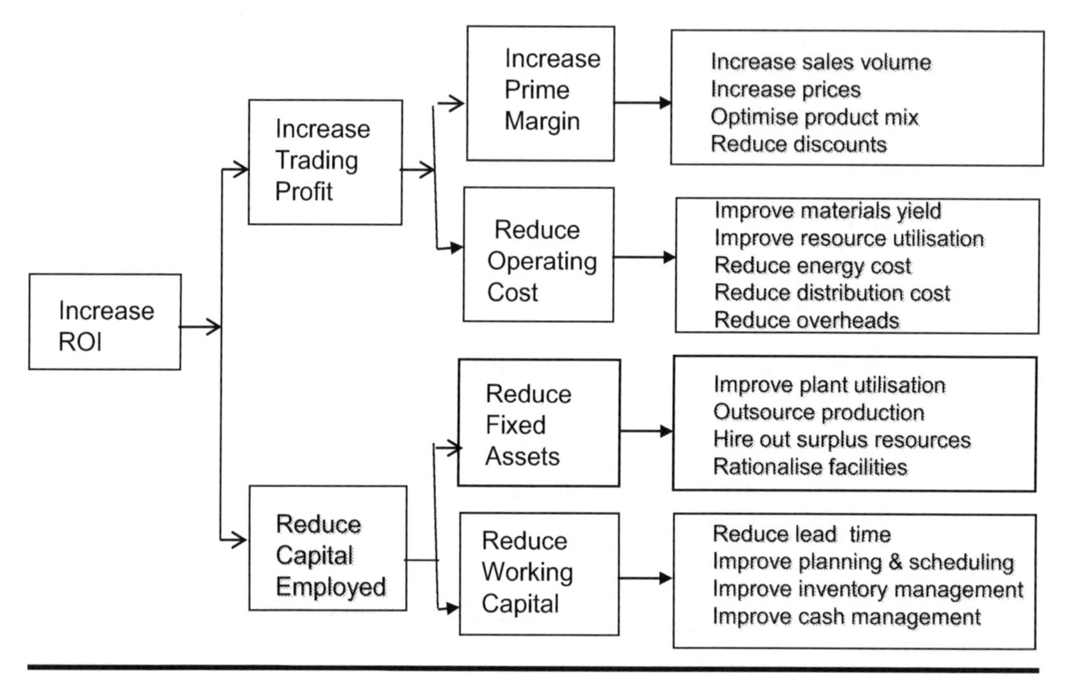

Figure 10.10 Tree diagram.

10.2.9.6 Final Thoughts

The tree diagram is a useful tool for analysing a complex problem into components. It is recommended that the tree diagram is used in a brainstorming and team-building session.

10.3 Summary

Consider yourself as a detective in the Analyse phase. You search for clues to solve the problem. The challenge is to analyse potential causes in a reliable manner to find significant evidence leading to a solution. You investigate the significance of root causes finding out which ones are more important than other ones.

Some of the tools of the Measure phase (e.g. flow process chart) are also used in the Analyse phase. When DMAICS (define, measure, analyse, improve, control and sustain) Lite is applied to small and medium-sized enterprises, the Measure and the Analyse phases are combined.

GREEN TIPS

- Start the Analyse phase to search potential causes and then identify root causes.
- When you find measures to eliminate root causes, you find a solution.
- When the primary cause is related to defects, select Six Sigma tools; when the primary case is related to time, select Lean analysis tools.

Notes

1 In Unilever Plc, the methodology was known as PAMCO (plant and machine control).
2 In GlaxoWellcome it was called CAPRO (capacity analysis of production).

Chapter 11

Tools for Improvement

Invention, strictly speaking, is little more than a new combination of those images which have been previously gathered and deposited in the memory; nothing can come of nothing.

—Joshua Reynolds, circa. 1780

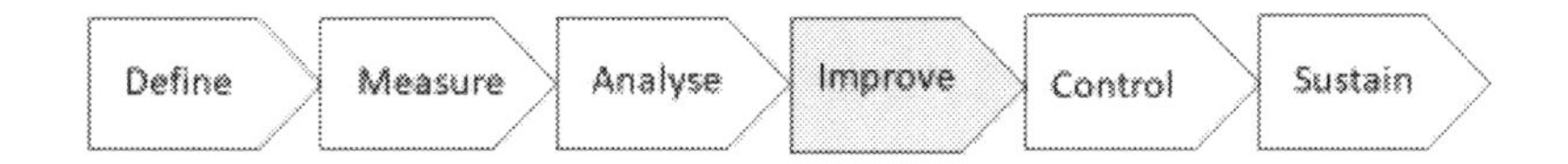

11.1 Introduction

The earlier stages of the project, in particular the Analyse stage, have pinpointed the areas for improvement. During the Improve phase, the ideas and solutions are put to work. Various options are then compared with each other to determine the most promising solution. Some experiments and trials may be required to validate the best solution. Finally, this solution is usually piloted on a small scale in the business environment.

The objectives of the tools of the Improve phase are to help the team to develop a solution for improving process performance and to confirm that the proposed solution will meet or exceed the quality improvement goals of the project.

The key deliverables of the Improve phase are as follows:

1. Proposed solution: a solution for reducing variation or eliminating the special causes of the problem in the process.

DOI: 10.4324/9781003268239-13

2. Validate solution: Process improvement that has been piloted in a real business environment.

11.2 Description of Tools for Improvement

The important tools for improvement should include

I1: Affinity diagram
I2: Nominal group technique
I3: Single minute exchange of dies (SMED)
I4: Five S's
I5: Mistake proofing
I6: Value stream mapping
I7: Brainstorming
I8: Mind mapping

The improvement stage also depends on techniques including brainstorming, design of experiments (DOE), quality function deployment (QFD) and failure mode and effect analysis (FMEA).

11.2.1 I1: Affinity Diagram

11.2.1.1 Definition

An affinity diagram is an improvement tool to generate creatively a number of ideas and then summarise logical groupings among them to understand the problem and then to lead to a solution.

This is also known as the KJ method, identified with Kawakita Jiro, the Japanese scientist who first applied it in the 1950s.

11.2.1.2 Application

An affinity diagram is used to categorise verbal information into an organised visual pattern. It is used in conjunction with brainstorming when problems are uncertain, large and complex, thereby enabling the user to create a discipline out of chaos.

It is a tool to overcome 'team paralysis' which is created by the generation of a large number of options and a lack of consensus. As part of brainstorming all ideas are recorded on sticky notes or index cards. The ideas on the notes are then clustered into major categories. It aims to be a creative, rather than a logical process. Hence the affinity diagram is regarded more as an improvement tool rather than an analytical one.

11.2.1.3 Basic Steps

1. Clarify the chosen problem or opportunity in a full sentence.
2. Collect the current data available on the problem or opportunity. This is usually done by brainstorming within a group.
3. Record each piece of data or idea onto a card or Post-it note and place them at random onto a board.
4. Sort ideas simultaneously into related groups.
5. Arrange the group affinity cards, usually less than ten in each group, in a logical order.
6. For each grouping, label header cards and draw broad lines around the group affinity cards.

11.2.1.4 Worked-Out Example

The following example is adapted from Schmidt et al. (1999, p. 125).

Consider the following problem as clearly stated:

What are the barriers to a quality improvement programme?

Figure 11.1 shows an example of an affinity diagram related to this problem.

11.2.1.5 Training Requirements

The knowledge and application of an affinity diagram are best derived by 'on-the-job' training during a brainstorming exercise on an actual problem. The facilitator should have experience in applying a number of affinity diagrams to direct the team and gain a consensus of grouping.

11.2.1.6 Final Thoughts

It is important to note that an affinity diagram is an improvement tool used in conjunction with a brainstorming exercise. It should not be confused with CEDAC (cause and effect diagram assisted by cards), a hybrid of the cause and effect diagram, which also uses cards or Post-it notes.

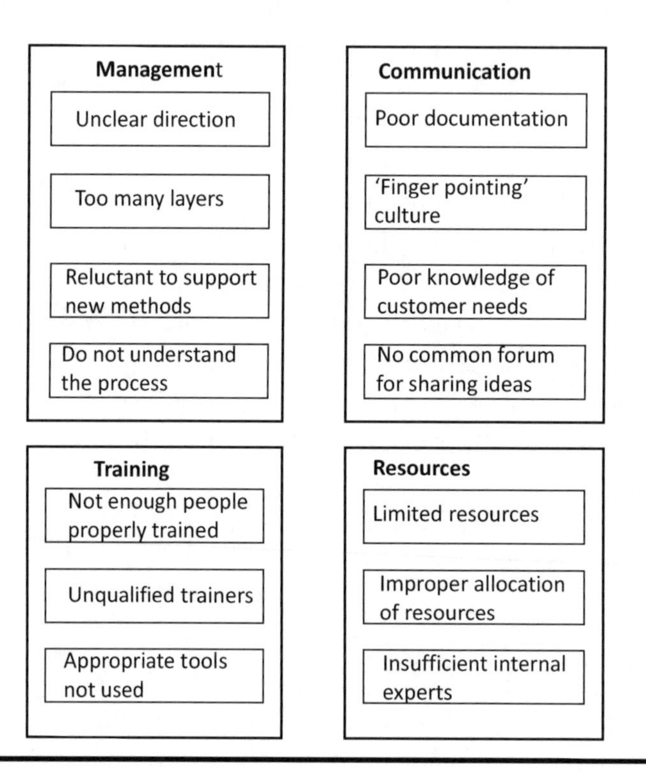

Figure 11.1 An affinity diagram.

11.2.2 I2: Nominal Group Technique

11.2.2.1 Definition

Nominal Group Technique (NGT) is an improvement tool to derive an importance ranking in a team's list of ideas arising from a brainstorming exercise.

This is also known as the Weighted Multivoting Technique.

11.2.2.2 Application

NGT is widely used to arrive quickly at a consensus on the relative importance of ideas or issues in a group working environment. It allows every team member to rank issues without any pressure from others. Thus the quieter individuals exercise the same power as the more dominant team members.

The application of NGT usually follows a brainstorming session. Incidentally, you may have recognised a similar process being applied during the voting procedure for the Eurovision Song Contest. (Now you have a good reason for watching it next time.)

11.2.2.3 Basic Steps

1. Assemble the team and generate a list of issues, problems or ideas to be prioritised. This could be done either by a brainstorming session or each member may be asked to write down their ideas.
2. Write the list on a flip chart or board. Refine the list by eliminating duplicate or similar statements.
3. Assign labels (A, B, C etc.) to the final list of statements and record them on a flip chart or board.
4. Each member records the labels of each statement in a rank order with the highest number allocated to the most important statement. For example if there are five statements then 5 is the most important and 1 is the least important ranking.
5. Aggregate the rankings of all team members, and the statement with the highest point would have the highest priority.

11.2.2.4 Worked-Out Example

Consider the case of a private school where several complaints have been made about the low morale of staff, but they are of an informal and unstructured nature and the Board of Governors have little to work with. Table 11.1 shows the team's list of complaints.

Table 11.1 List of Problems

A	Not enough teamwork and consultation
B	Information coming from an 'in crowd'
C	No consistent procedures
D	Dogmatic approach of the Head
E	Lack of feedback from reports made to parents and pupils
F	Strong hierarchy and work demarcation
G	Lack of training for support staff
H	More focus on fees than education
I	Lack of credit or recognition for support staff
J	Inadequate workspace except for the Head and 'cronies'

Table 11.2 Nominal Group Technique

Problem	*Team Members*					*Total*	*Priority*
	1	*2*	*3*	*4*	*5*		
A	7	9	9	6	6	37	
B	10	7	7	9	8	41	High
C	6	10	8	6	7	37	
D	9	8	10	8	10	45	High
E	5	3	6	7	9	30	
F	8	5	4	5	3	25	
G	3	4	5	4	2	18	
H	1	2	1	3	1	8	
I	4	5	2	2	5	18	
J	2	1	3	1	4	11	

Five members of the team have allocated a rank order to each problem listed in Table 11.1, and the results are shown in Table 11.2.

11.2.2.5 Training Requirements

Similar to the 'affinity diagram', the knowledge and application of NGT can be best derived by 'on-the-job' training in an actual group exercise. The facilitator should have the experience of directing a team and gaining a consensus.

11.2.2.6 Final Thoughts

NGT is not a 'scientific' approach, but it is simple and builds commitment to the team's choice through equal participation. It generates an atmosphere of fun and individual inclusion, building 'team spirit'. We recommend the application of NGT for solving in particular problems related to cultural and 'softer' issues.

11.2.3 I3: SMED

SMED is the name of an approach used for reducing output and quality losses due to changeovers and setups.

'Single minute' means that necessary set-up time is counted on a single digit.

11.2.3.1 Application

This method was developed in Japan by Shigeo Shingo (1985) and has proven its effectiveness in many manufacturing operations by reducing the changeover times of packaging machines from hours to minutes.

The primary application area of SMED is the reduction of set-up times in production lines. This process enables operators to analyse and find out themselves why the changeovers take so long and how this time can be reduced. In many cases, changeover and set-up times can be condensed to less than 10 minutes so that the changeover time can be expressed with a single digit, and it is therefore called 'single minute exchange of dies'.

SMED is considered as an essential tool in Lean manufacturing, and it is instrumental in the reduction of non-value-added activities in process times. Changeover loss is one of the six big losses that have been defined within the total productive maintenance (TPM). It is important to note that SMED is directly linked with the analytical process of OEE (overall equipment effectiveness).

With due respect to the success of the SMED method, it is fair to point out that the basic principles are fundamentally the application of classical industrial engineering or work study.

11.2.3.2 Basic Steps

1. Study and measure the operations of the production line to discriminate
   Internal setup (IS), the operation that must be done, which machine is stopped
   External setup (ES), the operation that possibly can be done while the machine is still running
2. Suppress non-value-added operations and convert IS operating into ES. The data from OEE and the preparations of pre-requisites (e.g. tools, changeover parts, pre-assemblies, pre-heating, mobile storage etc.) are reviewed to achieve results. Some internal set-ups are converted to external set-ups.
3. The next stage is to simplify the design of the machine, especially fillings and tightening mechanisms. Some examples of the design

simplification are U-shaped washers, quarter turn screws and cam and lever tights.

4. Balance the work content of the line and ensure teamwork. For example, in one automatic insertion machine, one operator sets up on the machine front while the other operator feeds components on the other side.
5. Minimise trials and controls. Use of mistake proofing or poka-yoke enables the standard way to be carried out each time.

11.2.3.3 Worked-Out Example

The following example is taken from Basu and Wright (1997, p. 97).

Consider the setup time reduction of a packing machine.

The internal and external set-up times have been measured. As shown in Figure 11.2, the total set-up time is reduced by overlapping external set-up times on the internal set-up time.

11.2.3.4 Training Requirements

The training of SMED is likely to be more effective with team members with a good understanding of work study or industrial engineering. The basic

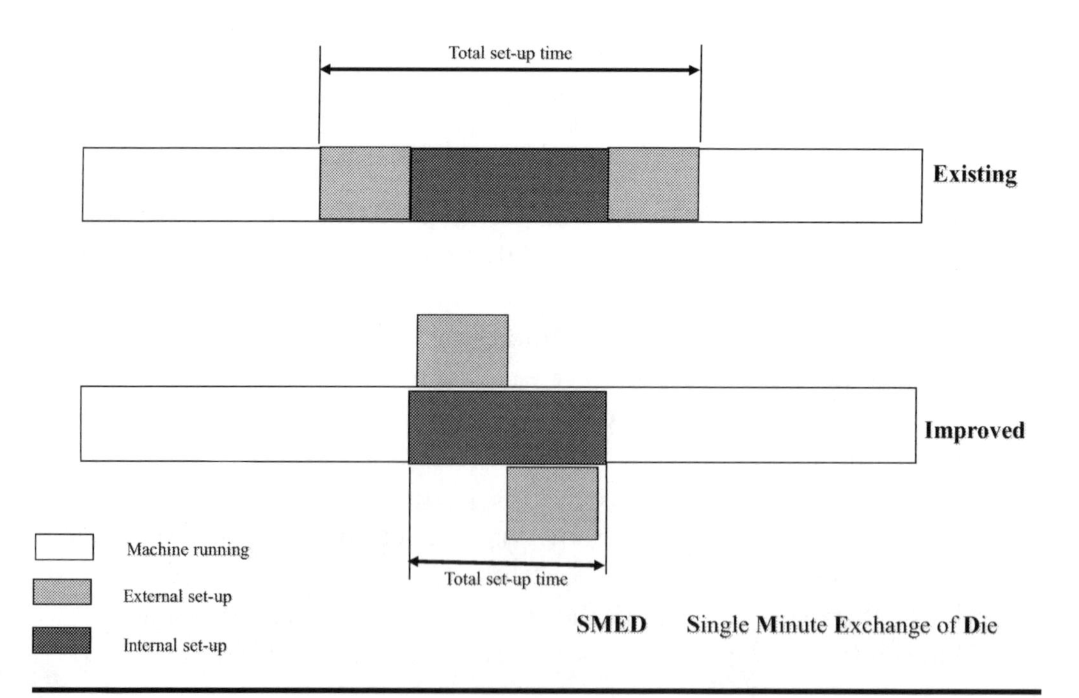

Figure 11.2 SMED: set-up time reduction.

principles should be explained to members with hands-on exercises being carried out in a room. This should be supplemented by a production line study in a factory environment. It is also important that the basics of OEE are covered during a SMED training session.

11.2.3.5 Final Thoughts

SMED can produce excellent results in achieving the reduction of set-up times but admittedly it is resource intensive. It should therefore be restricted to bottleneck operations.

11.2.4 I4: Five S's

11.2.4.1 Definition

The Five S's is a tool for improving the housekeeping of an operation, developed in Japan, where the Five S's represent five Japanese words all beginning with 's':

- Seiri (Organisation): Separate what is essential from what is not
- Seiton (Neatness): Sort and arrange the required items in an orderly manner and in a clearly marked space
- Seiso (Cleaning): Keep the workstation and the surrounding area clean and tidy
- Seiketson (Standardisation): Clean the equipment according to laid down standards
- Shitsuke (Discipline): Follow the established procedure

In order to retain the name 'Five S's', a number of English-language versions have evolved. These include:

- Seiri: Sort
- Seitor: Set in order/stabilise
- Seiso: Shine
- Seiketsu: Standardise
- Shitsuki: Sustain

11.2.4.2 Application

The Five S's method is a structured sequential programme to improve workplace organisation and standardisation. Five S's improves the safety, efficiency and orderliness of the process and establishes a sense of ownership within the team.

Five S's is used in organisations engaged in Lean Sigma, just-in-time (JIT), total productive maintenance (TPM) and Total Quality Management (TQM). This principle is widely applicable not just for the shop floor, but for the office too. As an additional bonus there are benefits to be found in environmental and safety factors due to the resulting reduced clutter. Quality is improved by better organisation, and productivity is increased due to the decreased time spent in searching for the right tool or material at the workstation. Consider the basic principle of a parent tidying a small child's room which is overflowing with clutter and sorting together various types of toys. The end product should be a neater, warmer, brighter and more civilised play environment which will encourage the child to utilise all toys and equipment more productively because all relevant pieces are together, space is enhanced and mess is reduced.

It is useful to note that the quality gurus of Japan like numbered lists (e.g. the Seven Mudas, the Five Whys, the Five S's). However, the exact number of S's is less important than observing the simple doctrine of achieving the elimination of wastes.

As the Five S's programme focuses on attaining visual order and visual control, it is a key component of visual factory management.

11.2.4.3 Basic Steps

1. Sort: The initial step in the Five S's programme is to eliminate excess materials and equipment lying around in the workplace. These non-essential items are clearly identified by 'red-tagging'.
2. Set in order: The second step is to organise, arrange and identify useful items in a work area to ensure their effective retrieval. The storage area, cabinets and shelves are all labelled properly. The objective of this step is, as the old mantra says, 'a place for everything and everything in its place'.
3. Shine: This third action point is sometimes known as 'sweep' or 'scrub'. It includes down-to-basics activities such as painting equipment after cleaning, painting walls and floors in bright colours and carrying out a regular cleaning programme.
4. Standardise: The fourth point encourages workers to simplify and standardise the process to ensure that the first three steps continue to be effective. Some of the related activities include establishing cleaning

procedures, colour coding containers, assigning responsibilities and using posters.

5. Sustain: The fifth step is to make Five S's a way of life. Spreading the message and enhancing the practice naturally involves people and cultural issues. The key activities leading to the success of Five S's include
   Recognise and reward the efforts of members
   Ensure top management awareness and support
   Publicise the benefits
6. The final step is to continue training and maintaining the standards of Five S's.

11.2.4.4 Worked-Out Examples

As Five S's is primarily a visual process, a good example of promoting its message would be to display pictures of a workplace with photographs showing both 'before' and 'after' depictions of the implementation of Five S's.

The following example is taken from Skinner (2001) to illustrate the benefits of a Five S's programme.

Northtrop Grumman Inc. in the US first deployed Five S's on a part delivery process. The work area assembled a variety of components into a single product.

Before Five S's, the area was not well organised, and the process was inefficient. With Five S's implementations, the area saw a huge 93% reduction in the space employees travel to complete tasks as well as a 42% reduction in the overall floor space.

The system has become a one-piece flow operation between assembly and mechanics, enabling everyone involved to know what the station has and what it needs.

11.2.4.5 Training Requirements

Five S's is a conceptually simple process, but it requires both initial and follow-up training to inculcate the methodology to all employees. The classroom training sessions should be followed by, as far as practicable, a visit to a site where visual changes due to Five S's could be observed. A second-best option is to show the members photography or videos illustrating the 'before and after' status of the workplace involved in a Five S's programme.

11.2.4.6 Final Thoughts

Five S's is a simple tool and should be considered for the housekeeping and visual control of all types of work areas, whether they are in manufacturing or service.

11.2.5 I5: Mistake Proofing

11.2.5.1 Definition

Mistake proofing is an improvement tool to prevent errors being converted into defects. It comprises two main activities: preventing the occurrence of a defect and detecting the defect.

Mistake proofing is also known as poka-yoke. The concept was developed by Shigeo Shingo and the term 'poka-yoke' comes from the Japanese words 'poka' (inadvertent mistake) and 'yoke' (prevent).

11.2.5.2 Application

Mistake proofing is applied in fundamental areas. Although poka-yoke was devised as a component of Shingo's 'Zero Quality Control' for Toyota production lines, it is very easy to understand and grounded in basic common sense.

The process of mistake proofing is simply paying careful attention to every activity in the process and then placing appropriate checks at each step of the process. Mistake proofing emphasises the detection and correction of mistakes at the Design stage before they become defects. This is then followed by checking. It is achieved by 100% inspection while the work is in progress by the operator and not by the quality inspectors. This inspection is an integral part of the work process.

There is an abundance of examples of simple devices related to mistake proofing in our everyday surroundings including limit switches, colour coding of cables, error detection alarms, a level crossing gate and many more.

11.2.5.3 Basic Steps

1. Perform Shingo's 'source inspection' at the Design stage. In other words, identify possible errors that might occur in spite of preventive actions. For example, there may be some limit switches that provide some degree of regulatory control to stop the machine automatically.

2. Ensure 100% inspection by the operator to detect that an error is either taking place or is imminent.
3. Provide immediate feedback for corrective action. There are three basic actions in order of preference:

Control: An action that self-corrects the error (e.g. spell checker)
Shutdown: A device that shuts down the process when an error occurs (e.g. a limit switch)
Warning: Alerts the operator that some error is imminent (e.g. alarm)

11.2.5.4 Worked-Out Example

Consider the situation leading to the development of a level crossing. This is a place where cars and trains are crossing paths and the chances of accidents are very high.

The possible errors that might occur would relate to car drivers, who might be thinking one thing or another or distracted while driving (source inspection).

Both the level crossing operator and the car driver should ensure safety features while the work is in progress (judgement inspection).

In order to prevent drivers from making mistakes when a train is approaching, traffic lights were installed to alert the driver to stop (warning).

The lights might not be completely effective, so a gate was installed when a train was coming (shutdown or regulatory function).

The operation of the gate was controlled automatically as the train was approaching (control).

With the above mistake proofing devices in place, an accident can only occur if either the control and regulatory measures are malfunctioning or the driver drives around the gate.

11.2.5.5 Training Requirements

There is no 'rocket science' involved in mistake proofing, and it may be perceived in a dismissive fashion: 'that's only common sense'. However, it is critical that there should be some basic training in the principles and applications of mistake proofing. Furthermore, the employees need to be empowered to make improvements in the process by using mistake proofing. A half-day workshop should meet these training requirements.

11.2.5.6 Final Thoughts

Mistake proofing is a simple tool in principle, but its execution is the difficult part. The contrast between mistake proofing and 'fool-proofing' however is critical. The essential difference is that in mistake proofing, operators are respected and treated as partners in solving problems.

11.2.6 I6: Value Stream Mapping

11.2.6.1 Definition

Value stream mapping (VSM) is a visual illustration of all activities required to bring a product through the main flow, from raw material to the stage of reaching the customer.

Mapping out the activities in a production process with cycle times, down times, in-process inventory and information flow paths helps us to visualise the current state of the process and guides us to the future improved state.

11.2.6.2 Application

VSM is an essential tool of Lean manufacturing in identifying non-value-added activities at a high level of the total process.

According to Womack and Jones (1998), the initial objective of creating a value stream map is to identify every action required to make a specific product. Thus the initial step is to group these activities into three categories:

- Those which actually create value for the customer
- Those which do not create value but are currently necessary (type one muda)
- Those which create no value as perceived by the customer (type two muda)

Once the third set has been eliminated, attention is focused on the remaining non-value-creating activities. This is achieved through making the value flow at the pull of the customer.

VSM is closely linked with the analytical tool of process mapping. Having established improvement opportunities at a high level by VSM, a detailed analysis of the specific areas of the process is effective with process mapping.

11.2.6.3 Basic Steps

1. The first step of VSM is to select the product or process for improvement.
2. Each component of production from the source to the point of delivery is then identified.
3. The entire supply chain of the product or process (e.g. through order entry, purchasing, manufacturing, packaging and shipping) is mapped sequentially.
4. The quantitative data of each activity (e.g. storage time, delay, distance travelled, process time and process rate) are then recorded.
5. Each component (i.e. activity) of production or process is evaluated to determine the extent to which it adds value to product quality and production efficiency.
6. These activities are then categorised as
   Value added
   Necessary non-value added
   Unnecessary non-value added
7. Areas of further analysis and improvement are then identified clearly.

11.2.6.4 Worked-Out Example

The following example is adapted from Womack and Jones (1998, pp. 38–43).

Consider a case containing eight cans of cola at a Tesco store.

Figure 11.3 shows a value stream map of cola, from the mining of Bauxite (the source of aluminium of the cans) to the user's home.

The quantitative data related to the activities in the value stream are summarised in Table 11.3.

It is evident from the details in Table 7.6 that value-added activities take only 3 hours compared to the total time (319 days) from the mine to the recycling bin. This proportion is surprisingly small when one considers the overall duration of the process.

11.2.6.5 Training Requirements

The basic principles of VSM are not new, but making sense of these ideas and applying them to practical problems clearly requires some training. There is no shortage of training consultants offering workshops and courses in the tools of Lean manufacturing including VSM.

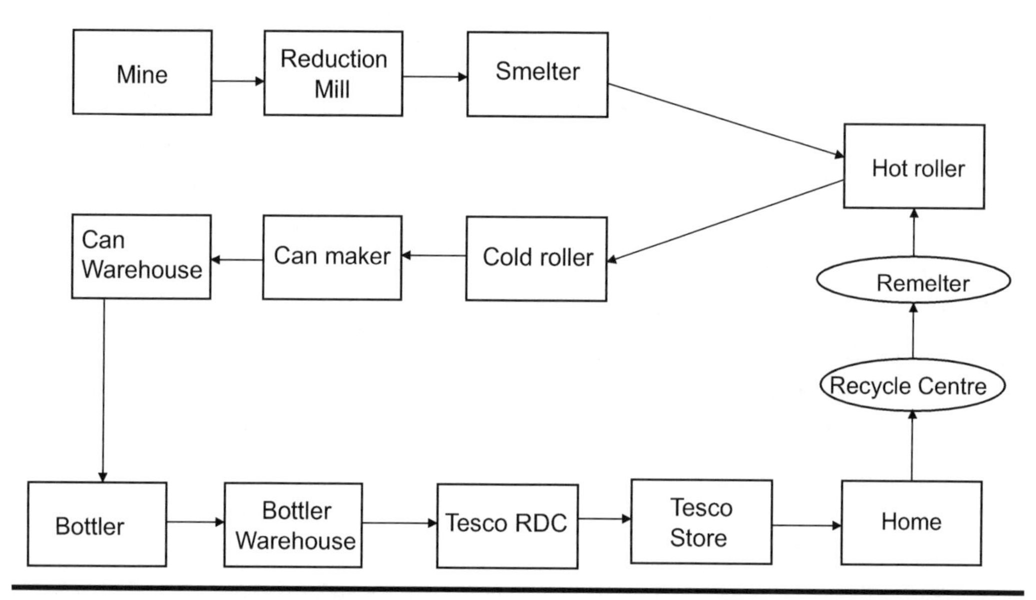

Figure 11.3 Value stream for cola cans.

Table 11.3 Quantitative Data of Cola Cans

	Incoming Storage	*Process Time*	*Finished Storage*	*Process Rate*	*Cumulative Days*
Mine	0	20 min	2 weeks	1,000 t/hr	319
Reduction mill	2 weeks	30 min	2 weeks	—	305
Smelter	3 months	2 hours	2 weeks	—	277
Hot rolling mill	2 weeks	1 min	4 weeks	10 ft/min	173
Cold rolling mill	2 weeks	< 1 min	4 weeks	2100 ft/min	131
Can maker	2 weeks	1 min	4 weeks	2,000/min	89
Bottler	4 days	1 min	5 weeks	1,500 min	47
Tesco RDC	0	0	3 days	—	8
Tesco store	0	0	2 days	—	5
Home storage	3 days	5 min	—	—	3
Totals	5 months	3 hours	6 months	—	319

We recommend that the team members should undergo a training workshop of at least half a day's duration for VSM. This training programme should be combined with other relevant tools like process mapping.

11.2.6.6 Final Thoughts

A complete value stream map quickly provides the visibility of the total process and is very effective in identifying non-value-added activities at a macro level.

11.2.6.7 Summary

Arguably the most difficult and certainly the most creative part of the Six Sigma and operational excellence initiatives is the Improvement stage. It is not rational to expect that the improvement tools described in this section will point out the obvious way forward. The solutions depend on the knowledge, innovative ideas and teamwork of the members. The tools and techniques are there to channel the ideas and analytical data towards improvement.

11.2.7 I7: Brainstorming

11.2.7.1 Definition

Brainstorming is an improvement tool for a team to generate, creatively and efficiently, a high volume of ideas on any topic by encouraging free thinking.

There are a few variations on the brainstorming process, of which two methods are more frequently used. First is the structured method (known as the 'round robin') where each member is asked to put forward an idea. The other technique is unstructured and is known as 'free-wheeling', in which ideas are produced and expressed by anyone at any time.

11.2.7.2 Application

Brainstorming is employed when the solution to a problem cannot be found by quantitative or logical tools. It works best by stimulating the synergy of a group. One member's thoughts trigger the idea of another participant, and so on. It is often used as a first step to open up ideas and explore options, and these are then followed up by appropriate quality management tools and techniques.

It has the advantage of getting every member involved, avoiding a possible scenario where just a few people dominate the whole group.

There are some simple ground rules or codes of conduct to observe:

- Agree to a time limit with the group
- Accept all ideas as given and do not interpret or abbreviate
- Do not evaluate ideas during the brainstorming process
- Encourage quantity rather than quality of ideas
- Discourage the role of an expert
- Keep ideas expressed in just a few words
- Emphasise causes and symptoms as opposed to solutions
- Write clearly and ensure the ideas are visible to everyone
- Have fun

11.2.7.3 Basic Steps

1. Clearly state the focused problem selected for the brainstorming session.
2. Form a group and select a facilitator, agree on a time limit and remind members of the ground rules.
3. Decide whether a structured approach or a free-wheeling basis will be used. For a larger group, a structured approach will allow everyone to get a turn and subsequently this could be switched to the free-wheeling method.
4. Write clearly on a flip chart or a board any ideas as they are suggested. The facilitator will motivate and encourage participants by prompting them, 'What else?'
5. Review the clarity of the written list of ideas, allow them to settle and discard any duplication.
6. Apply filters to reduce the list. Typical filters could include cost, quality, time and risk.
7. Ensure that everyone concurs with the shortlist of ideas.

11.2.7.4 Worked-Out Example

Consider the following focused statement for brainstorming:

What are the key selection criteria of a family holiday?

The five members of a family generated 26 ideas or issues. These were then filtered by a budgeted cost of US $4,000 for the whole family and the following key criteria were derived:

- Two weeks in August
- Seaside resort

- Indoor and outdoor recreational facilities
- Not near a nightclub
- Rich local culture
- Opportunities for sightseeing

11.2.7.5 Training Requirements

The application of brainstorming does not require any formal training in a classroom. A facilitator with some previous experience in the process can conduct a successful brainstorming session after briefing the team with the ground rules.

11.2.7.6 Final Thoughts

Brainstorming is a very useful tool for generating ideas in a group. Follow the ground rules with particular emphasis on two points:

- Do not dominate the group
- Set a time limit of, say, half an hour for the entire session

11.2.8 I8: Mind Mapping

11.2.8.1 Definition

Mind mapping is a learning tool for ordering and structuring the thinking process of an individual or team working on a focused theme.

According to Buzan (1995), the mind map 'harnesses the full range of cortical skills—word, image, number, logic, rhythm, colour and spatial awareness—in a single, uniquely powerful technique'.

It is a graphic tool to express 'radiant thinking' comprising four key characteristics:

1. The subject or theme is presented as a central image or key word.
2. The main components of the subject root out from the central image as branches.
3. Each branch contains a key word printed on an associated line.
4. The sub-components of each branch are also represented as branches attached to higher-level branches.

A mind map is arguably comparable to the cause and effect diagram where the effect represents the central image of the mind map. Each of the branches of the mind map are the causes in the cause and effect diagram.

11.2.8.2 Application

Mind maps have been applied for both individual and group objectives.

The applications in individual areas included

- Note taking
- Multidimensional memory device
- Creative thinking

Buzan (1995) claims that the mode of note taking involved in making a mind map saves more than 90% of the total time required for the conventional linear method of making notes. Instead, the mind map version involves noting, reading and reviewing only relevant key words.

Another way of thinking, the mnemonic device, involves the use of the imagination and association in order to produce multidimensional memorable images. The use of the memory mind map activates the brain to become mnemonically alert and thus increases the memory skill level.

The mind map is suited to creative thinking because it uses all the cortical skills commonly associated with creativity, especially imagination, association of ideas and flexibility.

The group mind map becomes a powerful tool during a group brainstorming exercise. The mind map becomes a 'hard copy' of the emerging group consensus and at the same time reflects the evolution of ideas through the branches and sub-branches radiating from the central image or theme.

In recent years, group mind maps have been successfully used by universities (including Oxford and Cambridge) and multinational companies like Boeing Aircraft Corporation, EDS, Digital Computers and British Petroleum.

11.2.8.3 Basic Steps

The mind map is intended to increase mental freedom and thus it is important not to introduce rigid disciplines. However, there is a need for a structured approach, otherwise freedom may be mistaken for chaos.

Buzan (1995) suggests six 'mind map laws' and offers three recommendations to supplement these 'laws'. The 'laws' are as follows:

1. Use a central image and emphasise that image by using variations in size of printing and colour.
2. Use arrows when you want to make connections within and across branch patterns.
3. Be clear and use only one key word per line.
4. Develop a personal style, while maintaining the mind map 'laws'.
5. Use hierarchy and categorisation in the form of basic ordering of ideas.
6. Use a numerical order simply by numbering the branches in the desired order.

The three recommendations are designed to help you implement the 'laws'. The recommendations are as follows:

1. Break mental blocks
   a. Add blank lines to your ongoing mind map
   b. Ask questions to stimulate a block-breaking response
   c. Add images to your mind map
   d. Maintain awareness of your associational capacity
2. Reinforce your mind map
   a. Review your mind maps
   b. Do quick mind map checks
3. Prepare
   a. Your mental attitude
   b. Your materials
   c. Your workplace/environments

The above 'laws' and recommendations are applicable to both individual and group mind maps. It is important to designate a facilitator to process a group mind map starting with a central image.

11.2.8.4 Worked-Out Example

An example of a mind map for late delivery is illustrated in Figure 11.4.

In this mind map late delivery is the central image. The main causes are noted by key words on radial lines and sub-causes are shown on branches emerging from these radial lines.

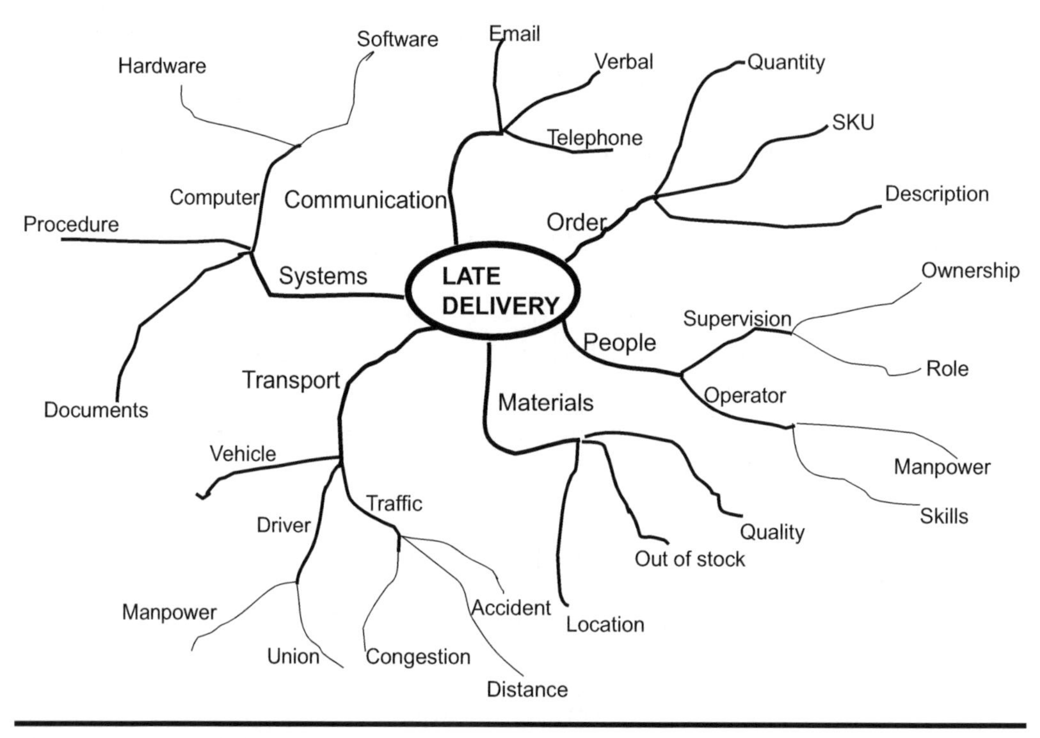

Figure 11.4 A mind map.

After the completion of the mind map, the importance of root causes can be weighted by assigning a number 1 to 100 according to its importance.

11.2.8.5 Training Requirements

The basics of mind mapping can be grasped by reading Tony Buzan's *The Mind Map Book* (1995). However, it would be useful if a facilitator with previous experience in mind mapping conducts a number of trial exercises before applying it to a real-life problem.

11.2.8.6 Final Thoughts

Mind mapping has various individual and group applications. It is particularly useful as a mnemonic or analytical tool for developing personal choices. It gives the brain a wider range of information on which to base its decision.

11.2.9 I9: Force Field Analysis

11.2.9.1 Definition

The force field analysis diagram or simply force field diagram is a model built on the concept by Kurt Lewin (1951) that change is characterised by a state of equilibrium between driving forces (e.g. new technology) and opposing or restraining forces (e.g. fear of failure).

In order for any change to happen, the driving forces must exceed the restraining forces thus shifting the equilibrium.

11.2.9.2 Application

A force field diagram is a useful tool at the early stages of change management leading to improvement. It is often used to

- Investigate the balance of power involved in an issue or obstacle at any level (personnel, project, organisation or network)
- Identify important players or stakeholders—both allies and opponents
- Identify possible causes and solutions to the problem

11.2.9.3 Basic Steps

According to Lewin (1951) three key principles are involved in the concept of change management by the force field analysis diagram:

1. First, an organisation has to unfreeze the driving and restraining forces that hold it in a state of apparent equilibrium.
2. Second, an imbalance is introduced to the forces, by increasing the drivers or reducing the restrainers or both, to enable the change to take place.
3. Third, once the change is complete and stable, the forces are brought back to equilibrium and refrozen.

The force field diagram is constructed by a team and the following basic steps are suggested:

1. Agree on the current problem or issue under consideration and the desired situation

2. List all forces driving changes towards the desired situation
3. List also all forces resisting changes towards the desired situation
4. Review all forces and validate their importance
5. Allocate a score to each of the forces using a numeric scale (e.g. 5 = most important and 1 = least important)
6. Chart the forces by listing the driving forces on the left and restraining forces to the right
7. Decide how to minimise or eliminate the restraining forces and increase the driving forces
8. Agree on an action plan

11.2.9.4 Worked-Out Example

1. The issue identified is how to increase the usage of purchase orders in a pharmaceutical company. The driving and restraining forces were identified by the team and also rated on a scale of 1 to 5 (5 = most important and 1 = least important) and represented in a force field diagram as shown in Figure 11.5.

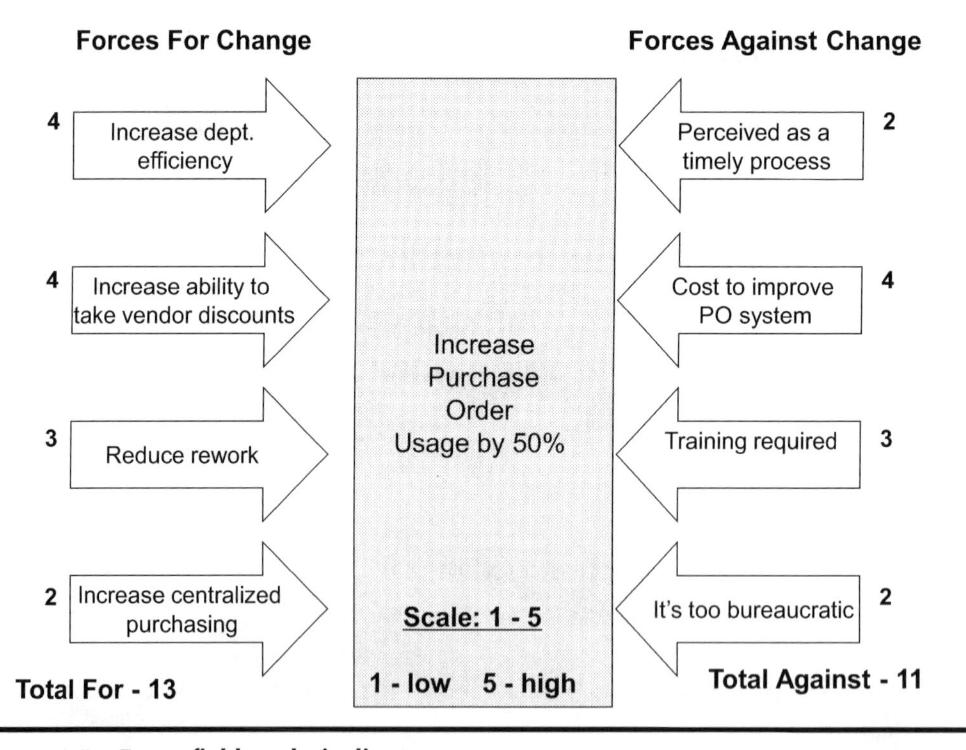

Figure 11.5 Force field analysis diagram.

The driving forces show a total of 13 against a total score of 11 by restraining forces.

11.2.9.5 Training Requirements

The knowledge and application of a force field diagram are best derived by 'on-the-job' training during a brainstorming exercise on an actual problem or project.

The facilitator should have the experience of change management in applying a number of force field diagrams to direct the team and gain a consensus of identifying forces and allocating scores.

11.2.9.6 Final Thoughts

The force field analysis diagram is a powerful tool in the Improvement stage of Green Six Sigma and also in tempering the mindset of any change management project. It could also provide new insights into the evaluation and implementation of corporate strategies.

11.3 Summary

In this chapter main tools for the Improve phase are included so that the outputs of the Analyse phase are used to generate solutions of the performance problem. During the Improve phase possible solutions are tested to select the best solution to the problem. An implementation plan may be developed to make the necessary transition from Improve to Control.

Similar to the Analyse phase, if defect reduction is your goal then your bias would be Six Sigma–orientated tools, and if time is your objective your bias would be Lean-orientated tools. However, you should tackle all-time and defect issues simultaneously with the Green Six Sigma toolset.

GREEN TIPS

- Consider standard work procedures of Improve for sustaining improved performance.
- Force field analysis is a powerful tool to achieve consensus for a sustainable change.
- It is important that possible solutions are tested at the Improve phase before moving to the Control phase.

Chapter 12

Tools for Control

That is what learning is. You suddenly understand something you have understood all your life, but in a new way.

—Doris Lessing, *The Four-Gated City*

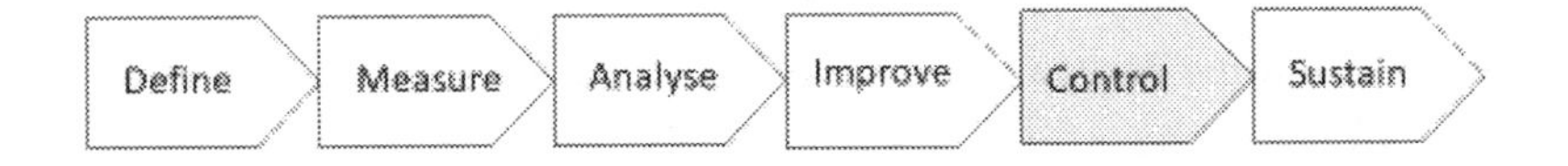

12.1 Introduction

Operational excellence is a long-term process. It can take an organisation several years to put the fundamental principles, practices and systems into action in a learning culture that will sustain the benefits gained. Managers who have steered their way through the challenges of a Six Sigma or a Lean Sigma programme over recent years are probably proud of the results. However, such a manager has only just embarked upon the path of success.

The objective of the Control phase is to implement the solution, ensure that this solution is sustained, and share the lessons learned from the improvement projects throughout the organisation. With this approach the projects start to create excellent returns. Thus the best practices in one part of the organisation are translated quickly to result in implementation in projects carried out by another part of the organisation.

DOI: 10.4324/9781003268239-14

The main deliverables of the Control stage are

1. Project documentation—A close-out report to record the key aspects of the project
2. Leverage of best practice—Transfer of key learnings from your project that may be adopted in other projects
3. Sustained solution—A fully implemented process that is supported with a Control plan to ensure that it is sustained over time

12.2 Description of Tools for Control

The key tools for Control should include

C1: Gantt chart
C2: Activity network diagram
C3: Radar chart
C4: PDCA (plan-do-check-act) cycle
C5: Milestone tracker diagram
C6: Earned value management (EVM)

Some of the tools from the early stages can be used during the implementation of the project (e.g. M7: Control Charts). In order to ensure the sustainability of the results, a number of qualitative techniques (see Chapter 15) are applicable at the Control stage:

- Balanced Scorecard
- European Foundation of Quality Management (EFQM)
- Sales and operations planning

12.2.1 C1: Gantt Chart

12.2.1.1 Definition

A Gantt chart is a simple tool which represents time as a bar or a line on a chart. The start and finish times for activities are displayed by the length of the bar and often the actual progress of the task is also indicated.

A Gantt chart is also known as a bar chart.

12.2.1.2 Application

The most common form of scheduling is the application of Gantt charts. The merits of Gantt charts are that they are simple to use and they provide a clear visual representation of both the scheduled and actual progress of activities. The current time is also indicated on the graph.

Gantt charts are also used to review alternative schedules by using movable pieces of paper or plastic channels. The charts can be drawn easily by standard software tools such as PowerPoint or Excel. However, a Gantt chart is not an optimising tool and therefore does not determine the 'critical path' of a project.

12.2.1.3 Basic Steps

1. Identify the key activities or the tasks related to the project and describe each activity by selective key words.
2. Prepare a scheduling board, and depending on the duration of the project, draw vertical lines to divide the board into monthly, weekly or daily intervals.
3. Arrange the activities in a sequence of estimated start dates and post them on the extreme left-hand column of the board.
4. Estimate the start and finish dates of each activity and draw horizontal bars or lines along the time scale to reflect the start and duration of each of the activities.
5. On completion of each activity, show the actual start and duration of the activity by a bar of a different colour.
6. Include a 'Time Now' marker on the chart; review and maintain the chart until the end of the project.

12.2.1.4 Worked-Out Example

Figure 12.1 presents an example of a Gantt chart showing the planned and completed activities of a FIT SIGMA programme.

12.2.1.5 Training Requirements

The application of Gantt charts does not require extensive training. The team members are usually experienced in the use of Gantt charts. A briefing session in front of the scheduling board should be adequate.

Activity	Week 1	Week 2	Week 3	Week 4	Week 5	Week 6	Week 7	Week 8	Week 9
A	2	2							
B			2	2					
C			6	6					
D					3	3			
E					1	1	1	1	
F									2
Total	2	2	8	8	4	4	1	1	2

Figure 12.1 A Gantt chart (*source*: Basu and Wright, 2003).

12.2.1.6 Final Thoughts

A Gantt chart is a simple but very effective visual tool for planning and monitoring the progress of a quality programme. We recommend its application at the Control stage of the programme.

12.2.2 C2: Activity Network Diagram

12.2.2.1 Definition

An activity network diagram is a control tool to determine and monitor the most efficient path, known as the critical path, and a realistic schedule for the completion of a project. The diagram is represented graphically showing a brief description of all tasks, their sequence, their expected completion time and the jobs that can be carried out simultaneously.

An activity network diagram with some variations is also referred to as project evaluation and review technique (PERT), critical path method (CPM), a precedence diagram and finally, as network analysis.

12.2.2.2 Application

The activity network diagram was extensively used in most projects during the 1960s and 1970s. As the larger projects became more and more complex comprising numerous tasks, its popularity by manual methods started to diminish. However, with the advent of software systems such as Primavera and MS Project, its application at the higher level of the project has increased significantly. It offers a number of benefits:

- The team members can visualise the criticality of major tasks in the overall success of the project.
- It highlights the problems of 'bottlenecks' and unrealistic timetables.
- It provides facilities to review and adjust both the resources and schedules for specific tasks.

12.2.2.3 Basic Steps

There are normally two methods applied for the construction of an activity network diagram: the 'activity on arrow' method and the 'activity on node' method. The former has been used most widely, and the steps for the application of the arrow technique can be outlined as follows:

1. Assemble the project team with the ownership and knowledge of key tasks.
2. List the key tasks with a brief description for each one.
3. Identify the first task that must be done, the tasks that can be done in parallel and the sequential relationship between tasks.
4. Draw arrows for each task which are labelled between numbered nodes and estimate a realistic time for the completion of each of these tasks.
5. Avoid feedback loops in the diagram. Unlike Gantt charts, the length of the arrows does not have any significance.
6. Determine the longest cumulative path as the critical path of the project.
7. Review the activity network diagram and adjust resources and schedules if appropriate.

For more detailed information on the activity network diagram, see Wild (2002, pp. 403–450).

12.2.2.4 Worked-Out Examples

Consider a project of writing and submitting the draft manuscript of a technical book such as this one to a publisher. Table 12.1 lists all the activities which constitute the project including their dependent relationship and estimated duration.

The activity network diagram is shown in Figure 12.2.

The critical path is A, B, E, F, G, J, K and M with a total project duration of 22 weeks.

Table 12.1 List of Project Activities for Production of a Technical Book

Resource	*Activity*	*Description*	*Predecessor*	*Duration (Weeks)*
Author	A	Prepare proposal	—	2
Publisher	B	Approve proposal	A	4
Author	C	Preliminary research	A	2
Author	D	Detailed research	C	10
Author	E	Write Chapters 1–3	B	3
Author	F	Write Chapters 4–6	E	3
Author	G	Write remaining chapters	F	4
Admin	H	Type Chapters 1–3	E	2
Admin	I	Type Chapters 4–6	F, H	2
Admin	J	Type remaining	G, I	3
Author	K	Compile full draft	D, J	2
Author	L	Obtain copyright clearance	B	12
Author	M	Submit manuscript	K, L	1

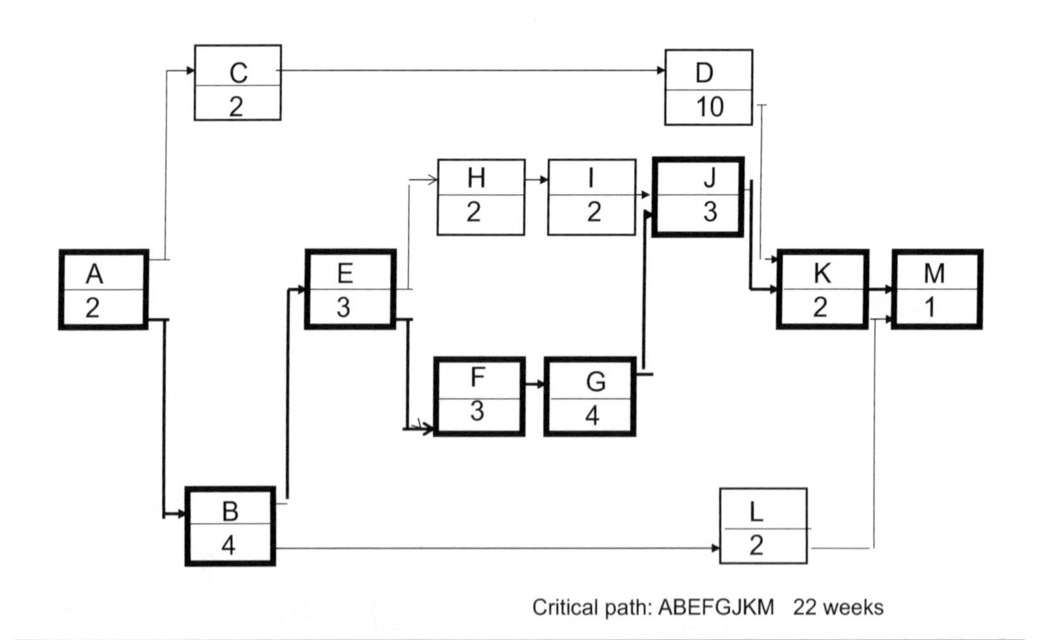

Figure 12.2 Activity network diagram.

12.2.2.5 *Training Requirements*

The basic principles of an activity network diagram are not difficult to follow and can be covered in a half-day workshop with practical exercises.

A detailed analysis of an activity network diagram can be complex when the variable estimates of duration, 'float' or 'slack', resource levelling and the probability of occurrence are considered. We suggest that the advanced applications should be treated for academic research.

12.2.2.6 *Final Thoughts*

With the flexibility, simulation and milestone facilities now available on software for Gantt charts, we recommend that an activity network diagram should be used primarily for the high-level mapping of tasks to identify the critical path.

12.2.3 C3: Radar Chart

12.2.3.1 *Definition*

A radar chart is a polar graph to show using just one graphic the size of the gaps in the performance levels of key performance indicators.

A radar chart is also known as a polar graph and, because of its appearance, as a spider diagram.

12.2.3.2 *Application*

A radar chart is a useful visual tool to display the important metrics of performance at the Control stage of a quality improvement programme. The other benefits of this chart include

- It highlights strengths and weaknesses of the total process, programme or organisation.
- It can define full performance in each category.
- It can act as a focal point to capture and review the different perception of all stakeholders of the organisation related to relevant performance metrics.
- Given a range of rating (say on a scale of 1–5), it can drive a total or average score of all entities.

However, a limitation of a radar chart is that it tends to provide just a snapshot of the performance levels at any given time.

12.2.3.3 Basic Steps

1. Select and define the performance categories. The chart can handle 10 to 20 categories.
2. Some performance metrics are likely to be easily quantifiable and expressed as a percentage. Other metrics may be qualitative and not represented by numbers.
3. Normalise all performance metrics in a scale of 1 to 5 with appropriate guidelines according to the following grades:
   1—Poor
   2—Fair
   3—Good
   4—Very good
   5—Excellent
4. Construct the chart by drawing a wheel with as many spokes as the performance categories and marking each spoke with 0 at the centre and 5 on the rim.
5. Connect the ratings for each performance category and highlight the gaps.
6. Use the results for consolidating strengths and improving weaknesses.

12.2.3.4 Worked-Out Example

The example of a radar chart is adapted from Slack et al. (2012, p. 42) (Figure 12.3).

Consider the following five aspects of operations performance, which as a whole will affect the customer service when a product is delivered to the customer:

Quality: Doing things right
Speed: Doing things fast
Dependability: Doing things on time
Flexibility: Ability to change
Cost: Doing things cheaply

Table 12.2 shows the actual performance of the supplier and the expectation of the customer for these five aspects of customer service on a scale from 1 to 5, where 5 is high performance and 1 is low performance.

(5= Excellent, 1= Poor)

Table 12.2 Supplier performance and Customer expectation

Five Aspects	Supplier	Customer
Quality	2	4
Speed	5	5
Dependability	4	4
Flexibility	2	5
Cost	4	4

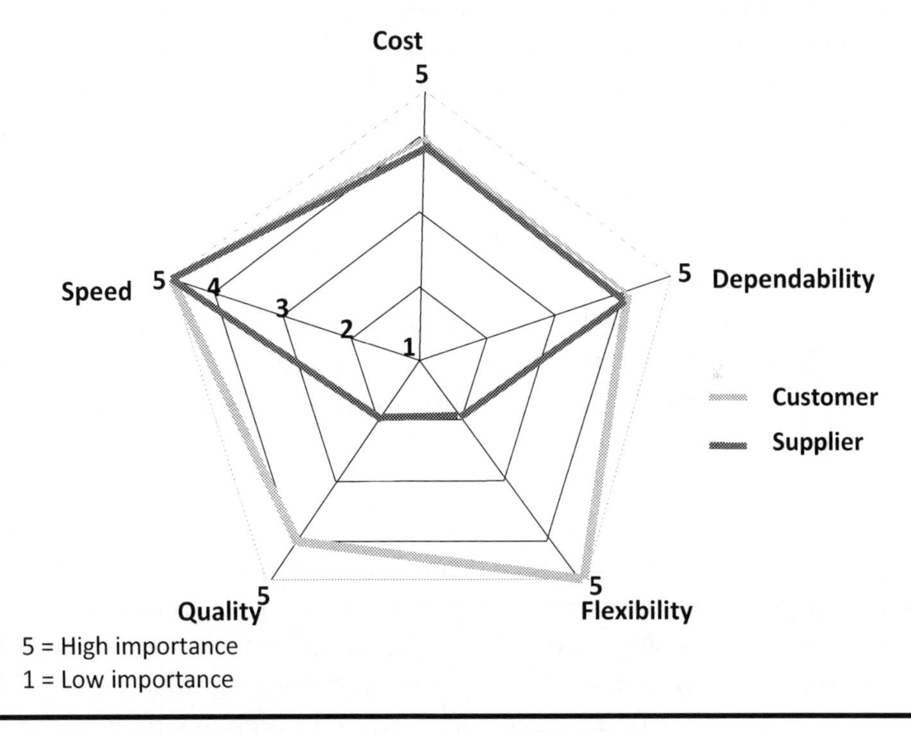

Figure 12.3 Radar Chart showing performance objectives.

12.2.3.5 Training Requirements

The application of a radar chart does not require any significant classroom-based training. However, it does necessitate a good understanding of how to rate all performance categories. This understanding and the selection criteria of the metrics should be explained in a period of about an hour as part of a training programme.

12.2.3.6 Final Thoughts

Given the availability of appropriate data, a radar chart can be drawn quite simply using easily accessible software tools. We recommend its application to monitor the progress of a programme by a snapshot of all key performance indicators.

12.2.4 C4: The PDCA Cycle

12.2.4.1 Definition

In a central process, the actual results of an action are compared with a target or a set point. The difference between the two is then monitored and corrective measures are adopted if the disparity becomes large. The repeated and continuous nature of continuous improvement follows this usual definition of Control and is represented by the PDCA (plan-do-check-act) cycle.

This is also referred to as the Deming Wheel, named after W.E. Deming. Another variation of PDCA is PDSA (plan-do-study-act).

12.2.4.2 Application

The application of the PDCA cycle has been found to be more effective than adopting the 'right first time' approach of concentrating on developing flawless plans (Juran, 1999, p. 41.3). The PDCA cycle means continuously looking for better methods of improvement.

The PDCA cycle is effective in both doing a job and managing a programme. The extent to which the PDCA cycle is applied to the job level depends on the self-control of the operators. Education and training enhance the self-control capacity of workers. At the programme level, the PDCA cycle acts as a process of repeatedly questioning the detailed working of the operations and thereby helps to sustain the improved results.

The PDCA cycle enables two types of corrective action—temporary and permanent. The temporary action is aimed at results by practically tackling and fixing the problem. The permanent corrective action, on the other hand, consists of investigating and eliminating the root causes and thus targets the sustainability of the improved process.

12.2.4.3 Basic Steps

1. P (Plan) Stage: The cycle starts with the Plan stage, comprising the formulation of a plan of action based on the analysis of the collected data.

2. D (Do) Stage: The next step is the Do or implementation stage. This may involve a mini-PDCA cycle until the issues of implementation are resolved.
3. C (Check) Stage: The next step is the Check stage where the results after implementation are compared with targets to assess if the expected performance improvement has been achieved.
4. A (Act) Stage: At the final Act stage, if the change has been successful then the outcome is consolidated or standardised.
5. If the change has not been successful however, the lessons are recorded and the cycle starts again. Even if the change is successful, the results are sustained by going through the PDCA cycle over and over again.

12.2.4.4 Worked-Out Example

The following example is taken from Juran (1999, pp. 32.8–32.10).

In this example, a healthcare organisation in the US was following the traditional reliance on extensive internal and external inspection to maintain quality standards. This resulted in a medical record system whose size, complexity and format were wasteful and cumbersome.

The aspects of the PDCA cycle were applied to their internal quality assurance procedures, and the medical record procedures were simplified.

Figure 12.4 shows the PDCA cycle.

12.2.4.5 Training Requirements

The importance of the training for the PDCA cycle, especially for first-time workers, has been well recognised at the early stage of the quality movement and thus the concept of the quality circle was born. During the late 1950s, about 100,000 transcripts of radio broadcast text for quality circles were sold in Japan.

The training of the PDCA cycle, although simple in principle, should be inculcated in everyone within the organisation on a continuous basis. This tool then becomes most effective when it becomes a way of life in the organisation.

12.2.4.6 Final Thoughts

The PDCA cycle is more than just a tool; it is a concept of continuous improvement processes embedded in the organisation's culture. The most

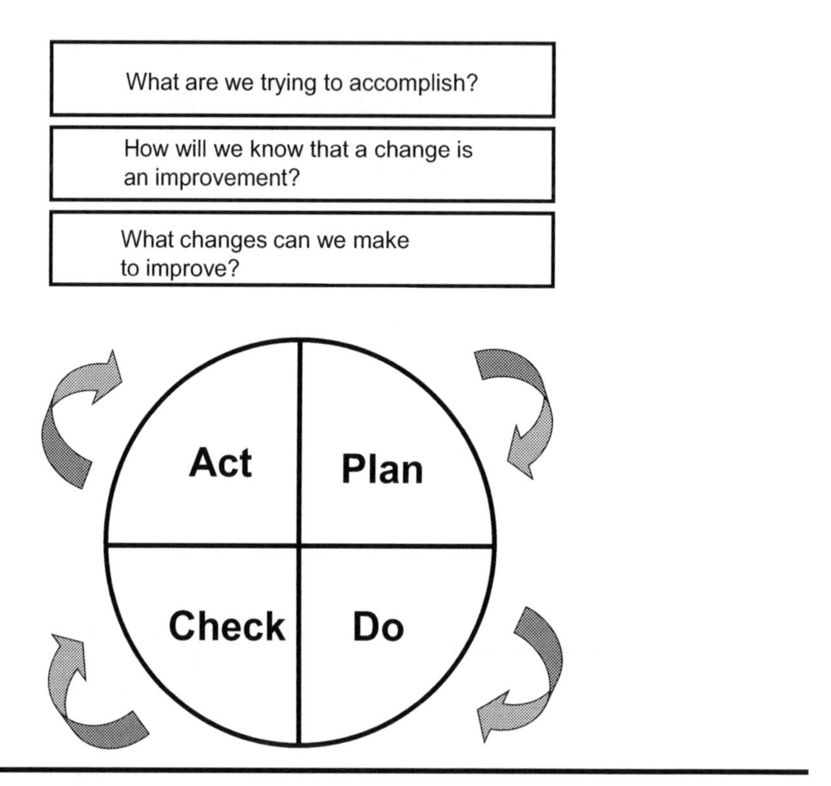

Figure 12.4 PDCA cycle.

important aspect of PDCA lies in the Act stage after the completion of a project when the cycle starts again for further improvement.

12.2.5 C5: Milestone Tracker Diagram

12.2.5.1 Definition

A milestone is a key event selected for its importance in the project. A milestone tracker diagram is used to show the projected milestone dates and the best estimated date on the week of the progress review on a single chart.

The purpose is to assess the achievement or slippage of the progress. Thus the milestone tracker diagram is also called the milestone slippage chart.

12.2.5.2 Application

A milestone tracker diagram is used in a large and long-running project to review the probability of achieving the key dates of the project. However, its principles can be applied to projects of shorter dimension.

The manager of the project or programme faces a problem when team members report their work as being 90% or 95% complete, leaving the last 10% or 5% continuously out of reach. The first requirement is to identify a set of key events or milestones in the plan that allows no compromise. A milestone is achieved when the objectives or targets of the key event are delivered in full. The dates of these milestones should not be far apart. A simple rule is that when the target passes from person to person or from department to department, the event must constitute a milestone. However, the workplace of a person or department may also contain several milestones.

12.2.5.3 Basic Steps

1. Determine the key events of the project as milestones.
2. Establish the planned delivery dates of the milestones and agree to the dates with the project manager and sponsor.
3. Draw two axes at right angles to each other and mark these axes in a suitable scale of weeks or months. The total period should allow the milestones to extend under the worst possible conditions.
4. Choose the x-axis as the 'Estimated Finish Week' of milestones and y-axis as the 'Week of Progress Review'. The axes should be of equal value. Note that the y-axis is in the reverse direction as compared to a conventional graph.
5. Draw a diagonal line from the top of the y-axis (zero) to the maximum duration point on the x-axis.
6. During every week of progress review, plot the estimated delivery dates of each milestone.
7. At any time during these progress reviews, the project manager can draw the 'best-fit' straight line to connect these revised estimates. If these straight lines are projected they will intersect the diagonal line at points which can be taken as the predicted delivery dates of milestones.

12.2.5.4 Worked-Out Examples

Figure 12.5 shows an example of a milestone tracker diagram where the progress of two milestones has been illustrated.

The planned finish dates of Milestone 1 and Milestone 2 are Week 4 and Week 8, respectively.

Figure 12.5 shows that Milestone 1 is on track while Milestone 2 has experienced some delays and thus its predicted finish date is Week 9.

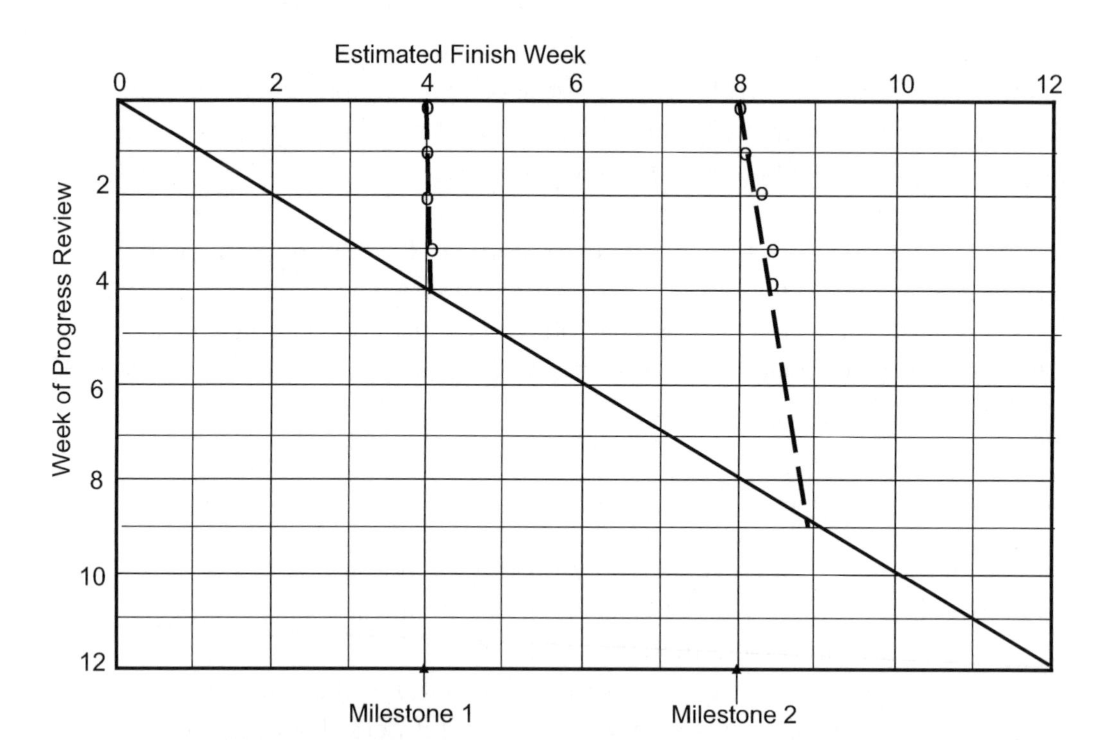

Figure 12.5 Milestone tracker diagram.

12.2.5.5 Training Requirements

The basic principles of a milestone tracker diagram are easy to follow and may not require any specific training in classrooms. The definition and guidance provided in this section of the book should enable the project team to apply this tool in practice.

12.2.5.6 Final Thoughts

The milestone tracker diagram is a useful, though not essential, tool for reviewing the progress of key events in a project. It is a better name than the 'milestone slip chart', something of a misnomer because it implies that milestones are expected to slip.

12.2.6 C6: Earned Value Management

12.2.6.1 Definition

EVM or earned value analysis is a project control tool for comparing the achieved value of work in progress against the project schedule and budget.

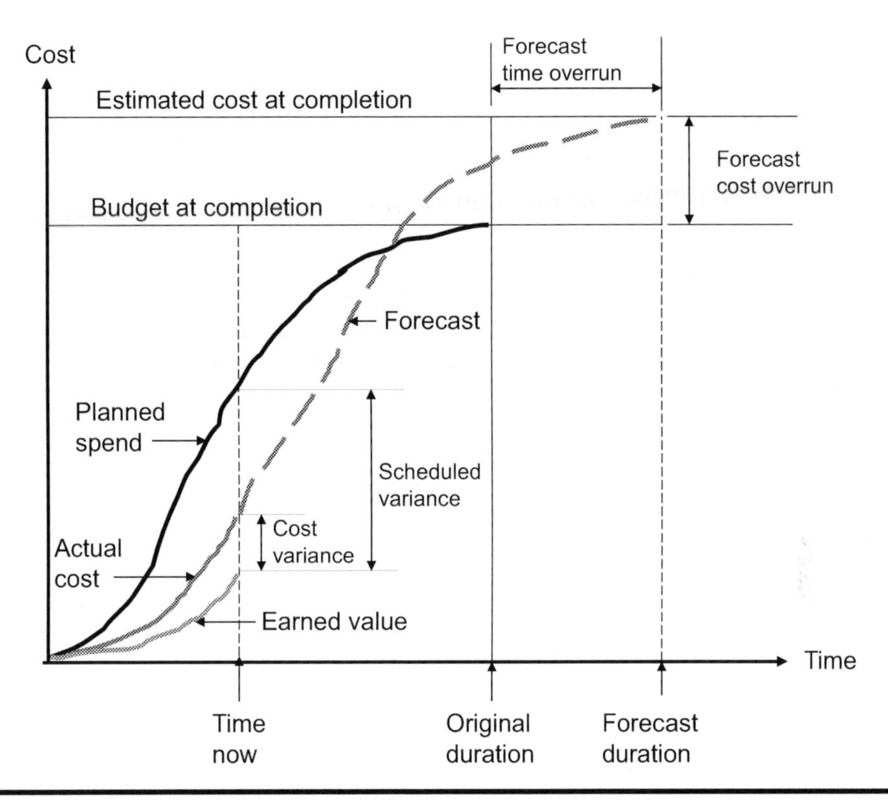

Figure 12.6 Earned value management.

It can be performed at the single activity level and by aggregating the results up through the hierarchy or work breakdown structure.

There are a few useful terms related to EVM as shown in Figure 12.6 and defined as follows:

Time Now: The reference point used to measure and evaluate the current status.

Earned Value: The value of useful work done at Time Now. It is also known as the budgeted cost of work performed (BCWP). It is typically calculated at activity level by multiplying the Budget at Completion for the activity with the percentage progress achieved for the activity.

Budget at Completion: The total budget for the work to be carried out.

Original Duration: The planned overall duration of the activity or project.

Planned Value (Spend): The planned rate of spend against time though the life of the project. It is also known as the budget cost of work schedule (BCWS).

Actual Cost: The cumulative costs incurred at Time Now. It is also known as the actual cost of work performed (ACWP).

Cost Variance: (CV): The difference between the Earned Value and the Actual Cost at Time Now (CV = Earned Value – Actual Cost).

Scheduled Variance (SV): The difference between the Earned Value and the Planned Value at Time Now (SV = Earned Value – Planned Value).

Scheduled Performance Index (SPI): The ratio between the Earned Value and Planned Value at Time Now (SPI = Earned Value/Planned Value). It is used to predict the final outcome of the project time.

Cost Performance Index (CPI): The ratio between the Earned Value and Actual Cost at Time Now (CPI = Earned Value/Actual Cost). It is used to forecast the Estimated Cost at Completion (EAC).

EAC = Actual Cost/CPI + (Budget at Completion – Earned Value)/CPI.

12.2.6.2 Application

EVM has been applied effectively in well-structured projects, especially engineering projects, where there is a defined work breakdown structure and various levels of hierarchy.

It is not a progress control tool in itself, as it can only highlight a need for corrective action by indicating trends. At each milestone review (i.e. Time Now), it provides some useful pointers of project control including

Variance analysis: It shows current status in terms of cost and schedule.

Estimating accuracy: It enables predictions of cost at completion and completion date.

Efficiency: It provides performance indices identifying areas requiring corrective action.

However, there is a danger of placing too much reliance on EVM, because the results are likely to be flawed for a number of reasons including the following:

- Forecasts depend on reliable measurements of the amount of work performed that can be difficult to achieve in some cost types.
- A considerable amount of clerical effort is needed to maintain the database and carry out calculations.
- Difficulty is likely to be greatest for projects containing a higher proportion of procured equipment and materials.
- It does not take into account risks and uncertainties.

- If the methodology of EVM is not fully understood by the sponsor and members of the project, it becomes difficult to get everyone's co-operation.

12.2.6.3 Basic Steps

1. Establish a sound framework of planning and control for implementing EVM. The key features of this framework include
   a. A detailed work breakdown structure
   b. A detailed cost coding system to match the work breakdown structure
   c. Timely and accurate collection of cost data
   d. A sound method of quantifying the amount of work done at each review date
   e. A well-trained administration
2. At each review date, determine for the work package
   a. Authorised current budget at completion (BAC)
   b. Percentage of work complete (ACWP)
   c. Planned spend (BCWS)
3. From the data in Step 3, calculate
   a. Earned Value = Current Budget × % Complete = BCWP
   b. Cost Variance = BCWP – ACWP
   c. Scheduled Variance = BCWP – BCWS
   d. CPI = BCWP/ACWP
   e. SPI = BCWP/BCWS
4. From the data in Step 4, calculate
   a. Estimated Cost at Completion (EAC) = BAC/CPI
   b. Estimated Time at Completion = Original Duration/SPI
5. Analyse the results and trends and take corrective action

12.2.6.4 Worked-Out Example

Consider the example of a small project which has a BAC of US $80,000 and duration of 6 weeks. The collected data at the end of Week 3 ('Time Now') are shown in Table 12.3.

Earned value at Week 3 (Time Now) = US $80 × 0.20 = US $16
CV = Earned Value – Actual Cost = 16 – 30 = –14

Table 12.3 Data collected at the end of Week 3

Project Weeks	*Planned Value (US $)*	*Actual Cost (US $)*	*Percentage (%) Complete*
1	10	8	10
2	20	18	15
3	35	30	20
4	60		
5	70		
6	80		

SV = Earned Value – Planned Value = 16 – 35 = –19
SPI = Earned Value/Planned Value = 16/35 = 0.46
CPI = Earned Value/Actual Cost = 16/30 = 0.53
Forecast Duration = Original Duration/SPI = 6 × 35 = 13 weeks
Estimated Cost at Completion = Budget at Completion/CPI

$$= \text{US } \frac{\$80 \times 30}{16} = \text{US } \$150{,}000$$

The project is currently overspending and in addition has fallen behind schedule.

12.2.6.5 Training Requirements

The principles of EVM are conceptually simple but detail rich. In the worked-out example, we have looked at only one activity, but the Earned Value and its related parameters would have to include all project work scheduled to be complete at 'Time Now'. The Earned Value should include all work actually finished as well as the completed portion of all work in progress. Therefore, it is important that the administrator is trained thoroughly in a classroom environment, say in a full-day workshop. The members of the project team will also benefit from a half-day appreciation workshop.

12.2.6.6 Final Thoughts

The detailed requirements of EVM should not deter the reader from its merits for implementation. We recommend that EVM should be applied with appropriate training in larger organisations.

12.3 Summary

The tools for DMAICS have been grouped according to their most appropriate or frequently used area in the DMAICS cycle. This is not a rigid taxonomy, the purpose of this classification is primarily for presenting the tools in a systematic way. The same tool (e.g. cause and effect diagram) can be used in Define, Measure, Analyse or Improve stages of the programme. The tools are not enough to make an improvement, and they need adequate training, resources, management support and an environment which is conducive to change and to their application. A tool is as good as a person who trained to use it the right way for the appropriate problem. In this chapter specific tools of project management are included to ensure that the process of the solution from the Improve phase is stable and consolidated.

GREEN TIPS

- The main purpose of the Control phase is to prevent the process of the solution from the Improve phase from decaying.
- Find as many mistake proofing solutions as you can to eliminate defects.
- Update flow diagrams and FMEAs.

Chapter 13

Tools for Sustainability

We cannot solve our problems with the same thinking we used when we created them.

—Albert Einstein

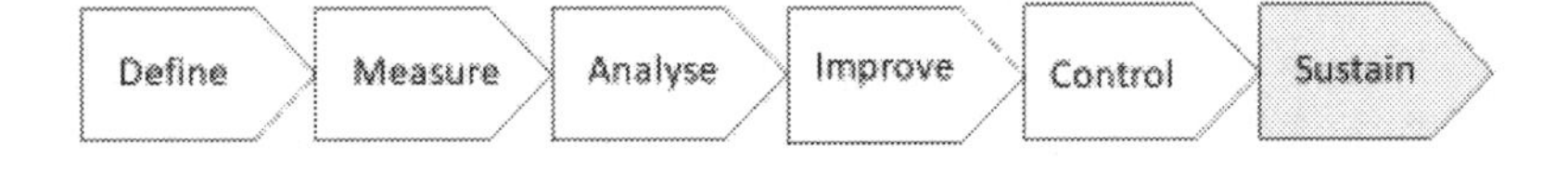

13.1 Introduction

Sustainability in a general sense is the capacity to maintain a certain process or state over a longer period, perhaps indefinitely. In the context of preserving the environment, it relates to uniting 'the needs of the present without compromising the ability of future generations to meet their own needs' (United Nations, 1987). In the context of Green Six Sigma, the sustainability of project outcomes includes both the longer-lasting stability of performance deliverables and environmental standards. The Sustain phase is a unique extension of DMAIC (define, measure, analyse, improve, control) to DMAICS (define, measure, analyse, improve, control and sustain) in Green Six Sigma methodology.

The objective of the Control phase is to implement the solution and ensure that this solution is sustained. Building on the start of process sustainability, the deliverables of the Sustain phase are threefold:

1. Sustain solution—Ensure that processes are stable and performance metrics are sustained

DOI: 10.4324/9781003268239-15

2. Sustain continuous improvement culture—Inculcate the culture of continuous improvement by appropriate training and knowledge management
3. Sustain environment—Include additional processes to measure and monitor environmental standards related to climate change

The reports of quality improvement programmes, including Six Sigma and Lean Six Sigma, have shown that there is a substantial failure rate in the implementation of each practice (Asif et al., 2009b). There are also examples of, metaphorically speaking, 'the operation was successful but the patient died'. This is where the Sustain phase of Green Six Sigma plays a crucial role.

Regarding the effective and sustainable implementation of Green Six Sigma programmes, useful insights were obtained from both the failed and successful quality improvement programmes. These examples suggest the need for a user-orientated implementation design to institutionalise the cycle of 4Ts (viz. Top management, Team, Tools and Training) as shown in Figure 13.1.

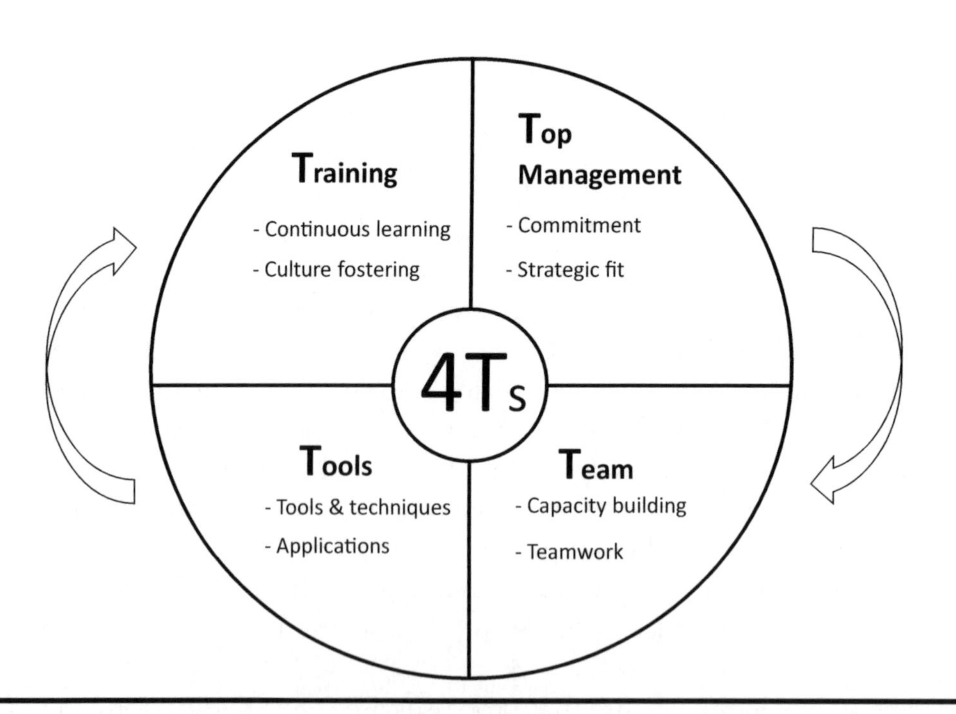

Figure 13.1 Sustaining Green Six Sigma programme.

The role of top management is to ensure a sustained commitment to the programme and the fitness with the organisation's business strategy. It is vital that the Green Six Sigma programme is aligned with the business strategy and corporate policy of the organisation. When integrated with the business strategy, Green Six Sigma becomes a competitive advantage.

The second key component of 4Ts is the Team. Adequate financial and people resources will build the Team. The user-orientated approach of Green Six Sigma promotes a greater buy-in by employees and the spirit of teamwork.

In order to make it in a structured process, we require appropriate tools and techniques as discussed in Chapter 8 through Chapter 15. The specific tools for the Sustain phase are described in this chapter.

The whole programme is underpinned by a structured and assessed training and education plan. The distinctive feature of Six Sigma programmes, including Green Six Sigma, is the training and certification of Black Belts, Green Belts and Yellow Belts.

13.2 Description of Tools for Sustainability

In order to ensure the sustainability of the results, a number of qualitative techniques (see Chapter 15) are applicable at the Sustain stage:

- Balanced Scorecard
- European Foundation of Quality Management (EFQM)
- Sales and operations planning
- Knowledge management

Key tools for Sustainability are

S1: Material flow analysis diagram
S2: Carbon footprint tool

13.2.1 S1. Material Flow Analysis

13.2.1.1 Definition

Materials flow analysis is a quantitative process for determining the flow of materials through the total enterprise or an ecosystem. When the study of materials flows is applied to an economic sector or a nation or a region it is also called material flow accounts.

13.2.1.2 Application

Materials flow analysis (MFA) has many application areas of which three areas are most prominent:

- On a national or regional scale. In this type of application material exchanges between an economic sector and the nation or region are analysed. This process is also known as materials flow accounts or material flow accounting.
- On a corporate level. This process involves the supply chains of a number of companies in the corporate group. The goal of MFA within a company is to calculate the balance of the input and output of materials so that materials are more efficiently used.
- In the life cycle of a product. In this application MFA is used to compile the life-cycle inventory to optimise the balance and waste reduction.

In each application of MFA, the central focus is the principles of a circular economy by recycling and waste reduction.

Basic Steps

1. Select the domain of MFA whether the analysis is for a nation, region, corporation or product life cycle.
2. Make precise use of the term 'material' whether it is a transforming material or a finished product.
3. Establish the unit of analysis, e.g. bottles of whisky or kilograms of transforming materials.
4. Measure the input and output of materials in each stage of the process or supply chain.
5. Present the quantitative data in a material flow diagram.
6. Analyse the material balance and losses in each stage.
7. Apply the principles of a circular economy to optimise the usage of materials.

13.2.1.3 Worked-Out Example

The MFA diagram in Figure 13.2 paints a picture of the scale and nature of Scotland's whisky production by calculating all the raw materials used to

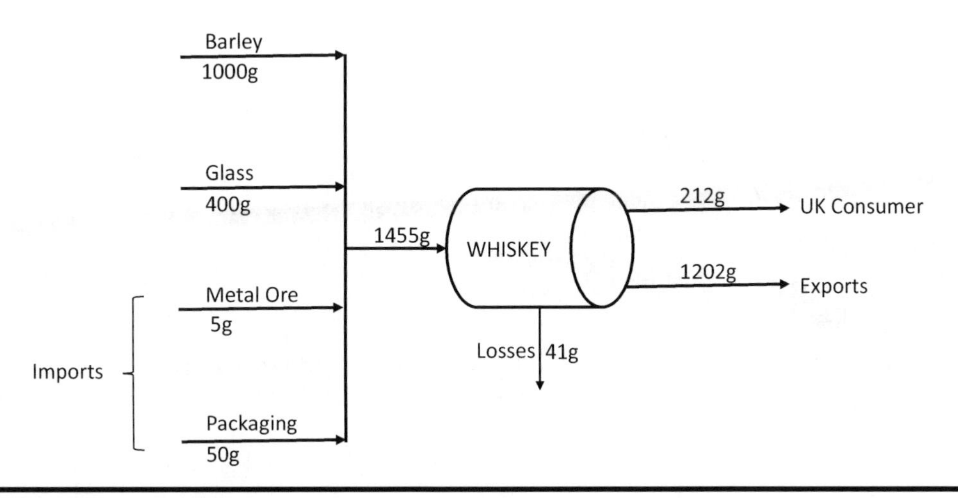

Figure 13.2 Material flow analysis diagram.

make whisky, local consumption and exports. The units are proportionate for material balance but not the total.

The example presented demonstrates how resources from Scotland are combined with a flow of imported materials to make a bottle of whisky. The materials used to produce the whisky have been measured against UK consumption and as exported material. The process also identified 2.8% wastes for further reduction.

This also enables users to explore Scotland's material flow accounts and compare Scotland's material consumption with additional indicators such as population, gross domestic product and carbon footprint data. Moving towards a circular economy is one of the solutions that will maximise value from the goods we already have in circulation while relieving pressure on finite natural materials.

13.2.1.4 Training Requirements

It is important to understand how to construct a material flow analysis diagram and recognize its subtle differences with material flow accounts. Workshops on MFA also assist in team building which is critical to collect and verify materials data from multiple sources. The training programmes for Black Belts and Green Belts should also include MFA.

13.2.2 S2. Carbon Footprint Tool

13.2.2.1 Definition

Carbon footprint is the amount of greenhouse gases released into the atmosphere as a result of the activities of a particular individual, organisation, service, product or community expressed as carbon dioxide equivalent.

There are three categories of scopes of carbon footprint as follows:

- Scope 1 covers the greenhouse gas emissions that are made directly, e.g. running a car.
- Scope 2 covers the emissions it makes indirectly, e.g. energy bought for heating.
- Scope 3 covers the emissions by suppliers to make the products along the supply chain.

13.2.2.2 Application

The application of carbon footprint was first promoted by BP Plc in an attempt to move public attention away from the activities of fossil fuel companies and onto individual responsibility. It has become a very useful tool for many applications including

- For an organisation to determine the degree of carbon offsetting by planting trees
- To determine the impact on climate change for a new strategy, especially a sourcing strategy
- To determine the carbon footprint per capita for a country
- To develop and validate climate change initiatives by governments, local governments and organisations
- To assess the impact of individual travels and lifestyles on climate change
- As a regulatory for larger companies to comply with streamlined energy and carbon reporting (SECR)

For calculating personal carbon footprints, several free online carbon footprint calculators exist. A high-level guide is shown in Appendix 1.

13.2.2.3 Basic Steps

1. Select the area for measuring carbon footprint (e.g whether a project, department or enterprise)
2. Define the boundary of calculation (e.g. which locations should be included in the inventory)

3. Develop likely greenhouse gas inventory (e.g. only CO_2 or also other greenhouse gases)
4. Determine consumption values for each component of emissions (e.g. quantify the energy consumption)
5. Calculate your carbon footprint (e.g. by using software or standard-compliant emission factors)
6. Develop a climate strategy and reduction targets (e.g. introduce key performance indicators [KPIs] for climate change)
7. Report your carbon footprint (e.g. prepare a report with a profile of carbon footprint, strategy and risk analysis)

13.2.2.4 Worked-Out Example

A large service company in the UK has been publishing their Carbon Footprint Report since 2008. The report is comprehensive containing a summary and also detailed emissions of different areas of activities. The following tables (Table 13.1, Table 13.2 and Table 13.3) show the performance summary and details of two areas.

13.2.2.5 Training Requirements

It is important to explain the purpose and method of calculating carbon footprints to all employees in an organisation to enhance the awareness and

Table 13.1 CO_2 Emissions: Performance Summary

	Year 2019	Change from 2018
Net emissions	20,252 tonnes CO_2e	+9.7%
Employees	3,290	+12.9%
Intensity per employee	6.16 tonnes CO_2e	–2.9%

Table 13.2 CO_2 Emissions: Car Travel

	Mileage	*Emission (Tonnes CO_2e)*	*Percentage of Total Emissions*
Commuting	12,177	2,735	13.5%
Company car	474	70	0.3%
Rental car	408	92	0.5%

Table 13.3 CO_2 Emissions: Wastes

Type of Wastes	*Treatment*	Volume (Tonnes)	Tonnes CO_2e
Organic	Compost	247	1.5
Paper	Recycled	246	5.2
Glass	Recycled	19	0.4
Plastic	Recycled	15	0.3
Mixed	Incinerated	165	3.5

urgency of climate change initiatives. The team with the responsibility of preparing and calculating greenhouse gases inventory will gain competence with experience.

The carbon footprint is a vital tool of the Sustainability stage of Green Six Sigma.

13.3 Summary

The Sustain phase is the substantive and distinctive feature of Green Six Sigma. The tools in this chapter (e.g. MFA diagram and carbon footprint tool) and techniques in Chapter 15 (e.g. Balanced Scorecard, sales and operations planning [S&OP], EFQM and knowledge management) are included to ensure the sustainability of the process, performance and environmental standards. The Control phase ensures the stability of the process, and the Sustain phase ensures the sustainability of the process meeting environmental standards.

GREEN TIPS

- Apply the MFA diagram and carbon footprint tool for each process.
- Apply Balanced Scorecard, S&OP, EFQM and knowledge management for the total Green Six Sigma programme.

TECHNIQUES FOR GREEN SIX SIGMA

Chapter 14

Quantitative Techniques

Product and service quality requires managerial, technological and statistical concepts throughout all major functions in an organization.

—Joseph M Juran

14.1 Introduction

This chapter addresses the advanced techniques of building process knowledge quantitatively, by analysing data, predicting outcome and measuring success. These techniques are based upon the quantitative and statistical analysis of data or results derived from other tools. There are a considerable number of quantitative techniques and we have selected the most frequently used techniques in Six Sigma and operational excellence programmes as follows:

Q1: Failure Mode and Effects Analysis (FMEA)
Q2: Statistical Process Control (SPC)
Q3: Quality Function Deployment (QFD)
Q4: Design of Experiments (DOE)
Q5: Design for Six Sigma (DFSS)
Q6: Monte Carlo Technique (MCT)
Q7: Inventive Problem-Solving (TRIZ)
Q8: Measurement System Analysis (MSA)

DOI: 10.4324/9781003268239-17

14.2 Selection of Techniques

Advanced techniques often generate a high expectation for results, but there is also a greater danger of using such a technique in a blinkered manner. For each technique, we have pinpointed its specific application areas as well as its benefits and pitfalls. Another reason for grouping these techniques together is that these are also so-called Six Sigma techniques. While selecting these techniques there are some common factors which should be considered with great care by the potential users of these techniques:

- There must be resources of appropriate skills and motivation in the organisation to gain expertise in these techniques.
- The fundamental rigours in purpose, measurement and use must be adhered to.
- The potential benefits, cost and difficulties in using the technique must be assessed.

14.3 Structure of Presentation

The following section covers the details of each technique under a common structure of presentation as follows:

- Background
- Definition
- Application
- Basic steps
- Worked-out examples
- Benefits
- Pitfalls
- Final thoughts

14.3.1 Q1: Failure Mode and Effects Analysis

14.3.1.1 Background

The technique of FMEA was developed by the aerospace industry in the 1960s as a method of reliability analysis. An early application was found at AlliedSignal Turbochargers. In 1972, the Ford Motor Company used FMEA to analyse engineering design and ever since, Ford has refined FMEA through continuous use including its application to Six Sigma. FMEA has proven to be a useful technique of risk analysis in defence industries.

14.3.1.2 Definition

FMEA is a systematic and analytical quality planning technique at the product, design, process and service stages assessing what potentially could go wrong and thereby aiding faulty diagnosis. The objective is to classify all possible failures according to their effect measured in terms of severity, occurrence and detection and then find solutions to eliminate or minimise them.

14.3.1.3 Application

There are five basic areas where FMEA can be applied. These are Concept, Design, Equipment, Process and Service.

Concept: FMEA can be used to analyse a product, system or its components in the conceptual stage of the design.

Design: FMEA is applied to analyse a product before the mass production of the product starts.

Equipment: FMEA can also be used to analyse equipment before it is procured.

Process: With regard to process, FMEA is applied to analyse the manufacturing, assembly and packaging processes.

Service: FMEA can also be applied to test industry processes for failure prior to their release to market.

14.3.1.4 Basic Steps

FMEA involves a 12-step process:

1. Form a team and flow chart the relevant details of the product, process or service that is selected for analysis.
2. Assign each component of the system a unique identifier.
3. List all the functions each component of the system performs.
4. Identify potential failure modes for each function listed in Step 3. (A failure mode is a short statement of how a function may fail to be performed.)

5. The next step describes the effects of each failure mode, especially the effects perceived by the user.
6. The causes of each failure mode are then examined and summarised.
7. Current controls to detect a potential failure mode are identified and assessed.
8. Determine the severity of the potential hazard of the failure to personnel or system in a scale of one to ten.
9. Estimate the relative likelihood of occurrence of each failure, ranging from highly unlikely (1) to most likely (10).
10. Estimate the ease with which the failure may be detected. A scale of 1 to 10 is used.
11. Determine a risk priority number (RPN) for each failure, which is the product of the numbers estimated in Steps 7, 8 and 9. The potential failure modes in descending order of RPN should be the focus of the improvement action to minimise the risk of failure.
12. The recommendations and corrective actions that have been put in place to eliminate or reduce failures are monitored for continuous improvement.

14.3.1.5 Worked-Out Example

A worked-out example of a process FMEA from the Ford Motor Company is shown in Figure 14.1. The procedure was used by the human resources department to analyse their 'internal job posting' process. It is interesting to note that the highest RPN was attributed to 'HR Input'.

14.3.1.6 Benefits

The major benefits of FMEA include the following:

- FMEA can be a powerful change agent for identifying weaknesses and risks in a product, process or service and suggesting methods for improvement.
- It is an effective analytical technique to capture the objective components derived from a group work or brainstorming.
- FMEA facilitates the relative weighting of a potential failure before the action is committed at a conceptual or early stage of an operation.

Failure Modes and Effects Analysis (FMEA)

Process or Product Name	Internal Job Posting	Prepared by:	Page .... of ....
Responsible		FMEA Date (Orig).................	(Rev) ....

Process Step/ Part Number	Potential Failure Mode	Potential Failure Effects	Seventy	Potential Causes	Occurrence	Current Controls	Detection	RPN	Actions Recommended	Responsi-bility
Finance sign-off	Does not happen or happens slowly	Stop process	8	Absence of controller, relations with controller, experience of manager; rules/ procedures/ bureaucracy	8	Department/ budget headcount limits	1	64		
Grading ratio	Prevent hiring at appropriate grade	Delay in commencing hiring process	8	Ratios out of date	7	Grading office/HR function	1	56		
Recruitment request	Delay in receiving request from manager	Ditto	8		1	Company policies	1	8		
Job description	Delayed	Delay in advertising	8	Lack of knowledge by manager, HR rules/procedures	4	HR function	1	32		
Advertising	Does not reach people	No/too few applicants, quality of applicants, diversity of applicants, re-advertise, delay to process	7	Company policies—HR/headcount	10	HR/headcount	1	70		
Application	None/too few, inconsistent information, lost, manager refusal to sign	Inability to shortlist/quality of shortlist, delay to process	7	Advertising process, headcount/backfill problems/poor paperwork/ understanding of process/restrictions on job	7		5	245		

Figure 14.1 An example of FMEA (adapted from the Ford Motor Company).

Pre-Screening	Screen already limited, number of applicants inconsistent	Reduce already low number of applicants	2	HR policies	4	HR policies	1	8		
Shortlist/ interview	Coordination, no show	Delay to process	5	Poor communication, availability	2	HR function	3	30		
Selection	No suitable candidates, candidate rejects offer	Delay to process, re-advertise	8	Advertising process, alternative offer	1	None		8		
Feedback	None or variable	Disengagement from process	5	Not priority for HR/ manager	3	None		15		
Re-advertise	Make same mistakes as first advert	Ditto	7	Ditto	10	Ditto	1	70		
HR input	Inflexible, inconsistent, inefficient	Delays process	7	Turnover, conflicting priorities, company policies	7	HR function/ company policies, not customer focused	9	441		
Releasability/ backfill	Delayed release	Delay in filling post	8	Lack of workforce/ succession planning, headcount restrictions, project deadlines/ departmental	4	20-day rule	1	32		

Figure 14.1 (Continued)

14.3.1.7 Pitfalls

FMEA is not an easy technique to handle and has many pitfalls including the following:

- Writing up the analysis is viewed as a tiresome task.
- The size and experience of the team often create difficulties and the meeting could become 'bogged down'.
- It can be viewed by technical staff as a catalogue of failures and merely a paperwork exercise to satisfy customer agreements.
- The determination of RPN from these factors is often received with doubt and uncertainty. It is viewed as a subjective assessment, as the criteria are not well understood.

14.3.1.8 Training Needs

The members of an FMEA team require 1-day classroom training followed by participation in at least one FMEA real-life exercise as observers. The initial reluctance by the manufacturing engineering function to take a leading role in the preparation of FMEA can be overcome by a 1-day 'hands-on' workshop.

14.3.1.9 Final Thoughts

FMEA is conceptually simple but detail rich. As the process involves steps of writing up modes, effects and causes of failures and assigning numbered ratings subjectively—albeit based on experience—FMEA has more sceptics than enthusiasts. Nonetheless, FMEA is effectively accomplished early in the design phase of a new product, process or even a project. Likewise, we do not recommend the use of FMEA when the design has reached a fixed state and changes will be much harder to effect.

14.3.2 Q2: Statistical Process Control

14.3.2.1 Background

The origin of SPC can be traced back to the work of Shewhart at Bell Laboratories in the 1920s. During the same period, the late 1920s, a British statistician named Dudding carried out work on statistical quality control along similar lines to that of Shewhart. During World War II, both American and British industries used SPC for the quality control of war materials.

Later in the 1980s, the Japanese led by Taguchi and stimulated by the teachings of Demurg effectively applied the SPC technique in quality programmes. The 1990s witnessed the resurgence of SPC following its successful application in Six Sigma.

14.3.2.2 Definition

In simple terms, SPC is the control or management of the process through the use of statistical methods and tools.

SPC is about control, capability and improvement and comprises some basic statistics (e.g. measures of control tendency and measures of dispersion), some tools for data collection (e.g. control charts) and analysis (e.g. process capability).

The seven 'basic tools' as shown are often included as SPC tools, although strictly speaking all of them are not in the SPC category:

1. Histogram
2. Pareto chart
3. Cause and effect diagram
4. Check sheet
5. Scatter diagram
6. Flow charts
7. Control charts

We described the seven 'basic tools' in previous chapters.

14.3.2.3 Application

SPC has four main areas of application:

1. To achieve process stability
2. To provide guidance on how the process may be improved by the reduction of variation
3. To assess the performance of a process
4. To provide information to assist with management decision-making

The application of SPC is potentially extensive, ranging from high-volume 'metal cutting' operations to non-manufacturing situations including services and commerce.

14.3.2.4 Basic Steps

The objective of SPC is to record the 'voice of the process', remove or reduce 'special' and 'common' causes of variation to the pursuit of continuous improvement. In this context the process means the whole combination of people, equipment, information, input materials, methods and environment that work together to produce output.

The basic steps of a SPC technique are

1. Collect data to a plan and plot the data on a graph such as a control chart or a line graph
2. Use the collected data to calculate control limits to determine whether the process is in a state of statistical control
3. Identify and rectify the 'special causes' of variation to stabilise the process
4. Assess the 'capability' of the process
5. Reduce, as much as possible, the 'common causes' of variation so that the output from the process is centred on a target value

This is an iterative process in pursuit of continuous improvement. A great deal of effort is required to stabilise the process to be in statistical control, and a great deal more to reduce the common causes of variation.

14.3.2.5 Glossary

Special causes of variation occur intermittently and reveal themselves as unusual patterns of variation on a control chart. These causes are assignable and are relatively easy to rectify. They include

- Incorrect material
- Change in machine setting
- Broken tool, die or component
- Keying in incorrect data
- Power failure

Common causes of variation arise from many sources, and they do not reveal themselves as obvious or unique patterns of variation; consequently, they are often difficult to identify. Some examples of common causes are

- Poor workmanship
- Poor workplace layout

- Poor quality of materials
- Bad condition of equipment
- Poor operating procedure

Variable data are the result of using some form of measuring instrument (e.g. scale, pressure gauge or thermometer).

Attribute data are the result of an assessment using go/no-go gauges or conforming/non-conforming criteria. An argument in favour of attribute data is that it is a less time-consuming task than that for variables and hence the sample size can be larger.

Arithmetic mean (χ) is determined by adding all the values and dividing the total by the number of values.

$$\chi = \frac{\Sigma\chi}{n}$$

where Σ means 'the sum of' and *n* is the number of values in the sample size.

Median is the middle value in a group of measurements when they are arranged in order (e.g. from the lowest to the highest).

Mode is the most frequently occurring value as part of a group of measurements.

Range (denoted by *R*) is the difference between the smallest and the largest values within the data being analysed.

$$R = \chi_{max} - \chi_{min}$$

Standard deviation (denoted by σ) is the measure which conveys by how much on average each value differs from the mean.

$$\sigma = \Sigma\sqrt{\frac{(x-x)^2}{n-1}}$$

Confidence level is the degree of confidence or certainty an event is likely to occur for a sample size. The different values of confidence level are given by the area under the curve of normal distribution. To define these values, two attributes are used as shown in Figure 14.2 (e.g. χ and σ).

The Central Limit Theorem states that if samples of a known size are drawn at random from a population then sample means will follow a normal distribution, the means of the parent and samples will tend to be the same

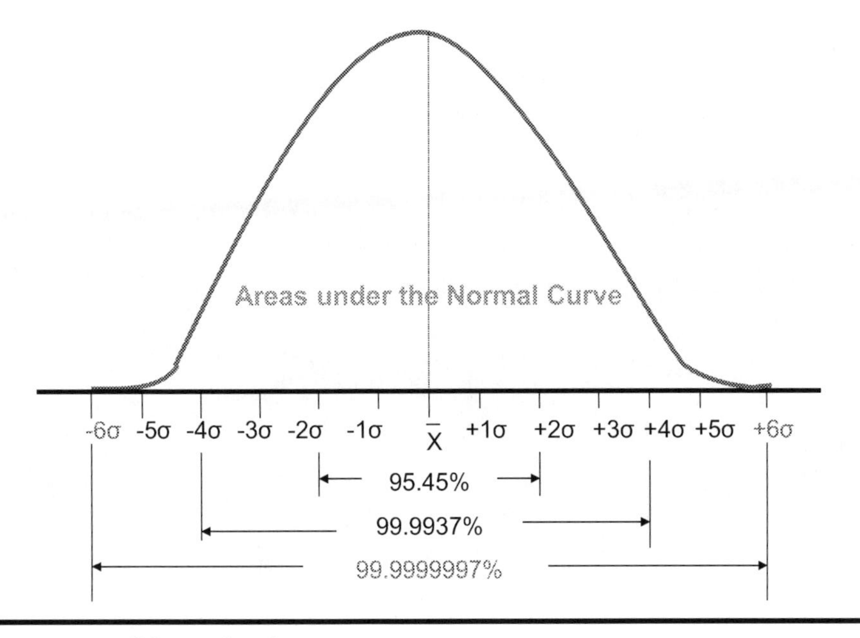

Figure 14.2 Confidence level.

and the standard deviation of the samples will be approximately that of the population divided by the square root of the sample size:

Population SD = Sample SD/SQRT (sample size)

Process capability refers to the ability of a process to produce a product that meets specification. For example, a highly capable process produces high volumes with few or no defects. The capability of a process is measured by one or a combination of four indices:

DPM = Defects per million

(e.g. for Six Sigma, level of process capability is 3.4 DPM)

σ level = Number of standard deviators between the centre of the process and the nearest specification

$$= \text{minimum}\left(\frac{(\text{USL} - \tilde{y}}{3\sigma,}, \frac{\tilde{y} - \text{LSL})}{3\sigma}\right)$$

where
USL = Upper specification level

LSL = Lower specification level
$\tilde{y}$ = Mean
σ = Standard deviation

C_{pk} = Process capability index
= Proposition of natural tolerances (3σ) between the centre of a process and the nearest specification

$$= \text{Minimum}\left(\frac{(\text{USL} - \tilde{y}}{3\sigma,}, \frac{\tilde{y} - \text{LSL})}{3\sigma}\right)$$

$$= \frac{\sigma \text{ level}}{3}$$

C_p = Process potential index
= Specification width
Process width

$$= \frac{\text{USL} - \text{LSL}}{6\sigma}$$

The C_{pk} index takes into account both accuracy and precision and is often defined as a 'process performance capability' index. For a Six Sigma capable process $C_p = 2.0$ and $C_{pk} = 1.5$.

14.3.2.6 Worked-Out Examples

14.3.2.6.1 Example 1

Determine mean, median, range, variance and standard deviation for the following ordered data set:

3,4,4,4,5,6,6,7,8,30
Mean χ = (3+4 + 4+4 + 5+6 + 6+7 + 8+30)/10 = 7.7
Median = Average of the two middle values 5 and 6 = 5.5
Range = 30 − 3 = 27

$$\text{Variance } S^2 = \frac{(3-7.7)2+(4-7.7)2+(30-7.7)2}{9} = 63.79$$

Standard deviation $s = \sqrt{63.79} = 7.99$

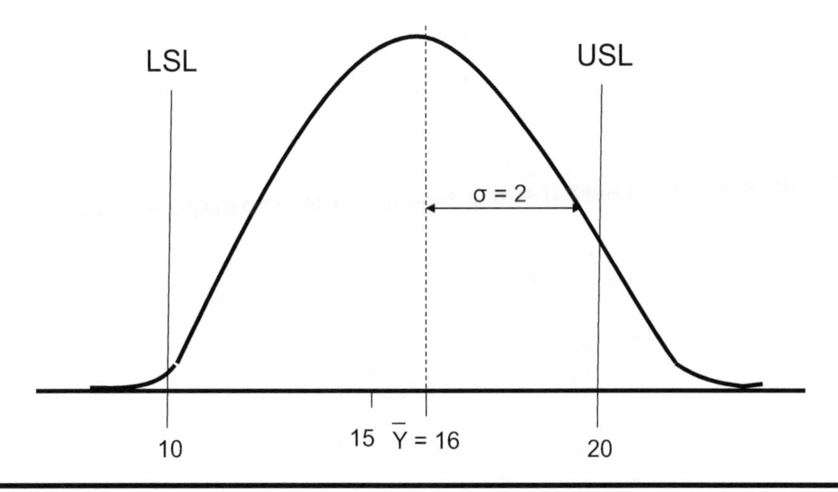

Figure 14.3 Process capability.

14.3.2.6.2 Example 2

Calculate C_{pk} and C_p from Figure 14.3.

Here $\sigma = 2$, $\tilde{y} = 16$, USL = 20, LSL = 10

$$C_{pk} = \text{Min}\left(\frac{(\text{USL} - \tilde{y}}{3\sigma,}, \frac{\tilde{y} - \text{LSL})}{3\sigma}\right)$$

$$= \text{Min}\left(\frac{20-16}{6}, \frac{16-10}{6}\right)$$

$$= \text{Min}\left(\frac{2}{3}, 1\right) = \frac{2}{3} = 0.67$$

$$C_p = \frac{\text{USL} - \text{LSL}}{6\sigma} = \frac{20-10}{6 \times 2} = 10/12 = 0.83$$

14.3.2.7 Benefits

The major benefits of SPC include

- The quantitative and statistical foundation of SPC provided the distinction between the Total Quality Management (TQM) approach (outside Japan) of the 1970s and the success of Six Sigma in the 1990s.

- SPC is an effective feedback system that links process outcomes with process outputs leading to continuous improvement.
- The fundamental principle of SPC (i.e. variation from common causes should be left to chance, but special causes of variation should be eliminated) is also a cornerstone of less quantitative approaches of continuous improvement.
- SPC can act as a focal point of a training and company-wide change programme.

14.3.2.8 Pitfalls

If it is not properly administered with top management support and a facilitator, SPC can be counterproductive. The major pitfalls include

- Poor understanding of the purpose of SPC within the company
- Often viewed as 'too much statistics' in shop floor environments
- Confusion between control and capability, or variation and standard deviation
- Confusion between seven basic continuous improvement tools and proper SPC tools

14.3.2.9 Training Needs

The basic understanding of SPC tools requires 1-day classroom training. However, a 3-day workshop is recommended for gaining proficiency in SPC technique. The Six Sigma Black Belt training programme includes an equivalent of 1 week's training in SPC tools and techniques.

14.3.2.10 Final Thoughts

SPC is a powerful technique, but it is basically a measurement process and can make a major contribution to operational excellence only when

- Management is committed
- A structured ongoing training programme is in place
- A mechanism is operational to eliminate 'special causes' and minimise 'common' causes of variation

14.3.3 Q3: Quality Function Deployment

14.3.3.1 Background

The QFD approach was developed by Mizuno and Akao and first applied by Mitsubishi Industries at the Kobe Shipyard. QFD describes a method of translating customer requirements (or the 'voice of the customer') into the functional design of a product or service. The technique has been used particularly by Japanese companies in the 1980s to achieve simultaneously a competitive advantage of quality, cost and delivery. The approach captured the attention of the West after the publication of *The House of Quality* by Hauser and Clausing (1988).

14.3.3.2 Definition

QFD is a technique that is used for converting the needs of customers and consumers into design requirements and follows the concept that the 'voice of the customer' drives all the company operations.

One specific approach of QFD, called the McCabe approach, proceeds by developing four related matrices in sequence:

- Product Planning Matrix
- Product Design Matrix
- Process Planning Matrix
- Production Planning Matrix

QFD is used to build in quality in the early stage of a new product development and helps to avoid downstream production and product delivery problems. A top-level view of QFD is shown in Figure 14.4.

In its simplest form, QFD involves a matrix in which customer requirements are rows and design requirements are columns. When the requirements matrix is expanded by the co-relation of columns, the result is called the 'house of quality'.

14.3.3.3 Application

In congruence with Figure 14.4, QFD is applied to deliver customer needs through four planning phases:

- Product planning
- Product design and development

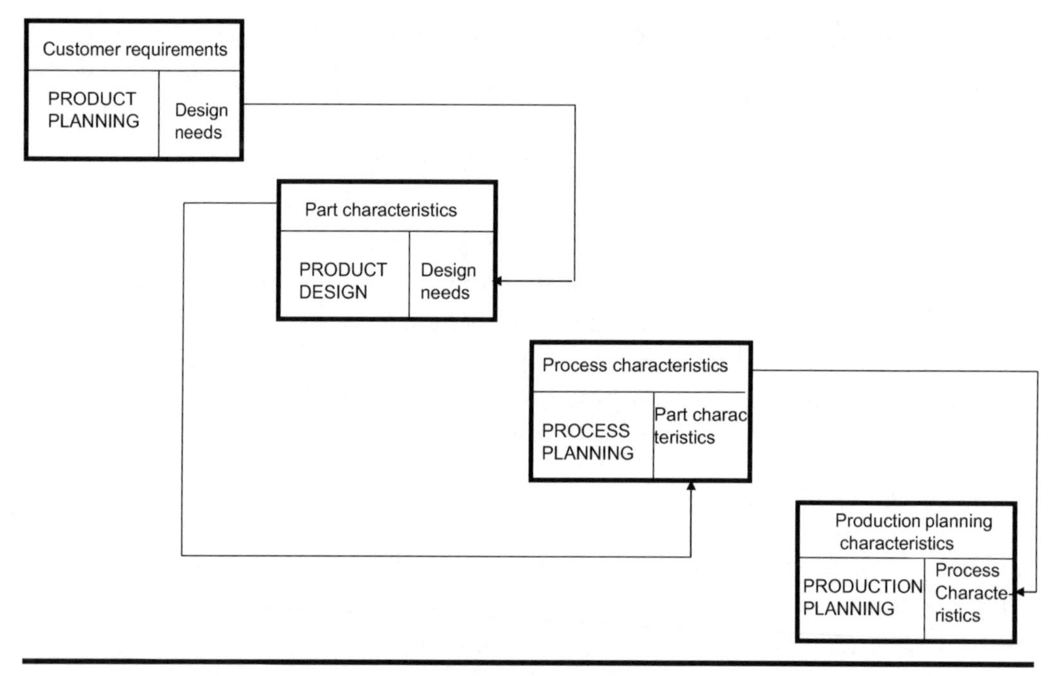

Figure 14.4 QFD approach.

- Process planning and development
- Production planning and delivery

During the product planning phase, the deliverables of QFD include customer requirements, competitive opportunities, design requirements and further study requirements.

The second stage involves product design which translates the design requirements from Stage 1 into component part design characteristics.

The third stage involves the selection of appropriate processes related to specific part characteristics. The deliverable of this stage will be a prioritised list of process characteristics which can be reproduced in production.

The purpose of the production planning stage is to ensure operations plan, training, maintenance and quality plans.

14.3.3.4 Basic Steps

The basic steps of how to develop a house of quality (see Figure 14.5) have been adapted from Mizuno and Akao (1994):

1. List customer requirements. The list of customer requirements includes the major elements of customer needs related to a specific product or

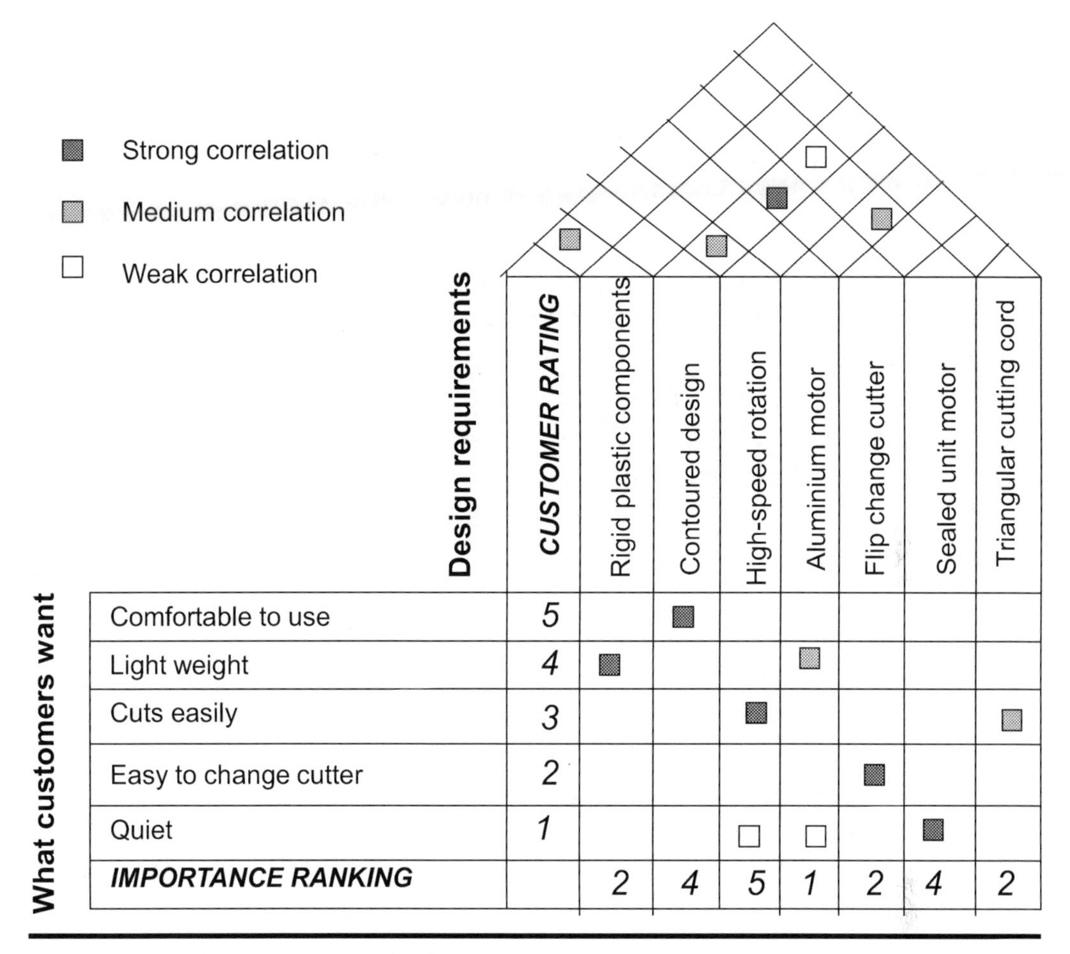

Figure 14.5 The QFD Matrix for an electric weed trimmer.

process. For example, customer requirements for a hotel room include a comfortable bed, a spacious room, a clean bathroom and a television.

2. List design elements. These design elements that relate to customer requirements are building materials and specifications. In the context of the hotel room, these are materials used for the mattress, lights in the bathroom, the size of the television and decoration of the room.
3. Demonstrate relationship between customer requirements and design elements. A diagram can be used to show these relationships. Furthermore, a score on a scale of 1 to 10 can be used; for example 9 means 'strongly related'.
4. Identify the co-relations between design elements. Positive or negative scores are assigned depending on whether the design elements are positively or negatively co-related. In the examples concerning hotels, tiles and television are negatively co-related.

5. Assess the competitiveness of customer requirements. This is an assessment of how your product (e.g. a hotel room) compares with those of your competitors. A five-point scale is used.
6. Prioritise customer requirements. 'Customer importance' is a subjective assessment, using a ten-point scale, of how critical a particular customer's requirement is. A focus group of customers usually assigns the rating. Target factors are set on a five-point scale (where 1 is for 'no change'). Sales point is established on a scale of 1 or 2 (where 2 is for 'high sales potential'). The absolute weight is then calculated by using the following formula:
   Absolute weight = Importance × Target value × Sales point
7. Prioritise technical requirements. Technical requirements are prioritised by assessing difficulty level, target value, absolute weight and relative weight. Difficulty level is assigned on a scale of 1 to 10, with 10 being 'most difficult'. Target value is set on a five-point scale as before. The value of absolute weight is the product of the relationship between customer and technical requirements and the 'importance of the customer' column. The value of the relative weight is the product of the column of relationships between customer requirements and technical requirements and the absolute weight of customer requirements.
8. Final assessment. The percentage weight factor for each of the absolute and relative weights is then computed. The technical elements with a higher share of relative weight are identified as key design requirements.

14.3.3.5 Worked-Out Example

Figure 14.5 shows a worked-out example taken from Waller (2002).

Here the customer requirements for an electric weed trimmer are

- Comfortable to use
- Light weight
- Cuts easily
- Easy to change cutter
- Quiet

The corresponding technical design elements are

- Rigid plastic components
- Contoured design
- High-speed rotation

- Aluminium motor
- Flip change cutter
- Sealed unit motor
- Triangular cutting cord

The strong correlations between the elements of customer requirements and design requirements are clearly indicated. For example there is a strong correlation between quiet operation (customer) and sealed unit motor (design). The roof of the 'house' shows high-speed rotation (design) has a strong correlation with sealed unit motor (design).

14.3.3.6 Benefits

The application of the QFD technique has achieved remarkable results, especially in Japan where it has been applied meticulously, to incorporate the 'voice of customers' into all stages of the design, manufacture, delivery and support of product and services. The key benefits of QFD include the following:

1. QFD seeks to identify the real needs of the customers and translates them into key design requirements.
2. This is a systematic and quantitative procedure which is used to build in quality into the upstream processes and new product development.
3. As the formation of a multifunctional team is a pre-requisite for QFD, this acts as a powerful catalyst to bridge the cultural gaps, especially between marketing and operations.
4. The use of QFD provides a structure for identifying those design characteristics for both products and services that contribute most or least to customer requirements.
5. QFD can be used to link the voice of the customer directly to internal processes. One of the most useful developments in this area is the 'policy deployment' as a measurement-based system for continuous quality improvement.

14.3.3.7 Pitfalls

Although QFD has been a highly successful technique in Japan, there are many difficulties which are experienced during the application of this practice. These pitfalls include the following:

1. It is often difficult to determine who the customer really is and identify their real needs, especially when in a new market the customer is not certain of their own requirements.

2. The development of matrices which are usually large can be a tedious and exhaustive process.
3. The condition that QFD cannot start until customer needs have been totally defined leads to a delay and loss of motivation.
4. Usually customer data are gathered by the marketing department, and engineers are eager to finalise the design specifications early. This leads to conflict and often delays the process.
5. Different chart formats are available, and there is often a lack of consensus to decide which format would suit the particular project.

14.3.3.8 Training Needs

The basic elements of QFD are not difficult to understand, but its application methodology varies according to the objectives of the project. The team members require attendance in a 1-day workshop. The team leader, unless initially supported by an expert or a consultant, must have hands-on experience of at least one QFD project under their belt.

14.3.3.9 Final Thoughts

When it is effectively used, QFD can bring customer focus to design and shorten the cycle time of the development of a product or services needing fewer changes. However, it is not a magic technique, and it requires considerable resources, understanding and attention to detail. It does not require any capital investment but the cost of training and at least six weeks of time should be taken into account to justify its application. When the application is repeated, the knowledge base is enhanced, cost decreases and benefit increases.

14.3.4 Q4: Design of Experiments

14.3.4.1 Background

If SPC can be termed as 'listening to the process', we can describe ideas implicit in the DOE as 'interrogating the process'. DOE has become the single most powerful technique in a Six Sigma programme. The origin of DOE dates back to the 1920s when R. Fisher applied complex statistical analysis in agricultural research. The work of Genichi Taguchi on experimental design in the 1970s is regarded as the basis of the current approach for DOE.

Taguchi views design from three perspectives: systems design, parameter design and tolerance design.

14.3.4.2 Definition

Design of experiments is a series of techniques that involves the identification and control of parameters or variables (termed 'factors') that have a potential impact on the output (termed 'response') of a process with the aim of optimising the design or the process. The experiment usually involves the selection of two or more values (termed 'levels') of these variables and then running the process at these levels. Each experimental run is termed as a 'trial'.

As a hypothetical example, consider that you wanted to bake a chocolate cake but you did not have a recipe to go with it. You need to know how much flour and sugar you require and how many eggs you should use. Furthermore, it would also call for information on the length of time and at what temperature you should bake your cake. A knowledgeable housewife or professional baker would probably resort to trial and error based on experience. However, the use of DOE will provide a systematic approach to finding out the best combination of variables to make your cake.

There are a number of methods of experimentation in DOE, of which the most commonly applied ones are

- Trial-and-error method
- Full factorial method
- Fractional factorial method

The trial-and-error method involves the step-by-step approach of changing one factor at a time, using the experience of the experimenter. This approach is easy to use and understand, but it is inefficient and time consuming.

The full factorial approach considers all combinations of the factors to find the best combination. For examples, three factors with two levels would need 2^3 or eight trials. Similarly, seven factors with two values will require 2^7 or 128 trials. This method is useful for a lower number of factors.

The fractional factorial method is applied when the number of variables or values is high. Typically for seven factors at two levels, the factorial method would need 32 trials which is a quarter of the full factorial method. This method changes several factors at the same time in a systematic way to ensure the reliability of results.

14.3.4.3 Application

DOE is an advanced technique which can be applied to both the design of a new product or process or to the redesign of the existing design or process. The technique is most effective for higher levels of variables and values.

The application areas include

- Product design and process design
- Minimum variation of a system performance
- Reduction of losses in a production line
- Achieving reproducibility of best system performance in manufacture

DOE has become an essential component of an advanced Six Sigma project and is particularly valuable in a DFSS project.

DOE helps to identify the effects of various types of factors as illustrated in Figure 14.6.

14.3.4.4 Basic Steps

The steps of a DOE would vary in detail depending on the methods of experimentation. The following basic steps have been simplified to describe the process of orthogonal arrays of the fractional factorial design:

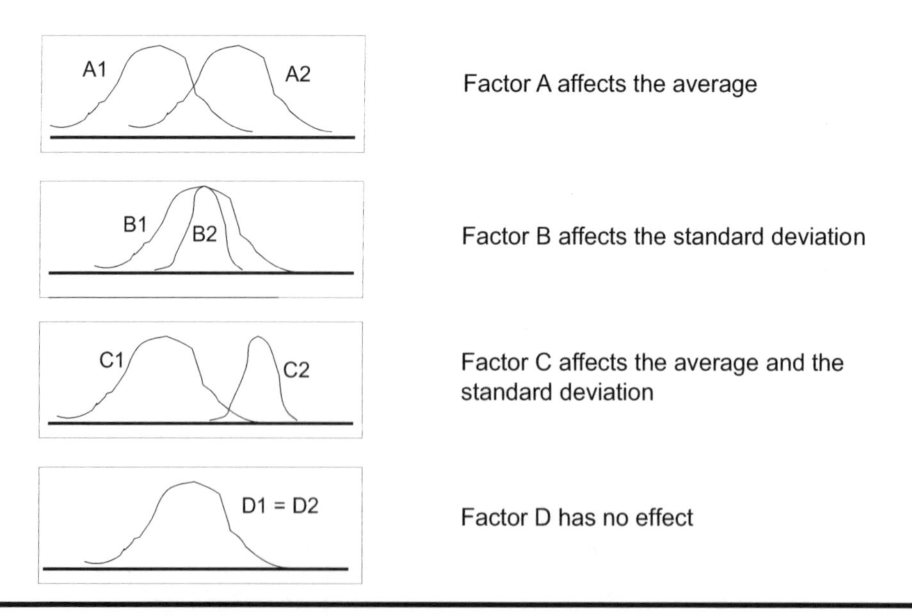

Figure 14.6 DOE identifies types of factors.

1. Define the project. The key information in the project definition report should include
   A background statement
   The purpose, scope and objectives
   Available statistical data
2. Develop an IPO diagram. An IPO (input-process-output) diagram is necessary to establish the critical factors and responses required in the experiment. This step should be undertaken by people who are knowledgeable about the process under investigation.
3. Select the factors to be optimised. It is useful at this stage to identify the factors to be optimised during the experiment.
4. Design the orthogonal array. The choice of the orthogonal array depends on member demands including the costs of the experiment, the number of factors and the number of levels to be studied. If there are more interactions to be studied, a larger design is required which is usually carried out by computer simulations.
5. Choose the levels of control factors and the sample size. The choice of levels is governed by the ease of measurement and the degree of difference between each level. The selection of the sample size and levels are also influenced by the economic consideration of the experiment. One way of reducing the sample size is to identify the highest and lowest level of the 'noise' and to control this.
6. Carry out the experiment. It is central that the experiment is carried out by a multidisciplined team with the rigour of project management principles. The accountability of key tasks such as data collection, leadership and interpretation of results should be clearly defined.
7. Analyse the data and confirm the results. The effective relative importance of each factor is determined by the use of variance. For more advanced experiments, Taguchi recommended 'signal-to-noise' ratios for situations requiring different input for a different output. As the orthogonal array is only a subset of the full factorial array, it is necessary to carry out a confirmation run.
8. Close the project. After the confirmation of the results, the project definition (Step 1) is reviewed to check if the objectives have been met. The methodology, results and conclusions are documented to signal the close-out of the project.

14.3.4.5 Worked-Out Example

This example is adapted from Dale (2000), and it concerns the part of the process used in the pharmaceutical industry in the manufacture of tablets.

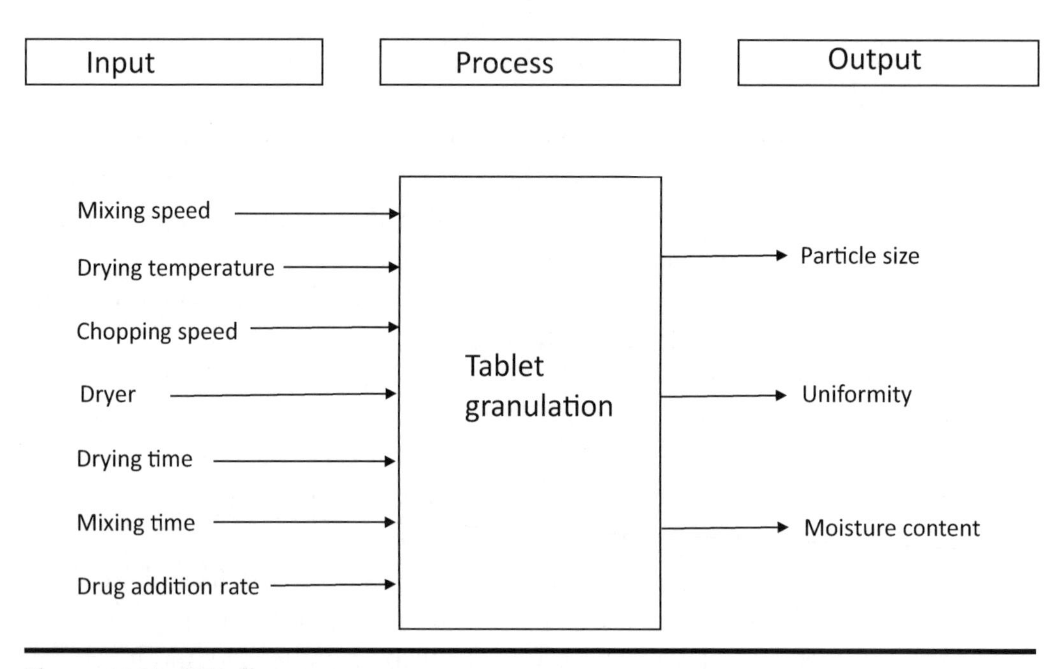

Figure 14.7 IPO diagram.

The objective is to produce uniform tablets in an optimum condition of manufacture. Figure 14.7 shows an IPO diagram of the process. A similar diagram was shown in Figure 8.1.

It was decided to run the first set of experiments to analyse the effect of seven input factors on the particle size on the response at two levels. Table 14.1 shows the experimental layout.

The results of experimental runs with a fractional factorial array (L_8) are shown in Table 14.2. The value of the particle size of each experiment is an average from an acceptable sample size.

The average of the experimental runs for the particle size is 3.96. The values of all responses for each level are calculated as shown in Table 14.3.

Comparisons have been made involving the relative difference between Level 1 and Level 2 of each factor as shown in Table 14.4. The comparative significance of each factor in affecting the response (i.e. Particle Size) is determined by the difference between levels.

From this analysis, it is evident that mixing speed (A), drug addition rate (G), the chopping speed (C) and mixing time (F) have the greatest effect in that order, while other factors have little significance on the response.

Table 14.1 Experimental Layout

	Control Factors	*Level 1*	*Level 2*
A	Mixing speed	High	Low
B	Drying temperature	High	Low
C	Chopping speed	Long	Short
D	Drier	Type A	Type B
E	Drying time	Long	Short
F	Mixing time	Long	Short
G	Drug addition rate	Fast	Slow

Table 14.2 Results of Experiments

Run	*A*	*B*	*C*	*D*	*E*	*F*	*G*	*Particle Size*
1	1	1	1	1	1	1	1	3.8
2	1	1	1	2	2	2	2	4.5
3	1	2	2	1	1	2	2	5.3
4	1	2	2	2	2	1	1	4.9
5	2	1	2	1	2	1	2	4.4
6	2	1	2	2	1	2	1	2.9
7	2	2	1	1	2	2	1	2.3
8	2	2	1	2	1	1	2	3.6

In this example, particle size is required to be as small as possible. The values of key factors (A, G, C and F) below the average (3.96) can now be used as the prediction of the result for a combination of factors that reflect their best effect on the output. In this case the optimal factor and levels are A_2, C_1, F_2 and G_1. The other factors can be set at the level where least cost is incurred. The prediction of the optimum combination is shown in Table 14.5.

As the orthogonal array is a subset of the full factorial array (e.g. 8 out of 128), a confirmation run is carried out to validate the predicted results.

14.3.4.6 Benefits

DOE has been recognised as the most important tool for gaining process knowledge. The traditional approach of cause and effect analysis has been

Table 14.3 Response Values

A_1	= 1/4 (3.8 + 4.5 + 5.3 + 4.9)	= 18.5/4	= 4.625
A_2	= 1/4 (4.4 + 2.9 + 2.3 + 3.6)	= 13.9/4	= 3.300
B_1	= 1/4 (3.8 + 4.5 + 4.4 + 2.9)	= 15.6/4	= 3.900
B_2	= 1/4 (5.3 + 4.9 + 2.3 + 3.6)	= 16.1/4	= 4.025
C_1	= 1/4 (3.8 + 4.5 + 2.3 + 3.6)	= 14.2/4	= 3.550
C_2	= 1/4 (5.3 + 4.9 + 4.4 + 2.9)	= 17.5/4	= 4.375
D_1	= 1/4 (3.8 + 5.3 + 4.4 + 2.3)	= 15.8/4	= 3.950
D_2	= 1/4 (4.5 + 4.9 + 2.9 + 3.6)	= 15.9/4	= 3.975
E_1	= 1/4 (3.8 + 5.3 + 2.9 + 3.6)	= 15.6/4	= 3.900
E_2	= 1/4 (4.5 + 4.9 + 4.4 + 2.3)	= 16.1/4	= 4.025
F_1	= 1/4 (3.8 + 4.9 + 4.4 + 3.6)	= 16.7/4	= 4.175
F_2	= 1/4 (4.5 + 5.3 + 2.9 + 2.3)	= 15.0/4	= 3.750
G_1	= 1/4 (3.8 + 4.9 + 2.9 + 2.3)	= 13.9/4	= 3.475
G_2	= 1/4 (4.5 + 5.3 + 4.4 + 3.6)	= 17.8/4	= 4.450

Table 14.4 Analysis of Factors

	A	*B*	*C*	*D*	*E*	*F*	*G*
Level 1	4.625	3.900	3.550	3.950	3.900	4.175	3.475
Level 2	3.300	4.025	4.375	3.975	4.025	3.750	4.450
Difference	1.325	0.125	0.825	0.025	0.125	0.425	0.975
Ranking	1	5	3	7	5	4	2

to hold all variables constant except one, but in practice it is not possible to hold all other variables constant. DOE has changed that by varying two or more variables simultaneously and obtaining multiple results under the same experimental conditions. In addition to this power of analysing practical conditions, the benefits of DOE include the following:

Table 14.5 Results of Optimum Combination

	Control Factors	*Level*
A2	Mixing speed	Low
C1	Chopping speed	High
F2	Mixing time	Short
G1	Drug addition rate	Fast
B2	Drying temperature	Low
E2	Drying time	Short
D1/D2*	Drier	Type A/Type B

* *Depending on lower operating cost*

1. A properly designed experiment enables you to use the same measurement to estimate several different effects.
2. It provides measurements of process performance and predictability.
3. It pinpoints the opportunities for improvement and indicates where to devote optimum results.
4. Experimental error is quantified, and a conformity run validates the conclusions.
5. It enables reproducibility of best systems performance in manufacture by minimising the variation.

14.3.4.7 Pitfalls

The main disadvantage of DOE is its apparent complexity which distracts potential users. In spite of its extensive deployment in Six Sigma projects, it is still viewed as an academic technique or guarded by 'Black Belt' experts. There are some practical pitfalls involved in DOE beyond this knowledge gap including the following:

1. The experiments are often run as an isolated intellectual exercise without a multidiscipline involvement or the rigour of project management.
2. For a practical problem with a large number of factors and corresponding responses, it is almost impossible to analyse results without an appropriate software (such as Minitab).

3. Although considerable process knowledge is gained by the experiment, savings are often difficult to quantify and implement.

14.3.4.8 Training Requirements

The training courses for DOE are usually covered over 1 week. During the course, the participants learn about statistical techniques for systematically manipulating many variables to discover the major factors affecting a selected result variable. The course is taught in a computer lab with each participant assigned to a desktop computer. The participants use statistical proprietary software such as Minitab, KISS, JMP or a customised Excel spreadsheet.

A DOE project is expected to be led by a trained Black Belt or Master Black Belt.

14.3.4.9 Final Thoughts

DOE is not a technique for an amateur enthusiast. If a DOE project is not properly run by a well-trained team, there is a danger of finding a solution which is 'exactly wrong rather than approximately right'. However, taking into account this cautionary note, DOE is a powerful technique to build quality into the upstream process by optimising critical characteristics. We recommend the use of DOE as an advanced technique for operational excellence invariably supported by a statistical software such as Minitab.

14.3.5 Q5: Design for Six Sigma

14.3.5.1 Background

The concept of DFSS was coined by Motorola who first applied it to the design and production of its pagers in the late 1980s. General Electric initially applied DFSS as a sequel to their Six Sigma programme to move the improvement process one step further, but the process actually turned out to constitute one step back—eliminating the flaws of the product and the process during the Design stage. Many other companies including Dow Chemical, Caterpillar and Seagate Technology have applied and developed the DFSS process further. In Six Sigma, DMAIC (define, measure, analyse, improve, control) has been accepted as a standard methodology. However, in DFSS there is little consistency among terms that define the process. The acronyms range from DMADV (define, measure, analyse, design, verify) to DMEDI (define, measure, explore, develop, implement) to IDDOV (identify, define, develop, optimise, verify) as well as some others.

One fundamental characteristic of DFSS is the verification which differentiates it from Six Sigma. The proponents of DFSS are promoting DFSS as a holistic approach of re-engineering rather than a technique to complement Six Sigma.

DFSS is also known as the application of Six Sigma techniques to the development process. Six Sigma is primarily a process improvement philosophy and methodology while DFSS is centred on designing new products and services. In practice, the difference between a formal DFSS and a Six Sigma programme can be indistinct, as a specific 'Black Belt' project may require DFSS to improve the capability (rather than performance alone) of an existing design.

14.3.5.2 Definition

DFSS is the system or process of designing and creating a component. This is done with the aim of meeting or exceeding all the needs of customers and the critical to quality (CTQ) output requirements when the product is first released. The goal of DFSS is 'right first time' so that there can be no manufacturing or service issues with the design after the initial release.

Most Six Sigma tools are applicable to DFSS, however the purpose and particularly the sequence of using these tools can be different from what might be expected within an ongoing operation. For example DOEs are often used in Six Sigma to solve problems in manufacturing, when in DFSS structured fractional factorial DOEs could be used as a control measure to achieve a high-quality product design. A large DFFS project requires the use of advanced techniques, such as QFD, FMEA and DOE.

14.3.5.3 Application

The primary application of DFSS as a technique is in the design and development stage of a product, process or service.

There are examples of the successful application of DFSS (Choudhary, 2003) in three environments:

1. Business transactions and services
2. Manufacturing processes and products
3. Engineering products

In the service industries, DFSS methodology bypasses the Measure and Analyse phases of DMAIC by creating a process that prevents the variations from emerging. This prevention methodology moves the five Sigma performances to yield Six Sigma results.

In manufacturing processes and products, DFSS encompasses the methodology of QFD and gets the designers and contractors working in concert to optimise the capability of the manufacturing process to attain a consistent quality for the product.

With regard to engineering products, in addition to applying the principles of concurrent engineering which entails the development of both products and processes, DFSS uses DOE for design optimisation.

14.3.5.4 Basic Steps

There are a number of process steps for putting DFSS into practice of which the most frequently reported methodologies are DMADV and IDOV (identify, design, optimise, validate). DMADV is often described as the next stage of DMAIC and thus may lead to a generic approach. In order to emphasise the distinctive characteristics of DFSS we adapted IDOV to show the basic steps of the process.

1. Identify Phase (Define and Measure): The Identify phase begins with the development of a team and a formal link of the design to the 'voice of the customer'. The fundamental rule is: do not start a DFSS project without the customer, management commitment and marketing involvement. The team members should have extensive experience in Six Sigma tools and techniques.

   The key tasks of this phase, also known as the Measure phase, are
   Identify customer and product requirements
   Identify CTQ variables and technical specifications
   Establish a business case
   Establish the roles and responsibilities of team members
   Plan the project and set milestones
   The most common and effective technique at this stage is QFD.
2. Design Phase (Analyse): The Design phase consists of developing alternative concepts, evaluating each option and selecting the best-fit concept. This is also known as the Analyse phase where design parameters (CTQs) are deployed. The key tasks of this phase include
   Formulate conceptual design
   Identify potential risks by using FMEA

Use DOE to determine CTQs and their influence on the technical requirements
Develop a bill of materials, formulation and procurement plans
Outline a manufacturing plan
DOE and FMEA are the most appropriate techniques for the Design phase.

3. Optimise Phase (Design): The Optimise phase uses the process capability information and simulation to develop and optimise detailed design elements. This is also called the Improve phase, and performance is predicted at this phase. The key tasks include
   Assess process capabilities of each of the design parameters to meet CTQ limits
   Use simulation to predict and design for robust performance
   Optimise tolerance and cost
   Implement design, commission and start-up
   The key tools used at this stage are SPC tools and simulation.
4. Validate Phase (Verify): The Validate phase comprises the testing and validation of the design. During this phase, also known as the Control phase, feedback of requirements and possible changes are shared with the design, manufacturing and procurement teams. The key tasks at this stage include
   Test prototype and validate
   Assess failure modes, performance and risks
   The FMEA technique is extensively used at this stage.

14.3.5.5 Worked-Out Example

The following flow diagram (Figure 14.8) is used as a worked-out example for DFSS. The diagram illustrates the main steps of developing a new consumer product by using DFSS.

14.3.5.6 Benefits

The proponents of DFSS and successful DFSS projects have established that DFSS helps fulfil the 'voice of the business' by fulfilling the 'voice of the customer'. The key benefits can be summarised as follows:

1. DFSS satisfied the voice of the business by:
   a. Increased sales volume by opening new markets with new designs and decreasing the cost of an existing design
   b. Decreasing development cost and capital investment

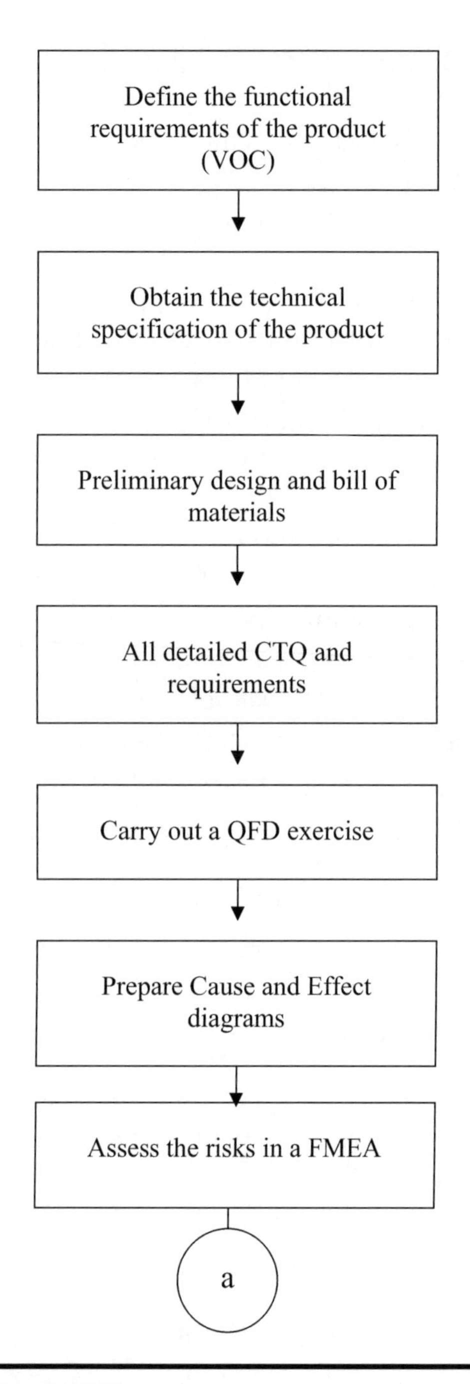

Figure 14.8 An example of DFSS.

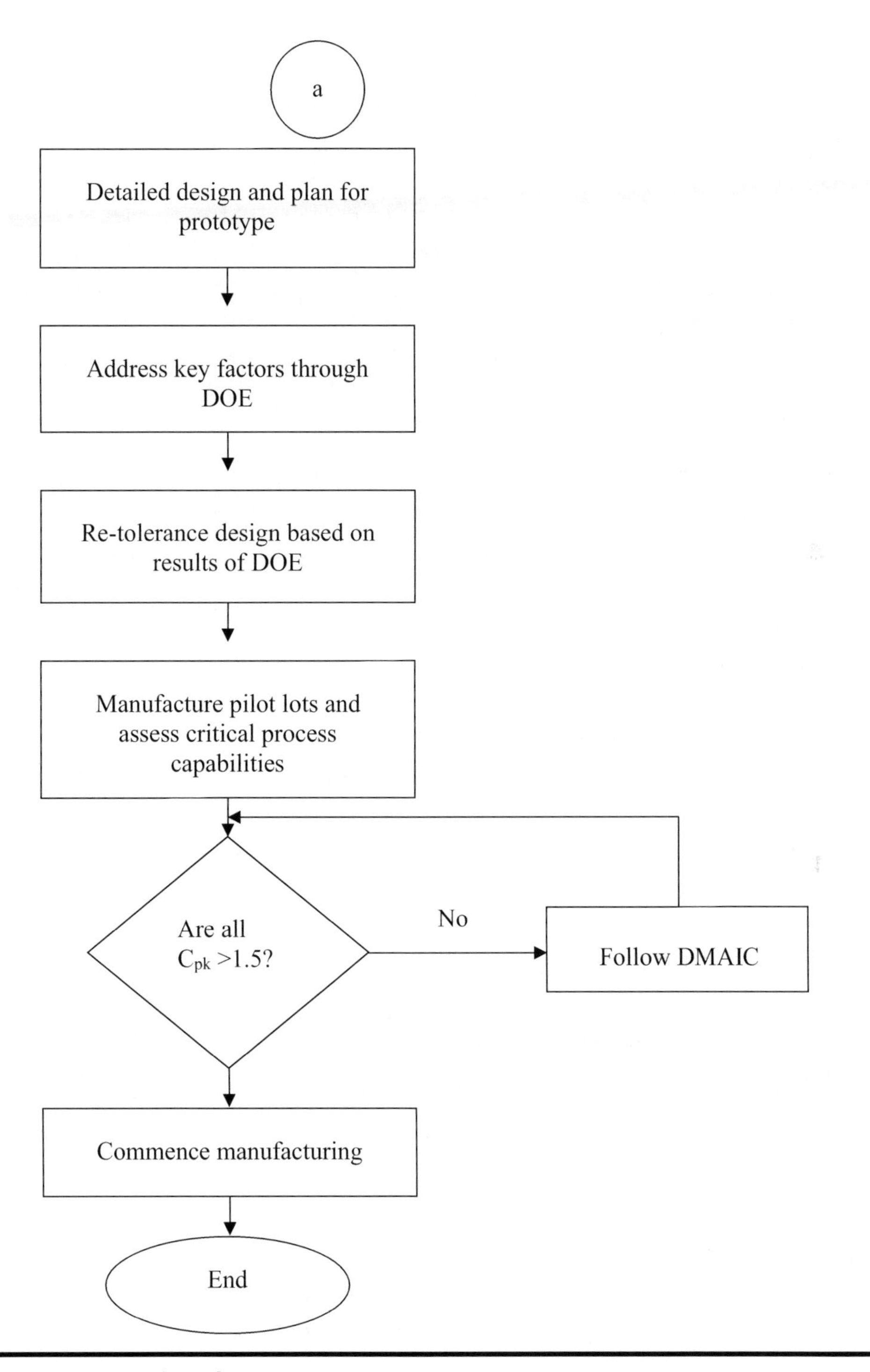

Figure 14.8 (Continued)

2. DFSS satisfies the voice of customers by:
    a. Improved design of existing products
    b. Generating value through new products
    c. Reducing the time to deliver new products
3. DFSS helps to improve organisation effectiveness by:
    a. Generating a discipline for product development excellence
    b. Creating a multifunctional synergy through active leadership to a common goal

14.3.5.7 Pitfalls

A major drawback of DFSS is the emphasis by its proponents on a 'pure-play' DFSS project. Thus it is often misunderstood by users and sponsors. Its links with other relevant techniques and methodologies, such as Six Sigma, DMAIC, QFD and concurrent engineering, are either confused or over-emphasised.

There are some practical pitfalls beyond the conceptual or philosophical issues such as follows:

1. The costs of the projects are more visible than the benefits for a long time along the project life cycle.
2. A DFSS project usually runs for a long period and the project close-out is not always well defined.
3. Based upon cost benefit alone, the justification of a DFSS is difficult, especially for a new product with an unpredictable forecast.
4. The success of DFSS depends heavily on the knowledge of the project team regarding advanced techniques such as QFD, DOE and FMEA.

14.3.5.8 Training Requirements

A DFSS programme requires that it is led by a trained Black Belt or Master Black Belt. The team members should also have a good training and hands-on experience in key Six Sigma tools and techniques, particularly in QFD, FMEA and DOE. If the trained team leader or members are not readily available, we recommend that a well-structured 1-week training programme is put into place before the start of a DFSS project.

14.3.5.9 Final Thoughts

DFSS is a longer-term, resource-hungry process, and it is expensive. Therefore, it should be deployed with care and on just a few vital projects, and specifically targeted towards the development of new products. Do not start a DFSS project without the customer, sales involvement, top management commitment and a team, preferably one with Six Sigma training. DFSS is a powerful technique, and its power should not be abused.

The proponents of DFSS believe that within the new few years as experience grows, DFSS will be used in design houses with the same familiarity as ISO standards. DFSS is primarily for new process or product development. Figure 14.9 illustrates the relationship between DFSS and DMAIC.

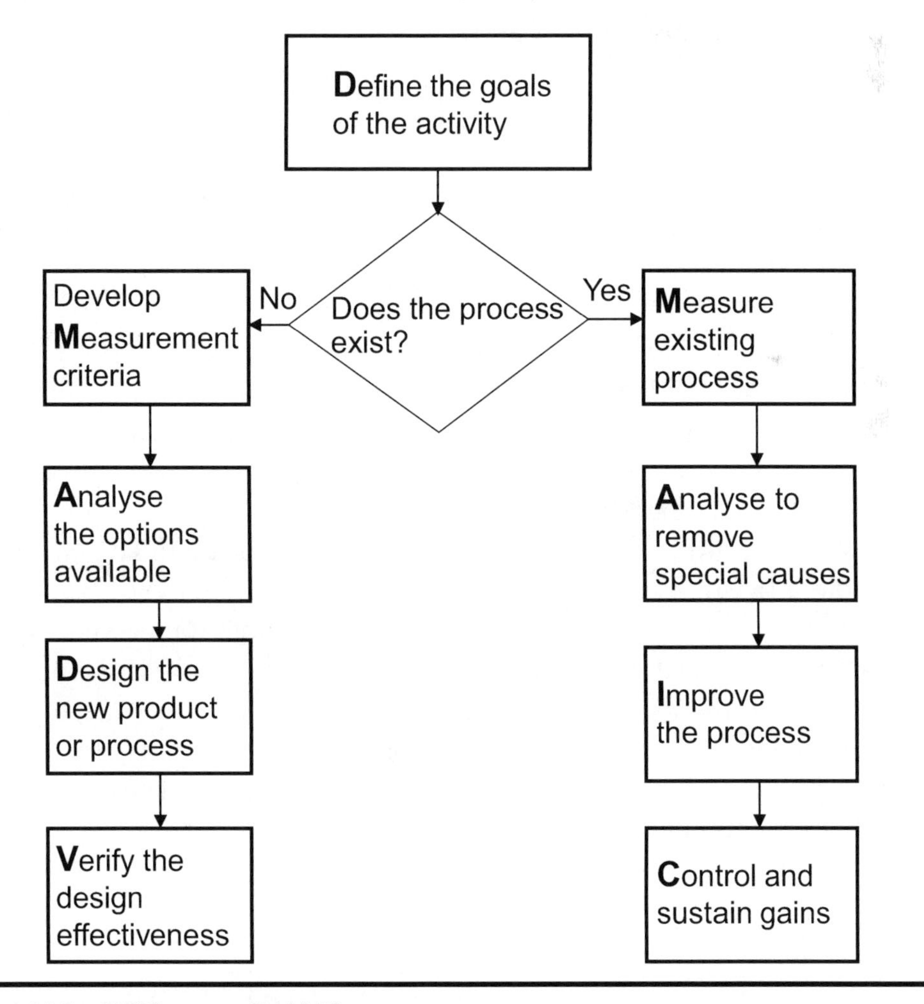

Figure 14.9 DFSS versus DMAIC.

14.3.6 Q6: Monte Carlo Technique

14.3.6.1 Background

The origin of the Monte Carlo technique goes back to a legendary mathematician observing the unpredictable behaviour of a saturated drunk. Each of the drunk's steps was supposed to have an equal probability of going in any direction. The mathematician wanted to estimate the average number of steps the drunk had to take to cover a specified distance. This was named the problem of 'random walk' and thus the application of random sampling was born, as the story goes. The method was found to have many practical applications, and subsequently it was named the Monte Carlo technique. The method requires a random number producing device, such as a roulette wheel, hence the name 'Monte Carlo'.

14.3.6.2 Definition

Churchman et al. (1968) define the Monte Carlo technique as 'simulating an experiment to determine some probabilistic theory of population of objects or events by the use of random sampling applied to the components of the objects or events'.

It often happens that an equation is so complex that it does not yield a quick solution by standard numerical methods. However, a stochastic process with distributions and parameters which satisfy the equation may exist. Instead of using the pure-play method it may be more efficient to construct the stochastic model of the problem and compute the solution. Thus an experiment is set up to mirror the features of the problem and a simulation is carried out by supplying random numbers into the system to obtain numbers from it as an answer. This is the basis of the Monte Carlo simulation process. The method does not mean a 'gamble', as the name may imply, but rather it refers to the manner in which individual numbers are selected from valid distribution functions. As shown in Figure 14.10, the output from a Monte Carlo simulation includes a histogram showing the relative probability of events and a cumulative probability chart ('S' curve).

14.3.6.3 Application

The Monte Carlo technique has been successfully used in scientific applications for over eight decades. The method has been a cornerstone of all

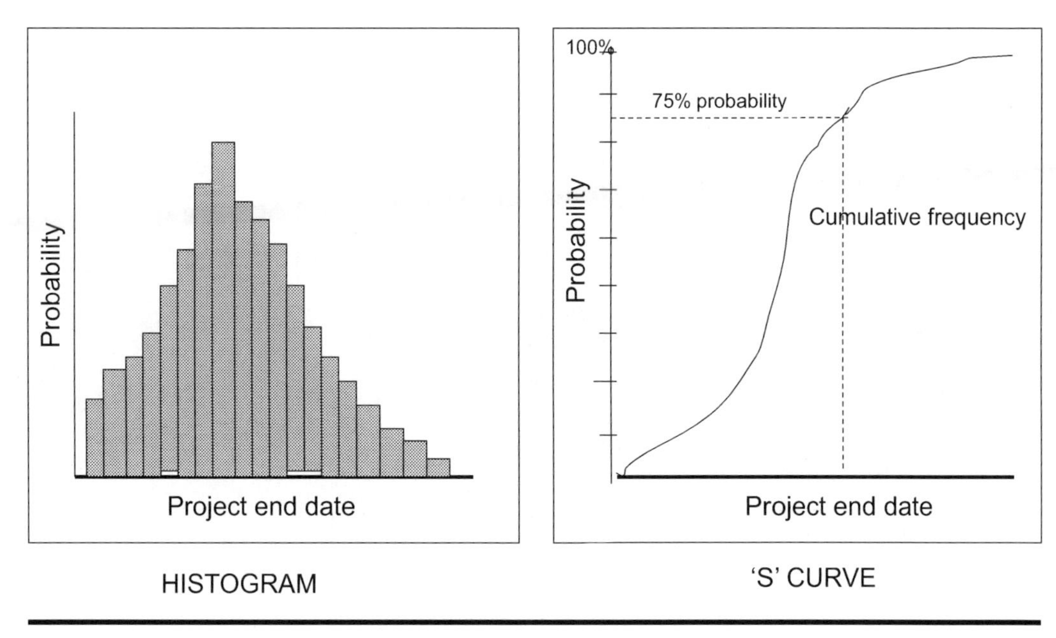

Figure 14.10 Output from Monte Carlo simulation.

NASA projects. It has been used extensively in investment appraisal and risk analyses in both scientific and industrial programmes.

There is an abundance of research publications containing the application of the Monte Carlo technique, particularly in solving queuing problems. It is evident that this technique can be used to solve any queuing problem for which the required data can be collected. Saaty (1959) demonstrates the successful application of the Monte Carlo technique in the optimisation of telephone calls, landing of aircraft, loading and unloading of ships, the scheduling of patients in clinics, customers and taxis at a stand and flow in production—to name but a few. It is important to realise that all these applications were accomplished before the advent of the computer.

Although the Monte Carlo technique is still a popular topic of academic and scientific research, with the availability of proven simulation software (e.g. PRM and Primavera Monte Carlo), the system is now in wider use in businesses and project management.

The applications of the Monte Carlo technique in project risk management are focused on two stages of the project life cycles, viz. feasibility and implementation. At the feasibility stage the simulation is the perfect tool for evaluating a pro forma of investment opportunities. We know that at the initial stage of a project there is uncertainty regarding input since pro forma

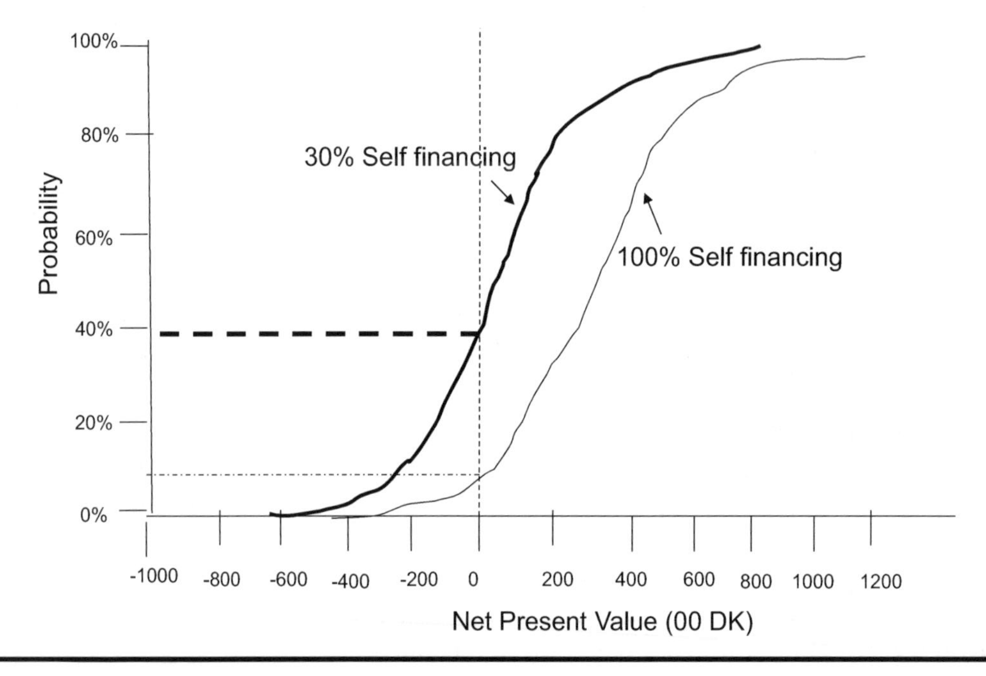

Figure 14.11 Monte Carlo in project appraisal.

is by definition an estimation of performance. We also know that the Monte Carlo technique will calculate answers that accurately reflect the uncertainty of input data. Figure 14.11 illustrates the probability curves, using Monte Carlo simulations, of the net present value for two options of a wind energy project in Denmark.

During the implementation stage of a project, the Monte Carlo technique examines the overall uncertainty of project completion times.

The technique is also very useful in assessing the probability and consequence of critical risk in a risk register during the implementation of a project. For example, having completed the preliminary assessment of probability/impact scores on the risk register for the project cost estimate and schedules, the data for high-score risk items are loaded in the Monte Carlo risk analysis software. This provides a quantitative evaluation of cost and schedule for high-risk items.

14.3.6.4 Basic Steps

1. Observe some basic rules on the use of this technique:
    a. The scope of each problem must be identified by studying each process individually to focus on the core issue and the aspects that are ignored.

b. The strength of the technique lies in its applicability to an equation which demands numerical solution but is not easy to solve by standard numerical methods.
c. If in a problem there are fluctuations due to time, apply the method to the parts which appear to cluster together in time.
d. The simulation should be applied several times to a problem with different sample sizes.
e. The technique follows a normalised distribution function. That means that we multiply the frequencies of the distribution functions by a constant such that the area under the normalised distribution function is equal to 1.

2. In practice, a Monte Carlo simulation is carried out by computer software. The input data requirements for each variable are the estimates for
   a. Minimum value
   b. Most likely value
   c. Maximum value

   By using random numbers and sampling techniques, the computer system will generate a probability distribution of each parameter. From the cumulative probability charts ('S' curves), the probability level is decided and the corresponding value of the parameter is selected.
3. The iterative process of the Monte Carlo technique is described as follows. (This can be done manually, although the use of a computer system is more sensible.)
   a. Given the value of Xi lies between a range of X_{min} and X_{max}.
   b. Start the ith iteration. Use a random number between 0 and 1 and the normalised distribution function of Xi to determine the single value of xi to be used in this iteration.
   c. Place the value of Xi in the appropriate 'bin' of a histogram. This histogram will become a frequency distribution of xi when all iterations are complete.
   d. Go to Step (b) and start the process over until the frequency distribution in Step (c) is complete for the range X_{min} and X_{max}.
   e. Plot the cumulative probability chart or 'S' curve for Xi (see Figure 14.10) and decide the probability level and corresponding value of Xi.

14.3.6.5 Worked-Out Example

This example is adapted from Churchman et al. (1968) and it concerns the home delivery of packages of goods purchased at a department store.

Consider the packages arriving for delivery from the store:

- Normal distribution
- Mean arrival rate: 1,000 packages per day
- Standard deviation 100

Similarly, assume the service pattern of delivery by trucks:

- Normal distribution
- Mean service rate 100 packages/day
- Standard deviation 10

Let us also assume, for the purposes of this example,

- Cost per truck: $25 per day
- Cost of overtime: $8 per hour

In practice, there could be a cost of delay, but in this exercise we assume that packages will be delivered on the same day and overtime may be required.

We run a Monte Carlo method of the delivery system 'on paper' with two fleets (10 and 12 trucks each) for five consecutive days. In a computerised simulation we could run the system for any length of time and for as many fleet sizes as desired. We begin by preparing a table as in Table 14.6.

In Table 14.6, Columns 1 and 2 are self-explanatory.

Column 3 refers to five successive numbers from the table of random normal numbers (see Appendix 2).

Column 4 shows the converted value of the number of packages arriving by taking into account the standard deviation and random normal number. This is equal to 1000 + 100 × Column 3.

Column 5 is the total requirement = Column 4 + Previous Column 8.

Column 6 is the next set of five consecutive numbers from the random normal number table. This represents the average number of deliveries per truck.

Column 7 is the converted value of total deliveries = Column 1 × (100 + 10 × Column 6).

Column 8 shows the leftover packages if no overtime is deployed. This is equal to Column 5 – Column 7.

Column 9 refers to the number of packages to be delivered at overtime rate. This is equal to Column 4 – Column 7.

Table 14.6 Manual Simulation of a Delivery System

1 Trucks in Fleet	2 Day	3 Random Table Value 1	4 Packages Arrived	5 Total Require-ment	6 Table Value 2	7 Number of Deli-veries	8 Leftover Packages	9 Number Delivered at Overtime	10 Cost of Overtime ($)
	1	2.455	1246	1246	–0.323	968	278	278	184
	2	–0.531	947	1225	–1.940	806	419	141	112
10	3	–0.634	937	1336	0.697	1070	286	0	
	4	1.279	1128	1414	3.521	1352	62	0	
	5	0.046	1005	1067	0.321	1032	35	0	
Total									296
	1	2.455	1246	1246	–0.323	1161	85	85	56
	2	–0.531	947	1032	–1.940	967	65	0	
12	3	–0.634	937	1002	0.697	1284			
	4	1.279	1128	1128	3.521	1623			
	5	0.046	1005	1005	0.321	1239			
Total									56

Column 10 is the cost of overtime. This is calculated as
[Column 9/Column 7] × $8.00
8 × Column 1

The total costs per week for each fleet can now be compared. Assuming a 5-day week, costs per week are

10 trucks: 10 × 5 × 25 + 296 = $1,546
12 trucks: 12 × 5 × 125 + 56 – $1,556

In this case the ten-truck fleet is more economical.

14.3.6.6 Benefits

1. The Monte Carlo technique is one of the most used simulation methods in both academic research and industrial applications. A major difficulty encountered in operations research is that of dealing with a situation so complex that it is impossible to set up an analytical equation. The Monte Carlo technique aims at simulating the operation systematically where the major factors and their interaction are studied.
2. This technique calculates the answers that accurately represent the input data by a large number of iterations. Experience over the last 80 years has proven that the solutions are also very close to real circumstances.
3. With the support of effective computer software, this technique has become an essential part of risk assessment in various stages of major projects.

14.3.6.7 Pitfalls

The Monte Carlo technique is not free from pitfalls. These include the following:

1. The methodology and algorithm of the Monte Carlo technique are perceived as very complex for practical managers. The theory of probabilities and the characteristics of random numbers are considered to be the domain of academics and statisticians.
2. Without the use of computer software, the iterations by a manual process can be laborious and prone to error.

3. Many practitioners tend to shy away from this technique with a notion that 'it is better to be approximately right than exactly wrong'.

14.3.6.8 Training Needs

The application of the Monte Carlo technique requires a good understanding of statistics, and how the technique actually works. It is also necessary for the user to gain hands-on experience in the use of the computer software. We strongly recommend that the modelling and simulation of a practical problem by the Monte Carlo technique should be given to someone who is trained and experienced in simulation techniques.

The project team members should receive an awareness training so that they are happy with the interpretation of the results obtained by the technique.

14.3.6.9 Final Thoughts

The Monte Carlo technique is an excellent aid to risk management, but it should be used with care, ideally by trained specialists and always with the support of an effective computer software.

14.3.7 Q7: TRIZ: Inventive Problem-Solving

14.3.7.1 Definition

TRIZ is the Russian acronym for Teoria Resheniya Izobreatatelskikh Zatatch (inventive or innovative problem-solving). It extends traditional systems engineering approaches and provides a powerful systematic method for problem formulation, systems and failure analysis.

There are 39 characteristics and 40 principles of inventive problem-solving. The contradictions generated by 39 characteristics are systematically eliminated by 40 principles to lead to the development of new innovations.

14.3.7.2 Application

There are two groups of problems people face: those with generally known solutions and those with unknown solutions. Those with known solutions can usually be found in books, technical journals or with subject matter experts. The other type of problem is one with no known solutions. It

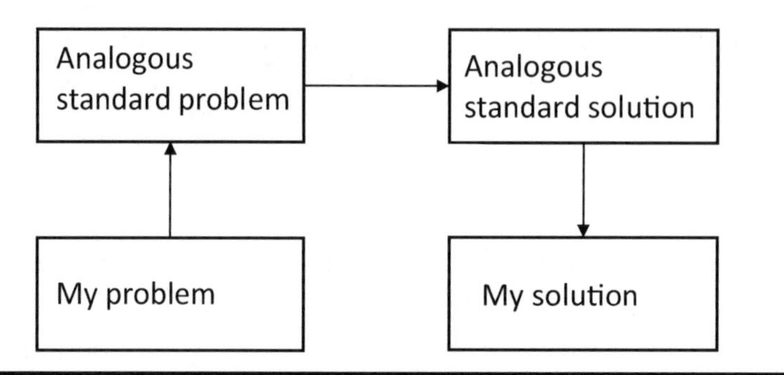

Figure 14.12 TRIZ general problem-solving model.

is called inventive problem and may contain contradictory requirements. Methods such as brainstorming and trial and error are commonly suggested. An alternative approach was developed by Genrich S. Altshuller (1994) born in the former Soviet Union in 1926, and this approach known as TRIZ should satisfy the following conditions:

1. Be a systematic, step-by-step procedure
2. Be a guide through a broad solution space to direct to the ideal solution
3. Be repeatable and not dependent on psychological tools
4. Be able to access the body of inventive knowledge
5. Be able to add to the body of inventive knowledge
6. Be familiar enough to inventors by following the general approach to problem-solving in Figure 14.12

Altshuller more clearly defined an inventive problem as one in which the solution causes another problem to appear (such as increasing the strength of a metal plate causing its weight to get heavier) and categorised the solutions into five levels:

- Level one: Routine design problems solved by methods well known within the speciality. No invention needed. About 32% of solutions fell into this level.
- Level two: More improvements to an existing system by methods known within the industry, usually with some compromise. About 45% of the solutions fell into this category.
- Level three: Fundamental improvement to an existing system by methods known in the industry. Contradictions are resolved. About 18% of the solutions fell into this category.

- Level four: A new generation that uses a new principle to perform the primary functions of the system. Solution found more in science than in technology. About 4% of the solutions fall into this category.
- Level five: A rare scientific discovery or pioneering invention of essentially a new system. About 1% of the solutions fell into this category.

TRIZ has been used in Russia to solve technical problems and develop thousands of patentable solutions. It is now gaining acceptance in major companies in the Western world and the Pacific Rim and also in Six Sigma problems. Samsung and Intel Corporation have recently reported successful application of TRIZ in innovation and projects.

14.3.7.3 Basic Steps

There are four basic steps of TRIZ as shown in Figure 14.12. These steps which follow a theme of increasing identity are as follows:

Step 1. Identifying my problem: Identify the engineering system being studied, its operating environment, resource requirement, primary useful function, harmful effects and ideal results.

Step 2. Formulate the problem: Restate the problem in terms of physical contradictions and identify problems that could occur. For example, identify it by improving one technical characteristic to solve a problem cause. Other technical characteristics may worsen resulting in secondary problems arising.

Step 3. Search for a previously well-solved problem. Find the contradicting engineering characteristics (from 39 TRIZ technical characteristics). First find the characteristic that needs to be changed. Then find the characteristic that is an undesirable secondary effect. State the standard technical conflict.

Step 4. Look for analogous solutions and adapt to my solution. Altshuller also extracted from the worldwide patents 40 inventive principles and the Table of Contradictions. These are hints that will help an engineer find a highly inventive (and patentable) solution to the problem. The Table of Contradictions lists the 39 Engineering Parameters on the x-axis (undesired secondary effect) and y-axis (feature to improve). In the intersecting cells are listed the appropriate Inventive Principle to use for a solution.

14.3.7.4 Worked-Out Example

A well-known example of TRIZ is the redesign of a metal beverage can. The first step is to identify the problem goal called the initial final result (IFR). The IFRs of our example include the design of a cylindrical metal container to hold beverage of a given volume that can support the weight of stacking to human height without causing damage to cans or the contents held by the container.

The standard technical conflicts are as follows. If we make the containers thinner more stress will be felt by the container walls. On the other hand, if we make the walls of the containers thicker, the containers would be heavier and the stacking of heavy containers could damage the bottom layer. Furthermore, heavy containers would be more expensive. This conflict of increasing strength and increasing weight could be reconciled by Segmentation, one of the 40 inventive principles. The answer is to change the smooth cylindrical surface of the container to a wavy surface made of many little walls. The new design will increase the edge strength of the cylinder while retaining its lighter weight.

14.3.7.5 Training Requirements

TRIZ is a highly specialised problem-solving tool. Although it is conceptually simple comprising four basic steps, it is very rich in details. Many consulting companies are guarding the 39 technical characteristics and 40 inventive principles along with the complex Contradiction Matrix to promote training workshops. The training workshops are usually covered in 3 to 5 days. During the course the participants learn about problem identification and formulation and search for solutions by applying the Contradiction Matrix.

14.3.7.6 Final Thoughts

TRIZ is conceptually a very powerful problem solving technique leading to new ideas and innovative solutions. The bottom-line question is 'if TRIZ is so good, why isn't everyone using it?' The answer is that because of its prescriptive approach with the baggage of 39 characteristics and 40 principles, it is difficult to sell and it also competes with techniques such as DFSS and QFD. Altshuller himself realized that the contradiction matrix was comparatively an inefficient tool and stopped working on it since 1973 and focused on the algorithmic approach of ARIZ with 76 inventive principles.

14.3.8 Q8: Measurement System Analysis

14.3.8.1 Background

Due to the problems caused in the calculation of measurement uncertainty, the International Bureau of Weights and Measures decided in 1977 to develop an international agreement on measurement uncertainty. This guideline was published in 1995 and is still known as the "Guide to the expression of uncertainty in measurement" (GUM).

Measurement is key and essential in Six Sigma. As a result the technique of measurement system analysis (MSA) was developed as an experimental and mathematical method of determining how much the variation within the measurement process contributes to overall process variability.

14.3.8.2 Definition

MSA is a technique where the same object is repeatedly measured by using different people or items of measurement equipment to quantify the amount of variation that comes from the measurement system itself. This is different from product or process variation.

14.3.8.3 Application

MSA is most commonly used in the Measure stage of the DMAIC process to test and improve the measurement systems before actual data collection.

The total observed variation has two components:

Total variation = Process variation + Measurement variation

MSA aims to qualify a measurement system for use by quantifying its accuracy, precision, and stability. The components of MSA are shown in Figure 14.13.

MSA assesses a measurement system for some or all of the following five characteristics:

Bias or accuracy: It is attained when the measured value has little deviation from the actual value. Bias is the difference between observed average measurement to the true or reference value.

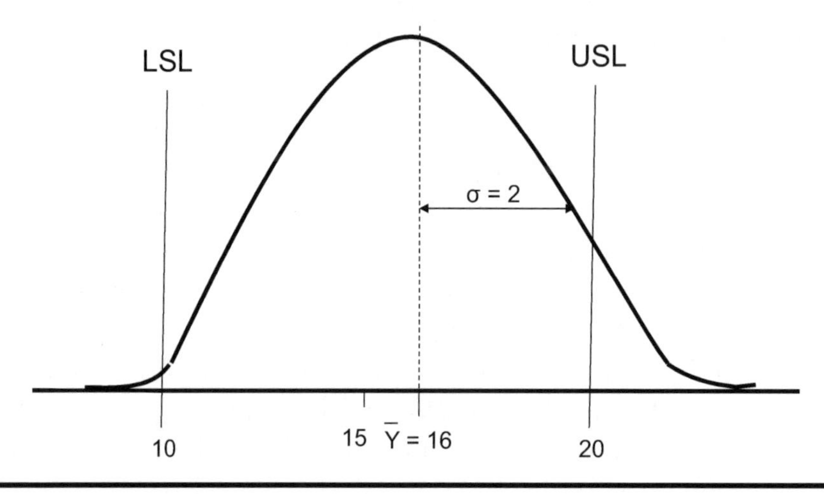

Figure 14.13 The components of MSA.

Linearity: Linearity is the difference in Bias value over the normal operating range of the measuring instrument or gauge.

Stability: Stability refers to the capacity of the measurement system to produce the same values over time when measuring the same sample.

Repeatability: Repeatability is the variation between successive measurements of the same part, same characteristic, by the same person using the same measuring tool or gauge.

Reproducibility: It is attained when other people or other instruments or locations get the same result.

The above five characteristics of MSA are illustrated in Figure 14.14.

Two important metrics used to assess the effectiveness of MSA are Fixed Effect ANOVA and Gauge R&R. Statistical software (e.g. Minitab and SPSS) are used to measure them from given data.

Fixed Effect ANOVA is used to compare independent groups of data and to determine what portion of the variation is due to operation differences and what portion is due to the measurement process.

Gauge R&R (Gauge Repeatability and Reproducibility). Calculations of the variance are made of the overall variation of the process plus the variation due to the repeatability and reproducibility. The results are then presented as a ratio of the repeatability and reproducibility and the overall variation.

Generally recognized criteria for gauge acceptability is when Gauge R&R is

- Under 10%: Acceptable gauge
- 10% to 30%: May be acceptable
- Over 30%: Gauge is unacceptable

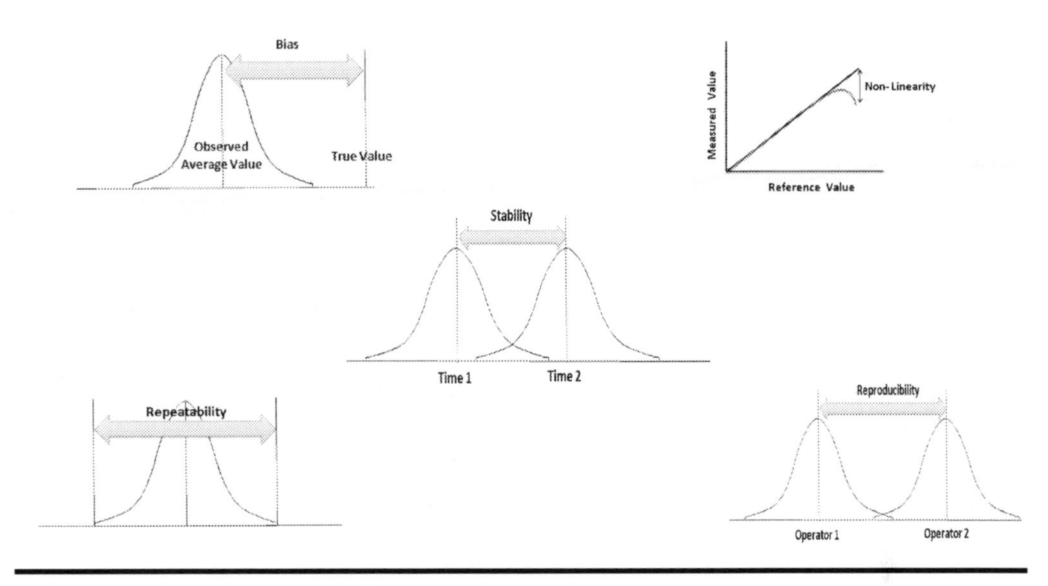

Figure 14.14 The five characteristics of MSA.

14.3.8.4 Basic Steps

1. Select the process or product for measurement.
2. Conduct an experiment where different people measure the same group of items repeatedly.
3. Collect and record data.
4. Analyse the data using Minitab or SPSS to evaluate ANOVA and Gauge R&R.
5. Improve the measurement process if necessary based on what you learn from the analysis.
6. Use the differences or variations as opportunities to eliminate problems with the measurement system.

14.3.8.5 Worked-Out Example

In a manufacturing industry, an engineer selected ten parts that represent the expected range of the process variation. Three operators (named Operator A, Operator B and Operator C) measured the ten parts, three times per part, in a random order. By using Minitab, the engineer performed a Gauge R&R study to assess the variability in measurements that may be from the measurement system.

The following graphs were also developed:

- Component variation graph
- R-chart by operator
- X-bar chart by operator

In a components of variation graph, the %Contribution from Part-To-Part > Total Gauge R&R. Thus, much of the variation is due to differences between parts.

The R-chart by operator shows that Operator B measures parts inconsistently.

In the X-bar chart by operator, most of the points are outside the control limits. Thus, much of the variation is due to differences between parts.

Gauge R&R value was 12.3%.

Although Gauge R&R value was 'may be acceptable', additional training was introduced to Operator B to improve Gauge R&R value.

14.3.8.6 Benefits

- MSA helps you to determine how much of an observed variation is due to the measurement system.
- It helps you to assess a measurement system for some or all of its five characteristics (e.g. Accuracy/Bias, Repeatability, Reproducibility, Linearity and Stability).
- It helps you improve the measurement system.

14.3.8.7 Pitfalls

- There is a risk of the inappropriate application of MSA for a measurement process of product of high tolerance.
- There is also a danger of inaccurate interpretation if adequate training were not given. It is often better to be approximately right than exactly wrong.

14.3.8.8 Training Requirements

Like other quantitative techniques (e.g. SPC), a basic understanding of MSA requires 1-day classroom training. However, a 3-day workshop is recommended for gaining proficiency in MSA technique including the application of Minitab or SPSS. The Six Sigma 'Black Belt' training programme includes an equivalent of 1 week's training in MSA techniques.

14.3.8.9 Final Thoughts

MSA is a powerful technique, but it is basically a measurement process and can make a major contribution to operational excellence only when

- A structured ongoing training programme is in place
- The users have a good understanding of statistical tools (e.g. ANOVA) and software (e.g. Minitab, SPSS)
- It is appropriately applied for products and processes requiring over three sigma tolerance limits

14.4 Summary

The quantitative techniques in Chapter 14 are primarily aimed at advanced applications by experts (e.g. 'Black Belts' in Green Six Sigma programmes). It is emphasised, by taking a Pareto analogy, that 80% of the problems can be addressed without the advanced techniques described in this chapter. We included DMAIC in this chapter in order to make it a grouping for a Six Sigma programme. The advanced techniques are more effective to improve the remaining and challenging 20% of the problems. The special emphasis on training is a pre-requisite for their successful applications.

GREEN TIPS

- Improve risk management by using FMEA.
- An advantage of SPC over inspection is that it emphasizes early detection and prevention of problems.
- QFD is best applied in product design driven by customer needs and values.
- Use DOE, DFSS, MSA and TRIZ in research and development projects.

Chapter 15

Qualitative Techniques

He that will not apply new remedies must expect new evils; for time is the greatest innovator.

—Francis Bacon

15.1 Introduction

Chapter 9 reported the advanced quantitative techniques of analysing data and improving process performance. In this chapter we deal with another category of techniques which we named 'qualitative techniques'. Although they are data driven, they depend heavily on logical reviews and judgemental assessments. There are numbers, but the improvements are not directly derived from a statistical process or a numeric solution. There are many such qualitative techniques, and we have selected the most relevant ones for Six Sigma and operational excellence programmes as follows:

R1: Benchmarking
R2: Balanced Scorecard
R3: European Foundation of Quality Management (EFQM)
R4: Sales and operations planning (S&OP)
R5: Knowledge management
R6: Kanban
R7: Activity-based costing (ABC)
R8: Quality management systems (QMSs; ISO 9000)
R9: Kazen

DOI: 10.4324/9781003268239-18

15.2 Description of Qualitative Techniques

The selection criteria of these techniques are the same as those for the quantitative techniques, except that for qualitative techniques there is less emphasis on specialist knowledge. However, qualitative techniques require specific training, in particular for S&OP, EFQM and ABC.

We followed the same structure of presentation as in Chapter 14, having pinpointed specific application areas as well as their benefits and pitfalls.

15.2.1 R1: Benchmarking

15.2.1.1 Background

The origin of benchmarking as it is known today, is credited to the Xerox Corporation who applied the improvement method based on comparing performances in the 1980s. The concept of benchmarking was well publicised by Camp (1989) based on the best practices of the Xerox Corporation.

It is to be noted that although the process was not formalised, many organisations, both in the private and public sectors, have been carrying out comparisons of the performance levels of various units for many years. The work of Camp (1989) was followed by numerous publications on benchmarking including those by Codling (1995) and Kartof and Ostblom (1994). Now various forms of benchmarking and sharing of best practices have become an accepted process for both performance improvement and knowledge management.

15.2.1.2 Definition

According to Karlof and Ostblom (1994), benchmarking is a continuous and systematic process for comparing your own efficiency in terms of productivity, quality and best practices with those companies and organisations that represent excellence.

Dale (1999) suggests three main types of formal benchmarking:

- Internal
- Competitive
- Functional

Internal benchmarking involves benchmarking between the same groups of companies so that best practices are shared across the corporate business.

Competitive benchmarking relates to a comparison with direct competitors to gather data on 'best in class' performance and practices.

Functional benchmarking is a comparison of specific processes in different industries to obtain information on 'best in school' performance and practices.

Internal benchmarking is the easiest one of the three to carry out while competitive benchmarking is often the most difficult one to put into place. Organisations are usually keen to share data in functional benchmarking when there is no direct threat of competition.

15.2.1.3 Application

A benchmarking process seeks to provide knowledge in a number of areas including the following:

1. What are the potential opportunities for improvement in our products or processes?
2. Who are the 'best in class' industry leaders in our competitive market?
3. How is our performance comparing with those of industry leaders?

The above three objectives could be provided respectively by internal, competitive and functional benchmarking. The application areas usually depend on the type of benchmarking.

Internal benchmarking is widely used by a multinational business with a number of subsidiaries in different countries or a national business which operates with some kind of branch structure of divisions. In such cases the business contains a number of similar operations that can be compared. GSK uses internal benchmarking to compare key performance indicators and to identify potential cost savings in its business. The Foods Division of Unilever have used internal benchmarking to compare Best Proven Practice for each of its major manufacturing sites.

Establishing a benchmarking partnership with other competitors can be mutually beneficial to both parties for the purpose of positioning the company in the market. External consultants and industry associations often play a key role to set up and conduct benchmarking on competitive issues and opportunities such as e-commerce, purchasing commodity-type materials and customer perceived quality.

The objective of functional benchmarking is to identify the best practice wherever it may be found. The purpose is to benchmark a part of the business which displays a logical similarity even in different industries. For example, a battery manufacturing company may want to benchmark its standard of direct delivery with Dell Computers, an organisation of acknowledged excellence in this area. Similarly, a pharmaceutical company may compare their production line changeover practices with Toyota Motors in Japan.

15.2.1.4 Basic Steps

1. Identify what to benchmark: Identifying what to benchmark is influenced by the knowledge of your own business. A SWOT (strengths, weaknesses, opportunities, threats) analysis may point towards the subject to be benchmarked. The subject could be products, production lines, customer service, working practices and so on.
2. Plan the benchmarking process: The preparation and planning should include
   a. Forming a team with their roles and responsibilities
   b. Selecting the measures of performance for the selected activity for benchmarking
   c. Determining a method of data collection
   d. Defining the scope and timeline for the exercise
3. Identify benchmarking partners: The participating units will vary depending on the type of benchmarking. Having identified a number of potential candidates, whether internal or external to the organisation, they must be contacted and briefed regarding the objective of the exercise. It is essential to establish a mutual trust with potential partners otherwise all efforts could be fruitless.
4. Collect data: The purpose of the fourth stage is to supply the information needed for the analysis. It is useful to draw up a questionnaire and test it by starting in your own organisation. The collection is often supplemented by interviews with partners.
5. Analyse data: A comparative analysis of the validated data is carried out to identify gaps. It is critical that the reasons for the gap with the 'best in class' are determined and understood. The trend of the gap should also be estimated over an appropriate time frame.
6. Implement and improve plan: Develop an action plan of closing the gap which the analysis stage has identified. The action plan is then implemented often requiring effective project management.

7. Review and repeat: More often than not, a benchmarking exercise is a continuous process. This should be conducted on a regular basis by sharing the results with benchmarking partners.

15.2.1.5 Worked-Out Example

The following example is taken from Welch and Byrne (2001).

General Electric Inc. with its global business of over US$120 billion per annum has been voted by Fortune as the 'most respected company'. GE is also known as the 'Cathedral of Six Sigma' and the high profile of the programme under the leadership of Jack Welch has been well publicised. GE licensed Six Sigma technology in 1994 from the Six Sigma Academy, rolled out the programme worldwide and achieved $2 billion savings in 1999.

GE Capital is the financial services arm of GE and accounted for approximately 40% of the group's turnover in 2001.

One heartening early success story at GE Capital relates to benchmarking and the sharing of good practice. GE Capital fielded about 300,000 calls a year from mortgage customers who had to use voicemail or call back 24% of the time because employees were unavailable. A Six Sigma team found that one of their 42 branches had a near perfect rate of answered calls.

The team carried out a benchmarking exercise between all branches. They analysed the systems, process flows, equipment, physical layout and stopping of the 'best in class' branch and then cloned it to the other 41 branches. As a result, a customer has a 99.9% chance of obtaining a GE person on their first try.

15.2.1.6 Benefits

The benchmarking techniques had delivered many tangible and intangible benefits. These include the following:

1. It enables process owners to identify what is to be improved and motivates them by the knowledge of what is achievable.
2. It enhances the communication, trust and partnership spirit between participating units by focusing on the high visibility of key processes and performance indicators.
3. It removes complacency even for high performers by providing a platform for improvement.

4. By focusing on processes rather than individuals, it helps to eliminate defensive and 'finger-pointing' practices.
5. It focuses attention on the details of best practice, thereby initiating a process of generating a learning culture.
6. It has a remarkable effect on strategy formulation, strategy implementation and leadership development.

15.2.1.7 Pitfalls

There are certain difficulties and pitfalls which must be recognised to make the benchmarking technique a success. These include the following:

1. Lack of trust in the process: Unless members at all levels of a participating company believe that the business can benefit, the exercise has little value.
2. Poor choice of metrics: The measures for comparison are often poorly chosen without due consideration for local variations. There is also a danger of number games being played or encouraging a 'league table' culture.
3. Inadequate process understanding: The exercise tends to focus on the outcome of a process rather than analysing the causes of the outcome.
4. A paper exercise remote from the source: Without the full participation of members from each unit, the exercise could be perceived as a mere 'paper exercise' by the centre. The centre is then accused of not understanding the variation and complexity of participating units.
5. Poor communication: If the participating units are not fully aware of the outcome of the project, data sharing will be guarded and the project is likely to fail.

15.2.1.8 Training Requirements

The team members of participating units require opportunities of team building to develop a mutual trust in sharing information. This is more fundamental than the knowledge of the key principles and steps of benchmarking. One method of achieving this is by encouraging the visits to selected sites involved in the process. This is further supplemented by e-communications and thus forming a virtual team.

15.2.1.9 Final Thoughts

Benchmarking exerts a powerful impact on learning organisations by leveraging best practices. It is a technique for knowledge management and not a cost-cutting method or a platform for number games.

15.2.2 R2: The Balanced Scorecard

15.2.2.1 Background

The basics of Balanced Scorecard were described earlier in Section 3.4.1. This section describes it in further details.

The concept of the Balanced Scorecard was first introduced by Kaplan and Norton (1992) in an article in the *Harvard Business Review*, 'The Balanced Scorecard—Measures That Drive Performance'. This generated considerable interest for senior business managers and led to the next round of development. The focus was shifted from short-term measurement towards generating growth, learning and value-added services to customers. This methodology was then published by Kaplan and Norton in a number of articles in the *Harvard Business Review* and in a book, *The Balanced Scorecard* (1996).

The senior executives of many companies are now using the Balanced Scorecard as the central organising framework for important decision processes. The rapid evolution of this technique has gradually transformed the performance measurement process into a strategic management system.

15.2.2.2 Definition

The Balanced Scorecard is a conceptual framework for translating an organisation's strategic objectives into a set of performance indicators distributed among four perspectives: Financial, Customer, Internal Business Processes and Learning and Growth (see Figure 15.1).

The indicators are aimed to measure an organisation's progress towards achieving its vision as well as the long-term drivers of success. Through the Balanced Scorecard an organisation monitors both its current performance (e.g. internal processes, finance, customer satisfaction) and its effort to improve and sustain performance (e.g. innovation and employee development). It is also balanced in terms of internal efficiency and external effectiveness.

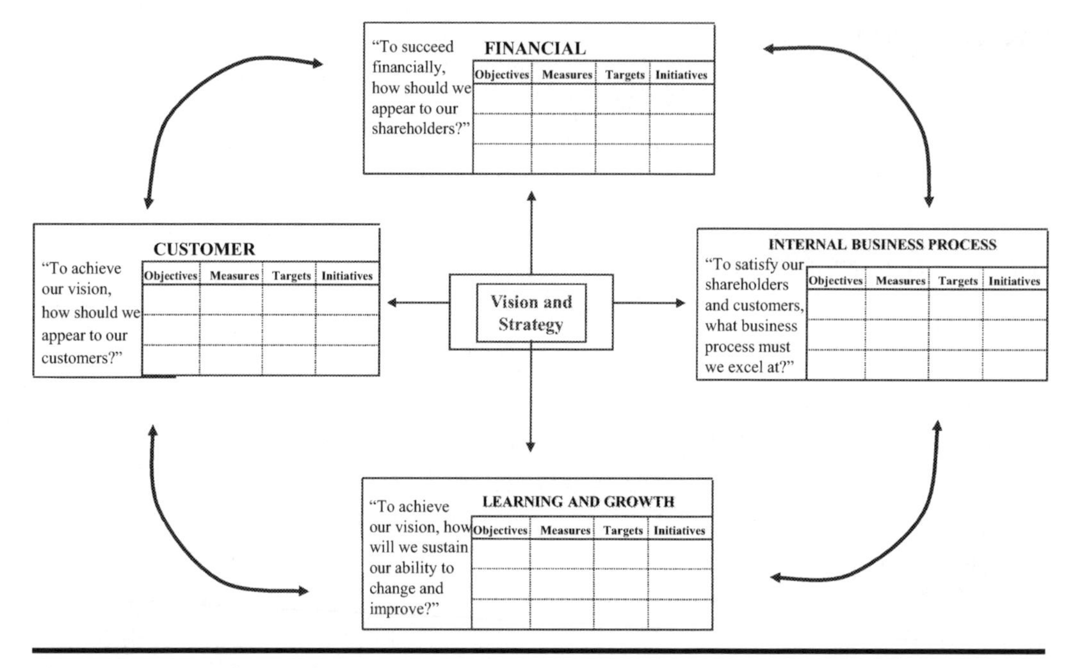

Figure 15.1 Kaplan and Norton's Balanced Scorecard.

For further details on the Balanced Scorecard see Kaplan and Norton (1996).

15.2.2.3 Application

The Balanced Scorecard has been applied successfully in several organisations around the world. The scorecard, with some customised changes, provides a management tool for senior executives primarily to focus on strategies and longer-term objectives. The organisations could vary from a large multinational business to a non-profit-making public service unit. The scorecard is sometimes named the 'Executive Dashboard'. The key performance indicators (KPIs) are reported as

- Current actual
- Target
- Year-to-date average

When the actual performance value is on or above target then the value is shown as green. If the actual is below the target but within a given tolerance, then the colour becomes amber. It is depicted in red when the value is below the tolerance limit of the target.

Another area of application is to assess the performance at the tactical operation level. Usually the top-level indicators (also known as 'Vital Flow') are designed in such a way that they can be cascaded to 'component' measures and the root causes can be analysed.

The published case studies by Kaplan and Norton (1996) provide examples of the application of the Balanced Scorecard in three areas: Chemical Bank, Mobil Corporation's US Marketing and Refining Division and United Way, a non-profit-making community service based in Rhode Island, US.

The application of the Balanced Scorecard has transformed methods of measuring a company's performance by financial indices alone. A recent publication by Basu (2001) has emphasised the impact of new measures on the collaborative supply chain. The Internet-enabled supply chain or e-supply chain has extended the linear flow of supply chain to collaborative management supported by supplier partnerships. This has triggered the emergence of new measures especially in five areas:

- External focus
- Power to the consumer
- Value-based competition and customer relationship management
- Network performance and supplier partnership
- Intellectual capacity

The design features and application requirements of the Balanced Scorecard are adapting to the collaborative culture of the integrated supply chain.

15.2.2.4 Basic Steps

As shown in Figure 15.2, Kaplan and Norton (1996) recommend an eight-step approach to introduce a Balanced Scorecard within an organisation.

Our experience suggests that the comprehensive approach of Kaplan and Norton is underpinned by three fundamental criteria leading to the success of a performance management system including the Balanced Scorecard. These are as follows:

1. Rigour in purpose
2. Rigour in measurement
3. Rigour in application

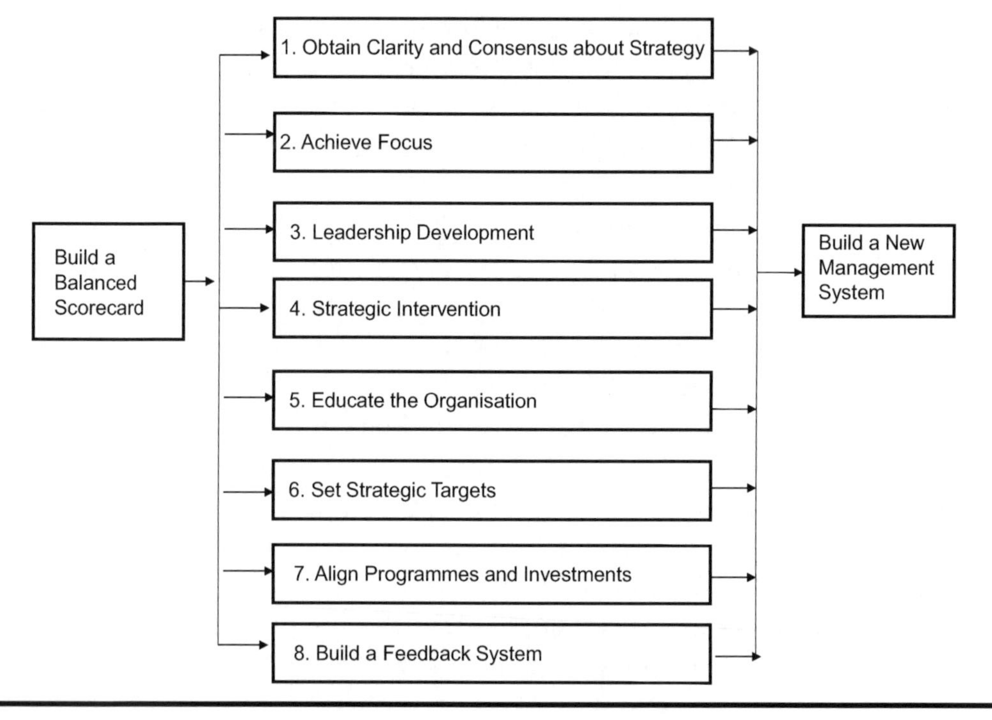

Figure 15.2 Kaplan and Norton's steps for implementation.

1. Rigour in purpose: Depending on the business objective, the metrics would vary in different industries. For example in a pharmaceutical company, order fulfilment and compliance with regulatory standards are critical while in a bulk chemical industry asset utilisation may be more important. The metrics should be derived in alignment with company objectives and an emerging area for the four inter-linked perspectives of the Balanced Scorecard. The metrics are clearly defined, validated and accepted by users during a pilot exercise.
2. Rigour in measurement: The success of established metrics will depend on the effectiveness of data collection and monitoring systems. This could vary from a manual process on a spreadsheet to a sophisticated data warehouse. Table 15.1 shows examples of monitoring systems depending on their application.
3. Rigour in application: The value of a well-designed and monitored Balanced Scorecard will be lost if the data are not used to improve and sustain performance. A process (such as S&OP) should be in place to review continuously the metrics and take action for performance improvement. Each measure should have a target both for the current year and the 'best in class' for the future. The measures are likely to

Table 15.1 Performance Monitoring System

Technology	*Tools*	*Application*
Local system	ERP	Local sites
	Excel	
Visual factory	Manual	Local system
	Multimedia	
Global system	ERP/SCM	Local sites
	Internet	Regional
	Data warehouse	Corporate

be modified or reaffirmed to reflect the active usage of the Balanced Scorecard.

15.2.2.5 Worked-Out Example

The following example is based upon the Balanced Scorecard developed by the Worldwide Manufacturing and Supply (WM&S) Division of GlaxoWellcome Plc in 1999, before the merger of the company with SmithKline Beecham to form GSK.

Four overall measures and targets were established to meet the four primary objectives of W&MS as shown in Table 15.2.

Table 15.2 Top Level Performance Measures

1. To support secure source of supply	1. Perfect order from suppliers
2. To support compliant and regulatory standards	2. Cost of poor quality
3. To support delivered point of sale	3. Perfect order to customers
4. To support 'best in class' cost	4. Cost of sales

For other supporting areas which are fundamental to the above objectives, additional measures were included in three perspectives:

5. Flexibility and adaptability
6. Growth and innovation
7. Health, safety and environment

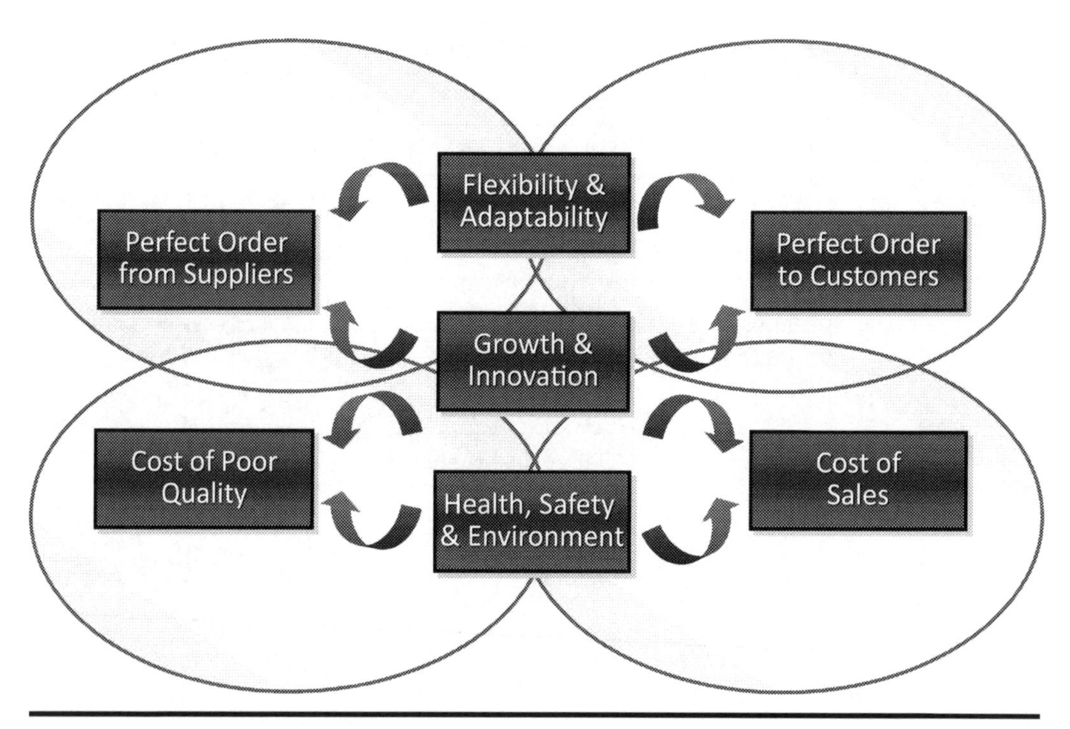

Figure 15.3 GlaxoSmithKline—Vital Few.

The seven 'Vital Flow' measures as described here and shown in Figure 15.3 sit at the top, and they are calculated from 16 component measures. For example, Perfect Order from Suppliers is derived from three components (viz. Order Fill, Vendor Managed Inventory and Quality Acceptance Rating) and the number of items.

As shown in Figure 15.4, there are, in addition, 33 supporting measures which do not directly feed into the seven Vital Few, but which provide important performance information for departmental reviews. There are also site-specific measures which were not part of the Balanced Scorecard.

The 56 measures were validated by a pilot exercise in three sites before the scorecard was rolled out. Following a continuous review of the benefits of the measures, the total number of measures was reduced to 27 metrics across the network.

15.2.2.6 Benefits

1. The Balanced Scorecard provides a sound basis to transform the approach of performance measurement to a management system.

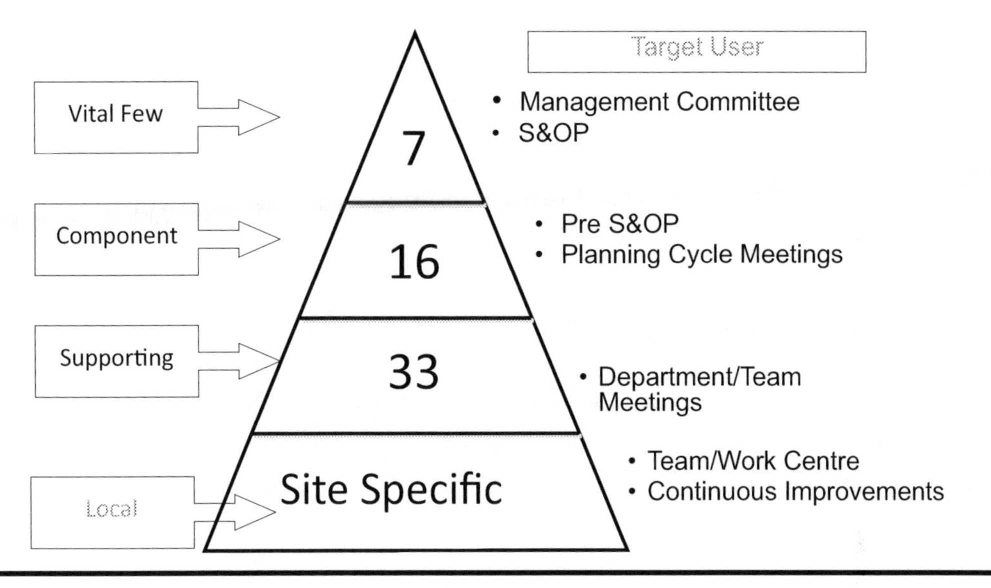

Figure 15.4 Balanced Scorecard measures and hierarchy.

2. It provides a tool for senior management to translate company strategies into longer-term goals.
3. It facilitates the cascading of high-level measures to component and supporting measures and thus to explore the root causes of variations.
4. The key measures are balanced covering all perspectives of the business and also for both strategic and tactical purposes.
5. It provides a framework to implement a customised performance management system and culture.

15.2.2.7 Pitfalls

1. If it is not properly administered it could promote a league table culture of number games.
2. Sometimes well-intentioned measures could remain unexplained and could be seen as a corporate scheme for 'Big Brother'.
3. A major challenge for the Balanced Scorecard is too many measures lead to bureaucracy and too few measures do not fit all.

15.2.2.8 Training Requirements

It is essential that the implementation of a Balanced Scorecard in an organisation is supported by training workshops. The workshop should contain the basic principles of the scorecard and the clarity and common definition of metrics. The need for local variation should also be considered seriously. The duration of such a workshop is usually 1 full day. The sharing of experience by visiting other sites is also part of the training requirements. There is a need for continuous review of requirements and metrics over the passage of time.

15.2.2.9 Final Thoughts

The Balanced Scorecard is a powerful and effective technique for both large and medium-sized businesses in all sectors. However, its effectiveness to a small enterprise of, say, less than 50 people should be reviewed carefully before its implementation.

15.2.3 R3: European Foundation of Quality Management

15.2.3.1 Background

The basics of EFQM were described in Chapter 3, Section 3.4.3. This section describes it in further detail.

The origin of the EFQM relates particularly to the Malcolm Baldridge Award and also to the Deming Prize. The Malcolm Baldridge National Quality Award (MBNQA) has been presented annually, since 1988, to recognise companies in the US that have excelled in quality management. The MBNQA criteria were based on seven categories:

1. Leadership
2. Strategic Planning
3. Customer and Market Focus
4. Information and Analysis
5. Human Resource Focus
6. Process Management
7. Business Results

The Deming Prize was awarded mainly in Japan during the 1950s and 1960s based upon ten examination viewpoints.

The EFQM was founded in the late 1980s by 14 large European companies to match the assessment criteria in Europe, and the EFQM Excellence Model was launched in 1991. The model was regularly reviewed, and an updated model was launched in 1999 which also included the RADAR (results, approaches, deploy, assess, review) logic.

15.2.3.2 Definition

The EFQM Excellence Model is a framework for assessing business excellence and serves to provide a stimulus to companies and individuals to develop quality improvement initiatives and demonstrate sustainable superior performance in all aspects of the business.

As shown in Figure 15.5 the model is structured around nine criteria and 32 sub-criteria with a fixed allocation of points or percentages as shown in Table 15.3.

The criteria are grouped into two broad areas:

1. Enablers: How we do things—the first five criteria
2. Results: What we measure, target and achieve—the second four criteria

The sub-criteria within an Enabler criterion have equal weighting. However, the weightings of Results criteria vary as

6a, 7a and 8b—75%
6b, 7b and 8a—25%

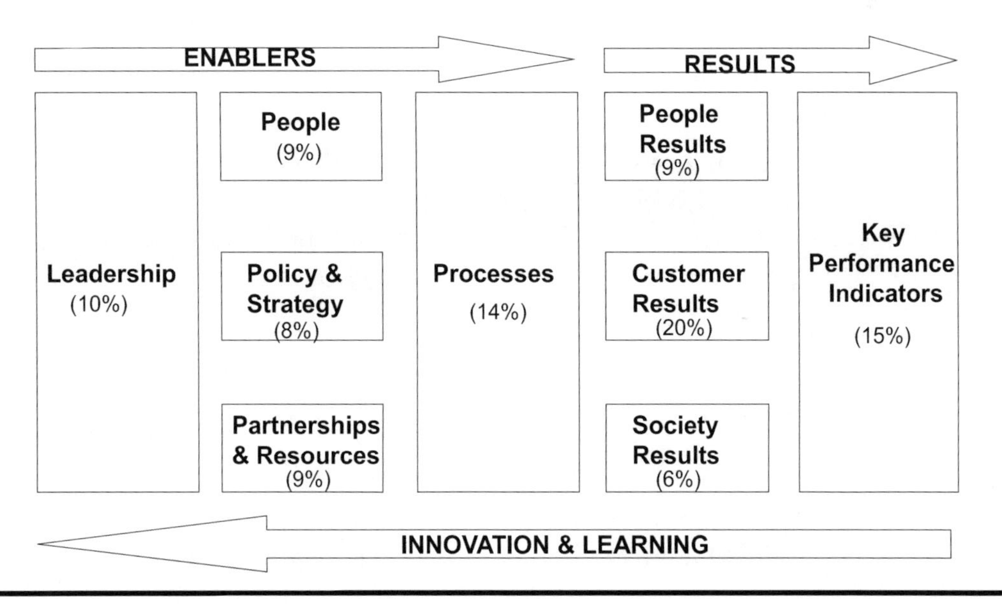

Figure 15.5 EFQM Excellence Model.

Table 15.3 EFQM Allocation of Points

Criteria	*Points*	*Number of Sub-criteria*	*Percentages (%)*
1. Leadership	100	4	10
2. People	90	5	9
3. Policy and Strategy	80	5	8
4. Partnership and Resources	90	5	9
5. Processes	140	5	14
6. People Results	90	2	9
7. Customer Results	200	2	20
8. Society Results	60	2	6
9. Key Performance Results	150	2	15
	1000	32	100

The scoring of each sub-criterion is guided by the RADAR logic which consists of four elements:

- Results
- Approach
- Deployment
- Assessment and Review

These elements and their attributes are applied to Enablers and Results as shown in Table 15.4.

The words on the RADAR scoring matrix reflect the grade of excellence for each attribute and what the assessor will be looking for in an organisation.

15.2.3.3 Application

The EFQM Excellence Model is intended to assist European managers to better understand best practices and support them in quality management programmes. The EFQM currently has 19 national partner organisations in

Table 15.4 RADAR Attributes

Elements	*Attributes*	*Applies to*
Results	Trends, Targets, Comparisons, Causes, Scope	Results
Approach	Sound, Integrated	Enablers
Deployment	Implemented, Systematic	Enablers
Assessment and Review	Measurement, Learning, Improvement	Enablers

Europe, and the British Quality Foundation is such an organisation in the UK. Over 20,000 companies, including 60% of the top 25 companies in Europe, are members of the EFQM.

The model has been used for several purposes, of which the four main ones are as follows:

1. Self-assessment: The holistic and structural framework of the model helps to identify the strengths and areas for improvement in any organisation and then to develop focused improvements.
2. Benchmarking: Undertaking the assessment of defined criteria against the model, the performance of an organisation is compared with that of others.
3. Excellence awards: A company with a robust quality programme can apply for a European Quality Award (EQA) to demonstrate excellence in all nine criteria of the model. Although only one EQA is made each year for company, public sector and small and medium-sized enterprise (SME), several EQAs are awarded to companies that demonstrate superiority according to the EFQM Excellence Model.
4. Strategy formulation: The criteria and sub-criteria of the model have been used by many companies to formulate their business strategy.

A survey of EFQM members in 2000 showed a high proportion of the usage of the model in self-assessment and strategy formulation (see Figure 15.6).

The use of the model originated in larger business; however, the applications and interest have been growing among the public sector and smaller organisations. To satisfy these needs, special versions of the model are available for public sector organisations and SMEs.

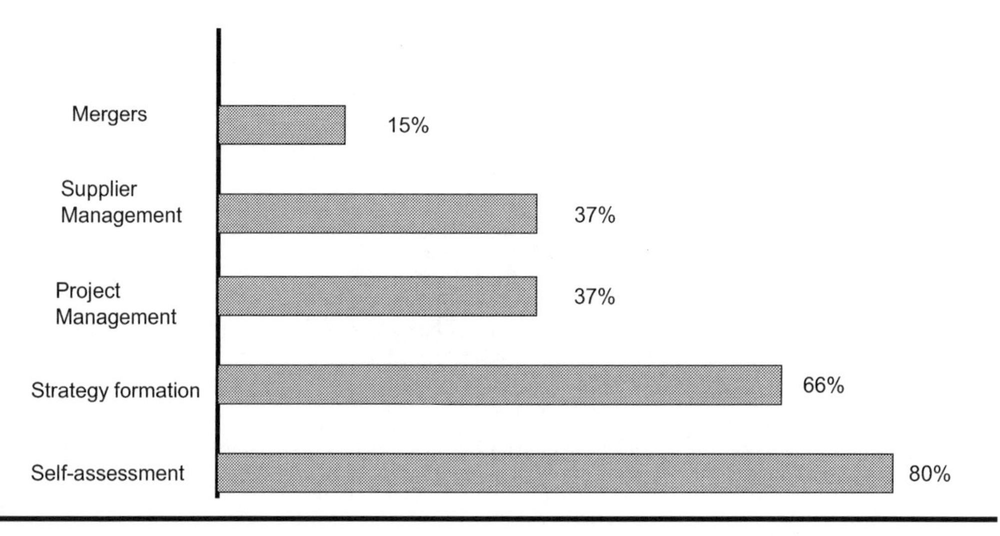

Figure 15.6 The use of EFQM.

15.2.3.4 Basic Steps

The organisations can use the EFQM model without the involvement of a third party, except during the training period. The sequence of steps could vary depending on the approach adopted by a company. The following steps are recommended with an emphasis on self-assessment:

1. Develop top management commitment: It is essential to gain top management support, otherwise the process will have a limited value. Educate senior managers with the benefits and resource requirements for the self-assessment process.
2. Plan a self-assessment process: Select the appropriate self-assessment approach and the Excellence Model. If the company is not aiming for an EQA then the criteria could be customised to companies' VMOST (vision, mission, objective, strategy and tactics). A roll-out plan is required starting from a pilot site.
3. Select assessors and training: Select a team comprising people with analytical and objective skills to carry out self-assessment. The team should be trained by EFQM-approved trainers.
4. Refine and communicate the self-assessment plan: It is important that the right message of self-assessment (that it is not another audit from the centre) is communicated to all sites. It is not an academic exercise by a third party, but it is a process of continuous health check.

5. Conduct self-assessment: There are a number of ways of conducting self-assessment. A proven process is assessment of all criteria over 2 days by a team comprising both trained assessors (see Step 3) and local members.
6. Develop action plan: Based on the assessment in Step 5, identify the improvement opportunities and agree on responsibilities and target dates for completion. Separate the activities which may require further investigation or investment.
7. Implement action plan: Set up an organisation to monitor progress and milestones. It is essential to secure senior management commitment.
8. Repeat and review: The self-assessment is a continuous process to achieve operational excellence. Steps 5, 6 and 7 should be repeated at least once a year.
9. Consider certification or awards: The company could consider an EQA if the pure-play EFQM Excellence Model is used. For a customised model the sites could be considered for their own company awards. For an EQA, a report of up to 75 typed A4 pages is submitted to an EFQM Partner Organisation (e.g. BQF) and then the company is audited by external assessors.

15.2.3.5 Worked-Out Example

The following example illustrates the scoring method of the EFQM model.

Consider an organisation, following an assessment based on its performance, by adopting the RADAR logic, scored percentage values for each of the 32 sub-criteria. The scoring summary sheet in Table 15.5 shows that an organisation was awarded 531 points out of 1,000 maximum points.

15.2.3.6 Benefits

1. It provides a holistic and realistic assessment of how good all the components and business processes of an organisation are.
2. It supports a balanced approach of assessing both qualitative (e.g. enablers) and quantitative (e.g. results) criteria covered in the model.
3. It provides a common language of enablers, results, assessment, scoring logic and certification for all types and sizes of organisation.
4. It brings the quality improvement initiatives into a single framework and creates a balance between different stakeholder groups.

Table 15.5 Summary Scoring Sheet

1. Enablers Criteria

Criteria number	*1*	%	*2*	%	*3*	%	*4*	%	*5*	%
Sub-criterion	1a	65	2a	55	3a	60	4a	50	5a	55
Sub-criterion	1b	55	2b	45	3b	65	4b	70	5b	45
Sub-criterion	1c	70	2c	60	3c	45	4c	50	5c	60
Sub-criterion	1d	65	2d	75	3d	60	4d	45	5d	50
			2e	65	3e	55	4e	55	5e	75
Sum		255		300		285		270		285
		/4		/5		/5		/5		/5
Score awarded		64		60		57		54		57

2. Results Criteria

Criteria number	6			%		*7*			%
Sub-criterion	6a	50	× .75	= 37		7a	60	× .75	= 45
Sub-criterion	6b	45	× .25	= 12		7b	50	× .25	= 13
Scores awarded				49					58
Criteria number	8			%		9			
Sub-criterion	8a	55	× .25	= 14		9a	70	× .5	= 35
Sub-criterion	8b	55	× .75	41		9b	40	× .5	= 20
Scores awarded				55					55

3. Calculation of Total Points

Criterion	*Scores*	*Factor*	*Points Awarded*
1. Leadership	64	× 1.0	64
2. Policy and Strategy	60	× 0.8	48
3. People	57	× 0.9	51
4. Partnership and Resources	54	× 0.9	49
5. Processes	57	× 1.4	79
6. Customer Results	49	× 2.0	98
7. People Results	58	× 0.9	52
8. Society Results	55	× 0.6	33
9. Key Performance Results	55	× 1.5	83
Total Points Awarded			531

Source: EFQM (1999).

5. The broad framework of the model allows its adaptation to a self-assessment checklist customised to individual company requirements.

15.2.3.7 Pitfalls

1. The EQA process requires a detailed report of up to 75 pages which is often viewed as a resource-intensive and bureaucratic process.
2. The generic nature of the model, unless it is moderated by a well-framed assessor, has a risk of misinterpretation.
3. The allocated weighting factors of percentages (e.g. 20% for customers) do not necessarily reflect the relative priority and mission of a business.
4. The European level of EFQM often faces conflict, in a multinational organisation, with MBNQA and ISO 9000.

15.2.3.8 Training Requirements

The success of EFQM as a driver of a quality initiative depends on two key factors, viz. top management commitment and properly trained assessors. It is vital that an organisation has a team of assessors (say 1% of the workforce) trained and licensed by an EFQM-approved organisation (e.g. British Quality Foundation). The senior managers and improvement team members should also receive 1-day awareness training by the company's own assessors.

15.2.3.9 Final Thoughts

EFQM or its adaptation to a self-assessment process is an essential technique for achieving and sustaining operational excellence. However, an organisation has to be at an advanced stage of its quality programme to be able to use self-assessment in an effective manner.

15.2.4 R4: Sales and Operations Planning

15.2.4.1 Background

The basics of S&OP are described in Chapter 3, Section 3.4.1. This section describes it in further detail.

The classical concept of S&OP is rooted to the MRP II (Manufacturing Resource Planning) process. In the basic S&OP, the company operating plan (comprising sales forecast, production plan, inventory plan and shipments) is updated on a regular monthly basis by the senior management of a manufacturing organisation. The virtues, application and training of the S&OP have been promoted by Oliver Wight Associates (see Ling and Goddard, 1988) since the early 1970s.

In recent years the pace of change in technology and marketplace dynamics have been so rapid that the traditional methodology of monitoring the actual performance against pre-determined budgets set at the beginning of the year may no longer be valid. It is fundamental that businesses are managed on current conditions and up-to-date assumptions. There is also a vital need to establish an effective communication link, both horizontally across functional divisions and vertically across the management hierarchy to share common data and decision processes. Thus S&OP has moved beyond the operations planning at the aggregate level to a multifunctional senior management review process.

15.2.4.2 Definition

The traditional S&OP is a senior management review process of establishing the operational plan and other key activities of the business to best satisfy the current levels of sales forecasts according to the delivery capacity of the business.

Ling and Goddard (1988) summarise a 'capsule description of the process':

> It starts with the sales and marketing departments comparing actual demand to the sales plan, assessing the marketplace potential and projecting future demand. The updated demand plan is then communicated to the manufacturing, engineering and finance departments, which offer to support it. Any difficulties in supporting the sales plan are worked out . . . with a formal meeting chaired by the general manager.

The outcome of the process is the updated operation plan over 18 months or 2 years (the 'planning horizon') with a firm commitment for at least 1 month.

The process is data driven. A report for each product family is prepared for the planning horizon, and it is usually divided into up to five sections containing 'a single set of numbers' for Sales Plan, Production Plan, Inventory, Backlog and Shipments.

15.2.4.3 Application

S&OP has become an established company-wide business planning process in the Oliver Wight MRP II methodology (see Wallace, 1990). It is now also known as integrated business management or senior management review.

The process has been developed and applied primarily for manufacturing organisations. The key members of all departments, such as R&D, Marketing, Sales, Logistics, Purchasing, Human Resources, Finance and Production, participate in the process but not in the same meeting. S&OP addresses the operations plan that deals with Sales, Production, Inventory and Backlog and thus it is expressed in units of measurements such as tons, pieces etc. rather than dollars or euros. The operation plan is reconciled with the business plans or budgets which are expressed in terms of money.

The S&OP or senior management eview process has been proven to be a key contributor to sustaining the performance level achieved through a TQM or Six Sigma programme (Basu and Wright, 2003, p. 97). The S&OP agenda, in addition to its main focus of establishing the operation plan, contains the reviews related to performance and key initiatives. This provides an effective platform for senior managers of all functions to assess the current performance and steer the future direction of the business.

With appropriate adjustments for the units of the products, the S&OP process can also be applied to service industries. This will encourage the managers in non-manufacturing sectors to review the demand, capacity, inventory and scheduling and enhance the synergy of different functions.

15.2.4.4 Basic Steps

The diagram in Figure 15.7 shows the five steps in the S&OP process that will usually be present, and the process can be adapted to specific organisation requirements.

New Product Review (Step 1): Many companies follow parallel projects related to the new products in R&D, Marketing and Operations. The purpose of this review process in Step 1 is to review the different objectives of various departments at the beginning of the month and resolve new product-related assumptions and issues. The issues raised will impact upon the demand plan and the supply chain at a later stage of the process.

Demand Review (Step 2): Demand planning is more of a consensus art than a forecasting science. Demand may change from month to month depending on market intelligence, customer confidence, exchange rates,

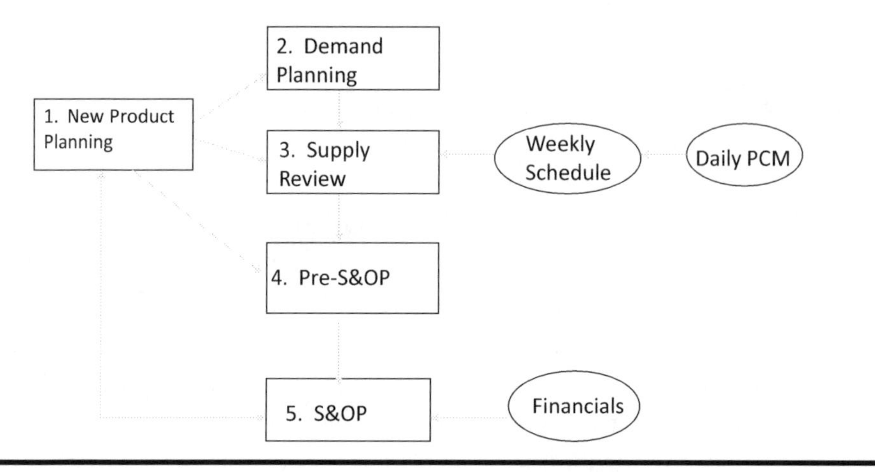

Figure 15.7 Sales and operations planning (S&OP).

promotions, product availability and many other internal and external factors. This review at the end of the first week of the month, between Marketing, Sales, IT and Logistics, establishes agreement and accountability for the latest demand plan identifying changes and issues arising.

Supply Review (Step 3): In the current climate of increasing outsourcing and supply partnership, the capacity of supply is highly variable, and there is a need to ensure the availability and optimisation of supply every month. This review, usually on the second week of the month, between Logistics, Purchasing and Production, establishes the production and procurement plans and raises capacity, inventory and scheduling issues.

Reconciliation Review (Step 4): Issues would have been identified in previous reviews of new products, demand and supply. The reconciliation step goes beyond the balancing of numbers to assess the business advantage and risk for each area of conflict. This review looks at issues from the business point of view rather than departmental objectives. This is also known as the Pre-S&OP Review and its aim is to minimise issues for the final S&OP stage.

Senior Management Review (Step 5): Senior managers or board members, with an MD or CEO in Chair, will approve the plan that will provide clear visibility for a single set of members driving the total business forward. The agenda includes the review of key performance indicators, business trends of operational and financial performance, issues arising from previous reviews and corporate initiatives. This is a powerful forum to adjust business direction and priorities. This is also known as the S&OP review.

In each process step the reviews must address a planning horizon of 18 to 24 months in order to make a decision for both operational and strategic objectives. There may be a perceived view that S&OP is a process of aggregate/volume planning for supply chain. However, it is also a top-level forum to provide a link between business plan and strategy.

15.2.4.5 Worked-Out Example

In addition it contains some useful data, such as production batch size (180), lead time (2 months) and stock target (3 months). The columns to the left of the line 'Today' show historical data and to the right is the information for the planning horizon in the future.

The data for the sales budget are taken from the annual business plan. 'Latest Forecast' represents what the sales and marketing teams are projecting based on the latest information. These data are updated every month. The stock target for each month is based on the sales forecast for the next 3 months, as the target is 3 months' stock cover. Due to a technical problem, production was suspended for 6 months and is resumed from this month. Therefore, a backlog of order or negative stock (–1,194) has built up in the current month.

The projected stock is calculated by using the following formula:

Stock this month = Stock last month – Sales + Delivery

For example, the projected stock for October =

–2,094 – 500 + 3,600 = 1,006

It is important to note that planned production should be in multiples of the batch size (i.e. 180) and the volume is available after 2 months' lead time.

Table 15.6 Shows a worked-out example of a product pack: Aquatic 500 in the unit of packs. The report is divided into four sections:

- Sales
- Stock
- QA Release
- Production

Table 15.6 Sales & Ops Planning

Month	*-6 Feb*	*-5 Mar*	*-4 Apr*	*-3 May*	*-2 Jun*	*-1 Jul*	*1 Aug*	*2 Sep*	*3 Oct*	*4 Nov*	*5 Dec*	*6 Jan*	*7 Feb*	*8 Mar*	*9 Apr*	*10 May*	*11 Jun*	*12 Jul*
SALES																		
Budget	500	500	500	525	600	600	600	550	550	525	525	525	525	525	525			
Latest Forecast	500	400	450	400	500	400	500	400	400	400	500	500	500	500	500			
Actual																		
STOCK																		
Target	1250	1350	1300	1400	1300	1400	1300	1400	1400	1500	1500	1500	1000	500	0			
Projected	956	556	106	-294	-794	-1194	-1694	-2094	1006	606	1906	1406	1806	1306	806			
Actual																		
QA RELEASE																		
Planned	0	0	0	0	0	0	0	0	3600	0	1800	0	900	0	0			
Actual																		
PRODUCTION START																		
Planned	0	0	0	0	0	0	3600	0	1800	0	900	0	0	0	0			
Actual																		

Year to Date: **Sales Budget:** 3225 **Actual Sales:** **Forecast performance**
Sales Forecast: 2650 **Stock performance**

15.2.4.6 Benefits

1. S&OP provides a practical up-to-date review of the operational plan of an organisation while meeting the business objectives of profitability, productivity and customer service.
2. It allows an excellent forum of senior managers of all functions to enhance the synergy to a common objective. The 'finger-pointing' culture is thus eliminated.
3. It is data driven and based on a 'single set of numbers' for all departments and thus helps to reconcile disputes and planning issues.
4. It can play an effective role in sustaining the high level of performance achieved by a TQM or Six Sigma related programme.

15.2.4.7 Pitfalls

1. If the S&OP process is introduced without proper training, the managers may be obstructive for fear of detail and lack of understanding of the process.
2. A critical success factor is that it must be supported and chaired by the CEO or general manager who can take a balanced approach related to past performance and future strategy and the degree of detail required.
3. The process will have limited value if all functions, especially Marketing, Sales, Logistics and Operations, are not involved at the appropriate stages of the business planning process.

15.2.4.8 Training Requirements

All key managers, including the CEO or general manager, should participate in a 2-day workshop on S&OP. It will also be beneficial for managers to sit in the S&OP meetings of other organisations where the process is fully operational.

There is a learning curve involved with S&OP. The first few meetings usually do not go well, but the process can become effective after 3 months. It is important to start S&OP as soon as possible.

15.2.4.9 Final Thoughts

S&OP is an excellent data-driven but people-based holistic process to establish and update business plans and sustain business performance in keeping with the changes in the company and its marketplace. It is essential that it is underpinned by good training and led by the general manager to ensure continuous and sustainable improvement.

15.2.5 R5: Knowledge Management

15.2.5.1 Background

The basics of knowledge management are described in Chapter 3, Section 3.4.4. This section describes it in further detail.

The concept and the terminology of knowledge management (KM) probably started in management consultancies. With the advent of the Internet they realised the benefits of sharing industry best practice and benchmarks of performance within the industry. However, a new product needs a name, and the name that emerged was 'knowledge management'. The term apparently was first used in its current context at McKinsey in 1987. Almost 400 years ago, Francis Bacon also stated that 'Knowledge is power'. The development of 'learning organisation' by Peter Senge (1990) further consolidated the understanding and application of KM.

15.2.5.2 Definition

The Gartner Group created a definition of KM, which has become the most frequently cited one (Duhon, 1998), and it is as follows:

> Knowledge management is a discipline that promotes an integrated approach to identifying, capturing, evaluating, retrieving, and sharing all of an enterprise's information assets. These assets may include databases, documents, policies, procedures, and previously un-captured expertise and experience in individual workers.

15.2.5.3 Application

There are three main areas of application for KM:

1. Accumulating knowledge
2. Storing knowledge
3. Sharing knowledge

We routinely solve many of the problems we encounter by just using the knowledge we have accumulated throughout our lifetime. But if we take this idea and apply it to more than one person, it gets even more effective. The accumulation of knowledge is the capture of the judgment and behaviour of a human or an organisation that has expert knowledge and experience in a particular field.

The best way of storing knowledge is information technology (IT). Now we have the benefit of powerful search engines (e.g. Google) which has become a way of life. We also have expert systems which are computer programs that use artificial intelligence (AI) technologies.

The sharing of knowledge will ensure that the specialized knowledge of employees does not leave with them, or go unutilized by other employees who would benefit from that knowledge. Also by sharing knowledge we can avoid the risk of 're-inventing the wheel'. In this way integrated information enables ongoing knowledge innovation processes.

15.2.5.4 Basic Steps

There are no prescribed steps of KM. It is very often evolved from continuous learning and sharing by IT and the Internet. The following four steps are suggested as a structured approach of benefiting from KM:

- Search data
- Capture data
- Process data
- Share and benefit

The search process starts within the organisation. In every organisation, there are multiple sources of knowledge. The sources of knowledge are identified and often data may be documented as a case study. Relevant best practices and published reports for external sources. The search process is helped by a solid understanding of the knowledge flow of the organisation.

The data must be stored and organised in a deliberate manner. By creating a system that is mapped, categorized and indexed, knowledge is more easily accessed and the organisational structure is increased. With advanced digital technology (e.g. data warehousing, website), data capturing and storage have become more effective.

The third step of processing data involves a deep analysis of the knowledge gathered in the previous two steps. The organisation must organise and assess the knowledge to see how best it can be folded into the structure of the organisation. It became clear that KM implementation would involve significant changes in the corporate culture. With the addition of the 'learning organisation' (Senge, 1990), the leadership of enabling these changes is usually given to the human resources department.

The last and the most important stage is sharing the benefits of KM. This step is also key in developing skills and management potentials of employees within an organisation. The common types of sharing the benefits of KM include

- Training and team-building workshops
- Internal newsletter and publicity materials
- Corporate websites
- Web seminars
- Case studies
- Personal development objectives

15.2.5.5 Worked-Out Examples

In May 1996, the National Committee of Inquiry into Higher Education was appointed to make recommendations on the purposes, structure and funding of higher education to meet the needs of the UK in the future. This committee produced the Dearing Report which highlighted the need for the UK to compete in increasingly competitive international markets by sharing knowledge between universities and research centres. Since the Dearing report, global competition has intensified, and high-level skills and knowledge have become ever more central to the UK's economic success.

A study by Cranfield et al. (2008) investigated how KM, following the Dearing Report, is perceived and implemented within seven higher educational institutes (HEIs) within the UK. The key findings of the study included the following:

- Although it was thought that academics are quite open to the idea of knowledge sharing, there is the further issue of creating opportunities for such interaction.
- Within HEIs there is the perception of an academic and administrative divide.
- The Russell Group Universities did not overtly prioritised KM but recognised the importance of KM in delivering research objectives.

Universities in general have a significant level of KM activities. It is evident from the cases that two out of the seven HEIs were engaging in KM in a systemic and institutional-wide way, and a further two had champions engaging in KM overtly within their faculty. These four institutions were therefore actively engaged in prioritizing 21st-century management tools.

15.2.5.6 Benefits

1. Increase customer satisfaction by cross-collaboration with customers and customer relationships management (CRM) systems.
2. Improve the decision-making process by obtaining rapid access to the knowledge base of the entire organisation.
3. Promote innovation and cultural changes by enabling access to the latest information.
4. Avoid redundant efforts and 'reinventing the wheel' by learning from previous experience.
5. Promote people development by continuous assessment and learning,

15.2.5.7 Pitfalls

1. Difficulty in efficiently capturing and storing business knowledge
2. Challenge to motivate people to share, reuse and apply knowledge consistently
3. Problems encountered to align KM with the business strategy and exiting processes

15.2.5.8 Final Thoughts

A major contribution of KM to Green Six Sigma is the continuous development of people skills. Along with the Balanced Scorecard, EFQM and S&OP, KM is an important technique to sustain the performance and culture of an organisation.

15.2.6 R6: Kanban

15.2.6.1 Background

The Toyota Motor Company of Japan pioneered the Kanban technique in the 1980s. As part of Lean manufacturing concepts, kanban was promoted as one of the primary tools of just-in-time (JIT) concepts by both Taguchi Ohno (1988) and Shingo (1988). Inspired by this technique, American supermarkets in particular replenished shelves as they were emptied and thus reduced the number of storage spaces and inventory levels. With a varied degree of success outside Japan, kanban has been applied to maintain an orderly flow of goods, materials and information throughout the entire operation.

15.2.6.2 Definition

Kanban literally means 'card'. It is usually a printed card in a transparent plastic cover that contains specific information regarding part number and quantity. It is a means of pulling parts and products through the manufacturing or logistics sequence as needed. It is therefore sometimes referred to as the 'pull system'. The variants of the kanban system utilise other markers such as light, electronic signals, voice command or even hand signals.

15.2.6.3 Application

Following the Japanese examples, kanban is accepted as a way of maximising efficiency by reducing both cost and inventory.

The key components of a kanban system are

- Kanban cards
- Standard containers or bins
- Workstations, usually a machine or a worktable
- Input and output areas

The input and output areas exist side by side for each workstation on the shop floor. The kanban cards are attached to standard containers. These cards are used to withdraw additional parts from the preceding workstation to replace the ones that are used. When a full container reaches the last downstream workstation, the card is switched to an empty container. This empty container and the card are then sent to the first workstation signalling that more parts are needed for its operation.

A kanban system may use either a single-card or two-cards (move and production) system. The dual-card system works well in a high up-time process for simpler products with well-trained operators. A single-card system is more appropriate in a batch process with a higher changeover time and has the advantage of being simpler to operate. The single-card system is also known as 'Withdrawal Kanban', and the dual-card system is sometimes called 'Production Kanban'.

The system has been modified in many applications; in some facilities although it is known as a kanban system, the card itself does not exist. In some cases the empty position on the input or output areas is sufficient to indicate that the next container is needed.

15.2.6.4 Basic Steps

1. Select the operation for the kanban system and decide whether a single- or dual-card system will be applied. We recommend the single-card system for its simplicity.
2. Determine the number of kanban containers to set the amount of authorised inventory. Use the following formula:

$$\text{Number of containers} = \frac{\text{Demand in lead time} + \text{Safety stock}}{\text{Size of container}}$$

3. Design and procure the standardised containers and kanban cards.
4. Develop and implement the workstation layout. Carry out a pilot run.
5. Train operators and activate the kanban system by following some basic rules:

   Each container must have a kanban card
   Each container must contain the exact quantity stated on the card
   The containers are pulled only when needed by the next down-stream station
   No defective parts are sent
6. Review the process regularly and aim to reduce the number of kanbans and the time period.

15.2.6.5 Worked-Out Example

The following example is based upon the experience of Level Industrial, the Brazil subsidiary of Unilever in Sao Paulo.

Lever Industrial was engaged in the batch production of industrial detergents comprising nearly 300 stock keeping units which varied from a 500-kilogram draw to a 200-gram bottle. After carrying out a Pareto analysis, the team selected three fast-moving products for a pilot kanban system. These products in total accounted for 18% of output.

The company adopted for each product, a simple single-card kanban system consisting of five stages as shown in Figure 15.8.

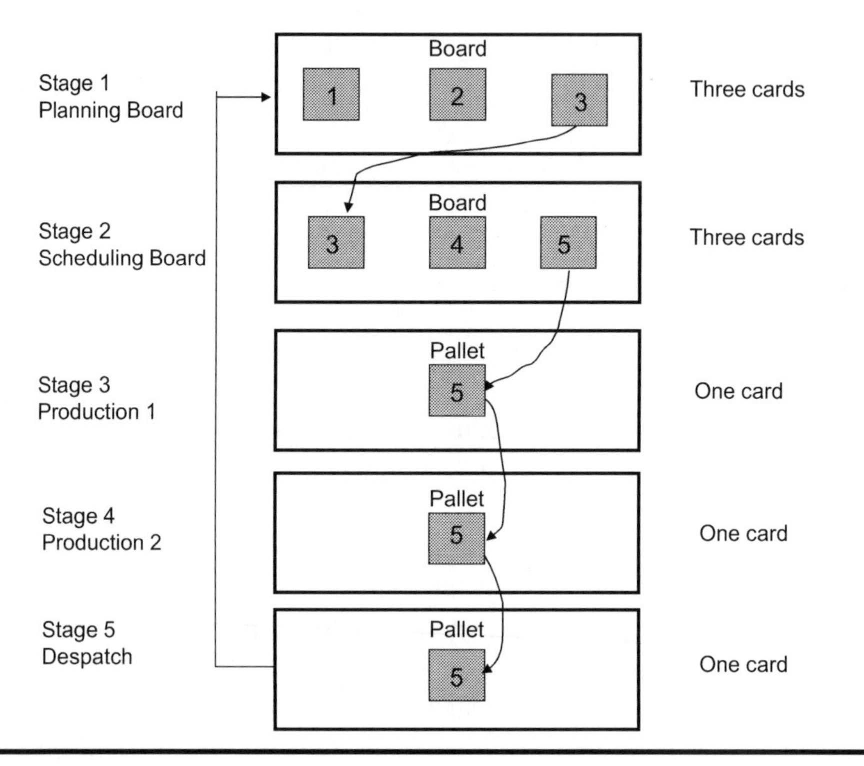

Figure 15.8 Kanban system.

Both the planning board and the scheduling board contain three cards each as a buffer between the variability of production cycle time and the availability of materials.

When the card arrives from the despatch (Stage 5) it is kept on the planning board and planning for the product starts. When the planning board is full with three cards, the third card is passed to the scheduling board and production scheduling is ensured. Similarly when the scheduling board is full, the third card is transferred to the pallet at the Production Station 1 and actual production begins.

When the pallet in Stage 3 (Production 1) is full, the card then moves to the next station (Production 2) in Stage 4, and then on to despatch in Stage 5. After the goods are despatched, the card returns to the planning board and the next cycle begins.

The pilot exercise was successful. It achieved an improvement in customer service which rose from 84% to an excellent 98% and inventory was also reduced. The kanban system was extended to nine additional key products. The manual system was retained for the five stages, although both the planning and stock adjustment processes were supported by MFG-Pro, the enterprise resource planning (ERP) system.

15.2.6.6 Benefits

1. Kanban enables only small inventory through the plant and pulling only when needed, thus allowing only a small quantity of faulty or delayed material.
2. Kanban minimises the negative aspects of inventory management including obsolescence, occupied space, working capital, increased material handling and poor quality.
3. Kanban uses standardised containers conducive to efficient material handling and lower costs.
4. It aims to create work sites that can respond to changes quickly empowering the operators to exercise their initiatives.
5. It facilitates the re-engineering of the process and works in harmony with JIT.

15.2.6.7 Pitfalls

1. It is an inflexible process, as the transfer batch is fixed. Therefore, it can cause additional stoppage periods.
2. Kanban is inappropriate for high mix, slow mover variants. It struggles with cyclic or seasonal demand.
3. It is perceived as a low-technology manual process and comes into conflict with the push MRP II/ERP systems. (However, there are good examples of a computerised MRP II system working hand-in-hand with a kanban call-off scheme.)
4. The application is visible as a solution to a part of the total operation and often not appreciated by employees who are not directly involved with the kanban system.

15.2.6.8 Training Requirements

Although the fundamental principles of the kanban technique are not complex, its training needs should not be underestimated. The two-card system requires a team of well-trained and motivated workers.

The operators should be trained for 1 day in a classroom environment, and this should be followed by 2 or 3 days of on-the-job training in the shop floor environment.

15.2.6.9 Final Thoughts

The attractiveness of a kanban system cannot be ignored even in an environment of flexibility and ERP systems. Select and apply kanban for fast-moving products containing the repetitive manufacturing of discrete units in large volumes which can be held steady for a period of time.

15.2.7 R7: Activity-Based Costing

15.2.7.1 Background

The classical method of costing allocates overhead costs to each product according to the amount of direct labour required to make that product. These days labour is a much smaller element of the product costs and thus the product costs are disproportionately distorted. Alternatively, the overhead cost is allocated according to the volume of the product. This could lead to increased consumer demand for under-costed low-volume product and decreased consumer demand for overpriced high-volume products.

These problems associated with the traditional overhead allocation methods were first highlighted by Kaplan and Johnson (1987) who demonstrated that historical methods of cost accounting have led to cost distortions. These distortions of both inaccuracy and inappropriateness have harmful effects on product profitability. The cost of quality is also affected by the traditional accounting system. Rabin Cooper in the US suggested the name 'activity-based costing' (ABC).

ABC seeks to correct these distortions by assigning indirect costs to products and services by using appropriate activity drivers that reflect resource consumption by the cost objectives. This also enables the quality cost associated with an activity to be more easily obtained.

15.2.7.2 Definition

ABC is an accounting technique that allows an organisation to assign more accurate product or service costs by understanding the activities that create cost and allocating overheads on those activities. Overhead costs and their causes are analysed so that they can be transferred, wherever possible, into direct costs.

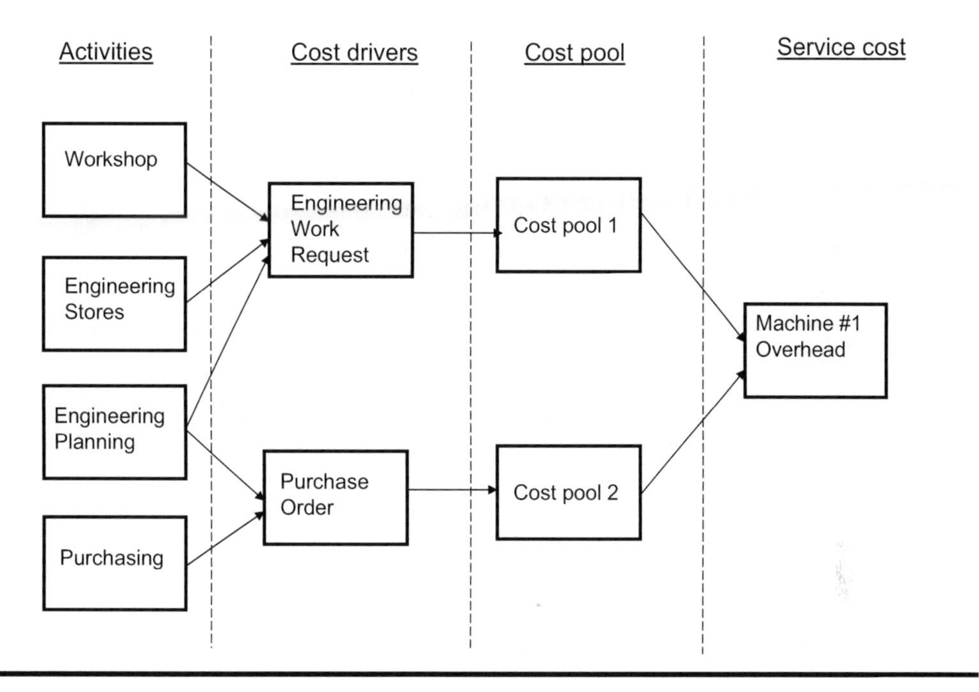

Figure 15.9 ABC terminology.

The following is some standard terminology used in ABC:

Activities: Activities are the types of work done and which consume resources in an organisation.
Cost Pool: A cost pool is the total amount of costs which may be derived from different departments, associated with an activity.
Cost Driver: A cost driver is an item that triggers an activity. Costs are assigned according to the number of occurrences associated with a cost driver.

The cost drivers are used to derive the cost of activities to cost pools. There is one cost driver for each cost pool.

The relationship between the key elements of ABC is illustrated in Figure 15.9.

15.2.7.3 Application

ABC is based on the concept that products give rise to activities which drive costs. The examples of such cost drivers in the procurement activity, for example, include purchase requisitions, quotations, purchase orders, invoices etc. In providing more accurate allocation of costs, ABC has also proved to be useful in identifying cost reduction opportunities.

Activity-based management (ABM) applies ABC-generated data for planning, controlling and cost-effectiveness. Since the early 1990s, many companies have attempted to apply ABC in their cost management initiatives. The UK companies with an active interest in ABC include British Airways, Unipart, Norwich Union and Cummins Engine.

A company is particularly suited to the application of ABC when

- The company produces a wide variety of customised products
- The company has high overhead costs

It is not the answer to all costing problems, however. It is just not appropriate, for example, for a business where simple products are produced. In fact it is now generally accepted that ABC can be used alongside a more traditional accounting system in the same company. The best approach has been to perform ABC for analysis purposes and retain the standard costing system for bookkeeping.

ABC can also be useful to a service organisation, provided that the activities are relatively homogenous, and its output can be defined. Examples of such applications are hospital and insurance services.

15.2.7.4 Basic Steps

The basic premise of ABC is that activities use resources and products consume activities. ABC uses many cost drivers. However, conventional methods typically use only one. Consequently, it is expected that the ABC method will increase the overhead allocation accuracy.

The basic steps for developing an ABC system are as follows:

1. Define and analyse activities: Activities are usually, but not necessarily, related to functional departments. However, one department may contain several activities.
2. Determine the cost driver for each activity: Cost drivers may be measured in terms of volume of transactions undertaken. For example for the quality control activity, the cost driver is given by number of inspections.
3. Identify cost pools: Activity costs with the same cost driver are collected in activity pools.

4. Assign cost products or services: The costs of activities in cost pools are assigned to products or services based on cost drivers.

15.2.7.5 Worked-Out Example

The following example is taken from Maskell (1996, p. 107).

Consider a customer service department where two cost reports are made for comparison purposes, as shown in Table 15.7.

It is evident that expediting, correction and issuing credits are non-value-added activities but they represent $520,000, nearly half of the total cost.

15.2.7.6 Benefits

1. ABC provides a means of increasing the accuracy of cost allocation for both manufacturing and service organisations.
2. It allows a better understanding of the cost of making a product and providing a service and thus helps to focus on priority activities for improvement.
3. ABC moves away from the notions of short-term fixed and variable cost and focuses on the variability of the cost in the longer term. Hence ABC has directed management attention from product costing to improving business processes leading to ABM.
4. The principles of ABC can be very effective in the analyses of customer profitability and cost of poor quality. Price estimates are also enhanced by ABC.

Table 15.7 Traditional versus ABC Reports

Traditional Cost Report	*Activity-Based Cost Report*
Salaries $920,000	Take Orders $600,000
Space $100,000	Expedite Orders $140,000
Depreciation $100,000	Correct Orders $120,000
Supplier $60,000	Issue Credits $160,000
Other $20,000	Amend Orders $60,000
	Answer Questions $40,000
	Supervise $80,000
Total $1,200,000	Total $1,200,000

15.2.7.7 Pitfalls

1. ABC has its limitations in high-technology industries and in allocating overheads where cause of these overheads are unknown.
2. The process of calculation in ABC is still viewed as complex. Empirical surveys in the UK (Bromwich and Bhimani, 1989) showed that companies expressed a great reluctance to change from their traditional cost accounting systems.
3. The process is time consuming, and its benefits are marginal in comparison to the effort associated with the detailed analyses required.

15.2.7.8 Training Requirements

The use of the ABC technique should be restricted to qualified management accountants. However, the project team members should benefit from attending the awareness workshop on ABC so that they can contribute to the identification of cost-saving opportunities.

15.2.7.9 Final Thoughts

ABC is conceptually simple and a useful technique for product pricing and profitability analysis. However, due to the complexity of calculation it should be used sparingly with support from qualified accountants.

15.2.8 R8: Quality Management Systems (ISO 9000)

15.2.8.1 Background

In this book, ISO 9000 represents the general area of accredited QMSs which relate to the organisation, procedures and processes for implementing quality management. In 1979 the British Standard Institute (BSI) issued the BS 5750 series of QMS standards. BS 5750 became ISO 9000 when in 1994 the International Organization for Standardization (ISO) created its now famous ISO 9000 by adoption of the BS 5750 together with parts of other national quality management standards. The ISO 9000 series of standards ran to around 20 different standards of which the main ones were

ISO 9001	For design, development and production
ISO 9002	For production, installation and servicing

ISO 9003	For final inspection and test
ISO 9004	For QMSs

In addition, ISO 14000 was published for environmental management standards.

The accreditation to ISO 9000 became very popular in the 1990s with government subsidies and customers asking their suppliers for confirmation of their accreditation. However, the 1994 version of ISO 9000 came into disrepute for four main reasons:

1. The numbering system left a lot to be desired. A company may be approved of ISO 9001, ISO 9002 or ISO 9003, but some customers still intended to audit them.
2. The government (e.g. Department of Trade and Industry in the UK) certified a large team of consultants and provided subsidies to promote accreditation and thus the standard of assessment could not be regulated.
3. The emphasis being on the maintenance of written quality procedures, it was viewed as 'institutionalising existing bad practice'.
4. The accreditation focused on one area of the organisation or process and therefore was not found to be a driver for improving the total business.

The vision for Phase Two to develop a single quality management standard and address these issues was conceived in 1996, and a new version is referred to as ISO 9001:2000 (see Figure 15.10). In the following sections we describe and analyse the new version of ISO 9000.

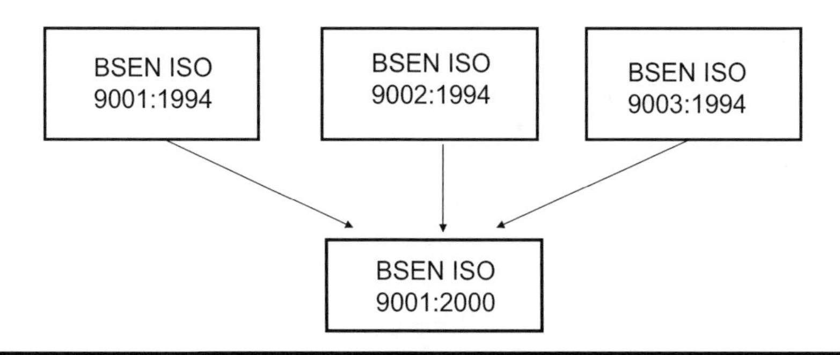

Figure 15.10 Link between ISO 9000:1994 and ISO 9000:2000.

15.2.8.2 Definition

ISO 9001:2000 is the updated QMS which specifies the requirements for an organisation to demonstrate its ability to provide products and processes that fulfil customer satisfaction.

As shown in Figure 15.11, ISO 9000:2000 contains significant changes from the 1994 standard and reflects the integration of six main areas:

1. Management responsibility: More emphasis on senior management involvement
2. Resource management: Less emphasis on paperwork and more on resources and business processes
3. Product realisation: Production and service under controlled conditions
4. Measurement, analysis and improvement: Requires measurement of processes
5. Customer focus: Requires measurement of customer satisfaction
6. Continuous improvement: Focuses the continuous improvement of both the processes and QMS

The first four are the fundamental requirements to achieve number 5 (i.e. customer focus) and number 6 (i.e. continuous improvement).

Nine guiding principles were identified to meet the requirements of ISO 9001:2000:

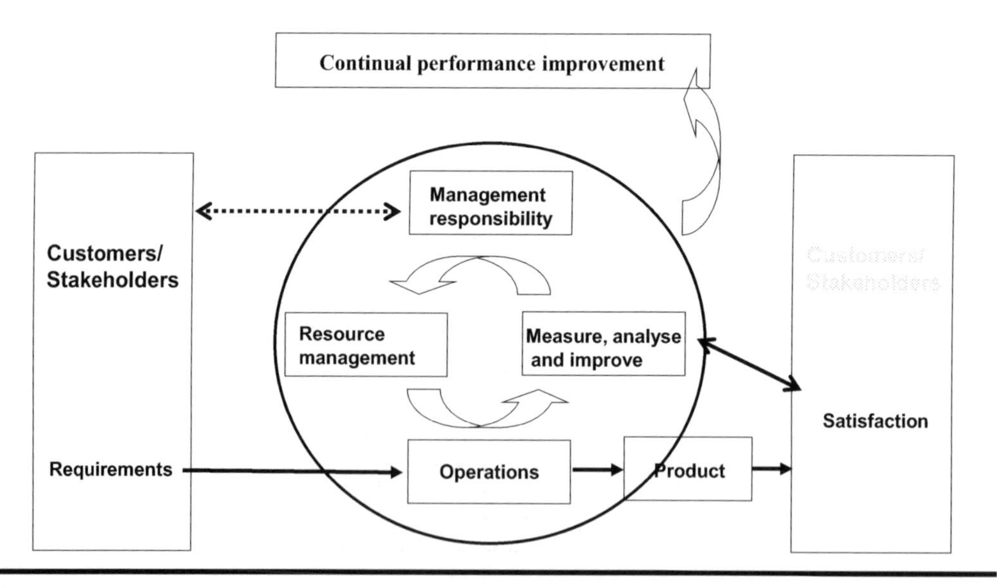

Figure 15.11 ISO 9000:2000.

- Role of leadership
- Involvement of people
- Systematic approach to management
- Customer focus
- Business process approach
- Factual approach to decision-making
- Continual improvement
- Mutually beneficial supplier relationships
- New requirements

15.2.8.3 Application

The 1994 version of ISO 9000 was extensively applied in both SMEs and large organisations. Many organisations have updated their accreditation to the new version of ISO 9001:2000.

It is a long and expensive process to gain ISO 9000 accreditation. It has been used for two primary objectives:

- To gain the customer's acceptance as a preferred supplier
- To demonstrate the quality system as a pillar in an organisation's approach to Total Quality Management (TQM)

However, ISO 9000 registration is not a pre-requisite for TQM. Many organisations, particularly in Japan, achieved excellent quality standards without the support of ISO 9000. It is also indisputable that the development and maintenance of procedures and control, as required by ISO 9000, help to sustain quality standards. As shown in Figure 15.12, ISO 9000 can act as a stopper that prevents the quality standard going in reverse.

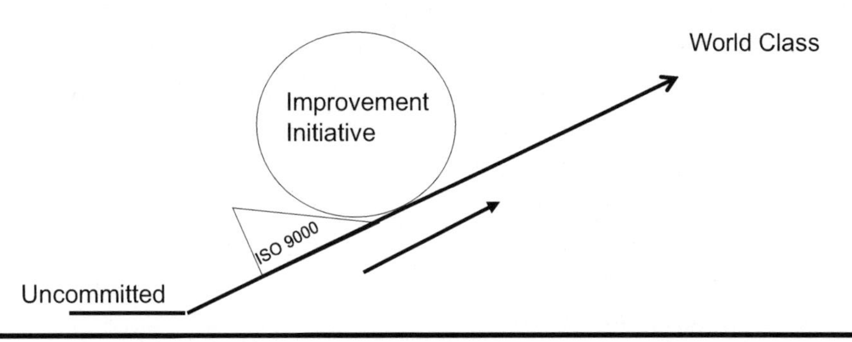

Figure 15.12 Quality improvement and ISO 9000.

The fragmented approach of the 1994 version still lingers on and appears to encourage the separation of a business into compartments where the requirements for ISO 9000 could be met. Organisations without a TQM or Six Sigma programme find it difficult to be involved in every function in a company-wide registration.

Many organisations or people are still confused about the relationship between TQM and ISO 9000. They are not alternatives, but they complement each other. As mentioned earlier, there are examples of successful TQM programmes without ISO 9000 (Sayle, 1991). Some partial improvements driven through the 1994 version of ISO 9000 would not lead an organisation to TQM. The new version ISO 9001:2000 has some potential to be an essential feature of a company-wide quality programme.

Another area of confusion is the relationship between ISO 9000:2000 and EFQM. If the new version of ISO 9000 is applied to the whole of the organisation, then it relates closely to EFQM. They have key principles in common (except perhaps the focus on supplies is less visible in ISO 9000) and the possible differences are in the scope of application. It is arguable that the differences, albeit limited, are promoted by the governing bodies of these two closely related systems.

15.2.8.4 Basic Steps

1. Prepare for the QMS: The organisation should state clearly the purpose, scope and benefits of going for ISO 9000. A steering committee should be established, ideally headed by the CEO.
2. Train a team: A QMS team is formed for a large and medium-sized organisation. Train the team with the fundamentals of ISO 9000:2000 so that the members can prepare the QMS document and conduct internal quality audits. Small organisations (of say less than 50 people) usually depend on external consultants.
3. Prepare the QMS document: The document for the QMS is determined by the nature of the business but follows the well-defined guidelines of ISO 9000:2000. It should cover the following checklist:
   a. Management responsibility
      i. Customer requirements
      ii. Quality policy, objectives and planning
      iii. Quality manual
      iv. Management review
      v. Control procedures

    b. Resource management
        i. Human resources
        ii. Business processes
    c. Product realisation
        i. Design and development
        ii. Procurement
        iii. Production and service delivery processes
        iv. Customer service
    d. Measurement, analysis and improvement
        i. Measurement and monitoring
        ii. Data analysis
        iii. Improvement
        iv. Control of non-conforming products
4. Pre-audit and identify gaps: It is important that an internal quality audit is conducted including the review of the QMS document by a qualified auditor. Involvement of the internal QMS team during this quality audit is also very important.
5. Registration for ISO 9000: When the QMS system is ready and supported by the pre-audit, a registration is sought in an accredited ISO 9000 certification body. The body will then supply an information pack and then the necessary terms are agreed with the accredited body.
6. Certification: The most appropriate time for assessment is decided after the QMS has been effectively running for 6 months. The assessment is carried out by a small team of independent assessors approved by the certification body. Any non-conformance with standards is rectified before a certification is awarded.
7. Review: The certification bodies usually follow a system of routine surveillance and revisit the organisation every 2 years at the invitation of the company.

15.2.8.5 Worked-Out Example

The following example has been adapted from a published case study by McLymont and Zukerman (2001).

Silberline Manufacturing of Lansford, US, completed a full transition to ISO 9001:2000 approximately 3 years after registering to ISO 9001:1994. The company is a global manufacturer of aluminium pigment and special products for coatings and the plastics industry.

The transition was led by a cross-functional steering committee and lasted for 1 full year from April 2000.

While no major changes were needed in product realisation processes, the site was organised around process flows that began with sales and ended with products delivered to customers. The process was then weakened until it was in control.

Rewriting the quality manual was the most frustrating part of the transition. The team abandoned the 20-element model of ISO 9001:1994 and based the new manual on the ISO 9001:2000 process model.

An ERP was already in place. All departments met monthly to discuss a 'red, yellow, green light' report which included output generated by the ERP system. The report was similar to a Balanced Scorecard format. The transit audit was conducted by BSI Inc., the US operation of the British Standards Institution. In spite of all of these trials, Silburn described the changeover to the new ISO 9001:2000 process as fairly seamless.

15.2.8.6 Benefits

1. The new version of ISO 9001:2000 reaches out to address company-wide quality issues and ensures that specified customer requirements are met.
2. Like the EFQM excellence model, it provides a framework of assessing a TQM or Six Sigma programme towards achieving operational excellence.
3. It has an international status covering all types of organisations across geographical and regional boundaries.
4. It establishes a discipline of improved controls and standard procedures, thus preventing the duplication and compromising of activities.
5. It provides an objective platform to enhance teamwork when it is applied as a company-wide programme.

15.2.8.7 Pitfalls

1. It still suffers from the chequered history of the 1994 version, and there are many 'agnostics' towards the system. It is being viewed as a bureaucratic process.
2. It is far from becoming a people process leading to self-assessments. The methodology is guarded by 'qualified' assessors.
3. The external assessors have often oversold the expectations of ISO 9000 and the certification has failed to deliver all the benefits.

15.2.8.8 Final Thoughts

The new version of ISO 9000:2000 can be used as an effective technique to improve process control and business performance in all types of organisations, whether large or small, manufacturing or service, private or the public sector. However, it has a long way to go to establish its credibility to make it people friendly and to remove the artificial demarcation with excellence models like EFQM.

15.2.9 R9: Kaizen

15.2.9.1 Background

Kaizen is the Japanese word for continuous improvement, literally meaning change and good or change for better. Kaizen was first practised in Japanese businesses after World War II, influenced by quality management experts and notably as part of the Toyota Production System.

15.2.9.2 Definition

Masaki Imai (Imai, 1986) defined kaizen as follows: 'Kaizen means improvement. Moreover it means improvement in personal life, social life and work life. When applied to the work place Kaizen means continuing improvement—managers and workers alike'.

The cycle of kaizen activity can be defined as PDCA (plan-do-check-act). This is also known as Deming cycle or PDCA. Other techniques used in conjunction with PDCA are 5 Whys and the fish-bone diagram for root cause analyses.

15.2.9.3 Application

There are two popular methods of applying kaizen, viz. Point Kaizen and System Kaizen. Point Kaizen is the most commonly used method. The measures are focused on an isolated area or a process. It is easy to implement. System Kaizen is applied in an organised manner to address the systems-level problems of the whole organisation.

Kaizen is for continuous improvement with incremental change. There are two versions of improvement by radical changes, viz. Kaikaku and Kakushin.

Kaikaku means a radical change during a limited time, whereas kaizen means an incremental change over a longer period.

There are four types of Kaikaku:

- Locally innovative implementation (e.g. AI)
- Locally innovative methodology (e.g. DMAIC)
- Globally innovative implementation (e.g. new robotic design)
- Globally innovative methodology (e.g. Green Six Sigma)

Kakushin looks at what you are doing now and not even trying to improve it. It only aims to do something radical.

Kakushin follows more from Kaikaku focusing on globally innovative products and processes.

Kaizen = Change + Good = Continuous improvement
Kaikaku = Change + Radical = Big improvement
Kaikashin = New + Radical = Radical innovation

It is important to note that Kaikaku and Kakushin are not part of kaizen, although they all are in the Lean Thinking family.

15.2.9.4 Basic Steps

1. Develop the ability to generate teamwork in kaizen. Management supports the kaizen process and lead by example.
2. Engage employees in problem-solving by using simple tools and techniques.
3. Generate the ability to move kaizen activities across organisational boundaries.
4. Specific kaizen projects with stakeholders (e.g. customers, suppliers etc.) take place.
5. The kaizen system is continually monitored and developed.
6. When a major organisational change is planned, its potential impact on the kaizen system is assessed.
7. People at all levels demonstrate a shared belief in the value of small steps and recognise incremental improvements.
8. Build the learning organisation by generating the ability to learn through kaizen activities.

15.2.9.5 Worked-Out Example

This case example (Shettar et al., 2015) was conducted in an automotive parts manufacturing industry in India. Kaizen was implemented in the hydraulic actuator assembly line. The operator had to pick up a heavy sizing tool and place it properly on the tube in the assembly stage for each assembly of the actuator. Due to the fatigue of the operator, he often missed the usage of the sizing tool causing high rejects.

Sizing tool material was changed from steel to aluminium and the weight was also reduced by 2.52 kg. The clamping of the piston rod was also changed from a manual tightening process by a pneumatic clamp. The process time was reduced by 30 seconds.

A culture of continuous improvement was followed, and further improvements were achieved by assembling two different sized actuators in the same area. Benefits achieved by kaizen included

- Reduction in cycle time
- Lower operator fatigue
- Lower rejects
- New pattern accommodated
- Increase in operator morale

15.2.9.6 Benefits

As part of the Lean Thinking toolset there are many generic benefits of kaizen, such as reducing waste, simplifying workplaces, improving safety and improving employee satisfaction.

As kaizen involves every employee making incremental small changes, the specific benefit derived from kaizen is that it changes culture to ensure problems stay solved.

15.2.9.7 Pitfalls

There are also some pitfalls of kaizen including the following:

- Localised kaizen events may lead to a shallow and short-lived burst of excitement that may be abandoned.
- It is important to have the full management commitment before something like kaizen would be well received.
- If kaizen is not properly rolled out with appropriate training, it may cause frictions in companies with a culture of closed communication.

15.2.9.8 Training Requirements

The kaizen methodology requires training staff and management to understand and adopt the kaizen philosophy. The focus of training is not on advanced tools and techniques but on communication and teamwork. This might require altering the usual process of work. Employees may need to take out time from work to undergo training. Also, new employees may need to undergo kaizen training in addition to their usual training.

15.2.9.9 Final Thoughts

Kaizen is in the camp of continuous improvement while Green Six Sigma is primarily for breakthrough. Hence there are some apparent conflicts between the approaches of kaizen and Green Six Sigma. However, kaizen fits in well with the lean processes and sustainable culture of Green Six Sigma and thus it can be considered as a qualitative technique of Green Six Sigma.

15.3 Summary

The techniques of problem-solving generally consist of use of a combination of tools. The qualitative techniques in Chapter 15 are applicable to most operations and organisations regardless of their size or whether they are in manufacturing or service, or in private or public sector. There are individual training requirements for qualitative techniques, but the focus must be on group working.

As the primary focus of Six Sigma is on data-driven solutions, qualitative techniques are often undervalued or overlooked. From the perspective of Green Six Sigma, the importance of qualitative methods to ensure sustainability of performance, processes and culture cannot be exaggerated.

GREEN TIPS

- Consider that kanban and ABC are more appropriate for improving a problem or process.
- Consider that benchmarking and QMSs are techniques to improve the performance of the whole organisation.
- Consider that Balanced Scorecard, EFQM, S&OP and KM are techniques to sustain the performance and culture of the whole organisation.

GREEN SIX SIGMA AND CLIMATE CHANGE

Chapter 16

Climate Change Challenges

What humans do over the next 50 years will determine the fate of all life on the planet.

—David Attenborough

16.1 Introduction

'Climate change' as a scientific term may well sound quite tame and vernacular, and possibly does not convey a sufficient sense of urgency. It becomes confusing when scientists additionally talk about ice ages and other natural changes to the climate which have occurred throughout the history of our planet. The term 'global warming' has also been used as another phrase for the same thing, but by contrast, this expression clearly imparts an unambiguous message concerning the imminent danger posed to humans and other living objects on earth.

What is beyond dispute is that the earth's temperature is rising, and even growth of 2°C would melt the ice sheets of the polar regions resulting in a sea level increase of many metres. I have to admit that in the context of discussions regarding our future well-being I do prefer the term 'global warming'. However as 'climate change' is the official term being used by the United Nations (e.g. United Nations Framework Convention on Climate Change [UNFCCC] and Intergovernmental Panel on Climate Change [IPCC]), I shall also use this phrase primarily to express the impact of global warming.

DOI: 10.4324/9781003268239-20

The world is producing 51 Gigatons or 51 billion tons of greenhouse gases every year. These greenhouse gases include carbon dioxide (CO_2), methane (CH_4), nitrous oxide (N_2O), water vapour (H_2O) and ozone (O_3) as shown in Table 16.1.

The share of greenhouse gases is also shown graphically in a pie chart in Chart 16.1.

Table 16.1 Greenhouse Gases

Greenhouse Gases	*Percentages*
Carbon dioxide	82
Methane	10
Nitrous oxide	6
Others	2

Source: US EPA (2019).

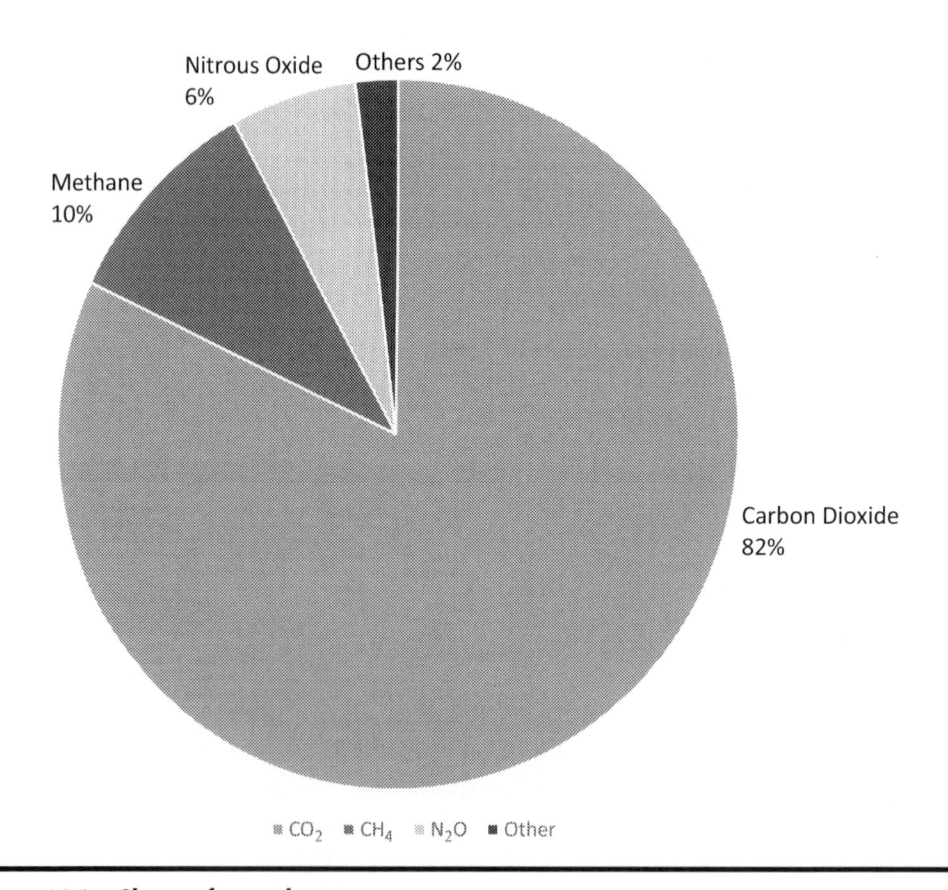

Chart 16.1 Share of greenhouse gases.

Let's consider the basic principle involved here. A greenhouse is made of glass to allow sunlight to pass through and warm the air inside. The heat is trapped by the glass and the interior becomes warmer and warmer. We experience this so-called greenhouse effect when our cars are parked in the sun and their interiors can become much hotter than the external temperature. Greenhouse gases work in a similar fashion but on a massive scale to increase the earth's temperature, hence this familiar term. Carbon dioxide, methane and other greenhouse gases stay in the atmosphere for a long time and trap the heat that would otherwise escape to the atmosphere, thus causing the earth's temperature to climb.

Carbon dioxide has been found to be the most important greenhouse gas related to global warming. It constitutes 82% of all greenhouse gases and stays in the atmosphere for a long time. Recent studies have shown that 75% of carbon will not disappear for between centuries to thousands of years while the other 25% will stay with us forever. Methane causes many times more warming, molecule for molecule, than carbon dioxide when it reaches the atmosphere, but methane does not stay in the atmosphere as long as carbon dioxide does. In addition, man-made (anthropogenic) activities are creating a serious global warming crisis that could last far longer than we ever thought possible—unless we act immediately.

Let us now examine how much the earth's temperature is rising and why this is causing serious concern.

As carbon dioxide commands the lion's share of greenhouse gases, the emission of all gases is expressed as a carbon dioxide equivalent (CO_2e).

16.2 The Earth's Temperature Is Rising

Greenhouse gas emissions, especially carbon dioxide, have increased gradually since 1840 and dramatically since 1950 as shown in Figure 16.1.

Sunlight passes through the thick layer of greenhouse gases without getting absorbed to reach the earth's surface and therefore warms up the planet. The earth radiates some of the heat energy back towards space and this hits the greenhouse gas molecules. This in turn makes the molecules vibrate faster, thus heating up the atmosphere. Only molecules of greenhouse gases, such as carbon dioxide, have the right structure to absorb radiation and thus to heat up the earth's temperature. Therefore, it is not surprising that the average global temperature since 1840 has also risen almost at the same rate as the increase of carbon dioxide (see Figure 16.2).

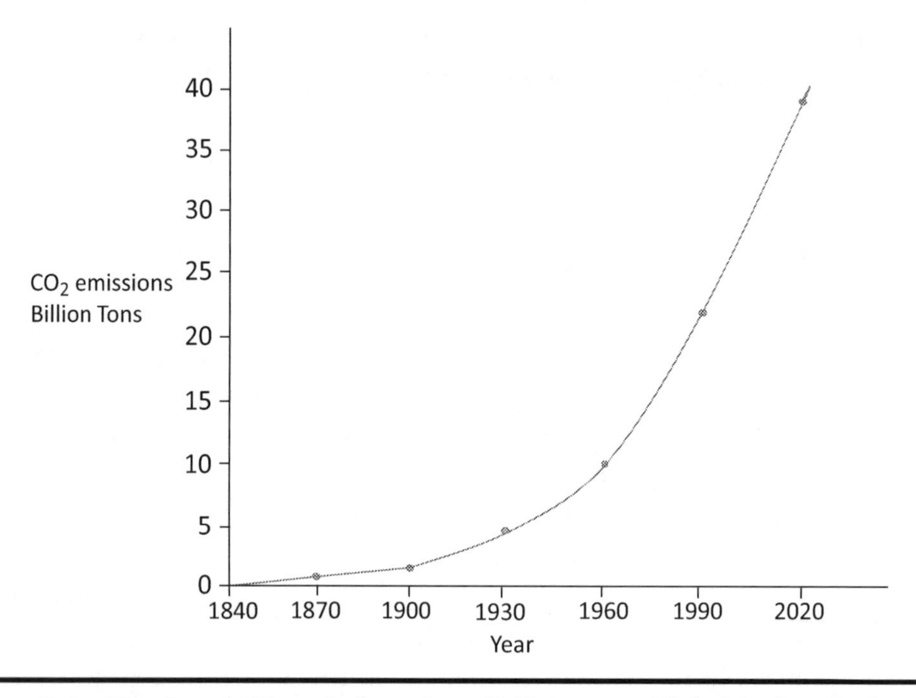

Figure 16.1 The rise of CO_2 emissions since 1840 (*source*: Global Carbon Project Report, 2019).

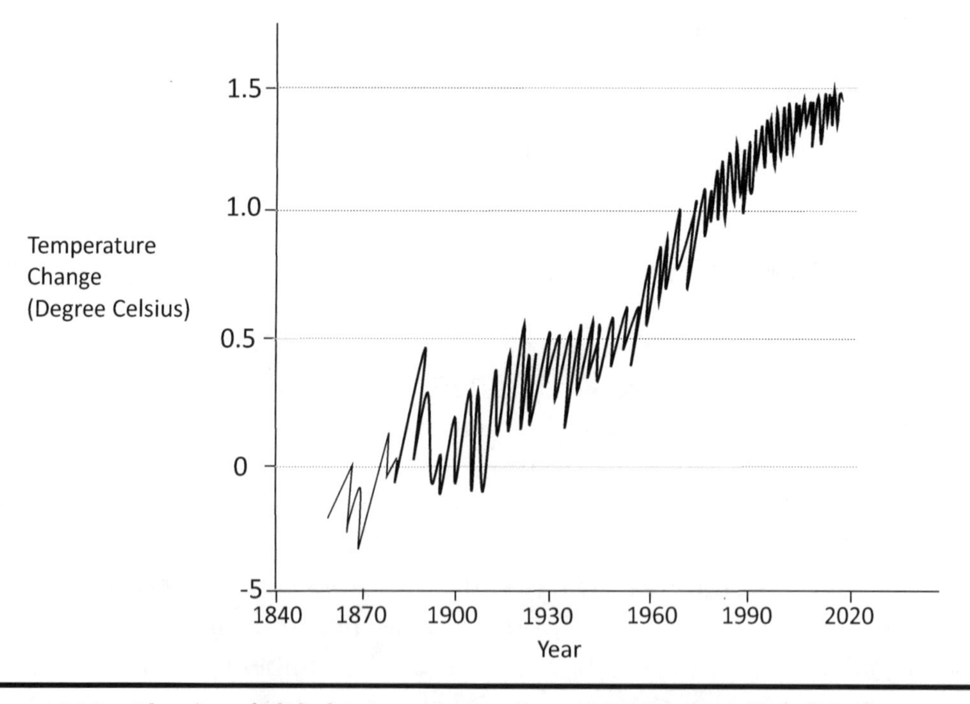

Figure 16.2 The rise of global temperatures since 1840 (*source*: Global Carbon Project Report, 2019).

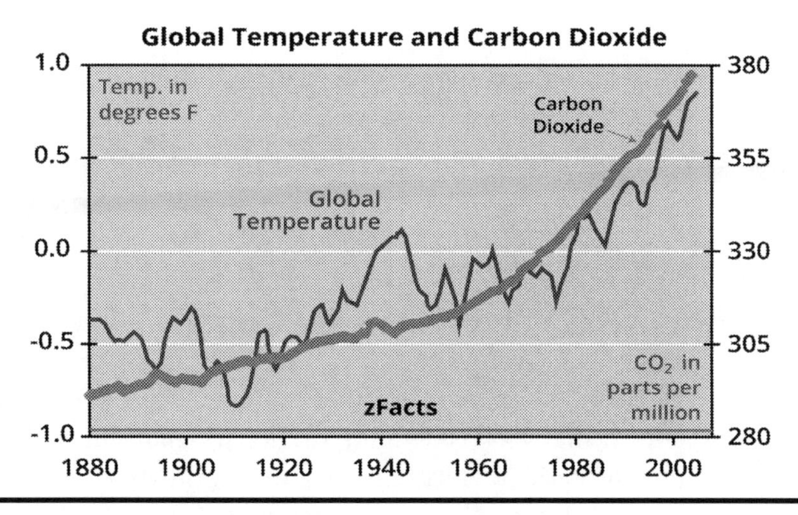

Figure 16.3 The rise of global temperatures and CO_2 emissions (*source*: Global Carbon Project Report, 2019).

The significant correlation between global temperatures and carbon dioxide emissions is clearly visible in Figure 16.3.

Scientists acknowledge (IPCC, 2021) that there is uncertainty about how much or how quickly the temperature will increase and exactly what effect these higher figures will have. The jury is out regarding the accuracy of the numbers, but there is no doubt that the earth is warming and that this is occurring as a direct result of human activities.

Scientists are debating the direct impact of global warming. However, IPCC reports are clear that global warming *is* the major factor involved in causing the harmful effects of climate change. There is growing evidence that global warming is making storms wetter and the occurrence of severe cyclones more frequent. A hotter climate also means that there will be more frequent wildfires. An alarming effect of global warming is that sea levels will rise mainly because the polar icebergs are melting. Rising sea levels will be even worse for the poorer people in the world such as those living in Bangladesh and the Pacific Islands.

The potential impacts on plants and animals also make for very bleak reading. The prediction in IPCC reports is that a rise of 2°C could destroy the geographic range of animals by 8%, plants by 16% and insects by 18%. There were 7,300 major disasters in the ten years between 2009 and 2019, resulting in 1.2 million deaths and wiping out $3 trillion (3.7%) from the global economy (IPCC, 2021). The scientific evidence would lead us to believe that, although the impact is gradual, at some point it will become catastrophic and irreversible. Even if the 'best case' scenario of that point is 50 years away, the inevitable conclusion from the data is that we must act now.

Let us start by exploring the root causes of global warming in more detail.

16.3 Greenhouse Gas Emissions by Country

The rate of greenhouse gas emissions from a country, especially carbon dioxide, depends upon many considerations. The two main factors are the generation and consumption of electricity and the population of the country. David Mackay of Cambridge University has produced a graph (Gates, 2021, page 6) which shows that the income per person of a country is proportional to the energy used per head. It was evident from the graph that richer countries (e.g. the US, Canada, Qatar) consume many times more energy per person than the poorer countries (such as Niger, Ethiopia, Haiti). Greater energy consumption means more carbon dioxide emission. However, it is not the richer nations alone who are causing the higher emission rates of carbon dioxide. Standards of living are going up in emerging economies with a rising demand for energy, cars, buildings and refrigerators. In addition, the global population is also rising.

Figure 16.4 shows the emission rate over last two decades for populous countries and regions. It is evident that emissions from advanced economies like the US and European Union have stayed fairly flat, but if we look at emerging economies, especially China, they are growing rapidly. Of course, undoubtedly it is good news that people are improving their standards of living thanks to globalisation, but the unfortunate consequence is that this is bad news for the planet we all live on. As standards of living are accelerating rapidly in developing countries like China, India, Brazil and Nigeria, both energy consumptions and carbon emissions are also increasing at a commensurate rate. It is important to note that in the future there will be an even faster growth in carbon dioxide emissions from these emerging economies.

Each country, depending on its power generation process, power consumption and population, emits differing amounts of greenhouse gases into the atmosphere. Table 16.2 below shows data which estimates carbon dioxide emissions (the main component of greenhouse gases) from the combustion of coal, natural gas, oil and other fuels, including industrial waste and

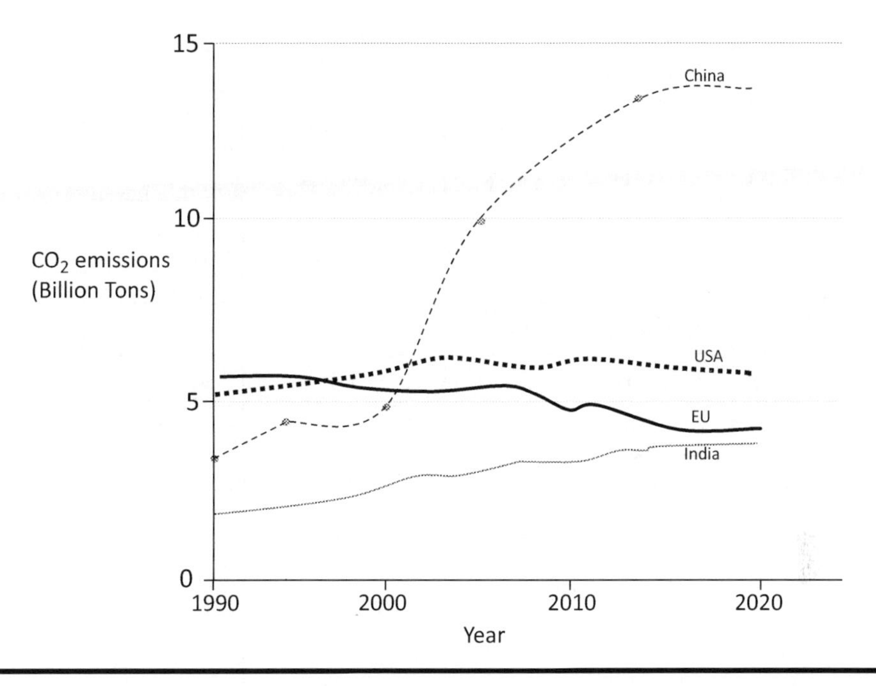

Figure 16.4 The CO_2 emission rates for countries and regions (*source*: Gates, 2021, page 41).

non-renewable municipal waste. It shows the ranking of the top 20 countries as the highest emitters of carbon dioxide in 2020.

Chart 16.2 shows graphically in a bar chart the top 10 carbon dioxide–emitting countries.

The top 20 carbon dioxide–emitting countries account for 79% of the total carbon dioxide emissions of the world, and the top 5 countries (China, US, India, Russia and Japan) are responsible for 58% of the total emissions.

A very useful way to show the impact of carbon dioxide in the atmosphere is by parts per million (PPM). This number tells us how many parts of carbon dioxide there are in 1 million parts of air. Figure 16.5 shows the rapid rise of carbon dioxide PPM levels along with the rise in carbon dioxide emissions in the atmosphere over the last 100 years.

Carbon dioxide concentrations are rising mostly because of the fossil fuels that people are burning for energy. Fossil fuels like coal and oil contain carbon that plants pulled out of the atmosphere through photosynthesis over the span of many millions of years; however now we are returning that carbon to the atmosphere within a time span of just a few hundred years. The current level of PPM in 2020 is 410, but this figure is growing.

Table 16.2 Top 20 Countries Emitting CO_2 in 2020

Rank	*Countries*	*CO_2 emissions (Giga Tons)*	*Percentage (%)*
1	China	10.06	28
2	US	5.41	15
3	India	2.65	7
4	Russian Federation	1.71	5
5	Japan	1.16	3
6	Germany	0.75	2
7	Iran	0.72	2
8	South Korea	0.65	2
9	Saudi Arabia	0.62	2
10	Indonesia	0.61	2
11	Canada	0.56	2
12	Mexico	0.47	1
13	South Africa	0.46	1
14	Brazil	0.45	1
15	Turkey	0.42	1
16	Australia	0.42	1
17	UK	0.37	1
18	Poland	0.34	1
19	France	0.33	1
20	Italy	0.33	1

Source: Earth Systems Science Data (2020).

The danger level of carbon dioxide could occur when we are exposed to levels above 5,000 PPM for a number of hours. A further critical point is that even higher levels of carbon dioxide can cause asphyxiation as carbon dioxide replaces oxygen in the blood. We do know that exposure to concentrations of around 40,000 PPM is immediately dangerous to life and health.

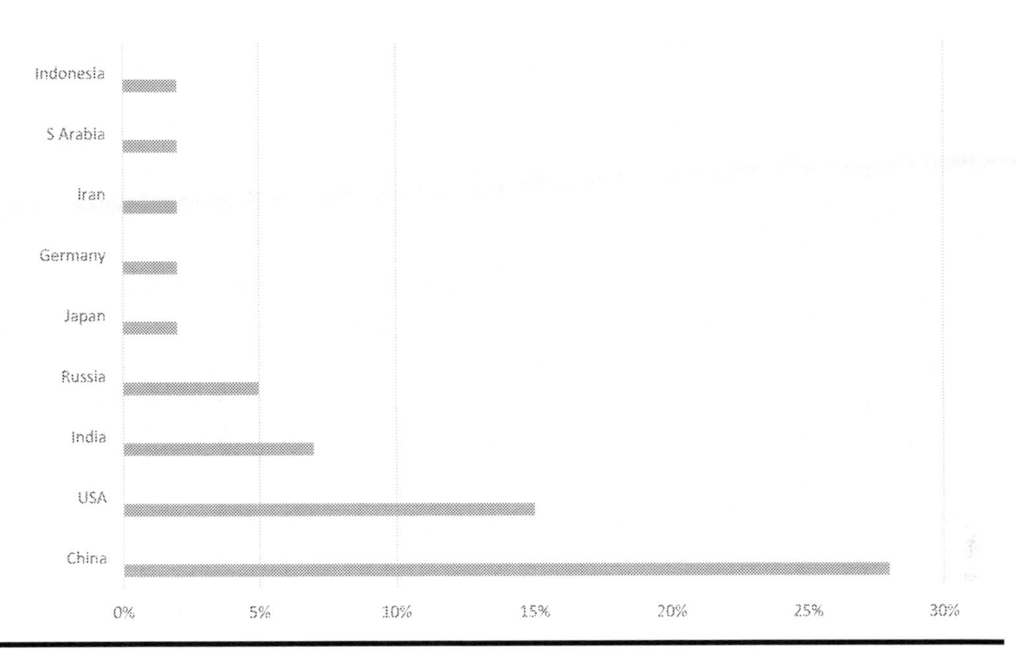

Chart 16.2 Top 10 CO_2-emitting countries.

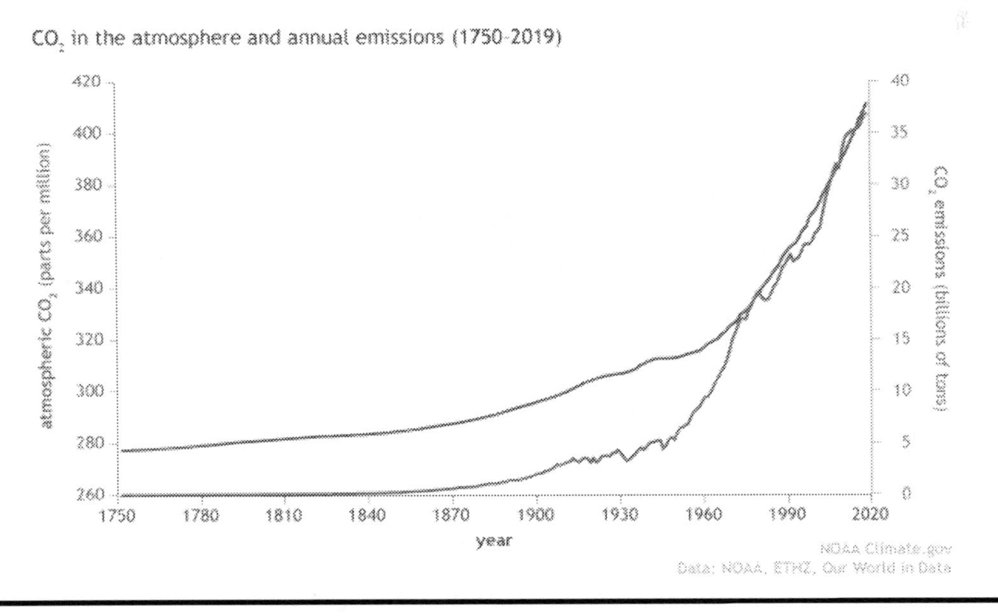

Figure 16.5 The CO_2 emission rates and CO_2 PPM.

16.4 Greenhouse Gas Emissions by Economic Sector

It is disturbing to learn that greenhouse gas emissions are anthropogenic, which means that these environmental changes are caused mainly by human behaviour, that is everything that humans do to live and prosper in a society. The share of emissions depends on the various sectors of the economy contributing to our lifestyle such as how we make things, how we grow our foods, how we are plugging in electricity, and how we are getting around.

We can see that there are some paradoxes involved. For example, while we would all agree that we need both cement and steel to build our infrastructures, making steel and cement alone accounts for approximately 10% of all emissions. There are some variations in the proportion of data for greenhouse gas emissions by economic sector depending on the sources and their chosen categories. For our discussion the categories and data from IPCC (2014a) have been chosen as shown in Table 16.3.

The share of greenhouse gas emissions by economic sectors is illustrated graphically in Chart 16.3.

From Chart 16.3, we can note that electricity and energy account for just over a third of all greenhouse gas emissions. Indeed, the burning of coal, natural gas and oil for electricity and heat is the largest single source of global greenhouse gas emissions.

Greenhouse gas emissions from the agriculture and land sector come mostly from agriculture (the cultivation of crops and livestock) and the

Table 16.3 Greenhouse Gas Emissions by Economic Sectors

Economic Sector	*Percentages*
Electricity and other energy	35
Agriculture and land	24
Industry	21
Transports	14
Buildings	6

Source: IPCC (2014a).

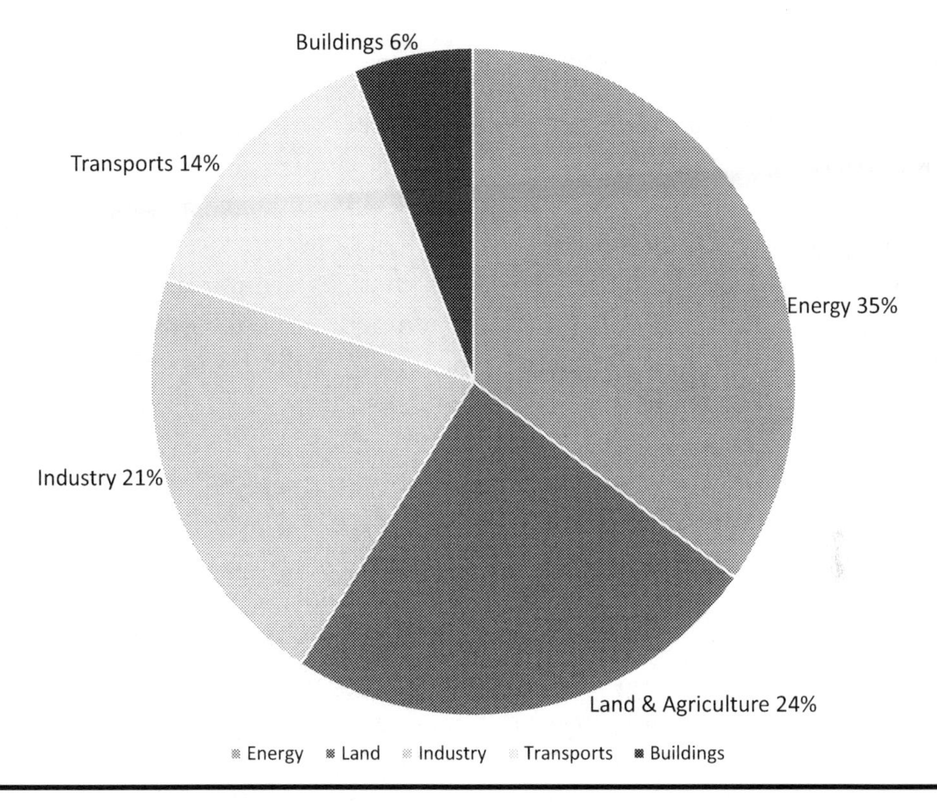

Chart 16.3 Greenhouse Gas Emissions by Economic Sectors.

practice of deforestation. The gas emissions from cattle and sheep are also a significant source in this sector.

Greenhouse gas emissions from industry primarily involve fossil fuels burned on-site at facilities for energy. This sector also includes emissions from chemical, metallurgical and mineral transformation processes not associated with energy consumption and emissions from waste management activities.

If we consider the transportation sector, greenhouse gas emissions primarily involve fossil fuels burned for road, rail, air and marine transportation. Almost all (95%) of the world's transportation energy comes from petroleum-based fuels, largely gasoline and diesel.

Greenhouse gas emissions from the energy sector arise from on-site energy generation and burning fuels for heat in buildings or cooking in homes. Emissions from the electricity and other sectors are covered in Chapter 18.

16.5 Summary

From the evidence-based analysis in this chapter some key points can be summarised:

- The global warming debate is over. There is incontrovertible evidence now that over the last 50 years our planet is warming between 0.5°C and 1.5°C. If no action is taken, the temperature will continue to rise and consequently sea levels will increase, with impacts on our weather cycle including more frequent cyclones, wildfires and incidences of flooding.
- The world is producing 51 billion tonnes of greenhouse gases every year; the biggest contributor is carbon dioxide (82%) generated mainly from the combustion of coal, natural gas, oil and other fuels.
- The world's most advanced economies (e.g. US, Japan, EU, UK and Canada) have been producing greenhouse gases. However the growth in emissions will come from the emerging and populous economies (e.g. China, India, Brazil, Russia, Nigeria and Indonesia).
- The largest economic sector of carbon dioxide emissions is the generation and consumption of energy (35%). Fossil fuels (e.g. coal, gas and oil) are the major sources of power plants and energy consumption in industries and houses.
- To avoid climate disaster, we must act now. We need to invest in future research for breakthrough solutions, but we have to deploy the tools we already have, such as solar and wind.

If we have a breakthrough process solution now, we must apply it immediately. We should also aim for a sustainable outcome. The UN (UN Foundation, 2020) projected 17 sustainable development goals (SDGs) to be achieved by 2030, of which climate change is the 13th goal. Sustainability focuses on meeting the requirements of the present without compromising the ability of future generations to meet their own needs.

Green Six Sigma (a hybrid of Six Sigma) is designed to provide both breakthrough process solutions and sustainable outcomes.

GREEN TIPS

1. Remember three numbers: 51 billion, zero and 2050.51 billion tonnes of greenhouse gases are added to the atmosphere every year. Our target is net zero emissions by 2050.
2. Carbon dioxide is the biggest contributor (82%) of greenhouse gases.
3. We must act now with urgency to avoid a climate disaster.

Chapter 17

International and National Climate Change Initiatives

We must now agree on binding review mechanism under international law, so that this century can be called the century of decarbonisation.

—Angela Merkel

17.1 Introduction

It was not until 1972 that environmental issues received serious attention from any national or international organisations. In June 1972 the first Earth Summit held in Stockholm adopted a declaration that set out principles for the preservation and enhancement of the human environment. The conference also proposed the establishment of stations to monitor long-term trends in atmospheric properties; however, at this stage, climate change was not the central preoccupation. Over the next two decades concerns for the global climate slowly gained traction and garnered international interest. From around the late 1980s global warming and the depletion of the ozone layer became increasingly prominent in the sphere of international public debate and a fixture on the political agenda, with the formation of the Intergovernmental Panel on Climate Change (IPCC) in 1988.

As a sense of urgency gained momentum and demands increased for stronger international action on climate change, the United Nations

DOI: 10.4324/9781003268239-21

General Assembly decided to convene the United Nations Conference on Environment and Development. This summit was held in Rio de Janeiro in 1992. The most significant event during the conference was the agreement of the United Nations Framework Convention on Climate Change (UNFCCC) which was signed by 158 states.

17.2 International Climate Change Initiatives

Against this backdrop, a number of international initiatives on climate change were established including

- The UNFCCC
- Kyoto Protocol
- International Carbon Action Partnership (ICAP)
- The Paris Climate Agreement
- The Conference of Parties (COP)
- IPCC

17.2.1 The United Nations Framework Convention on Climate Change

The UNFCCC is an international environmental treaty addressing climate change signed by 158 states at Rio de Janeiro in 1992. The United Nations established a Secretariat headquartered in Bonn to implement the Rio de Janeiro agreement.

The treaty established three categories of signatory states with differential responsibilities. The categories are developed countries, developed countries (Annex 1) with special financial responsibilities (Annex 2) and developing countries. Annex 1 countries are members of the EU and 13 Eastern European states in transition to market economies. This group is called upon to adopt national policies to limit greenhouse gas emissions. Annex 2 countries include all Annex 1 countries except the 13 East European states. This group is asked to provide financial resources to meet the costs of developing countries to reduce greenhouse gas emissions. Finally, the developing countries of the third category are required to submit progress reports of their individual commitments on climate change and emission targets for greenhouse gases.

The UNFCCC has been criticised for its apparent ineffectiveness in reducing levels of emissions of greenhouse gases since its creation. However, it should take the credit for being the major international forum of climate change and the source of other international climate change initiatives such as the Kyoto Protocol, ICAP, COP and IPCC, which are discussed next.

17.2.2 The Kyoto Protocol

The Kyoto Protocol was adopted in Kyoto in December 1997 by 192 countries. It extends the 1992 UNFCCC objective to reduce greenhouse gas emissions based on the scientific consensus that global warming is occurring and that carbon dioxide emissions as a result of human activity are causing this. The Kyoto Protocol applies to the six components of greenhouse gases including carbon dioxide (CO_2), methane (CH_4) and nitrous oxide (N_2O) as the main contributors.

The Protocol acknowledges that developed and developing countries have different capabilities in combating climate change and places the obligation more heavily on developed countries. The Protocol's first commitment period started in 2008 and 36 countries fully complied with it. The second commitment period, known as the Doha Amendment, commenced in 2012, under which 37 countries, including EU and Australia, have binding targets. However, there have been stumbling blocks along the way. Japan, New Zealand and Russia have participated in Kyoto but have not committed to the second-round targets. Other developed countries without second-round goals are Canada (which ceremonially withdrew from the Protocol in 2012) and the US (which has not ratified). Of the 37 countries with binding commitments, 34 have ratified and 147 states have accepted the Doha Amendment in principle.

Following the much-publicised departure of Canada in 2012 and the lack of ratification by many countries including the US, the Kyoto Protocol is considered to have failed to deliver. This resulted in the adoption of the Paris Agreement in 2015 as a separate instrument under the UNFCCC.

17.2.3 International Carbon Action Partnership

The ICAP was founded in Lisbon in 2007 by more than 15 government representatives as an international cooperative forum. From its Secretariat based in Berlin, ICAP coordinates the sharing of best practices and monitors the progress of the Emission Trading Scheme (ETS). ETS is an incentive-based approach for reducing emission pollutants, also known as 'cap and trade'. A maximum 'cap' or limit is set on the total amount of greenhouse gases that can be emitted by all participating members.

ICAP regularly publishes the status report of emission trading worldwide (ICAP, 2021). The partnership currently counts 32 full members and five observers. The UK launched its own domestic ETS following its departure from the European Union (EU) while the Chinese government officially announced the start of the first compliance cycle of its domestic ETS in January 2021, ending in December 2021.

ICAP advocates for a Paris Agreement that supports countries in using market mechanisms on a voluntary basis to help achieve their intended nationally determined contributions.

17.2.4 The Paris Climate Agreement

The Paris Climate Agreement (also known as the Paris Accord) is an international treaty on the climate crisis. Its stated aim is to radically reduce global carbon emissions and restrict the rise in the earth's temperature to less than 2°C. This treaty was signed by all 189 participating countries in Paris in December 2015 at COP 21.

There are four guiding principles for the agreement:

- To support parties in transferring part of their mitigation outcomes to other parties for compliance, leading to greater emissions reductions than can be achieved individually.
- To provide for a sound and transparent accounting framework for internationally transferred mitigation outcomes and build trust among parties.
- To encourage the development and use of robust monitoring, reporting and verification standards and ensure that the environmental integrity of parties' mitigation commitments is not undermined.
- To build upon the knowledge and institutions developed by countries and the UNFCCC.

The participating countries are also required to report every 2 years on how much greenhouse gas is being produced within their geographical boundaries. These reports are called 'greenhouse gas inventories'. Countries must also provide feedback on their efforts regarding climate change adaptation responses.

The treaty is bound by international laws. Although the participating countries are not bound by law to aim towards one universal goal, they are required to establish and review their own climate targets and plans every 5 years. These objectives must be set on a consistent basis. The plans and targets are known as Nationally Determined Contributions (NDCs). With 189 nations having ratified the landmark accord since 2015, the US is the only country in the world to have formally withdrawn from it in June 2017. However, the succeeding US administration of President Biden re-joined the Paris Climate Agreement in February 2021.

Notwithstanding the fact that, inevitably, some doubts arose when President Trump began the process of extracting the US from the Paris Agreement, the treaty has proved to be remarkably resilient. The key axis of the EU and China has remained intact. There are other successes such as the inclusion of 1°C as the aspirational target, a powerful movement towards 'net zero emissions' as well as a multiplicity of successful actions on climate by large businesses. However, even with all these positive steps, emissions have continued to rise globally. The UN Environment Programme (UNEP) reported that releases escalated from 50 billion tonnes in 2015 to 55 billion tonnes in 2019.

Despite the evident challenges, there is once again a sense of optimism around what can be achieved with President Joe Biden re-joining the Paris Agreement on his first day in office, and China's President Xi Jinping committing the world's largest emitter to a zero emissions target by 2060.

17.2.5 The Conference of Parties

The COP is the decision-making body of the UNFCCC and can produce some very important outcomes. For example, the Paris Climate Agreement was signed at the COP 21 in 2015. All states that are parties to the convention are represented and work together to make the relevant decisions that safeguard the successful implementation of the convention. The COP meets annually and reviews the national communications and emission inventories submitted by parties. COP dates back to March 1995 when the first COP meeting was held in Berlin, Germany, while the most recent, COP26 in 2021.

COP26 was due be held in Glasgow in 2020 but was rescheduled as a result of the Covid-19 pandemic and is due to occur in November 2021. It will take place in an atmosphere of anticipation, and expectations are high as President Biden has prioritised climate change, promising to convene a climate summit of the world's major economies within 100 days of taking office. All eyes will be on those countries that have not yet committed to long-term net zero targets. There will also be an appetite to see progress from those nations that have already dedicated themselves and that have already offered detailed plans for cutting emissions by 2030.

17.2.6 The Intergovernmental Panel on Climate Change

The IPCC is a highly respected body of the UN and was established in 1988. In conjunction with former US Vice President Al Gore, the IPCC won the Nobel Prize in 2007.

The group works to produce reports on climate change which are commonly held to be the official consensus of scientists, experts and global governments. These documents allow the IPCC to assess, summarise and provide an overview of knowledge, progress, impacts and the future risks of climate change. The reports are drafted and reviewed by the scientific community at various stages thus guaranteeing objectivity and authenticity.

17.2.7 Other International Initiatives

There are other international bodies involved in climate change initiatives, albeit with lower profiles, including Global Environment Facility (GEF) and Global Carbon Project (GCP).

The GEF was set up in 1992 at the Rio Earth Summit. GEF supports 184 countries in partnership with some private sector groups and civil society organisations to address global environmental issues. GEF also provides financial support for projects related to climate change, international waters, the ozone layer and environmental pollutions. The organisation has financed or co-financed more than 4,800 projects in 170 countries.

The GCP was established in 2001 as a global research project of Future Earth, a network of the international science community. Based in Canberra, its aim is to decelerate the escalation of greenhouse gases in the atmosphere. The primary focus of GCP research is carbon management and the group regularly publishes Global Carbon Project reports in English, Chinese, Japanese and Russian.

17.3 National Climate Change Initiatives

Both developed and developing countries, as part of UNFCCC and the Paris Climate Agreement, are contributing to climate change initiatives as well as submitting reports on greenhouse gas inventories. In addition, many states are proactively following national climate change strategies. In this section some of these national and regional initiatives are discussed.

17.3.1 European Union Emissions Trading System

The European Union Emissions Trading System (EU ETS) was launched in 2005 as the first large greenhouse trading scheme.

All installations must monitor and report their greenhouse gas emissions. A maximum cap is set on the total quantity of greenhouse gases that can be emitted by all participating installations. EU allowances for emissions are then auctioned off and can subsequently be traded. If emissions exceed the amount permitted by allowances, an installation must purchase allowances from others. Conversely, if an installation has performed well and has demonstrated success at reducing its emissions, it can sell its leftover credits. This is known as 'the cap and trade principle'.

In 2013 the EU ETS covered more than 11,000 installations of the power stations and factories of all 27 EU member countries. Subsequently the EU ETS was also extended to EU airports.

We examine in more detail how some key global economies are performing in terms of climate change, with a focus upon some of the worst 'sinners' in this area.

17.3.2 UK Climate Change Initiatives

British climate change initiatives are ambitious. The UK aims not just to meet the tenets of the Kyoto Protocol to reduce all greenhouse emissions by 12.5% from 1990 levels by 2012. In fact, new commitments will set the UK on a path to slash its carbon emissions by 78% by the year 2035. Furthermore, the 2008 Climate Change Act commits the UK government to cut national greenhouse gas emissions by at least 100% of 1990 levels (net zero) by 2050.

Achieving these targets would require more widespread use of electric cars, low-carbon heating and renewable electricity as well as cultural changes to food purchasing habits by a reduction in the consumption of meat and dairy products. Following Brexit, the UK entered its own UK-only ETS, although this is broadly similar to the EU ETS described earlier and the same ethos applies. Industries and power plants receive permits to emit greenhouse gases and can trade them at the market rate. Increasing renewable energy production, by wind, solar and nuclear power is one fundamental way that will allow the UK to meet its binding net zero target by 2050.

At the COP 21 meeting held in Paris, Britain joined 'Mission Innovation' with a group of 20 countries, pledging to 'double spending on clean tech R and D'.

17.3.3 US Climate Change Initiatives

The US is the second highest polluter of greenhouse gases but arguably holds maximum power and influence in global climate change initiatives. A look at the varying policies of US administrations towards green initiatives in this century reveals rather a 'swing of the pendulum' effect. It is evident that recent Republican presidents (viz. George W Bush and Donald Trump) have sought to protect fossil fuel industries while their Democrat counterparts (Barack Obama and Joe Biden) are more proactive supporters of global and national climate change initiatives.

The US, although a signatory to the Kyoto Protocol, has neither ratified nor withdrawn from this agreement. President Clinton in 1993 committed the US to reducing their greenhouse gas emissions to 1990 levels by 2020. However in 2001 George W Bush announced that the US would not implement the Kyoto Protocol.

Bush's successor adopted a different and more proactive approach to the pressing problem of climate change. In 2009 President Obama announced that the US would enter a 'cap and trade' system to limit global warming. He also established a new office, the White House Office of Energy and Climate Change, and appointed Todd Stern as the Special Envoy for Climate Change. In 2013 President Obama and Chinese President Xi Jinping formulated a landmark agreement to reduce carbon dioxide emissions and 2 years later in 2015, President Obama became a signatory of the Paris Climate Agreement. In the same year the US committed to reducing emissions to 26–28% below 2005 levels by the year 2025, a reflection of the US goal to convert the national economy into one of low-carbon reliance.

However 2017 saw another reversal of attitudes with a change of administration, when President Trump withdrew the US from the Paris Climate Agreement and appointed Scott Pruitt, a climate change denialist, as his Director of the Environmental Protection Agency (EPA).

Since taking office in 2021, the Biden administration has re-joined the Paris Climate Agreement and created a National Climate Task Force. President Biden has also proposed spending on climate change in his infrastructure bill, including $174 billion for electric cars and $35 billion for research and development into climate change initiatives.

17.3.4 China Climate Change Initiatives

China accounts for 28% of global greenhouse gas emissions and is the world's number one polluter due to an energy infrastructure heavily reliant upon the use of fossil fuels and coal. Furthermore, major industries including the construction and manufacturing sectors contribute heavily to the country's enormous levels of carbon dioxide emissions. China is already experiencing the severe impacts of global warming upon its agriculture, water resources and the environment.

However, as a signatory to the Paris Climate Agreement, the nation appears to be committed to climate change initiatives. President Xi Jinping announced in 2020 at the UN General Assembly that his country would hit peak carbon emissions before 2030, aiming towards attaining carbon neutral status by 2060. The National Leading Group to Address Climate Change was established by the Chinese government in June 2007. It was created in response to international pressure and constitutes a ministerial 'super group' designed to co-ordinate complex decision-making processes related to the key areas of climate change, emissions reductions and the conservation of energy. Furthermore, in terms of clean technology, China is a major world producer and exporter of solar panels, wind turbines and electric cars.

17.3.5 India Climate Change Initiatives

Another major problem area is India, the third largest global polluter of greenhouse gas emissions (7%). If India is to achieve an economic growth rate of over 8% then the country should also focus on its energy constraints. India is highly dependent on imported oil which constitutes 70% of its requirements.

The nation was on track to achieve its NDCs (nationally determined contributions) from 2015, but it is also one of the fastest growing economies in the world with a rapidly expanding climate footprint. The country's emissions have risen by a staggering 184% since 1990—in fact, China is the only nation whose emissions have increased by a larger rate. Prime Minister Modi played a prominent role in COP 21 in Paris and has said that he also expects developed countries to support international financing to India's climate change initiatives.

17.3.6 Russia Climate Change Initiatives

Russia ranks fourth in the greenhouse gas emission table (5%) and also has a high profile on the international political stage. The country has signed all

UN climate treaties including the Paris Climate Agreement and thus agrees to aim to reach net zero carbon emissions by 2050.

In 2020, a long-term strategy was presented by Russia on how to reduce greenhouse gas emissions by 2050. An internal review also claims that Russia's greenhouse gas emissions decreased by an encouraging 30% between 1990 and 2018.

However Russia's energy strategy is not clear. The country enjoys a major advantage, being fortunate in possessing one of the largest stores of solar, wind, geothermal and biofuels in the world. However the nation's 'Energy Strategy 2035' has also projected a substantial increase of Russian fossil fuel production, combustion and exports within the next 15 years.

17.3.7 Japan Climate Change Initiatives

Japan is fifth in the list of global polluters with its 3% share. The country is in many ways synonymous with key climate initiatives, as Japan acted as the host of COP 3 in Kyoto in 1997 when the Kyoto Protocol was adopted. However, the country's performance in emissions control prior to 2013 was less than satisfactory. Since the East Japan earthquake of 11 March 2011 Japan's dependence on coal-fired power has increased. The country's policy decision to continue the new construction of coal-fired power plants domestically and to support them financially abroad was criticised by the international community.

Matsushita (2020) recommends three key measures that Japan should adopt:

1. Set ambitious greenhouse gas reduction targets such as 40–50% reduction by 2030 and net zero emissions by 2050
2. Radically reform current coal-fired power policy, including stopping construction of new coal-fired power plants, and discontinue providing financial assistance for the construction of coal-fired power plants abroad
3. Introduce full-fledged carbon pricing

17.4 Summary

Despite the variety of records and achievements and discrepancies in these status updates, there is no doubt that, notwithstanding some pauses due to geopolitical factors, all major economies at least in theory are committed to international climate change initiatives. In addition, the major economies have signed all UN climate treaties including the Paris Climate Agreement.

The international community generally agrees on the science behind climate change and how to set greenhouse emissions reduction targets. However many experts have diverged on whether the Paris Climate Agreement will be enough to prevent the average global temperature from rising 1.5°C. The IPCC regularly assesses the latest climate science and produces consensus-based reports for countries. The/Its Green Strategy has become a positive force both in geopolitics and international businesses.

However the actual results achieved so far are not so encouraging. Current policies could result in a 2.9°C rise by 2100 (IPCC, 2021). In spite of a drop in greenhouse gas emissions in 2020 due the 'lockdown effect' of the Covid-19 pandemic, overall releases are rising, from 50 billion tonnes in 2015 to 55 billion tonnes in 2019 (UNEP, 2019). Regardless of the growth in renewables and clean technology, countries are on track to produce more than double the amount of fossil fuels by 2030.

Looking ahead, we must be optimistic in order to harness energy and expertise. It is encouraging that the desire for change is evident. We can expect a plethora of new announcements and initiatives on climate action, and we are beginning to witness the political momentum necessary to solve global warming problems. There is a new sense of optimism following the mobilisation of global efforts during the Covid-19 pandemic. In the immediate future, it is essential that we see concrete action including international and government plans and funding regarding climate change initiatives. However we need to make it happen at the sources of greenhouse emissions by taking grassroots action at power plants, factories, service centres, transport infrastructures and buildings.

The good news is that we have the technology and holistic processes to mitigate these climate change outcomes at the source. Let's now examine also how Green Six Sigma aims to address these key issues for our times.

GREEN TIPS

1. The aim of the Paris Climate Agreement, signed by 189 countries, is to restrict the rise of earth's temperature to less than 2°C by 2050.
2. Three top polluters of greenhouse gases are China, US and India.
3. IPCC is the most respected body for climate change reporting.

Chapter 18

Green Six Sigma and Clean Energy

I would like nuclear fusion to become a practical power source. It would provide an inexhaustible supply of energy without pollution or global warming.

—Stephen Hawking

18.1 Introduction

In preceding chapters I have addressed the existential challenges facing our planet caused by greenhouse gas emissions, as well as the extent of both global and national initiatives to mitigate these perils. I have also described some of the available tools and processes offered by Green Six Sigma and digital technology to accelerate the implementation of these initiatives. Although neither these strategies nor the wisdom of the carbon neutral target by 2050 are in doubt, it is important to note that in reality the reduction of greenhouse gas emissions can only be delivered by tackling its sources, such as power stations, factories, buildings, land usages and transport systems.

Table 16.3 in Chapter 16 indicated that the biggest source of greenhouse gas emissions is the energy sector, contributing 35% of total releases. This chapter analyses the challenges and opportunities in this sector as well as the role of Green Six Sigma to accelerate and sustain the desired outcomes.

DOI: 10.4324/9781003268239-22

Table 18.1 Green Six Sigma by Economic Sector

Economic Sector	*Percentage (%)*	*Chapters and Topics*
Electricity and other energy	35	Chapter 7: Clean Energy
Agriculture and land Industry	24 21	Chapter 8: Green Supply Chain
Transports	14	Chapter 9: Green Transports
Building	6	Chapter 10: Retrofitting Buildings

Before I address the issues of clean energy, Table 18.1 illustrates how the share of greenhouse gas emissions by economic sector has been grouped to describe the impact of Green Six Sigma in each sector; this is addressed in the later chapters of this book.

18.2 Guiding Factors of Clean Energy

There are a number of aspects that arise if we scrutinise articles or plans regarding clean energy and the shaping of its future direction. The good news is that even though the energy sector constitutes only 35% of greenhouse gas emissions, it represents much more than 35% of the potential solution. This is because energy generation and supply are at the start of the greenhouse gases value chain. Industry, transports and buildings, and even agriculture, are all driven by the supply of energy.

The first guiding factor is the basic demand for power to support the energy supply of a house, city or a country. Although this demand varies according to seasonality, degree of industrialisation and the time of the day, Table 18.2 is indicative of how much power we do need to support us.

The global requirement for electricity and energy is growing, and this is bearing in mind that currently some 860 million people (including 600 million in Africa) do not have access to any electricity at all.

We are highly dependent on fossil fuels (84.3% for combined oil, gas and coal) for the generation of primary power as shown in Table 18.3.

There are positive growths for renewables and nuclear sources of primary energy, but these advances are not enough meet the carbon target of 2050. Another worrying trend is that gas consumption is increasing.

In the UK as well, most of the primary energy is produced by fossil fuels, mainly natural gas (45%) and coal (9%). However, shares of renewables (25%) and nuclear (21%) are also relatively high.

Table 18.2 How Much Power Do We Need?

Location	*Power Supply*
Average household	1 KW
Medium-size city	1 GW
UK	35 GW
US	1,000 GW
World	5,000 GW

Source: Gates (2021).

Table 18.3 Fuel Share of Primary Energy in 2019

Fuel Source	*Share of Energy (%)*	*Percentage Change from 2028*
Oil	33.1	–0.2
Gas	24.2	+ 0.2
Coal	27	–0.5
Hydro	6.4	0.0
Renewables	5.0	+0.5
Nuclear	4.3	+0.1

Source: BP Statistical Review of World Energy (2020).

The space required for primary power generation is also an important guiding factor. Power density is a useful metric, allowing us to understand the amount of power generated per square metre of land and water for different energy sources, as shown in Table 18.4.

In terms of the required space, power plants run by fossil fuels have a clear advantage over renewables. However, one benefit of wind turbines is that they could be installed on otherwise unusable land and water.

Then there is the small matter of the cost advantage of an installation. Detailed cost-benefit analyses are beyond the scope of this book; however, we can discuss some of the principles involved. Zero-carbon solutions are in general more expensive than fossil fuel power plants. When we consider the cost of the environmental damage caused by fossil fuels or the government-imposed 'carbon tax', then renewable alternatives may seem more viable. Some protagonists of fossil fuels have suggested 'direct air capture' (DAC) as a feasible alternative. With DAC, emissions are blown over a device

Table 18.4 Power Density of Energy Sources

Energy Source	*Watts per Square Meter*
Fossil fuels	500–10,000
Nuclear	500–1,000
Hydro	5–50
Solar	5–20
Wind	1–2

Source: Gates (2021).

that absorbs carbon dioxide and the output is then stored for safekeeping. However, DAC is a very expensive and largely unproven technology.

A novel concept of the Green Premium is advocated by Bill Gates (Gates, 2021). The Green Premium is the difference between the cost of a green solution and the alternative or existing option. There are many such Green Premiums—some for electricity from various fuels, others for cement, steel and so on. The size of the premium depends upon first what you are replacing and second what you are substituting it with. In rare cases, a Green Premium can be negative—for example it may be cheaper to replace your gas cooker with an electric one.

18.3 How Clean Energy Solutions Can Reduce Greenhouse Gas Emissions

The primary objective of clean energy solutions is to generate carbon-free electricity. We know that the key suppliers of clean energy are renewable sources, which are natural sources that are constantly replenished. These constitute

- Wind energy
- Solar energy
- Hydropower
- Geothermal energy
- Tidal energy
- Biomass
- Hydrogen

Nuclear energy is also a source of carbon-free electricity although the material (e.g. uranium) used in nuclear power plants is not a renewable source. Nuclear energy is considered to be an effective way of producing power without the harmful by-products emitted by fossil fuels.

If we combine the energy from renewable sources and nuclear energy we get clean energy without any greenhouse gas emissions. The biggest share of the 2,790 gigawatts of clean energy in 2019 came from hydropower (42%) as shown in Table 18.5.

The share of clean energy is also shown in a pie chart in Chart 18.1.

The portion of hydroelectricity has been stable for several years. The share of wind and solar energy is growing while the segment of nuclear energy is in decline, especially after the 'Great East Japan Earthquake' in 2011. Thanks to wind and solar power, renewable energy experienced a record growth in 2019 accounting for 40% of the expansion in primary energy. This is undoubtedly a good sign for clean energy.

Hydropower that harnesses the strength of water is the oldest source of energy. In Ancient Greece flowing water was used to turn the wheels of flour mills. The most common type of hydroelectric plant is called an impounded facility. Here a large reservoir is created by building a dam and the flow of controlled water drives a turbine thereby generating electricity. This natural energy source has the advantage of being more reliable than wind or solar power. Furthermore, electricity can be stored for potential use at a time of peak demand. In another type of plant, called a diversion plant, a series of canals are used to channel a flowing river towards the turbines. A third type of operation, called a pumped storage facility, is where the plant collects energy produced from renewable sources by pumping water uphill from a pool at a lower level to a reservoir situated at a higher point.

Table 18.5 Share of Clean Energy in 2019

Source	*Share of Clean Energy (%)*
Hydropower	42
Wind energy	20
Solar energy	17
Nuclear energy	16
Others (Bio, Geothermal, etc.)	5

Source: BP Statistical Review of World Energy (2020).

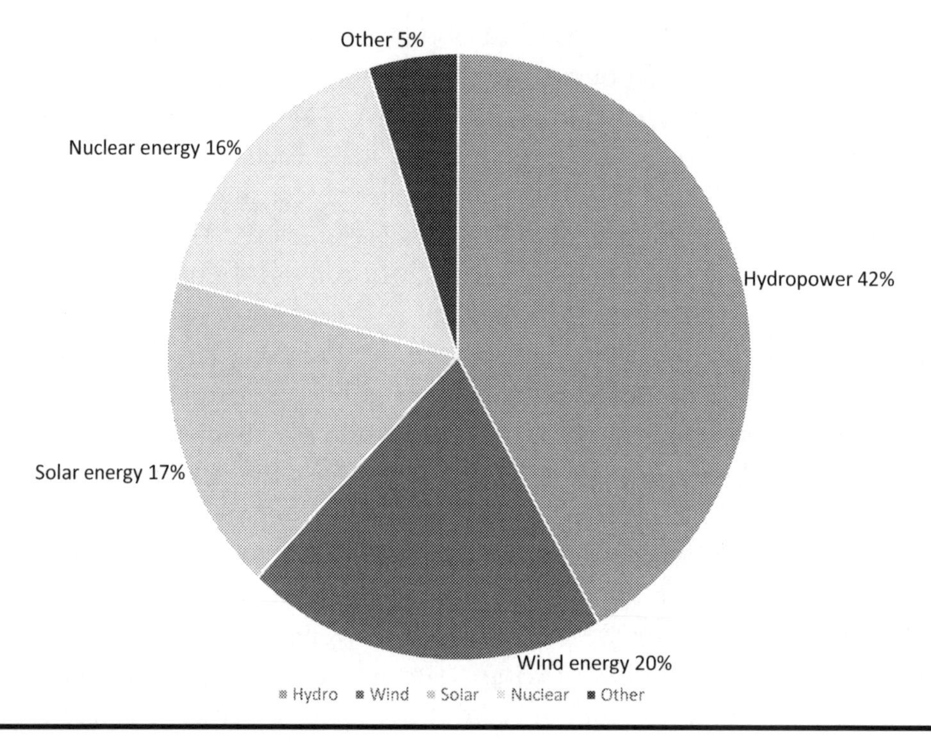

Chart 18.1 Share of Clean Energy.

Table 18.6 Hydropower

Rank and Country	*Hydropower (GW)*
1. China	341
2. US	102
3. Brazil	100
4. Canada	81
5. Russia	51

Source: EY Global Renewables (2020).

When a greater demand for electricity occurs, water from the reservoir is released to turn the turbines.

Table 18.6 shows ranking in the hydropower capacity of the top five countries.

Wind is a plentiful source of clean energy. It captures the natural air-stream in our environment and converts the wind motion into electricity by using wind turbines. The spinning blades of these wind turbines are connected to electro-magnetic generators that create electricity. This electricity is then fed into the national grid. However, there are some locations better suited for wind power than others. In general wind speeds are higher near the coast and offshore as there are fewer obstructions. A group of large structures of wind turbines, called wind farms, are familiar sights both in the outskirts of towns and at sea. There are also some domestic or 'off-grid' wind energy generation systems available.

The main advantages of wind energy are that it is a clean renewable form of power and has low operational and maintenance costs. However, as the wind speed varies both throughout the day and on an annual basis, wind energy presents an intermittency issue for the power grid. Fortunately, reliable storage systems, including battery storage, compressed air storage and pump storage are now available. The price of wind energy is declining and the share of renewable energy by wind is increasing.

Table 18.7 shows ranking in the wind power capacity of the top five countries.

Like wind, sunlight is one of our most abundant and freely available resources. Solar energy is absorbed in specially designed and manufactured solar panels to generate electricity directly. Solar panels are made out of photovoltaic cells (or PV cells) that convert the Sun's energy into electricity. A single PV cell can produce typically around 0.58 Volts. PV cells are sandwiched between layers of silicon. When hit by photons from sunlight each layer creates an electric field that creates the direct current (DC) needed to produce electricity. This current is then passed through an inverter to convert it into an alternating current (AC) which then can be connected to the national grid or used by a building with solar panels. Solar energy is carbon

Table 18.7 Wind Power

Rank and Country	*Wind Power (GW)*
1. China	288
2. US	122
3. Germany	63
4. India	29
5. Spain	27

Source: EY Global Renewables (2020).

Table 18.8 Solar Power

Rank and Country	*Solar Power (GW)*
1. China	205
2. US	76
3. Japan	63
4. Germany	49
5. India	38

Source: EY Global Renewables (2020).

free and renewable and thus has many benefits including its diverse applications for generating both electricity and heat, as well as its low maintenance costs. However, there are disadvantages such as the initial high cost, weather dependency and storage expenses.

Table 18.8 shows ranking in the solar power capacity of the top five countries.

Nuclear power is controversial with environmental activists primarily for the risks of radioactive waste storage and disposal that it poses. However, it is the only carbon-free energy source that reliably delivers electricity throughout every season almost anywhere in the world. The process of obtaining energy by splitting the atom is known as nuclear fission. The released energy is capable of generating steam and can then be used to turn a turbine and thus produce electricity. The most commonly used fuel for fission is uranium. In theory, fission power offers the exciting prospect of an almost inexhaustible source of energy for future generations. Nuclear fuels (uranium and plutonium) can be used to create nuclear weapons as well as nuclear reactors. Hence only nations that are part of the Nuclear Non-Proliferation Treaty are allowed to import nuclear fuels. In spite of the possible risks of radioactive wastes and misuse of nuclear fuels, nuclear energy has been found to be the cheapest and most continuously available path towards the goal of achieving the net zero carbon target.

Scientists at the Joint European Torus near Oxford, UK, announced on 9 February 2022 that they had generated the highest sustained energy pulse ever created by the fusion of atoms. The success of this breakthrough of nuclear fusion—the process that powers the Sun—promises to provide a near-limitless source of clean energy. It is a long way to go to see nuclear fusion in a full-size power plant, but it is a big step towards the supply of

Table 18.9 Nuclear Power

Rank and Country	*Nuclear Power (GW)*
1. US	97
2. France	61
3. China	48
4. Japan	38
5. Russia	29

Source: EY Global Renewables (2020).

sustainable green energy. Table 18.9 shows ranking in the nuclear power capacity of the top five countries.

China is often criticised as a leading emitter of greenhouse gases. However, the above data show that in fact the country ranks first in global terms in renewable power generation (hydro, wind and solar) and is in third place for the generation of nuclear power.

Having discussed the opportunities and challenges of clean energy initiatives in our battle against climate change, it is useful to describe a few well-known clean energy plants as case studies for successful clean energy endeavours.

One of the most famous renewable projects is the Three Gorges Dam in China, the largest hydroelectric dam in the world. Almost 100 years after it was envisioned and following two decades of construction the dam became operational in 2012. Built across the Yangtze River in western China the construction is 185 metres high and stretches over a length of 2 km. The reported cost of the project was US $17 billion, however the efficacy of its operations is clear. The water flows over 32 turbines capable of producing 700 MW of power. It is providing 1.7% of China's electricity, reducing carbon dioxide emissions by 10 million tonnes every year.

Walney Offshore Windfarm is located off the coast of Cumbria, England, and is the largest wind energy provider in the world. It was built at a cost of US $1.6 billion and began generating power from the beginning of 2011. The farm, stretching over 145 km^2, has a total of 189 wind turbines each standing 190 metres high. It has a total capacity of 600 MW which is enough to power 600,000 homes.

A further interesting example can be found in the Ivanpah Solar Electric Generating system, located in the Mojave Desert of Southern California, which is the largest solar energy facility in the world. The scheme was created jointly by Bechtel and BrightSource Energy with a total investment of US $2.4 billion and began operations from 2013. About 300,000 large (10 ft × 7 ft) mirrors (called heliostats) are digitally controlled to reflect sunlight to 140-metre towers fitted with solar panels. The concentrated sunlight heats water stored at the top of the three towers to create steam for turbines. The three plants together can produce 392 MW of electricity, sufficient to power 140,000 homes.

The largest nuclear plant in the world is the Kashiwazaki-Kariwa plant in Japan. It is owned by the Tokyo Electric Power Company and has a gross installed capacity of 8,200 MW. It has seven boiling water reactors and all units became fully operational from 1997. However, activities at the plant stopped in May 2012 following the Fukushima nuclear disaster. New safety guidelines from the Japan Nuclear Regulatory Authority are being implemented and all reactors of the plant are expected to be operational again from 2022.

18.4 How Six Sigma Is Helping Clean Energy Initiatives

There are good applications of Six Sigma and Lean Six Sigma in the renewable energy sector especially in wind energy. However, Six Sigma projects seem to be almost non-existent with no mention of them in the published papers related to hydropower or nuclear power plants. There are some academic publications on Six Sigma applications in thermal power plants (Kharub et al., 2018). The following case examples illustrate how Six Sigma can help in the renewable energy sector to save money and achieve sustainable improvements.

CASE EXAMPLE 18.1 SIX SIGMA IVANPAH SOLAR ELECTRIC GENERATING SYSTEM

As stated earlier the Ivanpah Solar Electric Generating system in California is the largest solar energy facility in the world and was a joint venture of Bechtel and BrightSource. Bechtel is a global project management organisation that has used Six Sigma to deliver many major projects to customers. Ivanpah was considered an appropriate venture for applying

Six Sigma because the scheme had a budget of US $2.4 billion over several years and Ivanpah had a lot of repetitive operations. The project was operationally challenging, requiring the design, construction and installation of around 300,000 large mirrors or heliostats.

Bechtel's project team joined with engineering and construction experts from BrightSource to form a Six Sigma group and all colleagues went through the Black Belt and Green Belt training programme together. The team identified five areas for process improvements by using Six Sigma: material handling, heliostat assembly, field transportation, heliostat installation and tower erection. In each of these spheres, especially in heliostat assembly, the team found new or more efficient ways of achieving results. For example, by using Six Sigma tools group members managed to redesign a stable process that would enable craftsmen to assemble 500 heliostats per day. The Six Sigma philosophy of 'continuous improvement' also carried over to the daily work of craftsmen.

All the processes involved in the heliostats including transportation had to be developed from scratch. This kind of complexity was a good fit for the Six Sigma approach. Bechtel used Six Sigma tools and processes to address the design, procurement and construction challenges of heliostats. The result was the development of new procedures that helped to meet sustainable performance goals and execute the project successfully. Six Sigma methodology was also instrumental for ensuring teamwork among project group members while simultaneously building trust and credibility with stakeholders.

CASE EXAMPLE 18.2 VESTAS INDIA, SAVED $10 MILLION THROUGH LEAN SIX SIGMA INITIATIVES

Vestas Wind Systems A/S is the largest manufacturer and installer of wind turbines in the world headquartered in Aarhus, Denmark. (Other notable large manufacturers are GE Renewable Energy, US, and Goldwind, China.) Vestas' subsidiary in India is based in Chennai. Vestas India saved US$10 million over 3 years through the implementation of Lean Six Sigma methodology in its wind turbine manufacturing and supply operations. The company engaged a local consulting firm, Millennium Global Business Solutions in 2006 for coaching 50% of its employees in Green

Belt and Yellow Belt training. The majority of them were schooled in the basic level of the Yellow Belt tools. Quality Director Dhananjay Joshi claimed, 'This enhanced collective competency as opposed to that of just a few employees at the senior level, as in the Black Belt methodology' (*The Economic Times*, 2009).

According to Joshi, employees trained as Yellow Belts implemented two to four projects per year based on Lean Six Sigma methodology. These projects were chosen from their own departments and their own domain of expertise. This enabled each Yellow Belt to take their own decisions based on available data, rather than being dependent on a team leader. The company spent Rs 50,000 (US $500) per employee on training while the savings per employee constituted up to Rs 5 lakh (US $5,000). Over 3 years (2006 to 2008) Vestas India saved an equivalent US$10 million and achieved sustainable performance to boot.

CASE EXAMPLE 18.3 BUSINESS INTEGRATION AT GE WIND ENERGY

General Electric (GE) is considered to be a real champion of the application of Six Sigma. As a leading manufacturer and supplier of wind turbines, GE has implemented the Six Sigma methodology very successfully in the renewable energy sector. Goel and Chen (2008) describe the re-engineering process of GE's wind energy division to integrate business operations across its globally dispersed acquisitions. The practice involved defining metrics and evaluating alternative processes through those metrics by deploying Six Sigma methodology. The application of Six Sigma ensured the sustainable best practices for process integration across global operations.

CASE EXAMPLE 18.4 IMPROVING THE QUALITY OF THE ROAD FOR WIND FARMS

An Indian organisation engaged in the installation of wind turbines utilised Six Sigma methodology for the development of sustainable wind farm roads. Gijo and Sarkar (2013) explain that the data-driven approach of Six Sigma identified the root causes and solutions for road damage. The implementation plan also ensured good teamwork and sustainable processes.

CASE EXAMPLE 18.5 SIX SIGMA IN RENEWABLE HYDROGEN ENERGY

In spite of the growth in renewable energy, it is envisaged that fossil fuels will continue to be used in the energy sector for the near future. As an alternative energy source, hydrogen also is regarded as a viable source of energy for renewable power systems. A research report by Apak et al. (2017) shows how Six Sigma methodology has been applied to hydrogen energy to boost energy efficiency. The authors also emphasize the importance of hydrogen energy in exploring potential future sources of sustainable, reliable and competitively priced energy. This study was an initiative to implement the Six Sigma methodology in a hydrogen power plant in Turkey with the aim of encouraging governments to support the use of hydrogen as a source of renewable energy.

18.5 How Green Six Sigma Can Help Clean Energy Initiatives Further

There is strong evidence of the application of Six Sigma/Lean Six Sigma methodology in solar power and wind power initiatives. Six Sigma has also been applied in thermal power plants. However, there is little testimony regarding Six Sigma methodology in hydro power and nuclear power projects.

Green Six Sigma should be very effective in all new clean energy projects whether they are for hydropower, solar power, wind energy or nuclear plants. The existing clean energy installations, especially hydropower and nuclear plants, should consider proactively the application of the Green Six Sigma approach. The case example of Vestas India clearly demonstrates that a modest investment to develop employees even up to the basic Yellow Belt level can deliver big savings and sustainable processes.

The distinctive additional contributions of Green Six Sigma include

- A comprehensive performance management system supported by a periodic self- assessment to monitor and sustain performance levels (see Chapter 15)
- A regular senior management review (e.g. sales and operations planning) to include Green Six Sigma projects (see Chapter 15)

- Application of carbon footprint tools and software (see Chapter 13 and Chapter 6)
- Application of material flow account (see Chapter 13)

18.6 Summary

We know that the energy sector is the top contributor of greenhouse gas emissions and its mitigation will come from all clean energy initiatives including nuclear energy. There is evidence of good progress within the realm of wind energy initiatives with the solar energy sector coming close behind. Although natural gas is the least of the carbon polluters among fossil fuels, the faster growth in natural gas consumption is not promising with regard to attaining the net zero carbon target. The big advantage of nuclear energy sources is that they can deliver more electricity day and night without interruption. Therefore, greater national efforts should be focused on developing the field of nuclear energy.

This chapter dealt with four main sources of clean energy but there are of course other energy supplies. One example would be geothermal, using deep underground hot rocks. High-pressure water can be pumped down into these rocks and the water with absorbed heat comes out from another hole; in turn this can spin a turbine to generate electricity. Another example is hydrogen which serves as a key ingredient of fuel cell batteries. We can use the intermittent supply of electricity from solar or wind farms to create hydrogen and then put hydrogen in fuel cells to generate electricity on demand.

Green Six Sigma is not a silver bullet to solve all clean energy demands. However, it can play a significant role in accelerating the initiatives at power plants and renewable farms as well as being instrumental to sustaining improved processes and environmental targets.

GREEN TIPS

- Emissions come from five economic sectors including a third from energy supply and we need solutions in all of them.
- For kilowatt think of a house; for gigawatt think of a city.
- Unfortunately 85% of our energy supply comes from fossil fuel, and we have to depend on natural gas for several years to keep our lights on.

- There has been good progress with the renewable energy both by wind power and solar energy.
- With the success of nuclear fusion at an Oxford Laboratory on 7 February 2022 there is justifiable hope for the supply of green energy by nuclear power stations.
- Green Six Sigma can play a significant role in the decarbonisation of the energy sector.

Chapter 19

Green Six Sigma and Green Supply Chain

When we walk away from global warming, when we don't advance and live up to our own rhetoric and standards we set a terrible message of duplicity and hypocrisy.

—John Kerry

19.1 Introduction

A supply chain, in simple terms, is a network between the suppliers to source the materials, the producer to convert them into products and the distributors to distribute the products to customers. The network could be complex involving many activities, people, facilities, information, processes and systems. In a typical supply chain, raw materials are procured (some local and some imported) and items are produced at one or more factories, transported to warehouses for intermediate storage and then transported locally and internationally to retailers or customers. With supply chain management the flow of materials and flow of information across traditional functional boundaries is seen as a single process. These flows are depicted in a simplified model in Figure 19.1. Thanks to ease of travel, the media and the 'World Wide Web' customers have never been more informed than they are today and they know what they want. This is especially true in service industries. As a result of the heightened expectations of customers,

DOI: 10.4324/9781003268239-23

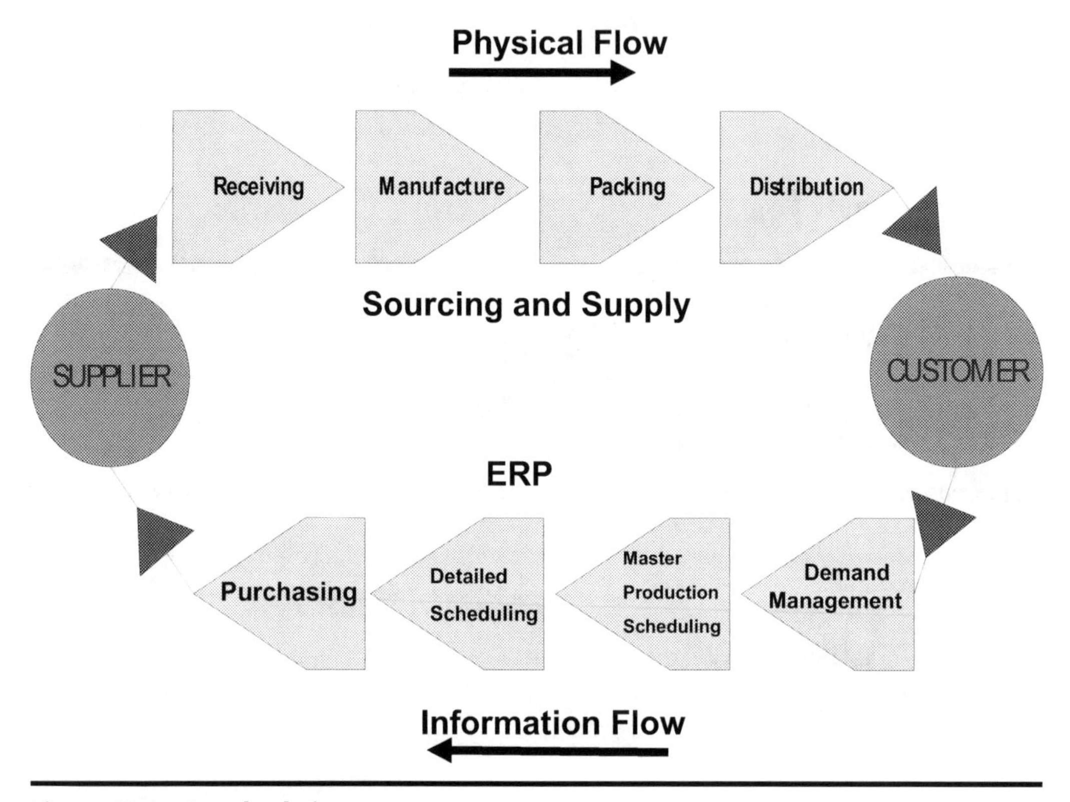

Figure 19.1 Supply chain management.

operations managers in service sectors have been forced to using the principles of supply chain management.

If we look at a supply chain in another way it is made of how we grow and procure things (land and agriculture), how we make things and provide services (industry) and how we get around (transport). As shown in Table 18.1 this chapter on the green supply chain includes agriculture and land and industry. The green supply chain in total accounts for 45% of greenhouse gas emissions. This chapter analyses the challenges and opportunities in the green supply chain and also the role of Green Six Sigma to accelerate and sustain the outcomes.

Environmental regulations are also changing the way in which supply chains are designed and managed. The problem is that the sheer number of rules, other influences such as changing consumer sentiment, and the complexity of global trade make it difficult for companies to decide exactly how they should respond to these pressures. Firms must also address the ethical requirements of their corporate social responsibility (CSR).

On a global scale, industrial pollution is one of the main contributors to the so-called greenhouse effect and global warming. It is important to note that environment and safety are not just social or political issues; they are vital ingredients contributing to the performance of an organisation. In manufacturing industries, there is much scope for environment and safety. The agriculture sector appears to present a picture of green pasture but a large proportion of methane gas is emerging from cattle and sheep farming activities. Apart from humanitarian reasons it is a truism that accidents cost money. Likewise many businesses and organisations are facing declining reserves of natural resources, increased waste disposal costs, keener interest in their human rights records, and tighter legislation.

These rising environmental pressures and social expectations can be turned to commercial advantage if a strategic approach is taken to develop a Green Thinking policy as described in the next section. Green Thinking is exactly on the same page as climate change initiatives. The strategic approach of Green Thinking entails complex longer-term considerations involving not just industry but environmental protection by regulations as an important international issue. Therefore, it is inevitable that no organisation can in the long term hope to avoid legislation and regulations designed to support the spirit of Green Thinking. In fact well-designed Green Thinking strategies can lead to sustainable business advantages.

Such a policy should be supported by a process to implement its objectives. Here Green Six Sigma can play its part and contribute its own specific role in this global challenge and opportunity. In the domain of quality and holistic processes, the specific methodology of Green Six Sigma (which as we know is evolved from Six Sigma methodology) could be such a process to support a Green Thinking strategy. The aim of this chapter is to review some of the critical issues and initiatives of Green Thinking under the following headings:

- Green Thinking and Climate Change Initiatives
- Why Green Six Sigma Is Relevant to Green Supply Chain
- Green Initiatives by Manufacturers and Suppliers
- Green Initiatives by Retailers
- Green Initiatives by Consumers
- Green Initiatives by Farmers
- How Green Six Sigma Can Help Green Supply Chain

19.2 Green Thinking and Climate Change Initiatives

The concept of Green Thinking is not new. Pearce (1992) analysed the relationship with Nature and the environmental strategies identified with the Green Movement. Elkington (1994) coined the phrase 'triple bottom line' as a new concept of accounting practice. The triple bottom line (TBL; also known as 'people, planet, profit') captures an expanded spectrum of values and criteria for measuring organisational success: economic, environmental and social. TBL is not without its critics, and it is now more aligned with CSR. Sustainability or sustainable development was first defined by the Brundtland Commission of the United Nations (1987) as 'the needs of the present without compromising the ability of future generations to meet their own needs'. As discussed in Chapter 17, this then led to the United Nations initiative of the United Nations Framework Convention on Climate Change (UNFCCC) environmental treaty addressing climate change signed by 158 states at Rio de Janeiro in 1992 and then other international and climate change initiatives

With this background the 3Es model (efficiency, environment and ethics) was developed as the three dimensions of Green Thinking. As shown in Figure 19.2, each aspect includes specific attributes (e.g. Efficiency relates to Lean Thinking, Energy Saving, the Green Economy and Green Logistics). Better training, greater awareness and improved communication on green issues characterise the Green Economy of an organisation. Green Logistics refer to buyer companies requiring a certain level of environmental responsibility in the core business practices of their suppliers and vendors. Ethics includes issues and processes related to Health and Safety, Fair Trading, Carbon Footprint and CSR. The most important dimension is Environment containing Pollution Control, Resources (including Energy), Conservation, Climate Change and Biodiversity. Climate change relates to the factors (such as CO_2 emission) causing the longer-term alterations in weather statistics. Major energy companies like BP (https://www.bp.com/en/global/corporate/sustainability.html) are focusing on new energy technology and low-carbon energy products in alternative sources, such as wind, solar, biofuels and carbon capture and storage (CCS). The variety of life on earth or its biological diversity is commonly referred to as 'biodiversity' where each species, no matter how small, has an important role to play.

The primary focus of Green Thinking is on the Environment and its sustainability while Efficiency and Ethics contribute to the quality and sustainability of Environment. Therefore, the three dimensions and their attributes

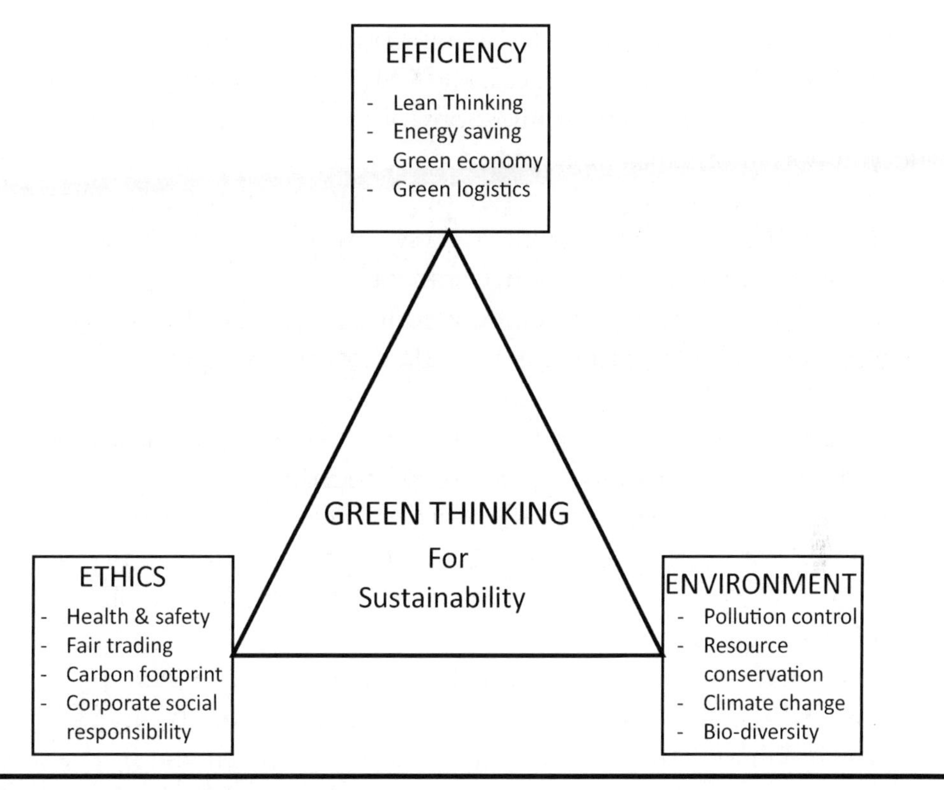

Figure 19.2 Green Thinking concept.

are interrelated. For example Energy Saving and Green Logistics in Efficiency are closely linked to Environment. Likewise the pillars of Fair Trading and Carbon Footprint within Ethics are intimately related to Environment.

The concept of Green Thinking is clearly incontrovertibly linked to climate change initiatives. Therefore, Green Initiatives, based on the concept of Green Thinking, also mean climate change initiatives.

19.3 Why Green Six Sigma Is Relevant to Green Supply Chain

As discussed in Chapter 2, the significant bottom-line results and extensive training deployment of Six Sigma and Lean Six Sigma must be sustained with additional features for securing the long-term competitive advantage of a company. If Lean Six Sigma provides agility and efficiency, then measures must be in place to develop a sustainable fitness. The process to do just that is FIT SIGMA (Basu, 2011). In addition the control of variation from the mean in the Six Sigma process (σ) is transformed to company-wide integration via FIT SIGMA methodology (Σ). FIT SIGMA is therefore synonymous with FIT Σ. Furthermore, the philosophy of FIT Σ should ensure that it is indeed fit for all organisations—whether large or small, manufacturing or service. Green Six Sigma is the recast of FIT SIGMA adapted specifically for Green Thinking or climate change initiatives with additional processes to ensure the sustainability of the environment.

Four additional features are embedded in the FIT SIGMA philosophy to create the sustainability of processes and performance levels:

- A formal senior management review process at regular intervals, similar to the sales and operational planning procedure
- Periodic self-assessment with a structured checklist which is formalised by a certification or prize, similar to the European Foundation of Quality Management (EFQM) award but with more emphasis on self-assessment
- A continuous learning and knowledge management scheme
- The extension of the programme across the whole business to ensure Green Thinking with the shifting of the theme from the variation control (σ) of Six Sigma to the integration of a seamless organisation (Σ)

The additional features of Green Six Sigma to ensure the sustainability of the environment are as follows:

- A formal extension of DMAIC (define, measure, analyse, improve, control) methodology to DMAICS (define, measure, analyse, improve, control and sustain)
- Inclusion of these features of FIT SIGMA to create the sustainability of the processes and performance levels
- Additional processes to ensure the sustainability of the environment with additional processes and tools including material flow account and carbon footprint tool

Green Thinking can leverage the contents of Green Six Sigma to achieve the specific environmental and business goals of supply chain stakeholders. For example, the value stream mapping of Lean Six Sigma (to pinpoint waste) can be adapted to create Green value stream mapping (to identify carbon footprints). Likewise, 'Voice of Environment' can represent the growing prevalence of environmental drivers not captured in 'Voice of Customer'.

19.4 Green Initiatives by Manufacturers and Suppliers

It is reasonable to state that manufacturing industries are major players regarding greenhouse gas emissions accounting for 21% of the total. Industries in the green supply chain must address health and safety in addition to the environment. But when the issues relate to health and safety, whether for products or workplaces, they apply seriously to both manufacturing and service organisations. Lack of safety in the product or in the workplace will inevitably cost money. Accidents mean lost production time plus time wasting inspections by government officials. They may also incur legal costs as well as the expense of correcting the situation. It has to be cheaper to do it right the first time. One of the better-known environmental standards is put forth by the International Organisation for Standardization (ISO), known as ISO 14001. Basu and Wright (2003) established that environmental protection relates to pollution control in two stages. Conventional restraints or 'first-generation pollution' controls are applied to pollution in air, water and regarding noise created in the manufacturing process. Such jurisdictions are usually regulated by legislation. There is also a 'second-generation pollution' which relates to the problems caused by the usage of certain products and chemicals over a long period. The most widespread example of such second-generation pollution is the contamination of land which permeates groundwater.

The primary focus of climate change initiatives is the reduction of greenhouse gases. Therefore, each major manufacturing company is expected to have a carbon neutrality strategy that is to achieve the net zero carbon footprints by 2050 or earlier. I propose a three-step carbon management strategy as follows:

1. Develop a total supply chain emission chart
2. Follow a carbon management action plan
3. Co-ordinate and share carbon management actions with key stakeholders

A typical total supply chain emission chart is shown in Figure 19.3. This chart is the starting point of measuring and monitoring carbon footprints and initiating a carbon management action plan.

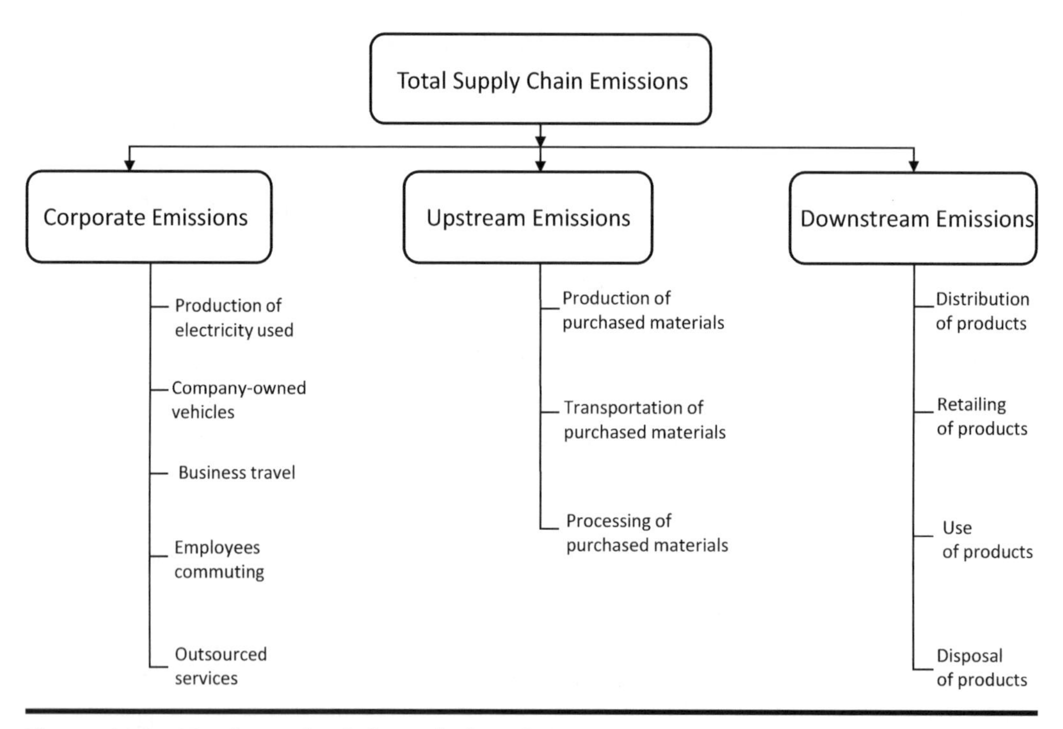

Figure 19.3 Total supply chain emission chart.

A carbon management action plan should follow the following hierarchy as proposed by Watt (2012):

- Avoid
- Reduce
- Replace
- Offset

Here are some tips to follow a carbon management action plan. Avoid carbon-intensive materials (e.g. steel and cement) and activities (e.g. air travel). Reduce waste and do whatever you do more efficiently (e.g. apply Green Six Sigma tools). Replace high-carbon energy sources with low-carbon energy ones (e.g. substitute plastics packaging). Offset those emissions that cannot be eliminated by these actions. Unfortunately, the process of offsetting is rather less straightforward. A carbon offset is the reduction of greenhouse gas emissions to compensate the emissions made elsewhere. Businesses can buy carbon credits generated by projects that are cleaning up our atmosphere.

We are still left with the environmental challenge with three major materials for the industry—steel, cement and plastic. When we produce steel or cement tons of carbon dioxide is released, but when we make a plastic around half of the carbon stays in the plastic. However, plastics can take hundreds of years to degrade. Additional research is needed specifically in these three heavily used materials beyond the scope of Green Six Sigma. The following Case Example 19.1 shows that some progress is made in manufacturing 'fossil-free steel'.

CASE EXAMPLE 19.1 SSAB DELIVERS 'FIRST FOSSIL-FREE STEEL'

Sweden' SSAB announced in August 2021 that its 'Hybrit technology' produced 'the world's first fossil-free steel' and started delivering it to the Volvo Group. The idea underpinning this 'Hybrit technology' is the use of fossil-free hydrogen rather than coal in steel production. The cost-effectiveness of the 'Hybrit technology' is not known, but the goal of SSAB is to market the technology on an industrial scale as early as 2026.

At this stage we can also follow the hierarchy of Avoid, Reduce, Replace and Offset. If we encourage companies to share their climate change initiatives with their key stakeholders up and down the supply chain, then we are likely to get radical responses. For example, Marks & Spencer may correctly decide to work with agricultural suppliers to ensure a future supply of low-carbon products. Unilever may correctly decide to influence consumer habits to use low-carbon ethically sourced products.

19.5 Green Initiatives by Retailers

Global retail giants Walmart and Carrefour as well as other supermarkets all over the world are responding to the pressures on packaging waste reduction and the additional environmental issues involved in the green supply chain. It is important to consider replacing all non-recyclable packaging materials as far as possible. Polyethylene terephthalate (PET) is a form of polyester and is recyclable.

Recent media reports are loaded with announcements on 'greening the supply chain' from large retail groups. Walmart, a US company and the world's largest retailer, unveiled its packaging scorecard to major suppliers such as Proctor & Gamble, Unilever and Nestle, which was designed to cut packaging. Walmart hopes that the scheme will reduce packaging across its global supply chain by 5% by 2013.

British supermarkets have taken the initiative in becoming more environmentally aware. ASDA supermarket, a subsidiary of Walmart in the UK, claimed, as an example, that taking pizzas out of cardboard boxes saved 747 tonnes of cardboard in a year. UK supermarket Sainsbury's announced in October 2006 that 500 of its own-brand goods would be presented in compostable packaging. 'Friends of the Earth', a non-profit organisation in the UK, gave a cautious welcome to Tesco's new environment fund of $100 million but said the supermarket giant still had a very long way to go if it was serious about greening its operations. Tesco would need to address a number of key areas if it was serious about reducing its environmental impacts. These include moving away from car-dependent stores, switching from its global supply chain, radically improving energy efficiency in its stores, and cleaning up its supply chains.

Perhaps the best example of a company devoting itself to acting upon environmental issues is the British retailer Marks and Spencer (M&S). Their website (https://corporate.marksandspencer.com/sustainability/plan-a-our-planet) details the evolution of their 'Plan A' scheme, launched in January 2007, which listed 100 commitments that the firm aimed to achieve within a 5-year time frame. Since then, M&S extended this commitment to encompass 180 goals that they wish to attain by 2035. M&S reset Plan A to Plan B to become net zero across own operations by 2035. Their proudly stated aspirations include 'working with our customers and our suppliers to combat climate change, reduce waste, use sustainable raw materials, trade ethically, and help our customers to lead healthier lifestyles' and that they have 'the ultimate goal of becoming the world's most sustainable major retailer'.

Even the airlines, the biggest polluters with their carbon dioxide emissions, have joined the green bandwagon. Richard Branson committed the next 10 years of profits for Virgin—around $3 billion—to fighting global warming.

There has been a stronger emphasis to introduce ethical, fair-trade, organic and bio products to the consumer. Case Example 19.4 of 'Carrefour Bio Coffee' illustrates that by promoting unbranded 500 g/1 kg coffee in bags as 'organic coffee to support fair trading' in 1997 sales increased by 80% in 4 years.

Such is the importance of a company displaying its eco credentials that there is a growing awareness that communicating their efforts to the consumer leads to favourable publicity and an enhanced reputation of the brand. Indeed, close perusal of any paper media advertising will reveal how many companies actually seem to be using newspapers and magazines to broadcast their environmentally conscious efforts instead of boasting about their products. In today's climate, ecological issues are such a current and hot topic that firms realise that they can tap into the public consciousness and portray themselves as 'green', thus superseding the traditional tenets of advertising, the focus on the goods themselves. This method is used not only by supermarkets but by manufacturers of white goods or motor cars (emphasis on energy efficiency of products), petroleum firms, energy companies and so on. The reader can also examine everyday objects to see this principle in action: the covers of recently published books (which inform us of their use of paper from sustainable forests), or the packaging of almost any food or cosmetic item to see details of the sustainability of their source materials.

19.6 Green Initiatives by Consumers

Consumers have both the power and responsibility to enhance the activities and effectiveness of the green supply chain. It is the consumer who pays for the end product or service, and it is the consumer who ultimately suffers or benefits from the resultant impact on the environment. Green initiatives from consumers could be manifested in three ways:

- Make your home green
- Feedback to retailers
- Reverse supply chain

'Make your home green' is becoming a conscious target of many consumers. This is effected in two paths. First consumers are attempting to minimise 'carbon emission' by making houses and household appliances more energy efficient and also by moving towards eco-friendly transport. Second, encouraged by local authorities, consumers are making commendable efforts in the recycling of household wastes by the provision of green (compostable) waste, paper and glass bins for the home; use of local recycling centres etc. The retrofitting of homes to 'make your home green' are discussed further in Chapter 10.

A reverse supply chain is a process of getting goods from the customers back to manufacturers. It is a relatively new trend in supply chain management that focuses on 'green manufacturing' to target recycling, recovery and remanufacturing systems. In these reverse networks, consumers bring products to a retailer or a collection centre. For example supermarkets in Germany have a bin where customers leave used batteries. Depending on the particular product, it can be refurbished, remanufactured or recycled; making sure the physical flow is efficient. It is estimated that 63 million personal computers worldwide became obsolete during 2003 and about 10 million electric waste products are dumped per year in Japan. By 2030, the obsolete PCs from developing regions will reach 400 to 700 million units, far more than from developed regions at 200 to 300 million units. The principle of 'circular economy' (i.e. reuse, repair, refurbish and recycle) is appropriate for domestic appliances minimising the use of new ones and the creation of wastes (Yu et al., 2010).

Mobile phones can be returned to the store where the new one is purchased. From there, the phones are resold and reused in other countries where the very technology that is being phased out in developed countries is being introduced. Many other products have the potential for second use, including computers, auto parts, printer cartridges, refillable containers and a host of other possibilities. In remanufacturing, reverse logistics introduces additional challenges to planning for a closed-loop supply chain. The planning, sourcing, making and delivery of the products are affected by the reverse flow of used products and materials for subsequent consumption in the manufacturing of new products. Reverse logistics play a key role as retail organisations tend to look at their reverse supply chains more closely to enhance customer satisfaction, cost/time efficiencies and supplier performance.

19.7 Green Initiatives by Farmers

Agriculture and land is a primary link of the green supply chain and farmers are its major stakeholders although it is unreasonable to make the farmers alone responsible for this sector. This sector, accounting for 24% of the total greenhouse emissions, covers a large range of human activities, from raising animals to growing crops and harvesting trees. With agriculture the main emission is not carbon dioxide but methane and nitrous oxide. Methane causes 28 times more warming than carbon dioxide, and nitrous oxide, though low in its share of all greenhouse gases, causes 265 times more warming. Scientists estimate that the emission from agriculture and animal farming is more than 7 billion tonnes of carbon dioxide equivalent per year. Another big challenge is deforestation and other uses of the land, which together is adding about 1.6 billion tonnes of carbon dioxide every year and destroying essential biodiversity.

The agriculture and land sectors offer more challenges and there are not many breakthrough solutions. Let us address some of these challenges. The first challenge is our food habits and the rapid growth in population. The world population has more than doubled within last 50 years to the current level at 7.9 billion people. As people are getting richer they eat more meat and dairy. Although there is little growth in meat consumption per person in the US, Europe and Latin America, it climbing rapidly in China. The supply of meat and dairy is coming from cows, sheep and pigs. The enteric fermentation in cattle is emitting methane equivalent to 4% of all greenhouse gas emissions. The excrements from farm animals also release a mix of powerful greenhouse gases—mostly nitrous oxide and some methane.

There is some progress with a compound which has to be administered to the cattle every day to reduce methane emissions by about 30%. This is not a winning area. We could try by applying the principle circular economy that is to replace and reduce. There is an active 'vegan movement' that we should just stop raising livestock for food only. There are some successful developments of plant-based meat and lab-cultivated meat. There is another way we can reduce emissions from the meat that we eat by wasting less of it—in the US more than 40% of food. Some good news is that the productivity of grain farming (e.g. rice, wheat and corns) has increased many-fold by two main factors—innovation (e.g. Borlaug's semi-dwarf wheat), mechanisation and fertilizers. However, a fertilizer is also a mixed blessing. The nitrogen from the fertilizer escapes into the air as nitrous oxide.

A large contributor (30%) of greenhouse gases in the agriculture and land sector is 'deforestation'. For example, Brazil's forests have shrunk by at least 10% since 1990 to clear pastureland for cattle. When a tree is removed the stored carbon gets released into the atmosphere as carbon dioxide. Also a tree in the rainforest can absorb 4 tonnes of carbon dioxide over 40 years. A popular process of carbon offsetting is the planting of trees. However, scientists have estimated that to offset the life-long emissions of one person we would need around 20 hectares worth of trees. Trees with higher carbon absorption capacity (e.g American Sweetgum tree, European Beech tree, Silver Maple tree, seagrass) should be preferred for new plantation.

19.8 How Green Six Sigma Can Help Green Supply Chain

The tools, techniques and processes of Green Six Sigma can provide an effective framework to implement the initiatives in all three dimensions of Green Thinking, viz. Efficiency, Environment and Ethics.

19.8.1 FIT SIGMA in Efficiency

The fundamentals of Lean Thinking are embedded in Green Six Sigma and these are underpinned by

1. Elimination of waste
2. Smooth operation flow
3. High level of efficiency
4. Quality assurance

The lean methodology as laid out by Womack and Jones (1998) is sharply focused on the identification and elimination of 'mudas' or waste; indeed their first two principles (Value and Value Stream) are centered on the eradication of waste. One important area of waste in processes is excess inventory. The cycle time or lead time reduction is another target area of waste reduction. Environmental misuses—including squandered energy—can cost companies thousands of dollars a year. Green Six Sigma tools (such as value stream mapping and the flow diagram) are there to analyse processes and identify wastes and non-value-added activities. The additional impact of Green Six Sigma on Efficiency is ensuring the sustainability of that efficiency by quality assurance and well-structured training programmes for Black Belts and Green Belts.

CASE EXAMPLE 19.2 GE JET ENGINE TESTING

A facility of General Electrics in Peebles, Ohio (where GE runs jet engine tests in the open air), is a leader within the company in applying Lean to address greenhouse gas emissions. The quantity of testing conducted is directly related to the amount of business the airlines are doing. More engine testing means more jet fuel consumption and more greenhouse gas emissions. These reductions have also resulted in significant cost savings.

Before Lean events, the engine had to be turned on three times in order to complete the balancing process. Lean methods helped Peebles develop a new balancing process that only required the engine be turned on once, which reduced both fuel consumption and greenhouse gas emissions. While the facility's old troubleshooting process required the engine to be running, the new procedure allowed troubleshooting to occur while the engine was not running. Through these changes the company reduced its use of fuel from 20,000 gallons to 10,000 gallons. There was also an overall cost savings of over a million dollars a year.

Source: US Environment Protection Agency (https://www.epa.gov).

The elimination of waste through Lean Thinking in any entity or form of consumption, whether products, processes, materials or utilities, is de facto in the domain of Green Thinking. Lean Thinking is also a key component of Green Six Sigma and thus Green Six Sigma can be a major driver of the 'Efficiency' leg of Green Thinking. In the well-known Toyota Production System (TPS), the forerunner of Lean Thinking, Toyota has developed a 5R programme (see Table 19.1) to reduce polluting wastes in order to maintain production with Green Thinking (Black and Phillips, 2010).

It is evident from the 5R programme that eliminating waste at the source by using Lean and related Green Six Sigma techniques helps the process of achieving zero waste as part of Green Thinking in Efficiency.

Table 19.1 Toyota's 5R for Green Thinking

5R	*Measures*	*Responsibility*
Refine	Expansion to reduce, reuse and recycle by changing design and raw materials	Operations departments
Reduce	Reduction of amount of waste generated at source	Operations departments
Reuse	Reuse within operations processes	Operations departments
Recycle	In-plant and outside use of generated waste	Operations departments
Retrieve energy	Recovery of energy from waste materials that cannot be refined, reduced, reused or recycled	Environment technology department

19.8.2 Green Six Sigma in the Environment

The impact of Green Six Sigma in the Environment is achieved by identifying and prioritising environmentally related projects as part of Black Belt/Green Belt training programmes. There are five areas under which environmental issues are addressed in identifying Green Six Sigma projects, namely,

1. Green procurement
2. Energy conservation
3. Resource conversation
4. Elimination of hazardous waste
5. Risk management

Green procurement is the selection of products and services that minimise environmental impacts. The rationalisation of the supply chain to significantly reduce delivery vehicle movements is such an example. Green procurement is part of Green Logistics of Efficiency but its focus on Environment is more significant in Green Six Sigma projects. Energy conservations is achieved through efficient energy use—in other words, energy use is reduced while achieving a similar outcome. In addition to industrial facilities, domestic housing and road transport are also two major areas of concern. The Resource Conservation Challenge is a national effort to protect natural resources (including energy) by managing materials more efficiently. 3M is a pioneer in the use of Lean Six Sigma methods and tools to minimise hazardous waste and improve operations and quality. The famous 3P programme (Pollution Prevention Pays) of the 3M Company brought about major savings including $2 million from the elimination of hydrocarbon wastes from a reactive casting process.

CASE EXAMPLE 19.3 3M COMPANY: ENVIRONMENTAL HEALTH AND SAFETY

When 3M instigated its 3P programme several years ago, the approach was to capture and control pollutions and emissions before they could damage the environment. This tactic, although effective, has been altered to a philosophy of prevention rather than containment. The 3P programme now aims to prevent pollution at the source by using different materials, changing the process, re-designing the plant and equipment, and recycling waste.

Since 2001, 3M's Environmental, Health and Safety Operations (EHS) organisation has deployed Lean Six Sigma to improve corporate EHS services and activities. EHS team members receive 2 weeks of Lean Six Sigma Green Belt training, and coaching is provided by Black Belts.

There are specific tools and processes of Green Six Sigma to contribute to the environment and sustainable climate change initiatives. In addition to the sustainability tools of Green Six Sigma, the rigour of Six Sigma and Green Six Sigma can be applied to verify the issues of the broader climate change debate. There are many Green Six Sigma tools in the book that are highly relevant for climate change projects. For example control charts can help analyse the range of variability in climate change or a cause and effect diagram can articulate the different causes of global warming. At a more advanced level of analysis, Six Sigma tools like design of experiments (DOE), measurement systems analysis (MSA) and statistical process control (SPC) can contribute a data-driven rigour to the debate of climate change. As a specific example SPC analysis of global temperature pinpointed special causes of variation affecting the temperature data (Dale, 1999). In this example three special causes are identified and analysed as the 'heat island' effect (urbanization), carbon dioxide emission due to human activity and the sunspot cycle.

Green Six Sigma is not the 'holy grail' to environmental challenges or the climate change debate, but it can contribute as a systematic and proven methodology in these challenges. There are many other initiatives from all stakeholders that are required as discussed earlier. Many such initiatives are in progress, such as the Renewable Energy Directive (EU target to produce 22% of electricity from renewable sources), Climate Change Levy (tax on greenhouse gases) and Carbon Reduction Commitment (UK target to achieve 78% reduction of carbon emission by 2035). There are more legislations in the pipeline including the regulation of waste in portable batteries and widening the scope of the Hazardous Waste Regulation. It is a good start, but it has a long way to go.

19.8.3 Green Six Sigma in Ethics

There are three key aspects of Ethics which could be supported and enhanced by Green Six Sigma, namely,

1. Product and occupational safety
2. Carbon footprint
3. Corporate social responsibility

A major incident in product safety, particularly in consumer and food products, can seriously damage the brand and business. Accidents in the workplace mean lost production time and are often accompanied by paying out both compensation to employees and legal costs. A carbon footprint is 'the total set of GHG (greenhouse gas) emissions caused directly and indirectly by an individual, organisation, event or product' (UK Carbon Trust, 2008). An individual organisation's carbon footprint is measured by undertaking a GHG emissions assessment. CSR is a self-regulating mechanism whereby organisations would monitor and ensure their adherence to the law, public interests and ethical standards. CSR also embraces the concept of the TBL covering environmental, social and economic bottom lines. A key role that Green Six Sigma can play in Ethics is by bringing together the specialists in health, safety and environment and other operations in Black Belt/Green Belt training.

CASE EXAMPLE 19.4. CARREFOUR BIO COFFEE

Carrefour is a global hypermarket retail chain organisation from France with a turnover of more than €100 billion making it second only to Walmart, the largest retail company in the world.

The first shipment of coffee beans was delivered in 10-kg sacks to Vitrolles, France, in 1970. The beans were roasted in store and sold in 500-g and 1-kg bags. In April 1997 Carrefour launched the 'organic' coffee brand under the name 'Carrefour Bio' to promote organic products and support fair trading. In 2001 it was decided to establish a green supply chain for 'Carrefour Bio'.

The organic coffee marketed under the name 'Carrefour Bio' is not indexed on the world coffee market. The purchase price is approximately 30% higher than the average cost in Mexico. The supplier is contracted to pay a guaranteed minimum price to producers. Thus producers can obtain up to 60% of the value of the coffee at current international rates.

Some 3,000 producers from 37 Mexican communities cultivate coffee using organic methods. A local infrastructure has been introduced to transport people between towns and villages (a 2-hour bus ride replacing what was previously a 2-day walk). A health scheme has been introduced providing free medicine and healthcare and a consortium has been set up to buy basic foodstuffs at cost price.

The coffee is cultivated by small farmers working for the Uciri cooperative in Mexico using an organic method of farming. Such cultivation helps prevent the land from becoming impoverished. The cultivation is carried out in accordance with French Organic Society standards without the use of organofluoridated fertilisers or chemical pesticides for tropical forest conditions. An organic fertiliser comprising sun-dried and hand-picked stoned cherries and animal waste is spread over the plants. This is the only plant treatment used by the farmers.

Cultivation methods are monitored by an organic certification body. An agricultural education centre has also been established catering for organic farming, animal breeding and bio culture.

The 'Carrefour Bio' coffee project appears to constitute a win-win initiative for the green supply chain. For Carrefour, sales for the product increased from 29.5 tonnes in 1997 to 54 tonnes in 2001. The fertility of the land has been protected. The average income per family of producers increased from €53 per year in 1985 to €1,524 per year in 2000. The local communities benefitted from the infrastructure and facilities for transport, healthcare and education. Finally, consumers are happy with an organic product at an affordable price.

By implementing environment management systems such as ISO 1401 or EMAS (Eco-Management and Audit Scheme) and adhering to a strong ethical culture by continuous training, the executive board of an organisation can be confident that brands or the business would not be tainted by prosecution or negative press. Training on the environment, safety and ethics can be embedded in Green Six Sigma training workshops.

19.9 Summary

'The scientific evidence is now overwhelming: climate change is a serious global threat, and it demands an urgent global response', concludes Nicholas Stern (2006).

It should be noted that this view has been disputed especially by fossil fuel lobbyists. However, irrespective of what we believe, the pressure is on for industry and nations to adopt a green approach to the supply chain. In this chapter we attempted to present a balanced view of various initiatives adopted by manufacturers and suppliers, farmers, retailers and also consumers. Every stakeholder has a role and responsibility in 'greening' the supply chain. We have shown that there are commercial benefits in reducing wastes (e.g. excessive packaging). Large retailers like Walmart, Carrefour and Tesco are probably facing disproportionate demands from environmental pressure groups and regulatory bodies but nonetheless are displaying visible efforts to respond to these demands.

Quality as a whole can play a major role in raising the awareness of Green Thinking and sustainable developments. In this chapter, mainly the impact of Green Six Sigma and Six Sigma programmes (which are part of the quality movement) are touched upon. Learning from the proactive focus on Green Thinking in the Six Sigma programmes of leading organisations like GE and 3M, the adaptive tools and processes of Green Six Sigma could be appropriately applied by all organisations in their Green Thinking initiatives. Edward de Bono (de Bono, 2016) suggests the inclusion of 'Green Hat' in his concept of 'Green Thinking Hats of Six Sigma'. In addition to following the fundamental tenets of Six Sigma, the additional features of Green Six Sigma should bolster the sustainable climate change initiatives of the stakeholders of the green supply chain.

GREEN TIPS

- Lean Six Sigma has been effective in improving the performance of the green supply chain. Green Six Sigma will add more value by ensuring sustainable processes and environmental standards.
- The amounts of methane emitted by cows and sheep account for 4% of global emissions.
- Each stakeholder of the supply chain has a role to play in reducing carbon dioxide emissions.
- Every organisation should have a carbon management action plan and follow a hierarchy as avoid, reduce, replace and offset.
- Toyota's 5R for Green Thinking is in alignment with the principles of circular economy in Green Six Sigma.

Chapter 20

Green Six Sigma and Green Transports

Global warming is causing the loss of living species at a level comparable to the extinction event that wiped out the dinosaurs 65 million years ago.

—Al Gore

20.1 Introduction

The various forms of transport that we use daily contribute only 14% of the total global greenhouse gas emissions. Thus this sector ranks fourth behind energy, land and industry. However, if you were to ask a random person which activities pollute most, that person probably would say burning coal for energy, driving cars and travelling by plane. Although transportation is not the biggest cause of emissions worldwide it is high on the list in the US and Europe where people drive and fly a great deal. In 2020 as the Covid-19 pandemic led to a sudden and drastic reduction in transport activities, it was noticeable that carbon dioxide transmissions also fell by 2.4%.

If we are going to attain net zero emissions, we will have to minimise all the greenhouse gases caused by transportation around the world.

DOI: 10.4324/9781003268239-24

20.2 Guiding Factors of Clean Transports

We cannot leave without transport, unless we plan to go everywhere on foot. Even if we reduce our non-essential travel we will still need transportation by land, sea or air to move goods for our survival. Farmers need to get their crops to the market. The benefits of globalisation cannot be realised without the movement of goods and people, both between nations and within countries. Therefore, in order to enjoy the benefits of transportation and travel without making our climate unliveable, we need to adopt green or eco-friendly solutions urgently.

In order to find such green solutions for transport-related emissions we need to analyse where these emissions are coming from. Although at present the US and Europe are responsible for a major part of such transport-related greenhouse gas emissions, these countries reached their peak in the past decade and in fact their emission rates have started to decline. At present, most of the growth in transport-related emissions can be traced to emerging economies as their growing populations are getting richer and buying more cars. For example in China, transport-related emissions have doubled over the past decade.

Table 20.1 shows the share of emissions from different types of transport such as cars, planes, ships etc. Passenger vehicles (cars, sports utility vehicles and motorcycles) constitute the biggest portion at 47% and trucks and buses contribute another 30%. Airplanes then add another 10%, as do cargo and container ships. Finally, trains account for the last 3%. As a whole, every year all forms of transports are emitting over 7 billion tonnes of carbon dioxide equivalent to the atmosphere. Our goal is quite simply to get every source of these transportation emissions to net zero.

Table 20.1 Emissions from Transports

Types of Transport	*Share of Emissions (%)*
Passenger vehicles	47
Trucks and buses	30
Cargo and cruise ships	10
Airplanes	10
Trains	3

Source: Gates (2021).

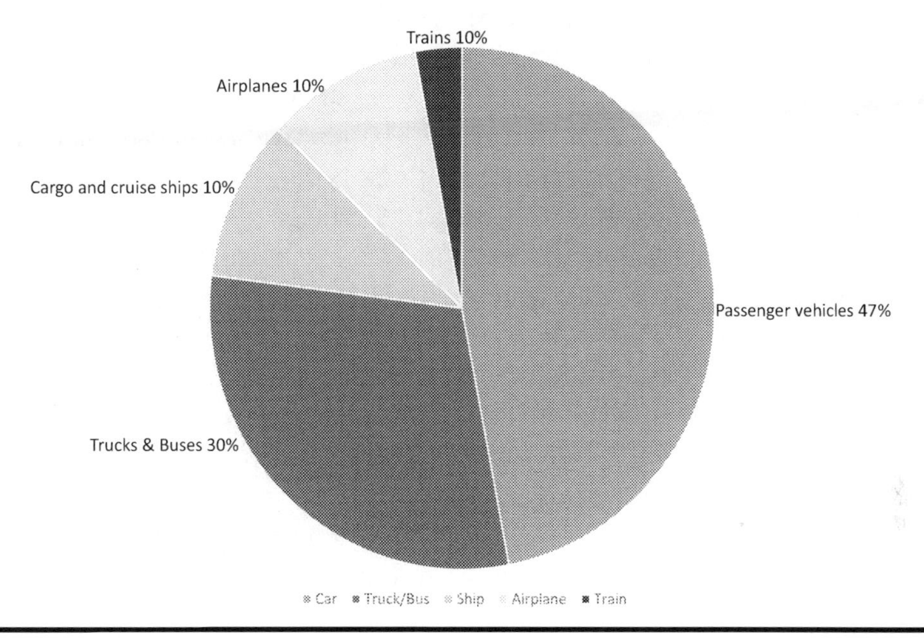

Chart 20.1 Emissions from Transports.

Chart 20.1 shows graphically in a pie chart the share of carbon dioxide emissions from different groups of transport.

There is also pollution from particulate matter or PM2.5. Inhalation of particulate pollution can have severe longer-term health impacts such as the age-specific mortality risk, particularly from cardiovascular causes. However, its damaging effects can be fatal regardless of age; in December 2020 a coroner ruled that the death of 9-year-old Ella Adoo-Kissi-Debrah, who lived very close to the South Circular Road in Lewisham, south-east London, was in part attributable to high levels of air pollution. This landmark ruling meant that Ella became the first person in the UK to have 'air pollution' recorded as a cause of death. In addition to it naturally occurring (e.g. volcanic eruptions and wildfires), PM2.5 is produced by human activities (e.g. emissions from car exhausts). Therefore, clean energy initiatives for passenger vehicles will also eliminate the man-made emission of PM2.5.

There are a number of methods that can be implemented to reduce greenhouse gas emissions from transportation, including the following:

- Do less moving around by transports. This requires less driving and reduced travel by air and sea. We should encourage more walking, cycling and working from home. During the Covid-19 pandemic a 'new normal' way of working, by using technology for virtual communication tools, has been established and proven to be effective.

- Use fewer carbon-intensive materials (e.g. steel and plastics) in making cars, trucks, buses, planes and ships. The less we use these materials in a car, the smaller its carbon footprint will be.
- Use fuels more efficiently by setting fuel efficiency standards. Cars have been designed with innovative components (e.g. catalytic converters) and more efficient engines to meet these standards.

These methods are all on the right track and will reduce the amount of carbon dioxide in the atmosphere—but they will not get us to zero emissions. The key way in which we can move towards net zero is by switching to electric vehicles (EVs) or using alternative fuels. These alternative fuels (e.g. ethanol for petrol and biodiesel for diesel) contain carbon, albeit in a negligible proportion. EV is the cleaner and greener option.

20.3 How Clean Transport Solutions Reduce Greenhouse Gas Emissions

Let us address in more detail how these methods of carbon reductions are being applied to the different types of transport listed in Table 20.1.

20.3.1 Passenger Cars

Top of the list is passenger cars contributing 47% of emissions by all types of transport. We now have two recognised solutions to this issue—electric alternatives and biofuels.

Fortunately we have the proven engineering and technology of making, using and maintaining EVs today. There are about one billion cars on the road currently and about 5% of these are electric. From Audi through to Volkswagen, all car manufacturers are delivering all-electric or hybrid cars. Hybrid vehicles are powered by an internal combustion engine and an electric motor–driven energy stored in batteries. The battery is charged through the internal combustion engine. However, as these cars also use petrol, they are not the preferred solutions for net zero carbon emissions. Instead, we should go for all-electric plug-in solutions.

Table 20.2 Electric Vehicles (EVs) in Europe

Country	*EV Shares of Total Cars (%)*
Norway	74.8
Iceland	52.4
Sweden	32.3
Netherlands	25.0
Finland	18.1
Denmark	16.4
Switzerland	14.3
Portugal	13.8
Germany	13.5
UK	11.3

Source: Pew Research Center Report, Washington (2021).

The fastest EV sales have been in Europe. As shown in Table 20.2 nearly three-quarters of the cars sold in 2020 in Norway and more than half in Iceland were electric. In ten other European countries the share of EVs was between 11 and 32%. Both in the UK and France the share of EVs was 11.3% in the same year. EVs in Europe as a whole experienced a compound growth rate of 60% per annum from 2016 to 2020. By contrast the share of the EVs in the US was only 2% and in China a little higher at 5.7%. The growth in EVs could be even greater but for three main reasons—higher costs, lack of battery charging points and a shortage of microchips. The elevated cost is due to the price of batteries. Fortunately for the consumer the cost of batteries is coming down, with an 87% drop since 2010. Lithium is the preferred choice of battery production for EVs due to its light weight and its excellent electricity conductance. These batteries can withstand higher temperatures and have longer battery lives compared to traditional lead-acid batteries.

The bar chart in Chart 20.2 shows the share of EVs in top 10 countries.

There are some government directives to encourage the use of EVs on the road. For example, the Mayor of London introduced congestion charges for passenger vehicles and EVs are exempted. All diesel and petrol vehicles will be gradually phased out by 2030.

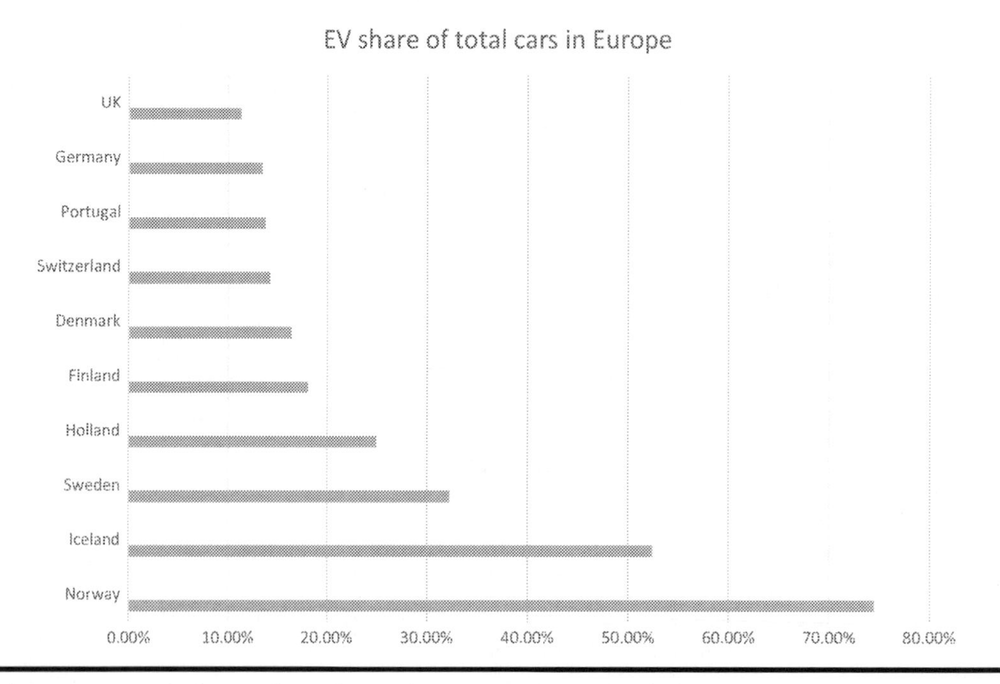

Chart 20.2 EV Share of Total Cars in Top 10 Countries in Europe.

It is estimated that as of July 2021, the cost of running an EV for an average family in the UK driving about 10,000 miles per year is approximately US $1,000 more compared to a similar diesel or petrol car. If the price of petrol rises then the extra cost for an EV will be less. The higher expense is negligible but with government incentives for road tax and congestion charges, EVs are an attractive option and are likely to persuade many car buyers.

As indicated earlier, another method of moving towards meeting the net zero emissions target is the use of alternative biofuels. There are traces of carbon in alternative fuels, but it is argued that these fuels use carbons that are already present in the atmosphere. The common form of alternative biofuel is ethanol. Ethanol fuel is actually ethyl alcohol made from sugar cane, corn or beet sugar. Fiat was the first car company to introduce a passenger car running only on ethanol in Brazil in 1978. However, ethanol poses a few problems. First it is less efficient than gasoline. Ethanol needs 1.5 times the volume of gasoline to produce the same energy. Therefore, the cost per litre for both petrol and ethanol fuel is in the same range although it varies from one country to another. However, the major challenge is that the amount of land needed to grow sugar cane or corn to produce a sufficient quantity of biofuel is so vast that it could impact the biodiversity of our environment. In spite of these problems, scientists are optimistic about alternative fuels and are working on converting biomass, such as trees, and left-overs from other processes (e.g. paper making) into carbon-free fuels.

20.3.2 Trucks and Buses

Medium-sized vans and city buses are reasonably lightweight and generally travel shorter routes. The advantage is that these vehicles can work on electric power. China has already electrified the local government fleet of some of its cities such as Shenzhen. Unfortunately, electrification is a less practical option for long-distance trucks and buses because they would need too many powerful batteries to travel over longer journeys without recharging en route. An expensive Tesla EV may travel 300 miles without charging but to cover the same distance an electric truck would need to carry many batteries. In order to compensate for this crucial extra weight, it would have to lose a quarter of its loads. To obtain the same amount of energy as from a litre of gasoline requires batteries 35 times heavier than a litre of gasoline. All in all, batteries for long-distance buses or trucks do not seem practical.

One possible solution under trial is so-called electrofuels produced by combining the hydrogen in water and the carbon obtained from carbon dioxide. But 'electrofuels' are at the moment four times more expensive and need electricity (which should be clean energy) to make them. Large organisations like Siemens Energy can see a huge potential for what they call the 'hydrogen economy'. Armin Schnettler of Siemens Energy has commented, 'I strongly believe that the next step of the global energy transition will be based on the hydrogen economy—transforming "green electrons" to "green molecules" via water electrolysis' (Whitlock, 2020). It is expected that by using renewable electrical energy like wind or solar power for 'green electrons' from the power sector this will unlock enormous environmental and business benefits across all sectors. However, we also need the infrastructure of special refuelling stations along the highways in order to make this work.

20.3.3 Airplanes and Ships

Batteries are unlikely to be powerful enough to move long-haul ships or planes. 'The best all-electric plane on the market can carry two passengers, reach a top speed of 210 miles per hour, and fly for three hours before recharging' (Gates, 2021, p. 143). The Airbus A350 is already carrying 350 passengers and flying 15,000 km non-stop for nearly 19 hours from New York to Singapore. Batteries are improving but it is hard to see their practical application in airplanes or ships in the near future. A possible solution for planes and ships (applicable to both cargos and cruises) could be to replace fossil fuels with electrofuels or advanced biofuels. Major oil companies, such as Exxon, are also serious about the future of biofuels and developing scalable biofuels derived from cellulose and algae.

Ship operators are now experimenting with available and affordable biodiesel. This is being seen as a 'drop-in' (with no major changes in ships' fuel storage and delivery systems) interim solution for reducing carbon emissions. Case Example 20.1 illustrates a successful trial of biodiesel on the Hurtigruten cruise line.

CASE EXAMPLE 20.1 NORWEGIAN CRUISE LINE HURTIGRUTEN TRIALS BIOFUEL ON A CRUISE SHIP

Hurtigruten AS is a Norwegian cruise line headquartered in Tromso, Norway. It also operates as a ferry service along the coast of Norway. This company claimed to be the first to test biofuels to power its 12,000 tonnes cruise ship MS *Polarlys* in 2019. The fuel for the test was a certified biodiesel that is free of palm oil.

Commenting on the development, CEO Daniel Skjeldam said, 'Biodiesel can in the long run potentially give a carbon dioxide reduction of as much as 95% compared to traditional marine fuels' (Warner, 2019).

The shipping industry consumes more than 330 million tonnes of fuel every year and is responsible for nearly 2% of global carbon dioxide emissions. By using biodiesel, which is sourced from waste cooking oil, corn, soya, wheat, tallow or palm, the industry can begin to reduce its carbon footprint. The move to biodiesel is in its infancy, but it has the potential to grow quickly and therefore transform the shipping industry.

20.3.4 Trains

Trains are gradually moving towards being all electric. Today, three-quarters of passenger rail transport activity takes place on electric trains, and the electrification of railways is on the rise although it varies between different countries. In China, Russia, Japan and India over 70% of trains are electrified while in the US, which has the largest rail network in the world, a mere

Table 20.3 Rail Electrification in Larger European Countries

Larger European Countries	*Electrification (%)*
Italy	72
Spain	64
France	58
Germany	53
UK	38

Source: Statistica Research Department, Hamburg (2021).

1% of trains are electric. At present in Europe only Switzerland can boast a 100% electric train network while larger European countries are behind with electrification schemes as shown by Table 20.3.

There are great opportunities for going green in the realm of rail transportation. Airplane and motor transport passengers should be encouraged to travel instead by green railways. Led by Japan, high-speed trains are cutting down the proportion of travel undertaken by airplanes and buses. The high-speed train network under construction in the UK (HS2) is expected to reduce carbon dioxide emissions by 400,000 tonnes per year. Each country should consider inter-city high-speed rail networks, with key routes such as Los Angeles to San Francisco and Washington to New York in the US, St Petersburg to Moscow in Russia, Mumbai to Delhi in India, or Melbourne to Sydney in Australia being ripe for such a change.

20.4 How Six Sigma Is Helping Green Transport Initiatives

The biggest player in green transport is EVs. Car manufacturers already have all the requirements to benefit from high-level Six Sigma methodology. The industry produces complex products requiring very high quality standards. The organisations are global, big and have many talented full-time employees. The largest car manufacturer, Toyota, constitutes the 'temple' of Lean Thinking or Toyota Production System (TPS) which is the integral part of Lean Six Sigma. Ford Motors has already demonstrated its commitment to Six Sigma by training its 350 senior executives including its CEO and more than 10,000 employees in various aspects of Six Sigma, producing both

Black Belts and Green Belts. The application of Lean Six Sigma techniques has enabled this multinational company to eliminate more than US$2.19 billion in waste over the last 15 years. However, it must be said that there were also some obstacles for implementation. These were an initial lack of commitment, finding time and money, and data needs. In spite of these impediments, Ford has completed about 10,000 Six Sigma projects since the early 2000s.

TPS has been blended with Six Sigma methodology to create a potent Lean Six Sigma programme. The outcome of such an effective programme has led to a high-performance culture and enabled the company to hold the position of being the top profitable car manufacturer. Toyota has revolutionised the hybrid car market over the last two decades. During this period according to Toyota, hybrid cars saved over 120 million tonnes of carbon dioxide compared to equivalent petrol cars. In 2022 Toyota is going to join the line-up of full-electric vehicles. It is a given that Toyota has applied TPS/Lean Six Sigma principles to manufacture and supply its Yaris or Corolla hybrid models and will also apply the methodology to bZ4X models.

There is evidence (Chaurasia et al., 2019) that car manufacturers in India and nearly all car manufacturers in Europe (e.g. Audi, BMW, Daimler-Benz, Volkswagen, Nissan) are engaged in delivering EVs or hybrid cars and also have applied some form of Lean Six Sigma to improve productivity and reduce waste.

Tesla Inc. has made a powerful brand name for itself under the leadership of Elon Musk. As shown in Case Example 20.2, their success was helped by the application of Lean Six Sigma in Tesla's manufacturing processes.

CASE EXAMPLE 20.2 ELON MUSK CAPITALISES ON LEAN SIX SIGMA

Tesla Inc. is one of Elon Musk's three major companies based in Palo Alto, California, manufacturing EVs. The Gigafactory in Nevada from Tesla's energy division will be the biggest factory in the world for lithium-ion batteries. Many are calling the Tesla Model 3 (costing around $50,000 in the US) the Model T of our time, a breakthrough new EV. As the only dedicated electric car manufacturer, Musk has engaged in clever strategies like First Principle Thinking and Lean Six Sigma to grow his company into a globally recognised brand.

Supported by the Six Sigma principles, as advocated by Peter Peterka of 'Global Six Sigma', the company is focused on a philosophy of enhancing efficiencies in their processes. It takes Tesla 3 days to build a car, starting from raw materials through to finishing the very last detail. The company embraced the Lean Six Sigma core principle, 'no extra processing that doesn't add value' and also DMAIC (define, measure, analyse, improve, control) methodology. Tesla's adaptation of DMAIC has been in three major steps:

- Identification of weakness (Define and Measure)
- Improvement of weakness (Analyse and Improve)
- Application of control measures (Control)

The success of Tesla was reflected when the company booked close to $14 billion in deposit-backed advance sales in under 3 weeks through the direct-to-consumer launch of the Model 3 in 2018. The Model 3 is the world's all-time best-selling plug-in electric vehicle (PEV).

There is also good evidence of the application of Six Sigma or Lean Six Sigma in rail infrastructure as Case Example 20.3 and Case Example 20.4 illustrate.

CASE EXAMPLE 20.3. LEAN SIX SIGMA AT NETWORK RAIL UK

Network Rail Limited is the infrastructure manager of most of the railway network in Great Britain. It is a public body of the UK Government and employs around 42,000 employees. In 2013 rail services were severely affected by regular delays in passenger trains and Bourton Consulting Group were asked to investigate. Their consultants identified five areas causing over 50% of the delays: Points, Track Circuits, Signalling, Seasonal Preparedness and Operational Procedures. Working with Network Rail, senior staff consultants introduced a Lean Six Sigma programme across 1,200 selected people.

Full-time leaders were trained as Black Belts, leaders of smaller-scale projects were schooled as Green Belts, while part-time project team members became Yellow Belts. All classroom training was supported by about 300 real projects. As a result, US $60 million in efficiency savings were achieved by Lean Six Sigma projects. Lean Six Sigma improvements were also recognised as a key contributing factor behind the 50% reduction of train delays.

CASE EXAMPLE 20.4. SIX SIGMA IN HS1

The Channel Tunnel High Speed Rail Link (HS1) is the first ever high-speed railway in the UK, capable of speeds of up to 300 kmph. It comprises 100 kilometres between London and the Channel Tunnel near Dover, with three stations and two depots. Rail Link Engineering (RLE) was appointed as project manager. The group was a consortium of construction companies formed between Arup, Bechtel, Halcrow and Systra. At the height of construction in 2001, the combined workforce of LCR (London and Continental Railway) and RLE was over 1,000, and at the closure stage the figure was 350. The completed US $5.8 billion project opened on time and within budget and was delivered for commercial operation to Eurostar on 14 November 2007.

Bechtel was the first major engineering and construction company to adopt Six Sigma, a data-driven approach to improving efficiency and quality, for major projects. On big rail modernisation schemes in the UK, including the HS1 project, Bechtel teams used Six Sigma to minimise costly train delays caused by engineering work, and in so doing reduced the 'break-in' period for the renovated high-speed tracks.

The introduction of Six Sigma to the HS1 scheme delivered both cost savings and programme benefits. The Six Sigma programme trained 23 Black Belts and around 250 Green Belts and Yellow Belts. A further 100+ senior managers were educated to act as Champions on improvement projects. In fact, over 500 such improvement projects were completed which in turn led to a cost saving/avoidance of at least US $40 million. These ventures covered and benefitted a wide range of activities across the whole HS1 undertaking including numerous architectural, civil and railway construction endeavours. Consequently this had the effect of ensuring timely third-party methodology approvals, facilitating procurement, accelerating drawing reviews and allowing the judicious generation of construction record documentation. It is evident that some of these improvement projects, such the reduction of lead time in methodology approvals and drawing reviews, also applied Lean Thinking concepts.

20.5 How Green Six Sigma Can Help Green Transport Initiatives Further

There is clear evidence of the application of Six Sigma/Lean Six Sigma methodology in the transport sector in Case Examples 20.2, 20.3 and 20.4. Green Six Sigma comprising both the toolsets of Six Sigma and Lean Six Sigma should be very effective in all new Green Transport projects whether they are for passenger cars, trucks and buses, trains, ships or airplanes. Thus any new and future Green Transport initiatives, especially in the manufacturing sector, should adopt the Green Six Sigma approach to ensure both cost benefits and sustainable outcomes.

As indicated in Chapter 18, the distinctive additional contributions of Green Six Sigma include

- A comprehensive performance management system supported by a periodic self- assessment to monitor and sustain performance levels (see Chapter 15)
- A regular senior management review (e.g. sales and operations planning) to include Green Six Sigma projects (see Chapter 15)
- Application of carbon footprint tools and software (see Chapter 13 and Chapter 6)
- Application of material flow account (see Chapter 13)

20.6 Summary

Although transports are not the biggest polluter compared to energy, agriculture and industry, their impact on the environment is arguably most visible. We are making good progress in electrifying passenger cars and trains, but we still face a major challenge to find zero-carbon solutions for long-distance haulage by trucks and buses as well as air travel. The growth in online shopping has also increased the volume of local deliveries by vans. We can of course aim to use electric vans but green solutions for trucks and buses are still a long way from being driven by hydrogen-powered internal combustion engines. The alternative low-carbon aviation fuels are still in development. A further problem is that there are not yet biofuels available in sufficient quantity suitable to run cargo ships or cruise ships.

Therefore, we need more investment and research to develop cost-effective and proven solutions for longer-distance transport whether by trucks, buses, airplanes or ships.

GREEN TIPS

- We are making good progress in electrifying passenger cars, but we still do not have practical solutions for the long-haul road transport.
- The alternative low-carbon aviation fuels are still in development.
- There should be further expansions of high-speed rails between neighbouring big cities in larger countries.
- Green solutions overlap each other. For example if we find a solution in hydrogen we may not need magic batteries.
- Building on the success of Lean Six Sigma in green transports, Green Six Sigma can add the extra value of sustainability.

Chapter 21

Green Six Sigma and Retrofitting Buildings

I have spent a fortune traveling to distant shores and looked at lofty mountains and boundless oceans, and yet I haven't found time to take a few steps from my house to look at a single dew drop on a single blade of grass.

—Rabindranath Tagore

21.1 Introduction

Buildings contribute only 6% of total global greenhouse gas transmissions, thus ranking last of all sectors. Some reports (RIBA, 2021) have estimated that the contribution of buildings to greenhouse gas emissions is in fact much higher. Regardless, the important point is that this is one area where we as individuals can start to do something—now.

The term 'buildings' covers both industrial and residential structures. Residential building operations account for 72% of the total emissions from this sector. This chapter focuses on the retrofitting of buildings (i.e. the addition or replacement of new features) to improve energy efficiency and reduce carbon emissions.

As with other sectors, if we are going to attain net zero emissions we will have to minimise all the greenhouse gases caused by building operations around the world.

DOI: 10.4324/9781003268239-25

21.2 Guiding Factors of Retrofitting Buildings

In building operations the main areas of energy consumption are heating, cooling, lighting and cooking. When considering lighting I have included the electricity used to run television, radio, computers and other home appliances. Energy consumption for lighting is higher in advanced economies compared to poorer countries.

Heating and cooling is an interesting area. In general, in hot countries in the tropics region we need more cooling devices (with electric fans and air conditioners) and in colder countries, naturally, we need more heating. However, many nations in the subtropical region (e.g. US, Japan, China, Australia) spend huge amounts of energy on both heating *and* cooling. Air conditioning is also essential in some industries including for cooling server centres containing thousands of computers. Table 21.1 shows the proportion of houses fitted with air conditioning in the top five countries:

It is likely that people will be adding to this global number of air conditioning (A/C) units as the population grows and heatwaves become more frequent. However these A/C units not only consume a large amount of electricity, they also contain fluorine (F-gas) which is a powerful contributor to global warming.

Domestic boilers and water heaters account for about a third of all emissions that come from domestic buildings. A/C units run on electricity which may also come from renewable sources but boilers and water heaters in our households mostly run on fossil fuels. Fortunately we have some proven technology already available to decarbonise our homes by retrofitting existing equipment. For example, we already have heat pumps, electric water heaters and heating systems and solar panels to provide domestic electricity.

Table 21.1 Top 5 Countries with Air Conditioning in Households

Country	*Households with Air Conditioning (%)*
Japan	85
US	84
S Korea	83
Saudi Arabia	62
China	60

Source: Gates (2021).

21.3 How Retrofitting Buildings Provides Solutions for Reducing Greenhouse Gas Emissions

The path to a net zero carbon strategy for domestic dwellings is likely to comprise the following steps that everyone can try to take:

1. Reduce energy losses and energy consumption
2. Replace fossil fuel boilers and water heaters
3. Seek to use renewable energy
4. Apply a circular economy

Let's look at each solution in more detail.

21.3.1 Reduce Energy Losses and Energy Consumption

There are many ways in which we can insulate our homes but first we should consider how to minimise heat loss from our buildings. In poorly constructed homes, the amount of heat loss varies according to the source of exit from the structure. For example we are squandering heat through external walls (40%), the roof (25%), windows and doors (20%), the ground floor (10%) and finally via draughts (5%).

To draught-proof our home, we should block up any unwanted gaps (with the exception of ventilators) that let cold air in and warm air out. This is the cheapest and most effective approach to save energy. There are various inexpensive ways of preventing draughts (e.g. self-adhesive foam strips for doors and windows, a letter box brush, chimney cap and silicone fillers around pipework). It is also important to make the house airtight as far as possible by replacing glass windows and external glass doors with double-glazed units.

It is easy to follow a number of energy-saving, common-sense actions including the following suggestions:

- Install a smart thermostat and set it at the desired temperature and timings
- Install a smart meter to monitor energy consumption
- Use energy efficient appliances
- Choose energy-saving LED electric bulbs
- Turn off standby appliances and computers when not in use
- Be smarter about hot water usage, for example, shower rather than running a long bath

Most newly built homes are fitted with loft insulation, and older homes can save up to 25% of potential heat losses by investing in loft insulation. This insulation will last over 40 years and the pay-back period is about 2 years. Mineral and wool sold in rolls like a blanket is the most common loft insulation material.

The most effective method allowing protection against heat loss from walls is so-called cavity wall insulation. This involves injecting an insulation material into the void between the inner and outer layers of brickwork. It works in the same way in which a thermos flask keeps a drink hot, by creating a layer around the house. Most newly built homes have cavity wall insulation as standard but older houses (such as those built before 1920 in the UK) do not have cavity walls. Here a possible solution is external solid wall insulation. There are many insulation materials (such as aluminium composite panel or rain screen cladding) which can be used for cladding external walls. However, following a number of fire tragedies (e.g. Grenfell Tower, UK in 2017), it is an essential prerequisite that cladding materials conform to building regulatory standards.

Appropriate floor insulation materials (e.g. expanded polystyrene sheets, polyurethane spray) can be used to insulate floorboards on the ground level of houses and to seal the gaps between floors and skirting boards. It is also important to insulate water tanks by using approved cylinder jackets and water pipes by polyethylene foam insulation.

21.3.2 Replace Fossil Fuel Boilers and Water Heaters

Heating systems in residential homes come in two parts—one section is for the supply of hot water to bathrooms, the kitchen and washing machines, while the other aspect is for hot water radiators. Many homes in the UK and rich countries are provided with a central heating system which serves both hot water and heating. The supply of energy in this system comes from fossil fuel boilers. There are two types of fossil fuel boilers. First, combination or 'combi' boilers provide hot water on demand for domestic use as well as central heating, and are a good choice for smaller homes. With the second type, a heat-only boiler (also known as a conventional boiler system), the hot water is stored in a hot water cylinder or storage tank. This system is suitable for a larger home with greater demand, where several people frequently need to use hot water at the same time. They also require space for a cold water feed tank, usually housed in the loft.

During July 2021 I interviewed 12 UK suppliers specialising in replacing fossil fuel boilers to find cost estimates for different possible scenarios. It should be noted that the figures that follow are indicative only.

In order to replace fossil fuel boilers and water heaters we have different options depending on the type and size of households. If the household is a one- or two-bedroom apartment then a simple solution could be to install a stand-alone electric water heater with a hot water cylinder for domestic hot water supply and plug-in electric convector heaters or electric storage heaters. A typical cost for this solution for a small apartment is around US $3,000 in the UK.

For a medium-sized household with three or four bedrooms fitted with a fossil fuel central heating system, an energy-saving alternative could be to replace the fossil fuel boiler with an appropriate electric combi boiler. As shown in Figure 21.1 this solution has the advantage that it will retain the existing radiators and can be retrofitted in the existing space. The cost of this type of solution could be US $4,000 to US $8,000 in the UK.

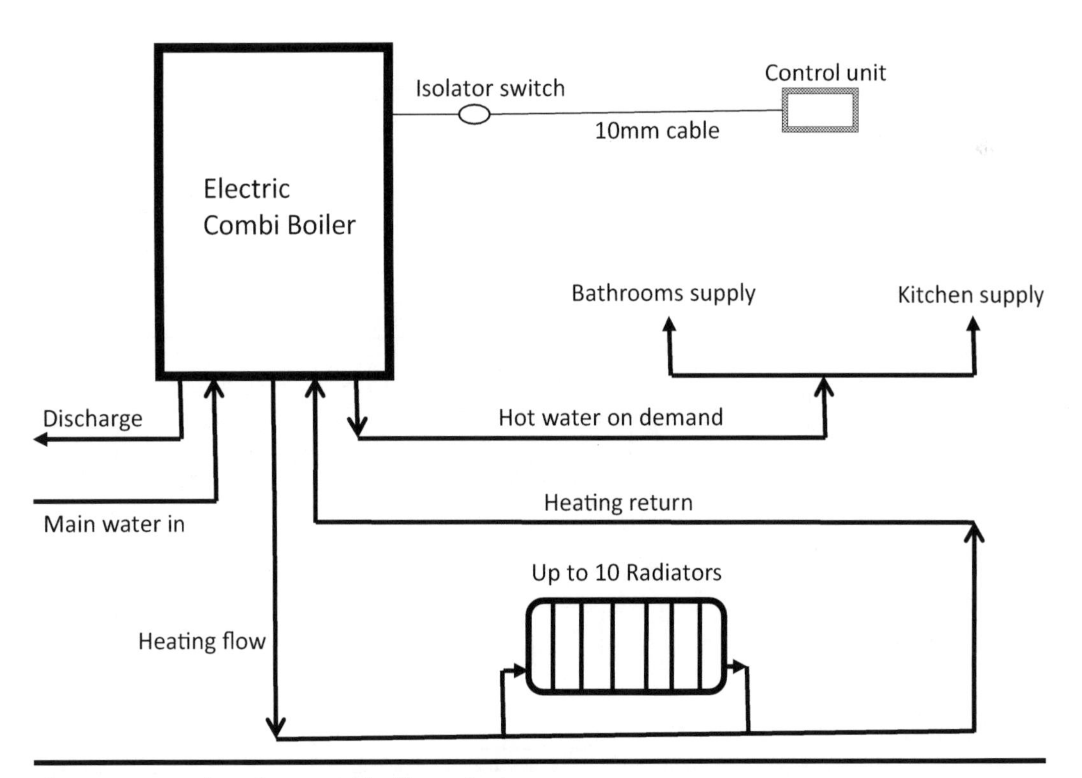

Figure 21.1 Electric combi boiler solution.

Retrofitting an electric combi boiler is a simpler solution, but it is limited to smaller properties and at present the running cost of a gas boiler is more economical. It is envisaged that a longer-term green energy solution for households is likely to be electric heat pumps.

A heat pump is either air or ground sourced. As shown in the schematic diagram in Figure 21.2 an air source heat pump (ASHP) works rather like a reverse fridge. The heat pump is installed outside and outside particles of air are blown over a network of tubes filled with a refrigerant to turn it into gas. The gas then passes through a compressor. The compressed hot gases pass into a heat exchanger surrounded by cool water in order to heat this water. This is circulated around the household to provide heating and hot water. Then the refrigerant condenses into cool liquid and starts the cycle again.

An ASHP costs from US $8,000 to US $18,000 to install in the UK. A Renewable Energy Incentive scheme is also operational in some countries including the UK. It works both as a heater or an air cooler. ASHPs have key advantages: energy bills are lower than comparable gas heating systems, and they are known to work efficiently in severely cold countries such as

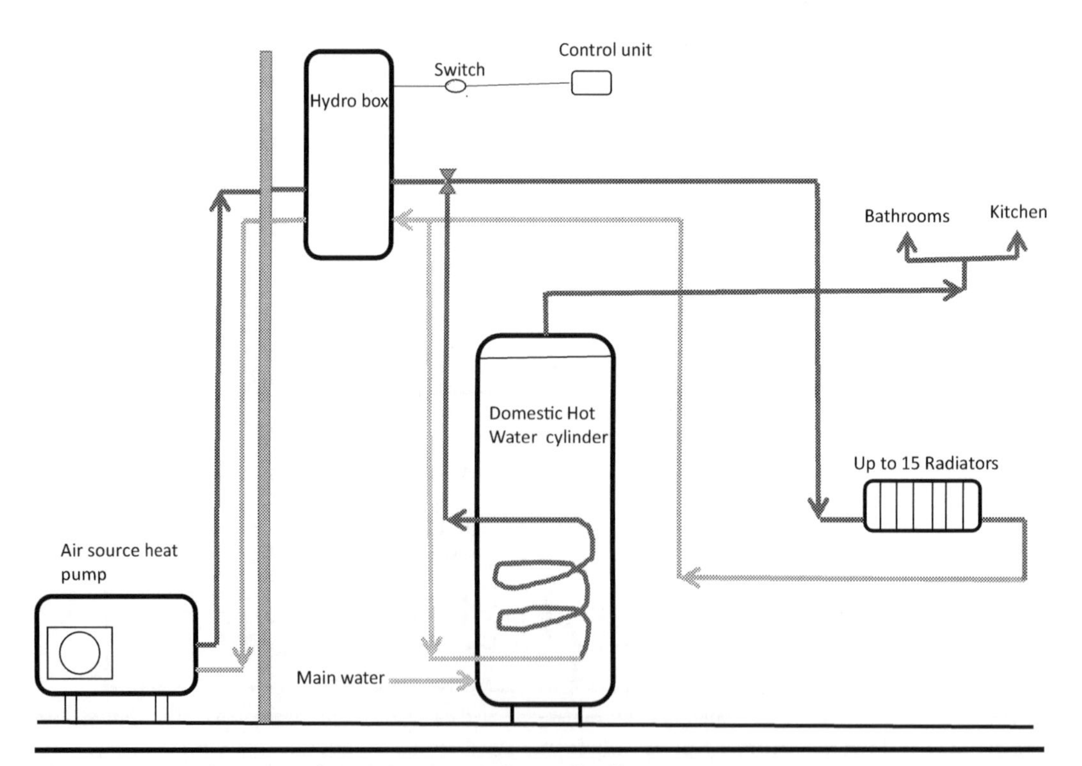

Figure 21.2 Air source heat pump: a schematic diagram.

Canada. However an ASHP also poses drawbacks including its noisy operations and requirement for larger radiators.

Ground source heat pumps (GSHPs) are also good low-carbon heating systems and they have higher efficiency rates and lower running costs than ASHPs. A GSHP makes use of the ground's constant temperature and uses that to both heat up homes and supply domestic hot water. A GSHP absorbs low-grade surrounding energy from the ground and then compresses and condenses this energy to a higher temperature. Heat is transferred to the heating and hot water system of the household as shown in a schematic diagram in Figure 21.3. The fluid from the heat exchanger then continues its circuit back to the submerged pipework to commence the cycle all over again.

By comparison with an ASHP, a GSHP has several advantages as it is not noisy and is both more energy efficient and cheaper to run. The main disadvantage of a GSHP is it is very expensive to install (in the region of between US $20,000 to US $40,000 in the UK) and it requires a lot of ground space. Households in the UK would also require planning permission. Thus an ASHP could be considered for a medium-sized house, and a GSHP is more suitable for a larger property with available land. Heat pumps are more attractive in countries requiring both heating and air conditioning (e.g. US, Canada, Japan).

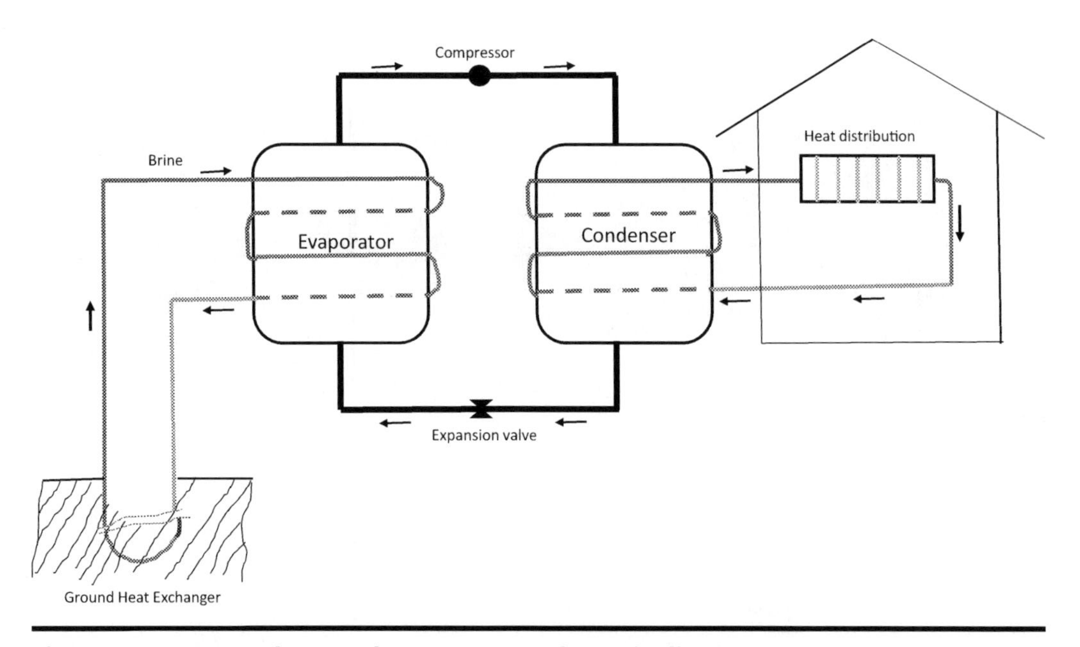

Figure 21.3 Ground source heat pump: a schematic diagram.

21.3.3 Seek to Use Renewable Energy

A stead forward approach for a household aiming to attain lower carbon emissions is to switch to a renewable energy supplier. However, herein lies the problem. Some major suppliers in the UK (e.g. E-On, Centrica, Pure Planet) are claiming to be renewable suppliers but while this may be true, their approach at present is likely to be offsetting carbon by planting trees all over the world. There are some suppliers (e.g. Octopus Energy, Ecotricity) that have a better reputation as truly renewable energy suppliers. Centrica have invested in EDF Energy's existing and future nuclear plants (e.g at Hinkley Point and Sizewell in the UK) and aim to increase their share of nuclear source electricity.

The most genuine method of using a source of renewable energy for a household is to consider domestic solar energy. As explained in Chapter 7, solar panels, also known as photovoltaic systems (PV systems), convert the sun's energy into electricity that can power our households. The system uses semiconductor technology to convert energy from sunlight into direct current (DC) electricity. This current is then passed through an inverter to convert it into an alternating current (AC). As shown in Figure 21.4 the system can be either grid connected or stand alone.

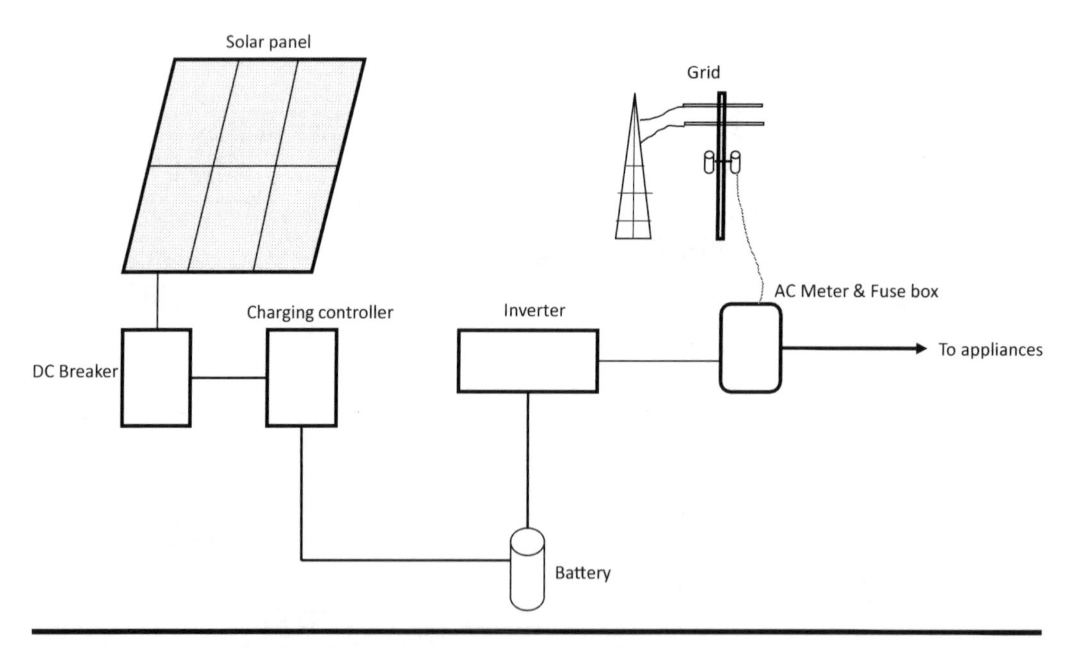

Figure 21.4 Domestic solar energy: a schematic diagram.

Grid-connected systems are linked to the local utility grid to ensure a continuous supply of electricity. When the domestic solar panel system generates more electricity than the household needs, the surplus energy can be exported back to the national grid. Likewise, if more electricity is needed the grid can supply this. Stand-alone systems are not connected to the grid but instead charge a solar battery system. These batteries store the electricity generated by solar panels. The stored electricity from these batteries will be used to operate household appliances. Stand-alone systems are operated in areas with local utility grids and are typically more expensive to include the cost of expensive storage batteries. The average domestic solar panel system, without storage batteries, costs between US $5,000 to US $10,000 in the UK.

Solar thermal collectors do not have PV cells and use sunlight directly to heat up water that is stored in a cylinder. The hot water from the cylinder can then be used for domestic heating needs. Solar thermal panels are roof-mounted, just like solar PV panels, but look slightly different, as instead of cells they have multiple pipes to heat water.

21.3.4 Apply a Circular Economy

As defined in Chapter 2, a circular economy is aimed at eliminating waste and the continual use of resources. The approach is to reuse resources, repair defects, refurbish facilities, rebuild products to original specifications, and recycle wastes to create a closed-loop system. The outcome is minimising the use of resource input, wastes, pollution and carbon emissions.

The application of a circular economy can benefit all sectors of our economy. In the context of retrofitting buildings two specific areas are discussed: the construction industry and household appliances.

According to a recent report (Retro-First, 2021), every year some 50,000 buildings are demolished in the UK, producing 126 million tonnes of waste and accounting for 10% of carbon emissions. RIBA (the Royal Institute of British Architects) has advocated that the demolition of buildings should be halted and buildings should be preserved and re-purposed. Materials should be salvaged and re-used whenever possible. RIBA argues that the construction of large buildings gobbles up fossil fuel hungry materials (e.g. steel, cement, aluminium and plastics) and hence developers ought to be more considerate and should be obliged to refurbish. The institute has requested that it should be allowed to force firms to calculate the total carbon impact of each project they wish to undertake. The Royal Institute of Chartered Surveyors, in partnership with RIBA and other organisations, is developing the first international standard for reporting carbon emissions across all areas of construction.

The campaign by these influential professional bodies is on the same page as the principles of the circular economy. The end goal of a circular economy is to retain the value of materials and resources indefinitely with little or no residual wastes. This requires transformational change in the way that buildings are designed, built, operated and demolished. Government intervention in both taxation and legislation should help. For example, retrofitting could be tax-exempt like new constructions and all new build proposals should indicate how much carbon will be emitted during the manufacturing and construction process.

The principles of a circular economy should also be applied to the 'replace or repair' policy of home appliances (e.g. washing machines, dishwashers, cookers, fridges). Their suppliers should be obliged to produce and supply spare parts for say 10 years so that appliances can be easily repaired. Suppliers should also introduce an exchange plan and offer to take the customer's old item away when delivering the new one. As shown in Figure 21.5

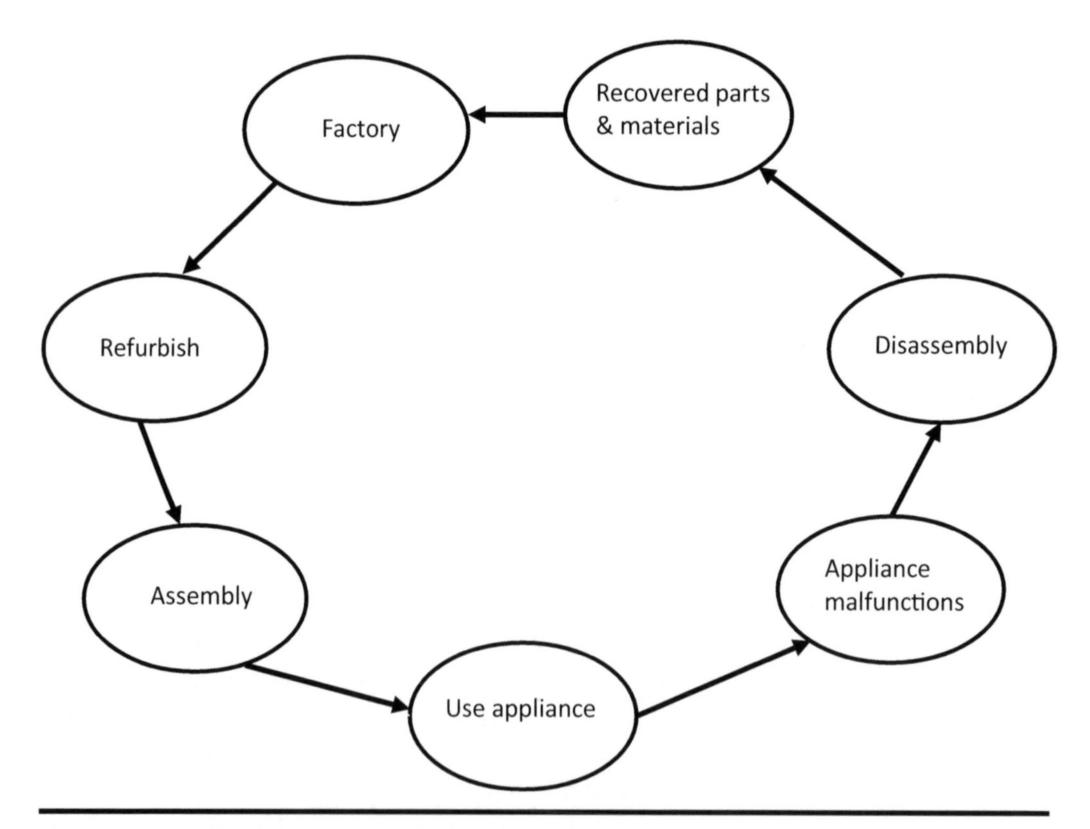

Figure 21.5 Circular economy for home appliances.

the cycle of the circular economy can work well and to the benefit of both suppliers and customers.

The recycling of products also applies to electrical and electronic items (such as televisions or computers). The Waste Electrical and Electronic Equipment (WEEE) directive regulates the manner in which manufacturers and retailers in European countries should behave regarding recycling.

21.4 How Six Sigma Is Helping Retrofitting Buildings Initiatives

There is no significant visible evidence of the application of any type of Six Sigma methodology in the manufacture and supply of electric boilers or heat pumps. However, there are some publications showing the application of Six Sigma research in delivering domestic solar energy as discussed in the following examples.

CASE EXAMPLE 21.1 SIX SIGMA APPROACH TO OPTIMISE SOLAR CELL MANUFACTURING

Domestic solar energy systems use silicon-based PV cells in solar panels. In general, silicon solar cell process uses either *p*-type or *n*-type doped silicon as the starting material. In a research paper (Prasad et al., 2016), boron-doped wafers were considered and phosphorus was used as the doping material. The paper describes the application of Six Sigma DMAIC (define, measure, analyse, improve, control) methodology for improving the throughput of the phosphorous process. The optimised process was implemented in production and as a result an effective gain of 0.9 MW was reported per annum.

CASE EXAMPLE 21.2 SIX SIGMA APPROACH TO DOMESTIC SOLAR ENERGY EFFICIENCY PROGRAMME

Mowris et al. (2006) argue that the increased emphasis on energy efficiency to reduce energy use and mitigate global warming requires rigorous evaluations based on the International Performance Measurement and Verification Protocol (IPMVP). Increased focus on customer satisfaction and resource efficiency to improve profitability has motivated businesses worldwide to adopt Six Sigma strategies. Research data in this paper show how Six Sigma strategies can provide a framework to measure and verify energy savings and performance metrics at critical steps in the supply chain of domestic solar energy systems (i.e. design, manufacturing, installation and service). The authors advocate a publicly funded energy efficiency programme to transform the renewable energy market, with particular reference to domestic solar energy systems, so that the cost to deliver energy efficiency products can be minimised. This research also indicates that this goal can be achieved by incorporating IPMVP and Six Sigma into programme design, implementation and evaluation. The Six Sigma approach encompassing IPMVP is applicable to assess energy efficiency in all forms of energy systems.

21.5 How Green Six Sigma Can Help Retrofitting Houses Initiatives Further

It is evident that Six Sigma methodology has not been effectively applied in initiatives related to retrofitting houses. Therefore, there are significant opportunities for Green Six Sigma in this virtually untapped sector. The question is, simply—is it worthwhile for manufacturers and suppliers of electric boilers, ASHPs, GSHPs and solar energy systems to adopt Green Six Sigma tools and techniques in order to be cost effective? The answer is certainly yes. The majority of these suppliers are small and medium-sized enterprises (SMEs), and they do not need to retain high-profile and expensive Six Sigma consultants. They can train their key employees using a week-long training sessions for Green Belt or a 3-day training session for Yellow Belt and then apply the 'fitness for purpose' Green Six Sigma approach. The available government grants and bursaries for adult learners (e.g. UK Government grants) should help SMEs.

With adequate training, Green Six Sigma can show breakthrough results in the supply chain of retrofitting building initiatives. The data-driven and process focused DMAIC methodology will improve the design, performance and manufacturing costs of boilers, heat pumps, solar panels and associated components. Any outcome will be underpinned by sustainable processes and reduced carbon footprints.

Thus the benefits of the circular economy as discussed in Section 21.3.4 are a fundamental tenet of Green Six Sigma culture.

21.6 Summary

In this chapter general principles, processes and the appropriateness of retrofitting building initiatives have been discussed. It is important to note the success of these household-related initiatives depend on us, the customers. The majority of action points are customer-driven 'pull' processes. All facilities and help are available in the marketplace now and are ready to apply. Some practical guidelines are included to manage energy economy and retrofit appropriate domestic energy systems. For example, one should consider combi electric boilers for smaller households and ASHPs for medium households. GSHPs should not be contemplated unless the property is large with adequate garden space. The pay-back period of a solar energy system is usually 8 years. However it is strongly advisable that professional surveys by multiple suppliers and the eligibility of government grants are properly investigated before selecting any retrofitting system for a household.

The supply chain of the retrofitting houses sector thus constitutes fertile ground for Green Six Sigma to reap sustainable rewards.

GREEN TIPS

- Ensure the insulation of all sources of heat losses in the house (e.g. loft, walls, floor and windows).
- Consider combi electric boilers in small houses and ASHPs for medium-sized properties.
- There is a long pay-back period for total solar energy solutions.
- Hydrogen solutions for home heating are not yet feasible.
- Government grants are necessary to implement climate change initiatives for retrofitting homes.
- Green Six Sigma can be very effective for SMEs in home retrofitting supply and installations.

Chapter 22

Green Six Sigma and Climate Adaptation

The measure of intelligence is the ability to change.

—Albert Einstein

22.1 Introduction

'Climate change adaptation is the process of adjusting to current or expected climate change and its effects' (IPCC, 2014a). Even taking into account that some progress has been achieved due to present climate change initiatives, it is a certainty that we will breach 1.5°C of global warming in the early 2030s and may remain unprepared. The global reality of the climate crisis could hardly be more serious now. July 2021 alone saw the effects of several chaotic weather events; over 200 people perished in floods in Germany, the Netherlands and Belgium, causing unprecedented scenes of devastation in northern Europe, while at least 50 people died and a further 400,000 people were displaced in central China after heavy downpours. In western India many properties with 125 people were perished by torrential monsoon. At the same time in western Canada and the US, a blistering set of heatwaves provided the tinder for wildfires to rage for months, while drought threatens areas from Algeria to Yemen.

As I write this chapter Covid-19 has killed over 4 million people and the UK government is advising us that in the long term, we must adjust and

DOI: 10.4324/9781003268239-26

learn to live with the virus because we have vaccines. Many of the lessons learned from the pandemic also apply to climate change. The world has made remarkable progress by working together in developing and testing vaccines in record time. The same degree of concerted effort is needed worldwide to mitigate and adapt to the consequences of global warming. By investing in climate change initiatives and R&D we can both rescue the post-pandemic economy and avoid a climate disaster.

The fundamental strategy of climate change preparation is 'Predict, Prevent and Preserve'. In terms of Prediction, scientists have already forecast the longer-term causes and effects of climate change. We also need continuous prediction of climate fluctuations to prepare for events such as cyclones, floods, heavy downpours and extraordinary heatwaves. Prevention is rooted in the elimination of the root causes (e.g. greenhouse gas emissions) as well as mitigation initiatives as discussed in previous chapters. Preservation lies in dedicated adaptation to the changes to our lifestyles, health and safety caused by climate change in spite of preventative initiatives. In this chapter, climate adaptation processes and actions are discussed under the following headings:

- Climate Adaptation in the Global Community
- Climate Adaptation in Clean Energy
- Climate Adaptation in Green Supply Chains
- Climate Adaptation in Green Transports
- Climate Adaptation in Retrofitting Houses
- Climate Adaptation and Infra-structure Projects
- Climate Adaptation and Innovation

In each section the ways in which Green Six Sigma can help by making additional contributions are discussed.

22.2 Climate Adaptation in the Global Community

At the time of writing there are less than 100 days until more than 190 world leaders will gather in Glasgow at the United Nations COP26 climate conference on 31 October 2021. This is an excellent opportunity to reboot climate change targets and initiatives. The UK Government calls the summit the world's 'last best chance'. Although I'll complete my manuscript before the conference, this book will be on the market after it takes place. Hence it may be presumptuous, but I'll suggest that the agreed communique should include the following:

1. There will be no new oil, gas or coal exploration projects beyond the end of 2021 as recommended by the International Energy Agency (IEA).
2. There will be an increase on the $100 billion (promised in the Paris Climate Agreement) by rich nations to support poorer countries in cutting emissions and a pledge to deliver this as soon as possible.
3. Annual audits by the United Nations of top 10 greenhouse gas emitters will be introduced in order to monitor the targets of the Paris Climate Agreement.
4. New targets will be set for each of the signed-up countries, e.g.
   No new diesel or petrol vehicles after 2030
   No fossil fuel domestic heating after 2035
   Reduction in methane emission
5. Introduce afforestation targets for each country as it is difficult to monitor voluntary carbon offsets or 'greenwashing'.

As host, the UK is pivotal to the success of any planned objectives. This could start by committing to end all new fossil fuel projects, and rolling out a nationwide clean energy programme with grants to accelerate decarbonised domestic heating.

22.2.1 How Green Six Sigma Can Help

As discussed in preceding chapters Green Six Sigma can accelerate most initiatives by its holistic approach and ensure the sustainability of both processes and environment. In the 1950s and 1960s the United Nations rolled out its Work Study to developing countries with the proactive support of the International Labour Organisation (ILO, 1978). Bearing in mind the success of the Work Study programme, the United Nations should consider a similar global agenda designed to roll out Green Six Sigma.

22.3 Climate Adaptation in Clean Energy

As noted in Chapter 7, the biggest source of greenhouse gases is fossil fuel power stations and good progress has been made in the field of obtaining clean energy supplies from renewable sources and nuclear power. However, the current programmes in clean energy all over the world suggest that some amount of carbon dioxide from other energy sources in the future is inevitable. There are two sources of alternative fuels that can be considered and have the advantage of having processes in place for adaptation. These low-carbon energy funds are biofuels and so-called green gases (e.g. biogas and hydrogen produced by electricity). We may also have to live with power stations run by natural gas which is considered to be less polluting than other fossil fuels.

We can either capture the carbon dioxide from the exhaust pipes of power plants and bury it in underground rocks or actively remove greenhouse gases from the atmosphere for storage. Major oil and gas companies (e.g. Exxon Mobile, Shell and BP) are investing heavily in carbon capture research projects. ExxonMobil has an equity share in about 20% of the world's carbon capture capacity. Shell has developed and patented carbon capture technology utilising a re-generable amine that offers cutting-edge performance.

22.3.1 How Green Six Sigma Can Help

Unfortunately, most of the business practices carried out by big companies contribute to waste and non-renewable energy. This is where Green Six Sigma can help oil and gas corporations improve their waste management. The DMADV (define, measure, analyse, design, verify) programme is a modified version of DMAIC, and this methodology will lead to more effective research in the realm of carbon capture technology.

22.4 Climate Adaptation in Green Supply Chains

Green supply chains will have a central role in supporting societal adaptation to the physical impacts of climate change, especially in more directly affected sectors such as manufacturing, agriculture, forestry, construction or transportation.

The supply chain discipline has progressed towards a common understanding of resilience to adapt to changing business conditions because of the emergence of new competitors, new products and altered political, economic and legal conditions. Adaptation plans for coping with the Covid-19 pandemic and climate change are also beginning to emerge. There is a growing emphasis internationally on engaging supply chain businesses in adaptation given their potential to develop technologies and innovative solutions, and enhancing the cost-effectiveness of certain adaptation measures. Large national and multinational corporations are among the key actors in this respect. Already, many of these corporations are purportedly taking steps to adapt their operations to climate change as Case Example 22.1 illustrates.

CASE EXAMPLE 22.1 UNILEVER SETS OUT NEW ACTIONS TO ADAPT TO CLIMATE CHANGE

Unilever has been leading the industry on sustainable sourcing practices for over a decade, and 90% of its forest-related commodities are certified as sustainably sourced. Unilever's climate adaptation strategy of 2020 clearly states that brands will collectively invest €1 billion over the next 10 years in projects focused on landscape restoration, reforestation, carbon sequestration, wildlife protection and water preservation. Unilever has been the industry leader in sustainable and ethical sourcing policies for over a decade. Unilever's ice cream brand Ben & Jerry's initiative is already underway to reduce greenhouse gas emissions from dairy farms and another brand, Knorr, is supporting farmers to grow food more sustainably.

Unilever aims to attain net zero emissions from all its products by 2039. To achieve this goal 11 years ahead of the 2050 Paris Agreement deadline, the company prioritises building partnerships with its suppliers who have also committed to their own science-based targets. This transparency regarding corporate carbon footprint is used as an accelerator and a system is set up for all suppliers to declare, on each invoice, the carbon footprint of the goods and services provided. Unilever has also introduced a pioneering agricultural code for all suppliers, built on green farming practices. The corporation will join the 2030 Water Resources Group, a multi-stakeholder platform hosted by the World Bank.

It is pivotal that all major organisations, both in the manufacturing and service sectors, publish their 'Net Zero Carbon Plan' and evaluate its progress every year as part of their annual report. This plan will also include the systems and processes in place to adapt to the consequences of climate change. A well-designed and executed blueprint is most likely to project the organisation as a leading company of environmental sustainability and act as a competitive advantage. There are a few sustainable business awards in different sectors of the supply chain which are independently judged by an expert jury representing a cross section of the business sector. These awards also act as a facilitator towards a more sustainable future.

A vital link in the foods supply chain is agriculture. Climate change presents significant risks for agriculture as farmers, whether in a rich country or a poor one, are certain to be adversely affected by any future changes in higher mean temperatures, drought and floods caused by climate change. Adaptation pathways planning should be in place to allow stakeholders to identify and evaluate adaptation options.

In July 2021, the Prince of Wales launched a booklet to translate 50 sustainability terms into a language which is more accessible to farmers. Bill Gates (Gates, 2021) suggested that CGIAR (the Consultative Group for International Agricultural Research) could help farmers with a wider variety of crops and livestock so that one setback does not have the effect of wiping them out. The National Food Strategy Report of the UK, published in June 2021, recommends that the UK government should "nudge" people towards plant-based foods in an effort to cut national meat consumption in the interests of the environment and animal welfare. We should also shore up our natural defences to protect agriculture—this is discussed further in Section 21.8.

22.4.1 How Green Six Sigma Can Help

Because Green Six Sigma focuses on reducing wastes and increasing efficiency, major corporations have implemented a variety of Six Sigma methodologies to improve their business processes. In doing so, their actions have significantly improved their waste management and production efficiencies. As a result, fewer resources were needed to create the same products. If less energy is required to manufacture products, this in turn leads to fewer carbon emissions and a reduced dependency on fossil fuel energy. With regard to climate adaptation activities, key stakeholders of a supply chain will also benefit from the application of Green Six Sigma to reduce wastes and ensure sustainable outcomes. With a 10% reduction of wastes in the supply chain sector, 2.9 billion tonnes of greenhouse gases will be removed from the atmosphere.

22.5 Climate Adaptation in Green Transports

The transportation industry is emitting 14% of greenhouse gases and the aviation sector is responsible for a large portion of these emissions. Research studies (Lee et al., 2021) have modelled the impact of climate change–imposed constraints on the recoverability of airline networks. The climate change adaptation activities of multinational airlines have been focused on evaluating the costs of disruptions in the form of flight delays and cancellations, as well as passenger misconnections. There is also some evidence of research on the field of aviation fuels and finding an alternative to fossil fuel–based kerosene. A study by Seyam et al. (2021) has found that the maximum overall thermal efficiencies of hybrid turbofan engines are achieved by using a fuel composed of 75% methanol and 25% hydrogen, which reduces carbon emissions by 65% compared to fossil fuels. However, there is a long way to go before we see commercial airlines running on this low-carbon fuel.

Both the aviation industry and oil and gas corporations are investing in innovative fuel concepts that may provide environmental benefits. While some of these are already being produced and used regularly in aircraft operations (e.g. Sustainable Aviation Fuels or SAF), others are still under research and development, such as Lower Carbon Aviation Fuels and Hydrogen. Air BP's SAF is called BP Biojet and is currently made from used cooking oil and other wastes (such as household wastes and algae). Traditional jet fuel is blended with SAF to make it suitable for long-haul flights. Air BP claims that SAF gives an impressive reduction of up to 80% in carbon emissions over the life cycle of the fuel compared to the traditional jet fuel that it replaces. In the continuing quest for carbon-free aviation fuel, research is ongoing to evaluate hydrogen as a possible solution in the future. However, several factors are currently against a possible use of hydrogen on commercial flights, such as on-board storage, safety concerns and the high cost of producing the fuel. We need more research to mitigate these challenges.

Hydrogen combustion engines are also the most likely solution for long-distance buses and trucks. Research on hydrogen-fuelled internal combustion to replace fossil fuels has been ongoing for decades. However hydrogen engines still have the problems of lower volumetric efficiencies and frequent pre-ignition events relative to gasoline-fuelled engines. Therefore we have to keep faith for a breakthrough in the path towards carbon-free long-distance road transports.

The progress regarding green passenger cars is encouraging. We are gradually adapting ourselves to using plug-in electric cars and perhaps all-electric adaptation could be achieved in European countries by 2030. In doing so we need to improve the infrastructure and replace fossil fuel service stations with electric charge point stations. In the UK Tesco implemented the availability of free electric charging bays at some 400 stores in a collaboration with Volkswagen and Pod Point, with a variety of power chargers all using completely green energy.

22.5.1 How Green Six Sigma Can Help

We need extensive research in the area of developing and improving carbon-free aviation fuel. More research is required to create safe and efficient hydrogen-fuelled internal combustion engines. Green Six Sigma has broad applications in an R&D context because R&D is fundamentally a series of problem-defining and problem-solving processes and there are appropriate Green Six Sigma tools on offer. For example, DFSS (design for Six Sigma) is a Green Six Sigma tool that scientists can effectively use to accelerate R&D projects on alternative fuels.

22.6 Climate Adaptation in Retrofitting Houses

As consumers all of us together can have a huge impact on the demand side of climate adaptation. We can influence the market by choosing the nature of the energy supply we have, what kind of transport we use, what sort of food we eat or what type of houses we live in. If all of us make individual changes toward carbon neutral products and services then this can add up to a huge total reduction in carbon emission. Fossil fuel–driven suppliers will have to change to adapt to customers' choices.

The processes for climate adaptation in retrofitting houses have an overlap with the climate mitigation processes described in Chapter 10. These mitigation processes will also apply to climate adaptation practices in every household and should include

1. Reduce energy losses and energy consumption
2. Replace fossil fuel boilers and water heaters
3. Seek to use renewable energy
4. Apply a circular economy by reducing wastes and repairing appliances rather than discarding and replacing them

We should also prepare against frequent flooding and excessive downpours. As short-term measures, households near potentially flood affected areas should arrange houses with quick-fit flood barriers at their doors and flood pumps should be installed in the local community. In order to attain viable longer-term solutions, national governments should invest in flood protection schemes and agile rescue and relief resources. Some examples of flood protection schemes are discussed in Section 22.7. There are also many examples of flood relief operations over recent years as Case Example 22.2 illustrates.

CASE EXAMPLE 22.2 FLOOD RELIEF IN BARBADOS

Barbados can be affected by tropical storms, causing serious flooding. In 2018 Storm Kirk deposited between 50 and 100 mm of rain in just 24 hours. The resultant flooding forced the closure of schools, universities and other public services. Flood barriers were manufactured in the UK and shipped out within 48 hours on a direct flight. A building company based in Barbados managed the installation works on the following day. It is expected that climate change will cause sea levels to rise and will undoubtedly lead to further flooding in Barbados.

In fact, supply chain professionals are undervalued in disaster relief operations. It is important that their expertise and resources are deployed to develop and deliver a quick response logistics support as Case Example 22.3 illustrates.

CASE EXAMPLE 22.3 HURRICANE KATRINA DISASTER RELIEF BY MASTERED SUPPLY CHAIN MANAGEMENT

Referring to the Hurricane Katrina disaster in New Orleans, Louisiana, in 2005, Walmart responded more quickly and was more effective in providing what was required than both the Federal Emergency Management Agency and Red Cross. While their speed of response was commendable, one could argue that Walmart was only doing what it does every day. Walmart delivered 2,500 containers to the region and set up satellite links for its stores that lost phone or Internet service so that they could stay connected to headquarters.

New design concepts for residential buildings are required to enable low carbon dioxide operations and to adapt to climate change. Furthermore, any climate adaptation strategy for dwellings must be cognisant of building regulations, new technologies and occupant needs. The good news is that our architects and builders do know how to build green buildings. One positive example is the Bullitt Center in Seattle which has been designed to stay naturally warm in winter and remain cool in summer. The center is also equipped with carbon neutral energy-saving technologies. While it is too expensive to build a perfect green residential house like the Bullit Center every time, we can still ensure that our energy efficient homes are constructed at an affordable cost. Architects are already designing eco-friendly energy efficient homes for the future to adapt to climate change. Some of these design features are summarised in Table 22.1.

As discussed in Chapter 10, rather than demolishing large office or residential buildings to build new multi-storeyed structures, builders should refurbish existing buildings to help them adapt to the consequences of climate change.

Table 22.1 Climate Adaption for House Design

Climate and Energy Causes	*Building Design Responses*
Heating system	Airtightness All electric Heat pumps Solar panels
Floods	Air bricks Door guards Solid flooring Community flood defence Sustainable drainage (SUD) system
Storms	Reinforcement of building and roofs Robust guttering and drainpipes Vigilant maintenance
Cold events	Roof insulation Cavity wall insulation Smart meters Double-/triple-glazed windows
Heatwaves	Shutters Smart glass windows Heat pumps as air coolers (if installed) Aerated concrete

22.6.1 How Green Six Sigma Can Help

Six Sigma methodology and tools are proven to be very effective in delivering the quality standards and project deliverables required of building projects. Green Six Sigma will be equally effective as Six Sigma for building projects and in addition it will ensure the sustainability of outcomes as well as the all-important sustainability of the environment. The paradigm and processes of the circular economy are embedded in Green Six Sigma and are also central to the adaptation of climate change in both household practices and refurbishing larger buildings.

22.7 Climate Adaptation and Infra-Structure Projects

The Economist published an article on 30 May 2020 about the cyclone 'Amphan' and its effect. On 16 May an anticlockwise spiral of clouds over the Bay of Bengal detected by satellite warned of an imminent disaster. Four days later Amphan made landfall, gusting at up to 185 km per hour along the coast of West Bengal and Bangladesh. The number of fatalities in Bangladesh amounted to 20 while the death tolls for earlier cyclones had been in their thousands. What was the reason for the considerable saving of lives this time?

The answer lies in the fact that Bangladesh has now developed a layered adaptation plan with an early warning system to evacuate people to concrete cyclone shelters. Seawalls were also built with international support to protect stretches of coastlines. This is an example of a successful climate adaption project to prepare for major cyclones.

The United Nations Environment Programme (UNEP) has been assisting over 70 projects on climate change adaptation in more than 50 countries. These multinational schemes are essential to enable developing countries to adapt to the consequences of climate change. I have chosen to focus upon two case examples as described in the following case examples.

CASE EXAMPLE 22.4 BRING BACK MANGROVES PROJECT

Mangroves are short trees that grow on salt waters along coastlines. UNEP research shows that mangrove ecosystems underpin global and local economies by supporting fisheries, providing other food sources and protecting coastlines from flooding and erosion. However coastlines are among the most densely populated areas on earth and developers are clearing mangrove forests to create space for buildings.

UNEP has developed guidelines on mangrove ecosystem restoration. Following these guidelines, Kenya and Madagascar in particular have recognised the contribution of mangroves to their own livelihoods and are actively participating in carbon monitoring, reforestation and education to ensure the incomes and quality of life of future generations. Other governments are also taking action. Cuba, Haiti, Puerto Rico, the Dominican Republic and Pakistan have all prioritised mangrove restoration.

CASE EXAMPLE 22.5 ELIMINATE ADDITIONAL MARINE PLASTIC LITTER BY 2050

The annual discharge of plastic into the ocean is estimated to be 11 million tonnes and at the current rate this figure is set to double by 2040. However, through an ambitious combination of interventions using known technology and established approaches, marine plastic litter entering the ocean can be reduced by 82% compared to current levels by 2040.

UNEP has launched the Osaka Blue Ocean Vision plan which voluntarily commits G20 countries to 'reduce additional pollution by marine plastic litter to zero by 2050 through a comprehensive life-cycle approach'. This vision will only be achieved by adopting more progressive policy targets. These include moving from linear to circular plastic production by incentivising reuse and designing out waste. The international trade in plastic waste would be regulated to protect both people and nature. Many countries outside the G20, though not all, have signed up to UNEP's Osaka Blue Ocean Vision and are investing in national recycling facilities. The next phase is expanding on the 'source to sea' approach (i.e. focusing on the root cause of marine plastics).

Developed nations are also investing in climate adaptation and infrastructure projects as Case Example 22.6 illustrates.

CASE EXAMPLE 22.6. UK PROGRAMME FOR FLOOD PROTECTION

In July 2020, the UK Government announced a US $6.2 billion programme to protect 336,000 properties from flooding by 2027. The programme includes the building of new flood and coastal defences and sustainable drainage systems (SUDs). These will support 25 areas at risk of flooding. The programme also includes investment in 'shovel-ready' flood defence schemes to benefit 22 areas across the country. In addition, the plan sets out proposed changes to the joint government and insurance industry flood compensation schemes. Longer-term projects of climate adaptation (e.g. storing water upstream to prevent flooding during heavy rainfall, then capturing this water for use during dry weather) are also included in the programme.

However in national projects it is often found that due to changes of leadership or political systems the project deliverables are different in reality from those intended in the original project plans.

For example, spurred by the devastation of Hurricane Ike in 2008, a $10 billion flood wall scheme to protect Galveston Island and the Houston Ship Channel ballooned to a budget of $32 billion in 2019.

In another example, the 'smart city' projects in China are supposed to assist planners with urban management and safety. The system is designed to help the municipal authorities monitor water levels in real time through sensors. In spite of the merits of these 'smart city' projects, when record rainstorms battered the Henan province during July 2021, the city of Zhengzhou situated in central China with a population of 10 million people suffered at least 66 deaths, including 14 in the local subway system and six in the Jingguang Road Tunnel.

Climate adaptation projects are not part of any political paradigm—they are real and designed for saving lives and the environment. We must be both optimistic and determined to deliver project goals.

22.7.1 How Green Six Sigma Can Help

In this area of climate adaptation projects, the impact of Green Six Sigma will be most significant. Case examples have demonstrated that Six Sigma methodology has ensured the quality and deliverables of these large schemes comprising many stakeholders. McKesson, which is the largest pharmaceutical distribution company in the US, proclaims on its website that it has applied Six Sigma to achieve the successful distribution of both the Pfizer and Moderna Covid-19 vaccines. Both vaccines required ultra-cold storage and thus their distribution across the US was an enormous undertak ing. There is also evidence that Six Sigma tools were applied in the major project comprising the roll out of Covid-19 vaccines to millions of people across India.

Green Six Sigma will contribute additional values of ensuring sustainable environmental standards to large and international climate adaption projects.

22.8 Climate Adaptation and Innovation

It is evident that many climate change initiatives, both for mitigation and adaption, require new materials and breakthrough solutions which we need now but do not have. For this we require innovation supported by dedicated research and development. In energy, transport, construction, software and just about any other pursuit, innovation is not just inventing new equipment or designing a novel process. It is also offering a completely fresh way of doing things. We have sufficient demand for innovation to combat the inconvenient consequences of climate change. The essential requirements of upscaling and testing any new product or process in a larger population cannot be denied and these will follow later, but we are now hungrier and there is a greater sense of urgency for the supply side of innovation. Although we have a few affordable zero- or low-carbon solutions today, unfortunately we do not possess all the necessary technologies or materials to meet the target of zero emissions globally. Therefore presented in Table 22.2 is my list of urgently required improved technologies and

Table 22.2 Improved Technologies and Materials for R&D

Advanced biofuels
Carbon capture and storage
Commercial graphene
Electricity storage for a long period
Geothermal energy
Hydrogen combustion engines
Plant-based meat and dairy
Tidal energy
Green hydrogen
Zero-carbon aviation fuels
Zero-carbon cement
Zero-carbon fertilizer
Zero-carbon plastics
Zero-carbon steel

materials for climate change initiatives—although environmental scientists may suggest more.

Most of these products are not new, but we do need them to be affordable and safe to use. In order to achieve this we need more research and greater R&D investment hypothecated to climate change research initiatives. President Joe Biden's 2022 budget proposal includes around $36 billion to fight global climate change of which only $4 billion is allocated to advancing climate research. Other rich nations should follow suit and consider a tax break for R&D investment to encourage private enterprises to invest in R&D for climate research. Large oil and gas corporations and fossil fuel producing countries must generously support R&D initiatives for zero-carbon alternatives to fossil fuels and also for carbon capture projects. The commercial future of fossil fuels as a major energy supplier is not bright in the longer term and their appropriate strategy should be to focus on advancing alternative solutions and the by-products of fractional distillation (e.g. lubricants, asphalt, naphtha, feedstocks etc.).

22.8.1 How Green Six Sigma Can Help

There is evidence (Schweikhart and Dembe, 2009) that Lean Six Sigma process improvement methodologies are well suited to help research projects become more efficient and cost effective, thus enhancing the quality of the research. There are specific tools for R&D projects (e.g. DFSS) which are also in the toolset of Green Six Sigma. It can be concluded that the application of Green Six Sigma will accelerate the quality and delivery time of much needed R&D projects for improved climate change solutions.

22.9 Summary

It is inevitable that in spite of all current mitigation plans to combat climate change, we will gradually feel the impact of global warming in the form of more frequent floods, heatwaves and droughts. We will have to live with the medium-term consequences and adapt ourselves with cost-effective measures. There is a sense of optimism after the election of President Joe Biden and expectations are also very high regarding the outcomes of the 2021 United Nations Climate Change Conference, also known as COP26.

In this chapter, some essential measures to adapt to the likely consequences of climate change in the different sectors of our lives (e.g. energy, industry, farming, transport and housing) have been outlined. These measures are based on the technologies and processes that we have already and which can be put into practice right now. Green Six Sigma tools and techniques can also help starting from today. We need both government and private investment in R&D to develop and innovate materials as well as suitable technology for climate change solutions that are affordable and safe to use.

GREEN TIPS

- In spite of global, national and individual efforts to prevent the consequences of climate change we have to prepare ourselves to adapt to more frequent floods, heatwaves and droughts.
- There must be more investments, both in the public and private sector, to develop new materials and solutions for climate change adaptations as listed in Table 22.2.
- Green Six Sigma can play a major role in improving the effectiveness of R&D projects for climate change solutions.

IMPLEMENTATION OF GREEN SIX SIGMA

Chapter 23

Case Studies

The bitterness of low quality is not forgotten

Nor can it be sweetened with low price

—Marquis De Lavant (1734)

23.1 Introduction

The first wave of Six Sigma, following the grand groundwork of Motorola, included AlliedSignal, Texas Instruments, Raytheon and Polaroid (to name a few). GE entered the arena in the mid-1990s and in turn was followed by many powerful corporations including SONY, DuPont, Dow, Bombardier, Ford and GSK. The success of GE Capital attracted many companies in the service sector, such as American Express, Lloyds TSB, Egg.com and Vodafone, to apply Six Sigma as a major business initiative. Some established companies like Unilever, ICI and BT have also applied Six Sigma tools and techniques but attempted to align them to existing strategic and continuous improvement programmes.

The ability to leverage the experience of successful Six Sigma players proved highly attractive both as a competitive issue and also to improve profit margins. The following case examples provide insights for organisations that are achieving success in their business performance through the use of Six Sigma or related operational excellence programmes. The objective of including these case examples is to offer a set of good proven practices through which a student or a practitioner can obtain a better insight of the application of tools and techniques and how the organisations have adopted the approach.

DOI: 10.4324/9781003268239-28

The case examples are taken from the real-life experiences of many organisations around the world. The examples reflect projects within respective operational excellence programmes (e.g. Six Sigma, Lean Six Sigma, FIT SIGMA and Total Quality Management). All these operational excellence programmes are completely aligned with the fundamental approach of Green Six Sigma. No effort is made to make them tidy to suit any subject area. Most of the case examples happened before the urgent need to combat climate challenges. It is hoped that these cases provide a useful resource to stimulate discussions and training of tools and techniques in Green Six Sigma with the further opportunity of ensuring sustainability.

23.2 Case Example for Large Manufacturing Organisations

Case examples are presented in two categories: applications in large manufacturing and non-manufacturing organisations. The examples of large manufacturing organisations in this section show variations in approach depending on the complexity of problems involved.

CASE STUDY 23.2.1 PRODUCT INNOVATION AT ELIDA FABERGE, UK

BACKGROUND

Elida Faberge, UK is a subsidiary of Unilever Group and manufactures and supplies personal care products like shampoo, deodorants and toothpastes. The company examined the way in which innovation takes place in the business and how it is managed and controlled. The purpose of this review was to develop a clearly mapped transparent process for new products delivery ahead of competition with the optimum use of resources. The relaunch of a major shampoo brand (Timotei) in March 1992 was the pilot project using this process. The relaunch involved four variants of shampoo in four sizes and two variants of conditioner in three sizes and incorporated the main technical changes of a new dispensing cap and changes in formulations.

APPROACH

A multidisciplinary task force was set up to develop the new process in readiness for a company-wide launch in April 1993. The innovation

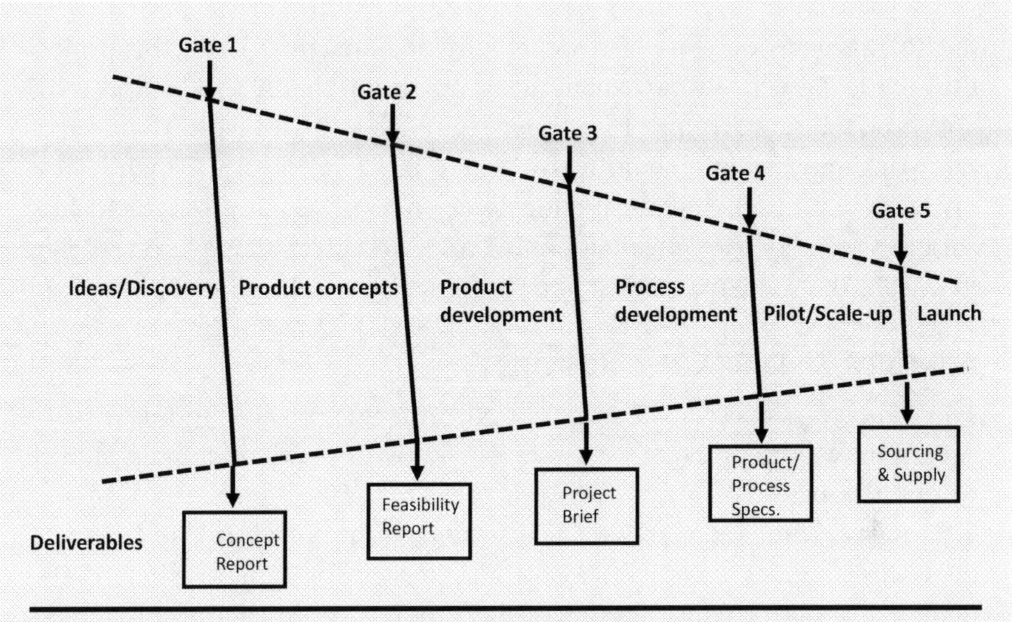

Figure 23.1 Product development phases and gates.

process used by Elida Faberge was based on the principles of 'phases and gates' (see Figure 23.1) and consisted of

- A logical sequence of phases and exit gates
- The use of empowered multidisciplinary core teams
- The use of gates as key points where 'gate keepers' review progress and decide whether a project may proceed or be terminated
- The monitoring of progress at each gate and for the overall project using key performance measures

The project team had a membership representing the key processes, such as Brand Development, Supply Chain, Quality Management and Customer Development. The members received additional training in project management and the use of statistical process control (SPC) tools. A project leader co-ordinated the team's efforts and reported the progress to the 'gate keepers'.

IMPLEMENTATION

Using the multifunctional team approach as described, the Timotei project was launched and managed to the implementation stage (Phase 4) by a team lead by a product manager. A second team was set up under the technical leadership of a development manager to see the project through to launch, in particular to set up supply lines, production trials and production start-ups. Two other teams were also involved—a separate packaging supply chain team, and a TQ team on the Timotei start-up production line. Technical audits were also set up at three packaging supply sites and corrective actions were issued. The sites were revisited to ensure that all corrective actions were satisfactorily achieved.

RESULTS AND LEARNING POINTS

The Timotei project was a success both in terms of sales and as a managed innovation process. No notable problems with materials supply were encountered.

However, the considerable sales success of the product overstretched the bottle supplier, so that a second supplier was brought on stream. The quality and performance ratings of the production line achieved targets.

The key learning points of the project included the following:

- The use of a multifunctional team allowed actions to be taken swiftly with shared ownership and resources at the team's disposal.
- The additional training of team members in project management and SPC tools ensured that the project was managed in a systematic, organised and disciplined manner.
- The effective use of phases (with multiple project leadership) and gates contributed to the development of the 'Innovation Funnel' of Unilever which was introduced to all product groups.

Source: Implementing Six Sigma and Lean (Basu, 2009).

CASE STUDY 23.2.2 LEAN MANUFACTURING AT ALUMINIUM INDUSTRY, US

BACKGROUND

Author and researcher James P Womack, president of the Lean Enterprise Institute, told attendees at a meeting of the Aluminium Association that aluminium producers are among the many basic industries that could do a lot more with less difficulty and lower costs if they did some lean thinking.

He described the labyrinth and costly processes associated with delivering an aircraft, automobile or drinks can, and suggested manufacturers think about the consumer first and work backward to gain efficiencies and to cut out the non-value-adding activities.

APPROACH

'Ask, from the customer's standpoint, what is of value among your activities? What is wasted? How can we eliminate the waste? It's so simple and yet so very hard to do. Most people are in love with their assets, technologies and organisation', Womack said. Womack cited a military aircraft program as an example. 'The typical subassembly goes through four plants, four states and 74 organisational handoffs between engineering, purchasing and fabricating operations. It goes 7,600 miles and takes darn near forever. It takes two to four years from beginning to flyaway condition'. Manufacturers need to identify the value stream for each of their products and document all the steps it takes to go from raw material to the customer.

IMPLEMENTATION

'Step 1, get the value right', said Womack. 'Your customer is not interested in your assets. He is interested in his value. Step 2, identify the value stream from start to finish, not just within the walls of your plant or company'.

'Organise the remaining value-creating steps so they impact the product in a continuous flow', he advised.

RESULTS AND LEARNING POINTS

'If, and only if, you can create flow, then you can move to a world of pull', Womack said. 'You put the forecast in your shredder and get on with your life. And you make people what they want. The customer says, "I want a green one". And you say, "Here's a green one". That's a very different world from your world of endless forecasts, always wrong, and the desperate desire to keep running, which makes you produce even more of the wrong thing because that makes the numbers look good in the short term'.

Source: Metal Center News, *May 1999, Vol. 39, No. 6, p. 123.*

CASE STUDY 23.2.3 SUPPLY CHAIN LOGISTICS AT NATIONAL STARCH, US

The National Starch and Chemical Company, based in Chicago, began measuring shipping performance on a corporate basis in the early 1990s. The initial results indicated that, at best, the Adhesives Division was achieving 80–85% on-time delivery performance. This is a business where it is essential to ship a multitude of products in a variety of containers with varying lead times from customers. The logistics department set an objective to reduce shipping delays from 15 to 5% and at the same time reduce overall inventories.

The approach the division took was to classify their products in specific containers by categories A, B and C. The categories represented volume of product and the customer base purchasing the product. The lead time for shipping the product and the inventory held were related to the A, B and C designations, with A being a shorter lead time with available inventory.

A number of local cross-functional teams were formed comprising sales, manufacturing, technical service, materials management and customer service. The task of the teams was to work on ABC product classification and delivery service standards. Essentially every product in each container was evaluated based on the volume and number of customers purchasing it. The Chicago plant reviewed over 200 products in approximately 15 different containers for more than 700 orders per month.

The next task was to develop a tracking system to report on delays against standard lead times and compare them with the date agreed with the customer.

After agreement was reached on the ABC products and their safety stocks, the order patterns of these products were reviewed and minimum, maximum and re-order levels were established. Employees were thoroughly trained, and the system was presented at each regional sales meeting. It was important to gain the support of sales staff since they needed to 'sell' it to the customers.

To ensure that the ABC and safety stocks reflected changing market demands, a cross-functional team reviewed the product A, B, C designations and inventory lists every month.

RESULTS AND LEARNING POINTS

Since the implementation of the ABC system, the improvement in shipping performance was dramatic as shown by the figures for shipping delays as follows:

Year 0 overall: 14.7%
Year 1 overall: 3.8%
Year 2 overall: 1.2%

There were other benefits in the supply chain operations and performance including the following:

- The sales team increased the amount of time used to work proactively with customers.
- There has been a reduction in finished goods and the raw materials inventory.
- 'Last-minute' expedited freight charges decreased.
- Communication improved between departments, especially between sales and logistics.

Source: Implementing Six Sigma and Lean *(Basu, 2009).*

CASE STUDY 23.2.4 PERFORMANCE MONITORING AT DUPONT TEIJIN FILMS, UK

Dupont Teijin Films is a global polyester films business with manufacturing sites in the US, Europe and Asia. The company was created following the acquisition of Teijin Films of Japan by Dupont. DTF is a market leader but was experiencing tough competition from new entrants. As part of the corporate Six Sigma programme, the Wilton Site of DTF in Middlesbrough, UK, started the deployment plan from 1999.

The main objectives of the programme included

- Increased capacity
- Improved material efficiencies

- Cost reduction
- Increased revenue by higher sales volume

The site project team followed a methodology of 'successful implementation' in three key categories—'Doing the Right Work' (Process); 'Doing the Work Right' (Efficiency) and 'Creating the Right Environment' (Education and Culture).

Within the category of 'Doing the Work Right' the team introduced

- Input metrics
- Output metrics
- Tracking profile

The input metrics included the number of Black Belts trained and people educated. The output metrics covered

- Money saved
- Number of projects per annum
- Quality index
- Critical to quality (CTQ) flowdown
- Cost of poor quality (COPQ)
- Strategic linking

The project team followed an internal self-assessment process every quarter based on a 'Do Right Work Checklist' comprising 24 questions in Customer Alignment, Business Alignment, Process Baselining and Project Selection.

RESULTS AND LEARNING POINTS

The Six Sigma programme has been running at the Wilton Site since 1998. The typical sources and relative magnitude of savings are given in Table 23.1.

The key learning point from DTF at Wilton is that training is a key deployment opportunity to influence the hearts and minds of your people. The structured continuous training programme has been vital to sustaining the Six Sigma programme for over 5 years.

Source: Implementing Six Sigma and Lean *(Basu, 2009).*

Table 23.1 Relative Magnitude of Savings

Source	Year 1	Year 5	Relative Value of Savings
'Low-hanging fruits'	40%	5%	$$
COPQ	40%	20%	$$$$
Key performance indicators	10%	10%	$$
CTQ flowdowns	10%	30%	$$$$
Strategic linking	0%	35%	$$$$$

CASE STUDY 23.2.5. TOTAL PRODUCTIVE MAINTENANCE AT NIPPON LEVER, JAPAN

The Utsunomia plant in Japan was commissioned in 1991 on a greenfield site by Nippon Lever to manufacture household detergents products and plastic bottles for liquid detergents. The factory was experiencing 'teething' problems primarily due to the poor reliability and lack of local support of the imported equipment. Many of the employees were new to factory work.

To improve this situation the company used the help of the Japanese Institute of Plant Maintenance (JIPM), an organisation which was working on total productive maintenance (TPM) with over 800 companies in Japan. TPM has been widely used in Japan, having been developed to support Lean/just-in-time (JIT) and Total Quality Management (TQM). It was considered to be appropriate for the Utsunomiya plant since TPM focuses on machine performance and concentrates on operator training and teamwork.

A TPM programme was launched at the Utsunomiya plant in July 1992 with the objective of zero losses:

- Zero stoppages
- Zero quality defects
- Zero waste in materials and manpower

Strong organisational support was provided by the Nippon Lever management in terms of

- A top management steering team to facilitate implementation by removing obstacles
- A manager to work full time supporting the programme
- One shift per week set aside for TPM work
- Training for managers, leaders and operators involving JIPM video training material

The programme launch was initiated at a 'kick-off' ceremony in the presence of the whole Nippon Lever Board and managers from other companies and suppliers' sites.

The initial thrust of the programme was the implementation of 'Autonomous Maintenance' following the JIPM's seven steps:

1. Initial clean-up
2. Elimination of contamination
3. Standard setting for operators
4. Skill development for inspection
5. Autonomous inspection
6. Orderliness and tidiness
7. All-out autonomous working

To implement the Seven Steps, 'model machines' (those giving the biggest problems) were chosen. This approach helps to develop operators' knowledge of a machine and ensures that work on the model can be used as the standard for work on other machines. It also enhances motivation, in that if the worst machine moves to the highest efficiency, this sets the tone for the rest of the process.

The improvements to the machines were made using kaizen methodology (small incremental improvements) and were carried out by groups of operators under their own guidance. Two means of support were given to operators—a kaizen budget per line so that small repairs and capital expenses could be agreed without delay; and the external JIPM facilitator provided encouragement and experience to work groups.

RESULTS AND LEARNING POINTS

By the end of 1993, substantial benefits were achieved within a year at the Utsunomiya plant including the following:

- US $2.8 million reduction in operating costs
- Reduced need for expensive third-party bottles
- Production efficiency increased from 54 to 64% for high-speed soap lines and from 63 to 80% for liquid filling lines
- A team of trained, motivated and empowered operators capable of carrying out running maintenance

The success of the programme at the Utsunomiya plant led to the introduction of TPM in two other factories at Nippon Liver (Shimizu and Sagamihara). Over the next few years the corporate groups of Unilever encouraged all sites outside Japan to implement TPM with remarkable successes achieved, particularly in factories in Indonesia, Brazil, Chile, the UK and Germany.

Source: Implementing Six Sigma and Lean *(Basu, 2009).*

CASE STUDY 23.2.6 SIX SIGMA TRAINING AT NORANDA, CANADA

Noranda Inc. is a leading international mining and metals company for copper, zinc, magnesium, aluminium and the recycling of metal. With its headquarters in Toronto, the company employs 17,000 people around the world, and its annual turnover in 2000 was $6.5 billion.

In August 1999, the Board of Noranda decided to embark upon a global Six Sigma project with an initial savings target of $100 million in 2000. There were some specific challenges to overcome. The company business was in a traditional industry with long serving history. Furthermore, Noranda is a 'de-centralised' company with multiple cultures and languages. Senior executives studied the experiences of other companies (GE, AlliedSignal, Dupont, Bombardier and Alcoa) and invited the Six Sigma Academy from Arizona to launch the training deployment programme.

The Six Sigma structure at Noranda focused on the training of the following levels:

- Deployment and project Champions
- Master Black Belts
- Black Belts
- Business analysts and validates
- Process owners
- Green Belts

The Six Sigma Academy was intensely involved for the first 3 months of the programme and then Noranda started its own education and training.

RESULTS AND LEARNING POINTS

The training accomplishments in 2000 were impressive:

- All 84 top executives followed a 2-day workshop.
- 90 Black Belts were certified.
- 31 Champions were trained.
- There were 17 days of Master Black Belt training.
- There were 3,000 days of Green Belt training.
- There were more than 3,500 days of training in 2000—and this has continued.

The key learning points from the experience of Noranda include the following:

- The rigorous deployment of structured learning schedules is key to the success of a company-wide Six Sigma programme.
- After the initial learning by external consultants, it is more cost effective and useful in the longer term to build up your own in-house training teams.

Source: Implementing Six Sigma and Lean *(Basu, 2009).*

CASE STUDY 23.2.7 TOTAL QUALITY AT CHESEBROUGH-POND'S, JEFFERSON CITY

The Jefferson City plant Chesebrough-Pond's was built in 1966 and was managed through the traditional functional structure. Each packing department had its own manager and supervisors responsible for each shift resulting in five reporting layers. Maintenance and industrial engineering were separate departments, and production planning was located with accounts in the office area. Raw materials and packaging suppliers were managed by the purchasing department in the head office at Connecticut, 1,000 miles away.

With the concentration of the purchasing power of the retail trade in the US (e.g. Walmart, K-Mart etc.) the plant had come under considerable pressure to deliver personal products more often, at shorter lead times and at a lower cost. Therefore, the factory decided it had to increase flexibility while at the same time reduce operating costs.

The management of the Jefferson City plant recognised that an organisational change was required to meet the demands of the business and launched its total quality programme in the early 1990s.

The process started with the writing and adoption of statements on plant vision, plant mission and plant values. These statements were communicated throughout the plant. The other key activities in establishing total quality were

- Understanding the supply chain
- Process management
- Problem-solving skills
- Setting goals and measuring progress
- Training

The Jefferson City plant displayed a plant vision, mission and values by posting them in the permanent (yearly) framed charts in primary production buildings.

The plant was divided into three focused manufacturing units replacing the traditional functional management structure. All employees were referred to as 'associates' and participated in self-directed work groups.

The first line supervisors were redeployed as 'coaches' of work groups formed by associates. Each of these work groups owned activities such as supplier contract, production scheduling and small projects.

All associates in the plant received at least 6 days' training per year in total quality tools and principles. Topics included Quality Characteristics, SPC, GMP, Safety, Scheduling and Waste Analysis. Even supplier employees visited the plant for 4 days' training in total quality.

RESULTS AND LEARNING POINTS

Within 3 years the plant achieved the satisfaction of customer expectations through improved quality and matching production on the basis of demand and lower costs. For example,

- First pass quality improved from 93 to 99%.
- Work-in-progress inventory reduced by 86% and total inventory by 65%.

As a result Chesebrough-Pond's at Jefferson City was recognised as one of its Ten Best plants in the US by *Industry Week* magazine in 1992.

Source: 'Our Search for the Best', Industry Week, *US, October 1992.*

CASE STUDY 23.2.8. SUSTAINING FIT SIGMA BY LEADERSHIP FORUMS AT PLIVA, CROATIA

Pliva is a generic pharmaceutical company with manufacturing sites in Croatia, Poland and the Czech Republic. It has a broad product portfolio including *Sumamed.* During the summer of 2004 Pliva embarked upon a Six Sigma programme called PEP (Pliva Excellence Process) and retained an American consulting company to train Black Belts and initiate projects. Although a good foundation for training 16 Black Belts was accomplished, underpinned by projects based on DMAIC (define, measure, analyse, improve, control), however a change of mindset and culture was lacking. In 2005 Performance Excellence Limited was engaged to relaunch the programme. The primary focus of this second wave included the certification of Black Belts with completed projects, the training of additional Black Belts and Green Belts and the management buy-in by using Leadership Forums. The following constitutes a detailed account and outcome of one such forum.

PEP or the Pliva Excellence Process came of age. Since the launch of this Six Sigma initiative in Global Product Supply in the summer of 2004, PEP demonstrated tangible results as reflected by the projected savings of US$13.4 million of which $4.6 million has already been realised. Perhaps more significantly PEP opened the door towards a self-supporting sustainable operational excellence competence and holistic culture in most business units of GPS, supported by 28 trained Back Belts and 108 qualified Green Belts. The time was right to leverage this success and good practice across and beyond GPS. A most critical success factor for a major change programme such as PEP was the better understanding of the process by and commitment of managers. With this backdrop in mind, the first PEP Leadership Forum was conducted by Zeljko Brebric, PEP Director, supported by Ron Basu, the retained consultant and mentor, and the PEP team.

The event took place in Zagreb and was attended by 24 managers from Pharma Chemicals, Local Product Supply Zagreb, Finance, Regulatory Affairs, Business Development, Medical Marketing Services and Human Resources. The primary objectives of the forum were to engage front-line managers in understanding Six Sigma/FIT SIGMA and PEP and to identify data-driven projects.

Zeljko Brebric presented an update of PEP and emphasised that PEP would be a continuous journey of achieving and sustaining operational excellence. The present strength of Black Belts and Green Belts would be fortified by additional stages of training, and the next wave of in-house Black Belt education would soon commence. Ron Basu then explained the quality movement leading to Six Sigma and FIT SIGMA and what they meant to Pliva. This was followed by presentations by Ron Basu and interactive discussions on DMAIC methodology, project selection, Six Sigma tools and how to make it happen. It was evident that PEP projects could span all business units of Pliva and should not just be confined to GPS manufacturing operations.

Managers from Pharma Chemicals and Local Product Supply who had involvement in PEP emphasised the advantage of a data-driven approach leading to sustainable results. Participants from Regulatory Affairs, Business Development, Medical Marketing Services, Finance and HR shared the potential benefits of extending PEP in their business units but also expressed concerns regarding possible extra workloads thus straining their resources.

RESULTS AND LEARNING POINTS

A product manager asked, 'In a simple sentence how would I say to Pliva senior management that PEP is not another fad?' Ron Basu replied, 'PEP has tangible results to show and further savings and improvements will be done by Pliva teams in a Pliva way'. Zeljko Brebric added, 'We also aim to build a team of self-sufficient competent managers based upon BB and GB education'.

Leadership workshops or forums were instrumental in gaining the acceptance of the change programme by members of both senior management and middle management.

Source: PEP News, *No. 5, January 2006, published by Pliva Group, Zagreb.*

CASE STUDY 23.2.9 LEAN SIX SIGMA CHANGES ABBOTT'S CULTURE

This case study is an outcome of an MBA dissertation at Henley Management College.

ADC UK is a division of Abbott Inc., which manufactures and distributes medical appliances, in vitro diagnostic devices and accessories. ADC was formed in April 2004 after a merger between Abbott Laboratories and MediSense Ltd. to become the third largest provider of blood glucose monitoring equipment in the world. Abbot Inc. has operated a succession of policies to improve its financial position. This has culminated in a Lean Six Sigma implementation in ADC UK from 2006.

A master (Six Sigma) Black Belt was employed to lead the Lean Six Sigma initiative called the Business Excellence Programme. This project was run in parallel with the closure of the original Abingdon site and consolidation to a purpose-built new facility at Witney in Oxfordshire.

All of the qualified Black Belts within the Business Excellence team ran approximately four projects at any given time, Green Belt training and mentoring. There had been no new Black Belts trained since the initial education programme. Plans were in place to start a Green Belt upgrading programme during 2008.

A single site case study approach was utilised to determine the success of this implementation and to ascertain whether the change process led to significant permanent cultural shifts. The effect of governmental regulations on the implementation was also considered. Most previous Lean Six Sigma initiatives have occurred in unregulated industries.

Data were collected by focused interviews with relevant local management and Lean Six Sigma trained employees. This information was combined with previously unanalysed statistics, and an internal white paper which correlated quality and regulatory issues with project initiatives at ADCUK.

RESULTS AND LEARNING POINTS

Analysis demonstrated that the implementation of Lean Six Sigma into ADCUK had not progressed at the expected rate. These results overall indicated that there were issues in several areas:

- Regulatory requirements
- Current information technology (IT) strategy and knowledge management
- Lean methodology use but less Six Sigma implementation
- Cultural change
- Parallel initiatives

Governmental regulatory issues ensured that this kind of implementation was slower and more complex than the textbook examples for a standard manufacturing company. Projects were shown to take approximately twice as long as in a non-regulated industry. The perceived slow implementation of Lean Six Sigma was creating some dissatisfaction.

It is recognised that issues in the regulatory environment and the many information streams are significant to the pharmaceutical world. Knowledge management strategy is crucial to a successful implementation. Fragmented IT systems are a hindrance explaining why Lean tools are used in preference to Six Sigma, as they rely less heavily on analysis of quantitative data.

Cultural change had been slow but appeared to be lasting. Furthermore, progress had been made with other initiatives including kaizen across the site providing added impetus. The Lean Six Sigma cultural change programme relies on the site management team to drive a true lasting cultural change, owned by all the employees. One significant aspect of cultural change is the inclusion of suppliers in training and delivering projects. At the time of writing this report there were 65 Green Belt projects of which 22 projects were led by suppliers.

Source: Sandra E Goodwin, ADC UK (2008).

CASE STUDY 23.2.10 EMPOWERMENT OF MIDDLE MANAGEMENT DELIVERS RESULTS AT NOVOZYMES

Novozymes is the world leader in industrial enzymes and biopharmaceutical ingredients. Its products are used in numerous industries such as biofuels, detergents, food and animal feed. Novozymes has more than 40% of the world market for industrial enzymes and is more than twice the size of its nearest competitor. In 2009, the company achieved revenue of DKK 8.5 billion based on a large portfolio of products and services worldwide.

During the period from 2001 to 2005, the yield of major products increased by only 5% as compared to the business plan seeking a 50% increase. On top of that the members of the Global Optimisation Network (GON) found that the quality of the project plans was poor. This in turn often led to discussion after the project because its scope and objectives were not well defined. The GON member from Quality Management recommended Six Sigma as the approach to take the improvement culture of Novozymes to the next level.

When presented with the idea of applying Six Sigma as a mandatory approach for improving productivity, Novozymes' top management explicitly stated that they would not support the implementation of Six Sigma because they did not see the instigation of a mandatory methodology as compatible with Novozymes' culture. Novozymes' top management was not only concerned about the cultural fit but just as much whether religious rigidity would destroy innovation and kill all initiative because of bureaucratic project models.

Despite the rejection of turning Six Sigma into a corporate concept, middle management within the Quality organisation started promoting Six Sigma and at the first wave approximately 120 employees throughout the organisation received training as either Green or Black Belts. The courses were popular, and word of mouth became enough to spread the interest of voluntarily participating in Six Sigma training. This indicated that there was a perceived value of Six Sigma even though the worth of implementing Six Sigma in Novozymes had not been evaluated or documented. The first Six Sigma training started in 2004 and evolved in 2007 into Lean Six Sigma, where a part of the statistical training was left out and replaced with fundamental Lean training.

However, it was evident that Novozymes' middle management believed in the value of Lean Six Sigma. This was supported by the fact that more and more employees were voluntarily joining the training programme. Despite this, the support or commitment of those at top management level was not clear. It appeared that they gave some tacit support and wanted to be convinced by tangible results. Novozymes' top management requested a study of the effect of the ongoing Lean Six Sigma effort. They thought it was important to know if the value was only assumed or if it could actually be quantified in any way that could justify the resources used for maintaining the training program and the continuation of the implementation.

The study was completed as an MBA dissertation at Henley Business School in 2009. The research was conducted as a single case study which included semi-structured interviews with members of top management, Black Belts and Green Belts.

RESULTS AND LEARNING POINTS

As a consequence of the study, Novozymes has now established a Lean Six Sigma strategy with top management commitment and a cross-functional team that coordinates activities. All production managers were trained as Green Belts and the cross-functional group was working on how a Black Belt programme should look at Novozymes.

It can be argued that top management did not commit to the initiative because they believed that dictating the means to achieve a goal was not compatible with Novozymes' culture. The implementation of Lean Six Sigma *was* however supported by the culture of Novozymes with a long tradition for working with continuous improvements. Employees were empowered to find their own means to meet the very clear objectives and goals that were cascaded down through all functions. This empowerment was found to be the reason why top management on the other hand would not stop the initiative and explains why middle management carried on with the deployment despite the absence of commitment from top management.

Source: Soren Carlsen, Novozymes A/S (Carlsen, 2009).

23.3 Case Examples for Services, SMEs and Projects

The case examples in this section cover organisations other than large manufacturing companies. The examples for services, small and medium-sized enterprises (SMEs) and projects underline the Green Six Sigma tenet of 'fitness for purpose'.

CASE STUDY 23.3.1 KAIZEN KEPT PRODUCTION IN SCOTLAND FOR ROSTI

Rosti Technical Plastics is a medium-sized company with 47 injection moulding machines and a range of ancillary processes including spray painting and ultrasonic welding. The production facilities are situated at Larkhall in Scotland.

Like many other manufacturing industries in the UK and other Western European countries, the viability of Rosti's Larkhall facility is under constant threat from low-cost manufacturing areas such as China and Eastern Europe. In order to remain competitive and secure the future of the Larkhall site, the company was required to reduce the proportion of labour content which went into their products. One such product was a cash cassette assembly of some 37 parts.

With the help of a local consulting firm the senior management team of Rosti carried out a Kaizen Event against a background of the introduction of the Euro note as currency in the European Union. The demand for the cash cassette was anticipated to increase twofold. A team of 14 people including a guest member from the Scottish Executive was brought together for this exercise. It contained four operators and a number of support people such as internal suppliers and engineers.

The week of the kaizen was the first time that any members of the team had been involved in such a close and intense work group. The strength and power of such a team being focused on a single area or problem for a full week was highly productive. The involvement of all the clusters of people in the process made it easy to gain acceptance of what was actually a very different manufacturing process. Many of the operators said they had never before been involved in deciding how their procedures should be set up and run and they were delighted at this opportunity to participate. The overall feeling of the group at the end of the week was one of exhaustion mixed with exhilaration at having achieved so much in so short a space of time.

RESULTS AND LEARNING POINTS

The Kaizen Event is a recommended process for Green Six Sigma, especially for SMEs. This exercise delivered remarkable results as shown in the following table:

Area/Problem	Goal	Actual
Productivity	50%	67%
Reduce space per cassette	30%	57%
Reduce work in progress	50%	83%
Improve purchased part quality	N/A	90%

Rosti moved on from this success to carry out several more events (including one in Poland) and they are now self-sufficient at running Kaizen Event programmes. Thirty months later the product (which was the subject of the case study) is still being manufactured in Larkhall (not China) thanks to the efforts of the kaizen team.

Source: Alex Morton, Carrick Lean Consultancy, Scotland (2010).

CASE STUDY 23.3.2 SIX SIGMA CUTS NETWORK RAIL'S WEST COAST MAIN LINE BILL

Britain's West Coast Route Modernisation (WCRM) upgrade project was at one point expected to cost more than US $10 billion—five times the original forecast. Skilled management and planning, though, brought this down to a more realistic US $8 billion, with projects in the programme from 2002 frequently meeting both budget and deadline.

'Six Sigma' methodology has been successfully used by manufacturers for many years and, among them, GE and Motorola have been at the forefront of its application. In recent years, many other product manufacturers and service providers have followed their lead for quality improvements, cost reduction, and customer satisfaction.

The DMAIC methodology of Six Sigma was relatively new to Network Rail, the sponsor of the WCRM project. However, the experience of Bechtel, a major contractor of the project, validated the fact that it can produce quality improvements and generate savings by reducing the cost and volume of rework.

The WCRM is a long-planned upgrade of the 724-km route between London and Glasgow, via Birmingham and Manchester. It is among Europe's busiest mixed-use railways, carrying more than 2,000 trains per day. It has 2,670 track-km and 13 major junctions.

Line speed was previously limited to 177 km/hr, but under WCRM, it is being raised to 200 km/hr. The first phase linking London with Manchester was completed in autumn 2004, while the second phase, to Glasgow, was completed by the end of 2006.

Large-scale construction projects such as WCRM have many special characteristics. Some elements constituted new construction without rail traffic operations (blockade approach), while other parts of it were major upgrades under partial line closure. The project generated large quantities of data on various construction activities with common elements of processes. These actions included design approval process, field reporting procedure, material acquisition and distribution, geometry measurements of constructed track, production tamping, and welding processes. All of these courses of action had to comply with Network Rail's (NR) project and railway standards for quality and output of work. These are measurable processes to which Six Sigma can be readily applied.

Recent experience on WCRM suggests that Six Sigma–based process improvement plans work and successfully deliver clear financial benefits for both project and client.

Bechtel brought the approach to WCRM. Both Bechtel and NR each made significant corporate commitments in training project staff to achieve the gains that were theoretically and practically possible by using Six Sigma. Contractors and sub-contractors were also encouraged to participate in process improvement plans and were delivering better results.

RESULTS AND LEARNING POINTS

Six Sigma is not a short-term activity. Rather, it requires the serious specialised training of staff and long-term resource commitment, which will definitely pay off in short-term process improvement plans as well as long-term operations.

Railway construction companies can achieve better profitability by drastically reducing the cost of rework, and by delivering the required quality the first time. Six Sigma improves service reliability and safety, and ultimately enhances the client's confidence and enriches relations between client and contractor.

The Six Sigma process at Network Rail can be summed up as identifying the process or activity that is not meeting targets; creating a process improvement team that directly affects the quality and quantity of infrastructure works; and assigning a champion who has a strong interest in the process and who can unblock any difficulties in the implementation plan.

Source: International Railway Journal *(June 2005)*.

CASE STUDY 23.3.3 SIX SIGMA IN RADIOACTIVE WASTE MANAGEMENT PROJECTS

Bechtel was the first major engineering and construction company to adopt Six Sigma, a data-driven approach to improving efficiency and quality in major projects. Here is an example of such an application.

The Hanford site in Washington State was established in 1943 to produce plutonium. During Hanford's years of defence production, significant amounts of wastes were created and nearly 54 million gallons of high-level radioactive and chemical excess remained in underground storage tanks. In 2001 the Bechtel Hanford formed a team called Environmental Restoration Contractor (ERC) with pre-selected contractors to improve the waste management process. Six Sigma methodology (DMAIC) was used to reduce redundancies and improve efficiency in this waste management process. Creating an efficient, cost-effective waste acceptance process entailed integrating dozens of waste-generating activities under five different projects. Each such scheme operated with different goals and regulatory restrictions. More regulatory issues resulted from the low-volume higher-risk activities.

In addition to deploying the training programme for Black Belts and Green Belts and the application of DMAIC in five different projects, approaches as appropriate to these five projects were adopted. Their purpose was to overcome challenges in four specific areas as follows:

- Quantifying the problem: Efforts were made to utilise the data available in a manner that fitted the task rather than to generate new information. This approach conforms to Green Six Sigma.
- Communication: The process owners from five projects were involved as key members of the process improvement team, encouraging their active participation.
- Application: It was impractical to try and adapt a set of tools derived from the manufacturing world to a project environment. Therefore the process was adapted to resemble a manufacturing industry and this allowed the acceptance of the appropriate toolkit. This is also in line with Green Six Sigma.
- Differing customer needs: To overcome this challenge, a secondary metric was designed specifically for each project to ensure that the improvements were not having a negative impact on the safety aspect of radioactive waste management.

RESULTS AND LEARNING POINTS

The Bechtel Hanford–led ERC team worked together with the regional US Department of Energy office to map out failure modes and effect analysis. The application of Six Sigma helped develop an improved process for regulatory compliance and decreased the amount of time needed to process waste.

The process improvements included the development of a complete process flow, the streamlining of waste management procedures and the implementation of waste management training courses at a general and project-specific level. The overall impact of implementing Six Sigma enhanced and sustained the waste management efficiencies of the US Department of Energy.

Source: Waste Management Conference, Tucson, Arizona (2002).

CASE STUDY 23.3.4. LEAN SIX SIGMA IN A MAJOR UK INSURANCE COMPANY

A major UK insurance company found itself at crossroads. Plans to grow its market share by 150% were threatened by market changes and economic downturns. Major efforts to streamline internal processes failed. Therefore, the insurer decided to initiate a Lean Six Sigma–based programme starting with an initial assessment called the Diagnostic X-ray.

The approach of the Diagnostic X-ray comprised three main activities:

- Enterprise value stream mapping
- Benchmarking
- Prioritising

The move taken by the X-ray team was to develop a value stream map of the organisation's processes and the costs associated with them. The team started by creating a business process fact base looking at all procedures from account set-up to claims handling, gathering data on 189 activities. The analysis showed that there were seven major steps in the value chain. By breaking each stage into the time spent and staffing level, the team found that the company was spending the same amount of time and money on every claim.

The next action was to establish valid benchmarks, both internal and external. The internal benchmarks were created by comparing how various business units carried out similar processes. The team found costly differences. For example, the placement process took twice as long in one business unit as it did in another. Benchmarking also helped the team to track its activity-based costing and evaluate risks.

In the final phase of the X-ray, the team decided which problems to pursue and in which order. This meant plotting the value of solutions against how difficult they would be to implement. With their priorities firmly established the Black Belts embarked on the Lean Six Sigma projects.

RESULTS AND LEARNING POINTS

The Diagnostic X-ray approach clearly conforms to the initial assessment of 'fitness for purpose' in FIT SIGMA and helps to align the resources to business objectives.

There are also other learning points conducive to favourable business results. First, the company set out to reprise some policies supported by activity-based costing. Within the first year the re-pricing alone earned the insurer an additional 10% of profits. Second, Black Belts worked on a series of targeted organisational changes, boosting profits by a further 10%. Third, over the next 30 months the company invested in technology upgrades justified by focused process improvements. Ultimately IT investments helped to deliver the final 30% profit increase.

Source: Bain & Co. (https://www.bain.com).

CASE STUDY 23.3.5 ENVIRONMENTAL MANAGEMENT DASHBOARD SLASHES A BANK'S CARBON FOOTPRINT

In 2007, a major UK bank, employing 140,000 people across 30 countries, reduced its carbon footprint by 210,000 tonnes of carbon dioxide which was 36% of its total footprint.

The tool responsible for this achievement was the Environmental Management Dashboard (the Footprint Tracker) by Building Sustainability Ltd (see Figure 23.2). The dashboard provided a flexible means to visualise the total environmental load with special focus on energy consumption, energy savings and energy generations by people, equipment, buildings and complete property portfolios. By measuring and presenting the actual cost of energy, energy consumption and carbon dioxide emission (carbon footprint), the occupants, engineers and managers were motivated to introduce energy efficient solutions.

The bank was using the dashboard system in a controlled experiment to measure the impact of carbon emissions on renewable energy sources and

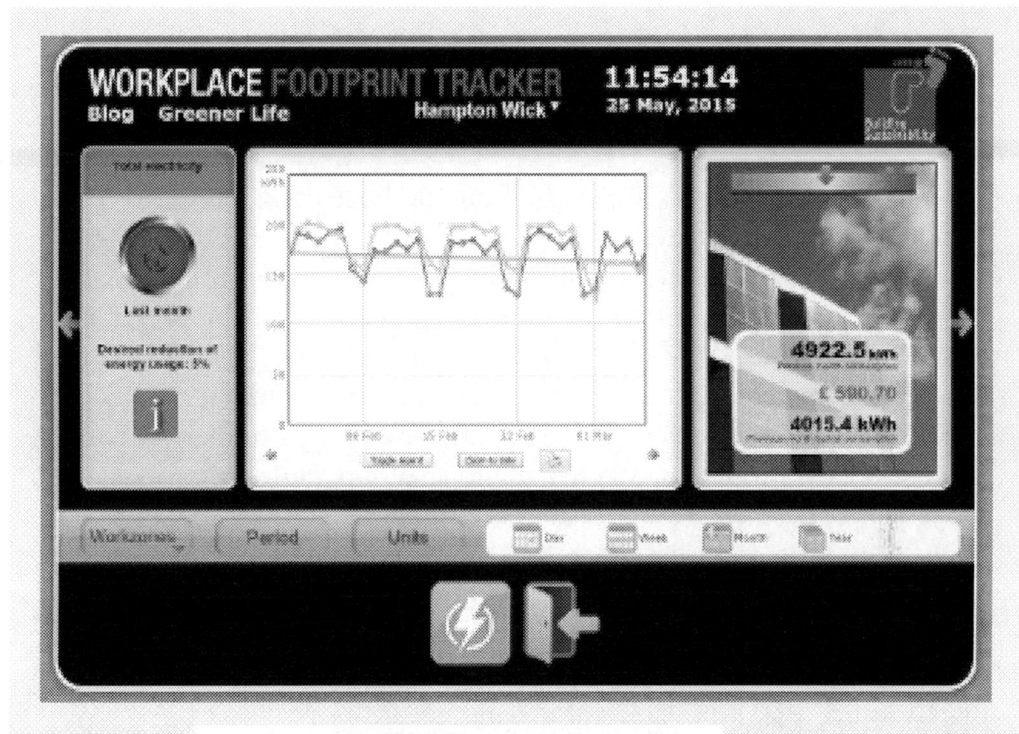

Figure 23.2 Environmental Management Dashboard.

on enabling employees to control their individual energy usage. The innovative new approaches employed included a digital lighting control system, rooftop photovoltaic cells and a solar thermal panel for water heating.

RESULTS AND LEARNING POINTS

The high profile and intuitive ease of the use of the carbon dashboard enabled the bank to demonstrate its environmental credentials and commitment to progressive technologies, as well as its corporate social responsibility to employees and customers alike. During the period of the 'credit crunch', all banks were unpopular to say the least, and this initiative was considered as an acceptable 'bonus' by the bank concerned. The visual and user-friendly system also validated the saying 'you cannot manage what you cannot measure'.

This case example conforms to FIT SIGMA in green thinking.

Source: Building Sustainability Limited (www.buildingsustainability.net).

CASE STUDY 23.3.6 A GIVSM APPROACH FOR IMPROVING SUSTAINABLE PERFORMANCE AT A SME

BACKGROUND

The organisation for this case study is one of the leading returnable packaging/manufacturing SMEs in the UK. It manufactures reusable transit packaging products such as pallies, lids and hog boxes, primarily used by industrial clients. The company's customer base includes retailers, healthcare companies, and transport operators including postal logistics, large manufacturing firms and global automotive manufactures (e.g. Honda, Jaguar, Land Rover). The company employs more than 50 people and uses batch prodction systems where small quantities of products are processed in batches. Such a system is manually operated and involves a large amount of non-value-added activities (wastes), which impact both environmental performance and operational efficiency. The company is beginning to come under pressure by its clients' requirements concerning carbon footprint reduction along the supply chain and by UK's commitment towards the 2015 Paris Agreement.

APPROACH

The company decided to initiate a project to increase its operational efficiency and improve its environmental performance. The methodology applied was GIVSM (Green Integrated Value Stream Mapping). The first step of this methodology consisted of mapping the flow diagram of the general packaging material manufacturing processes (see Figure 23.3).

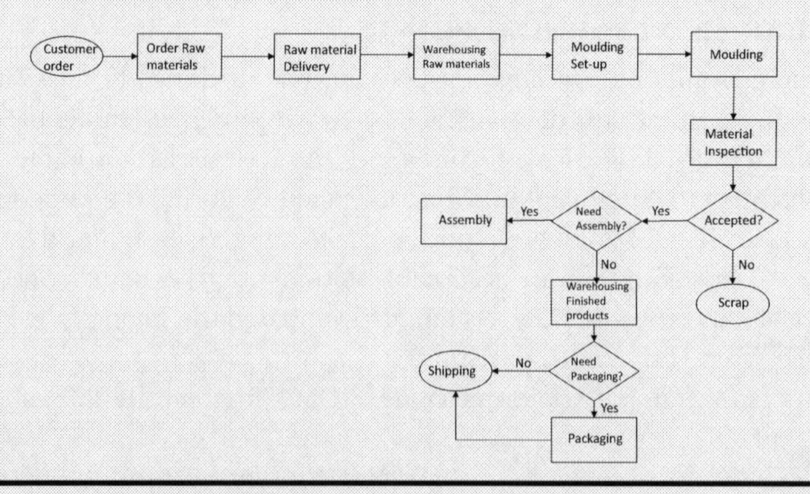

Figure 23.3 Packaging Material Manufacturing Process.

Table 23.2 Lean Wastes and Green Wastes

Lean Wastes	*Green Wastes*
Transportation Inventory Waiting Overproduction Overprocessing Defects	Raw materials Transportation Energy Scrap Emissions

There are two types of wastes in GIVSM as shown in Table 23.2.

The next step of the methodology involved analysing the causes of lean waste and green waste. This was done by doing root cause analysis (RCA) or using a fish-bone diagram. The RCA of various types of lean wastes points towards inefficiencies mainly in six categories, namely (1) materials, (2) methods, (3) machines, (4) management, (5) measurements and (6) (work) environment.

The process that generated the lean waste also generated the green waste. The key green wastes identified in the pally manufacturing process were CO_2e embedded in the polymer 'raw material' (1,438 tonnes CO_2e per year for producing 20,480 pallies). An average raw material 'inventory' of 3,362 pallies producing 243 tonnes CO_2e per year. This included carbon embedded in polymer along with the transportation and 'energy' required to store them. Product 'inventory' was the third largest carbon footprint hotspot accounting for 126 tonnes CO_2e per year. Other key green wastes were 'energy' consumed in the moulding process generating 114 tonnes of CO_2e for manufacturing 20,480 pallies per year. This was followed by 'transportation' or shipment of the raw materials from a Germany-based supplier to the UK manufacturing plant in 25 trips generating 32 tonnes of CO_2e footprint per year. As all defects were moulded back to the recycled polymer and used in manufacturing other components, it generated lowest 'scrap' green waste.

IMPLEMENTATION

As a part of the improvement strategy, a pilot test was run on a single customer order of pally production. Using the results obtained from this pilot study, the future state GIVSM was designed and developed. The company ensured a sequential flow of information in order which reduced the wait time from 5 hours to 20 minutes. The company uses an enterprise resource planning (ERP) system and that was used to automate the raw material procurement process. Owing to its substantial contribution to the company's carbon footprint, a strategic decision was taken to select a British supplier over the German supplier. The delivery of raw materials and warehousing problem was addressed by setting standard operating procedures. Resource levelling was used to ensure the availability of workforce to directly stack the raw material from truck-to-rack, thereby eliminating the wait time of 90 minutes. Instead of weekly revision, the production schedule was revised daily, which reduced machine downtime. Changing the procurement strategies towards more sustainable and low-carbon material as well as a local supplier allowed the company to reduce the raw material inventory from 3,362 to 1,500 pallies.

RESULTS AND LEARNING POINTS

This case study demonstrates that a traditional VSM can integrate both lean and green paradigms to launch a lean project and instigate appropriate improvements within a manufacturing packaging SME. This further shows that SMEs, who usually lack in capital resources, could use this simple GIVSM framework to achieve substantial improvements in their operational and environmental performance.

Source: Choudhary et al. (2019).

23.4 How Green Six Sigma Could Help More

The case studies described in this chapter illustrate how Six Sigma and Lean Six Sigma have been successfully applied in various organisations and projects. The tools, techniques and methodology of Green Six Sigma were not known then.

The additional input of Green Six Sigma will ensure the sustainability of the process and also sustainable environmental standards. The sustainability of the processes will be achieved by embedding the 'sustain' process of Green Six Sigma, such as performance management, knowledge management, self-assessment and senior management reviews. Finally, sustainable environmental standards such as material flow analysis, carbon footprint and circular economy will be achieved by Green Six Sigma tools.

23.5 Summary

In this chapter, case studies are included to offer a practical insight of their applications in both manufacturing and non-manufacturing organisations. We chose examples of both company-wide programmes and solutions to problems so that both the application of tools and techniques and the holistic approach of Green Six Sigma implementation can be illustrated. The case examples have been simplified to present the key learnings and approach of each example that may conform to Green Six Sigma methodology and objectives. Many case studies are not strictly related to climate change initiatives, but they reflect the key learning and approach of Green Six Sigma.

It is intended that the reader may attain a sharpened understanding of how Six Sigma and Lean projects have worked in practice, in compliance with Green Six Sigma methodology, as appropriate for each case.

GREEN TIPS

- Follow the key learning points and approach of each case study.
- Investigate further in more details, when appropriate, by following the reference of the specific case study.

Chapter 24

Implementation: Making it Happen

24.1 Introduction

When I read the summary of the UK Met Office report on climate change (Kendon et al., 2021) I was thinking of borrowing the title from Tom Clancy's novel, *Clear and Present Danger.* The report has shown that visible evidence of climate change is already here in the UK and that the 21st century so far has been warmer than the previous three centuries. The rate of sea level rise has been over 3 mm per year for the period 1993–2019. There can be no doubt looking at these figures that climate change is happening right now.

This is also reflected by the spate of climate-related disasters in July 2021 in geographical areas as widespread as Western Europe, North America, China and South Asia. In August 2021, a UNICEF report (*The Guardian,* 20 August 2021) noted that of the 2.2 billion children in the world, almost half were already at "extremely high risk" from the effects of both pollution and climate change, and that nearly every child alive was at risk from at least one risk such as disease, drought or air pollution and extreme weather events such as cyclones, flooding and heatwave. The report highlighted the fact that for those living in the 33 countries that constitute the most endangered areas—including sub-Saharan Africa, India and the Philippines—they face the consequences of at least three events at once.

The Intergovernmental Panel on Climate Change's (IPCC) Sixth Assessment Report (IPCC, 2021) also carries the same urgent message of the 'clear and present danger' of climate change. This report clearly states, 'Unless there are immediate, rapid and large-scale reductions in greenhouse

DOI: 10.4324/9781003268239-29

gas emissions, limiting warming to close to 1.5°C or even 2°C will be beyond reach . . .. Extreme sea level events that previously occurred once in 100 years could happen every year by the end of this century'. However, it is encouraging to note that the report also suggests that 'human action has the potential to determine the future course of climate'.

In the preceding chapters I described plans for both mitigating and adapting to the consequences of climate change. There is a Japanese proverb which says, 'Action without a plan is a nightmare and a plan without actions is a day dream'. It is clear that we need implementation plans urgently in order to avoid imminent catastrophe—we need to make it happen.

In this chapter, implementation plans are presented in two parts:

- Implementation of climate change initiatives
- Implementation of Green Six Sigma

As the primary domain of this book is Green Six Sigma, more details regarding implementation plans are included in the Green Six Sigma sections. This is where this book will add greater value. In the climate change initiatives section, high-level points for implementation are outlined.

24.2 Implementation of Climate Change Initiatives

Any implementation programme including climate change initiatives, whether international or national, should follow a structured plan based on the best practices of project management and change management (Basu, 2009). The implementation plan of Green Six Sigma, as described in Section 24.3, also follows the principle of change management. The success of the implementation of climate change initiatives at a national level, in particular, is linked to the way they are integrated with the economic and social policies of the national government in power. The leadership strategy/attitude to green initiatives of the national government also plays a critical role. There are many learning points that can be derived from earlier implementation projects related to climate change (de Oliviera, 2009; Sharp et al., 2011) which would be useful for other climate change initiatives. These leaning points include the following:

- Start with an immediate and recognisable threat such as frequent forest fires, hurricanes or flooding. It will help spur action, especially if the community has recently experienced the disaster.

- Begin with a project that can be integrated with an initiative that is already in progress to minimise the duplication of activities and resources.
- Recognise local values and be prepared to be flexible and respond to a community's needs. Reach out to the community and provide open communication.
- Involve elected officials of the local and regional governments early. Although they may not lead on climate change initially, they appreciate being involved and their support is crucial.
- Use outside resources and consultants based on their expertise and track records. Early stage community organising is the most important skill. Later on, technical experts can help with specific needs to provide the solution.
- Recognise mitigation can be a first step. Climate mitigation and adaptation are close cousins. If there are separate budgets for mitigation and adaptation for the same problem, combine them.
- Ensure that the project is funded adequately either from a budget or an emergency relief fund.
- Focus critically on data-driven communication, transparency and stakeholder management rather than media-driven public relations. An example of data-driven communication is to show that meat contributes 57% of greenhouse gases in an average diet. Two essential building materials alone (viz. steel and cement) are emitting 7 billion tonnes of greenhouse gases every year.
- Case examples of Lean Six Sigma in climate change initiatives have demonstrated the power of Lean Six Sigma in accelerating the project delivery. Green Six Sigma, in addition, will ensure both the sustainability of processes and the environmental standards, even in completed projects.

It is also recommended that the application of Green Six Sigma tools should be considered, and its approaches for 'fitness for purpose' and 'fitness for sustainability.

A stakeholder is a party that has interest in the project and can influence its outcomes. It cannot be emphasised enough that stakeholder management is crucial to the success of a climate change solution—both to deliver and also to initiate. It is important that key stakeholders are proactive either to activate or influence climate change projects right now.

24.2.1 International Community

The role of international community is arguably the most crucial driver of implementing any climate change initiatives. The year 2021 has been coined as a 'super year' for the environment. President Joe Biden and other G7 leaders have committed to a new partnership to build back better for the world. The upcoming summits, such as COP26 and the G20, are promising green recovery opportunities from the Covid-19 pandemic. Despite some failures of previous summits, experts are predicting tangible outcomes.

24.2.2 National and Local Governments

There are visible signs that the national, devolved and local governments of the UK have started the implementation of climate change projects. Communities across the UK are tackling the climate crisis with hundreds of local schemes ranging from neighbourhood heating to food co-ops, community land ownership projects and flood defences (*The Guardian*, 10 March 2021). In spite of the Covid-19 pandemic, major climate change projects have started in other countries as well. Political campaigns by Green parties are also driving changes.

24.2.3 Non-Governmental Organisations

There are tens of non-governmental organisations (NGOs), including Greenpeace and Friends of the Earth, across the world. These organisations have been campaigning, often by direct actions, over decades to influence the policymakers to achieve a greener planet. Now their campaign has changed from policy to actions.

24.2.4 Industry and Service Providers

This is a powerful sector to make changes happen. If the big four corporations of the oil and gas industry get together then alternative aviation fuels will be guaranteed within 5 years. If the largest private sector consumers of materials like steel, cement and plastics can cooperate then cleaner substitutes will soon be found. The good news is that some of the world's richest people like Bill Gates and Jeff Bezos are contributing billions of dollars to work to fix the climate crisis. Companies can also invest in their own R&D projects to develop innovative zero-carbon products. Many companies around the world have already committed to using renewable energy for a large part of their operations.

24.2.5 The General Public

Every member of the general public is both a consumer and a citizen and as such, there is much that each of us can do. As a consumer, a person can influence demand for products by fossil fuel energy and as a citizen an adult can vote to change a policy towards achieving climate change solutions. If all of us make individual adjustments in what we buy and use, as discussed in Chapter 10, it will also force the suppliers to change. We can also create demand, say, by buying electric cars and the market will respond accordingly. In a democracy, we can also use our voice to express a view and request change from those in power. As a citizen if we write to our local elected representative demanding actions for climate change then this can have a real impact.

24.2.6 Media

The media world of newspapers, books, television, radio and social media is playing its role by influencing the mindsets of people and affecting government policies on climate change paradigms. The impact of the movie, *An Inconvenient Truth* by Al Gore in 2006 and the *Blue Planet II* documentary on plastics pollution by David Attenborough in 2017 have been game changers. Progressive newspapers (e.g. *The Guardian*) and television channels (such as the BBC who showed *Blue Planet*) have been providing fact-based information to inform, educate and update the audience on both climate change challenges and possible solutions. Sky News have dedicated 30 minutes every day to broadcasting news exclusively about climate change. However, there are also some channels and social media platforms feeding 'fake news' to the deniers of climate change, and their impact and reach cannot be underestimated.

24.2.7 A New Measure of Sustainability

The standard yardstick of national wealth fails to account for the environmental costs of climate change. Polman et al. (2021) suggest that to combat climate change effectively, the world should abandon measuring economic progress in gross domestic product (GDP). The measure of GDP appears to count everything that raises spending as a good thing, but it does not measure the benefits of biodiversity or quality of education. Ultimately we need to find an answer to how we can measure a sustainable economy which decouples growth from environmental degradation. Some experts favour a metric, 'sustainable domestic product', based on the United Nation's sustainable development goals (SDGs).

Organisations are encouraged to include an environmental, social and governance (ESG) strategy and goals in their annual reports. This is often called sustainability. When producing and consuming everything from cars to food, we contribute to the environment, resource depletion, waste, pollution and deforestation to name a few examples. In the social context, organisations have a responsibility for their employees as well as their impact on the societies in which they operate. Governance can serve as a control mechanism in relation to corruption, tax evasion, executive remuneration and internal control. Knowledge sharing, listening to consumer demands and investing in green technology will all drive ethical governance.

24.2.8 Practitioners

The proactive role practitioners cannot be underestimated. In addition, engineers, project managers, management consultants and quality practitioners should also be proactively involved in climate change initiatives and promote the application of the Green Six Sigma approach.

24.3 The Implementation of Green Six Sigma

The implementation of Green Six Sigma presented here is part of the application of climate change initiatives. The implementation of Green Six Sigma and for that matter the instigation of any change programme is a bit like having a baby—it may be very pleasant to conceive, but the delivery of change can be a tricky process. According to Carnall (1999), 'the route to such changes lies in the behaviour: put some people in new settings within which they have to behave differently and, if properly trained, supported and rewarded, their behaviour will change. If successful this will lead to mindset change and ultimately will impact on the culture of the organisation'. The implementation of Six Sigma and Lean Six Sigma has been going on for decades. Thus learning from the proven pathways of both the successful and failed programmes is suggested here for consideration in implementing Green Six Sigma for climate change initiatives.

Next we outline proven pathways for implementing Green Six Sigma for organisations involved in climate change initiatives and potential organisations for climate change that are in different stages of Six Sigma awareness and development. As Green Six Sigma is an adaptation of Six Sigma and Lean Six Sigma principles, any reference to Six Sigma or Lean Six Sigma also relates to Green Six Sigma. We categorised three stages of development and also for SMEs:

1. New starters of Green Six Sigma
2. Started Six Sigma, but stalled
3. Green Six Sigma for small and medium-sized enterprises (SMEs)
4. Green Six Sigma for successful organisations

24.4 Implementation for New Starters

There are many large climate change projects waiting to happen where Green Six Sigma can be applied right from the start. The significant projects in this category include the decommissioning of a fossil fuel power plant, building of a new regeneration power supply system, working on a high-speed train project, setting up a new EV manufacturing plant, conceiving a major climate change research project (e.g. for carbon capture) and so on.

At the earliest stage the decision makers should understand the urgent need for a climate change initiative towards a net zero carbon objective and the necessity for an improvement programme underpinned by a data-driven holistic process of Green Six Sigma. The main concern will be the change required to the culture of the organisation and the absence of a proven structure for transformation of a culture. Management knows what they want but how do they convince their staff that they need to or want to change and encourage to buy in to the process? And how do they sell Green Six Sigma to stakeholders? You can take a horse to water—but how do you make it drink?

Here I provide a total and proven pathway for implementing a Green Six Sigma programme, from the start of the initiative via to the embedding of the change, right through to a sustainable organisation-wide culture. Note that both the entry point and the emphasis on each step of the programme could vary depending on the 'state of health' of the organisation.

The framework of a Green Six Sigma programme is shown in Figure 24.1 and described in the next sections.

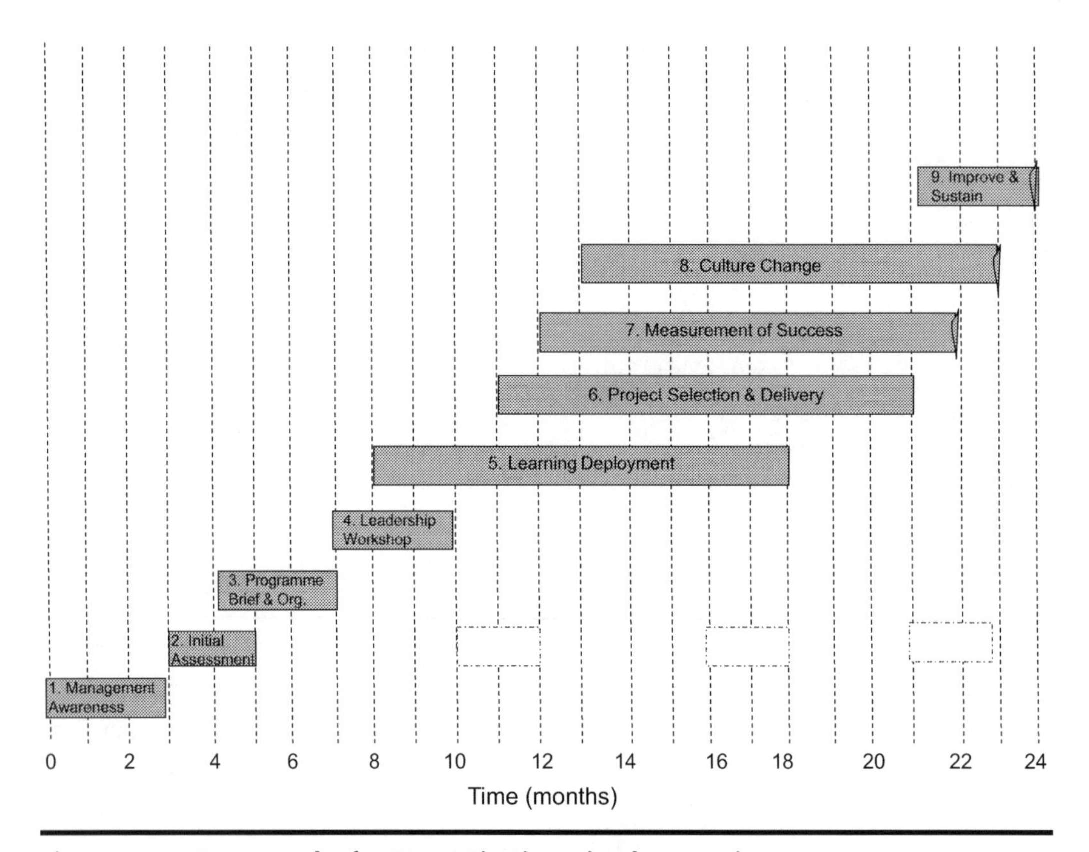

Figure 24.1 Framework of a Green Six Sigma implementation.

24.4.1 Step One: Management Awareness

A middle manager has been tasked by the CEO with leading a Green Six Sigma programme in a large organisation which has no previous experience of Six Sigma or Lean Six Sigma. The CEO has just read an article concerning Jack Welch's successes with Six Sigma at General Electric, and he is full of enthusiasm and has high expectations. The middle manager however is less enthused; in fact he does not want to participate at all and has grim forebodings of failure. He realises that the CEO is a powerful member of the board, but after all he is only one member. Meanwhile, in another organisation, the quality manager for a medium-sized company has attended a Six Sigma conference and has mixed feelings about the task ahead—optimism as well as some doubts. So which one of them has the correct approach? In fact, the author believes that both these managers are right to be concerned.

Learnings from previous Six Sigma and Lean Six Sigma programmes suggest that it is essential to convince the CEO and at least a third of the board regarding the scope and benefits of Green Six Sigma, prior to launching the programme. The success rate of a 'back door' approach without the endorsement of the key players cannot be guaranteed. If a programme is not company-wide and wholly supported by senior management, it is simply not Green Six Sigma. It may be a departmental improvement project—but it is not Green Six Sigma. In cricketing terms a CEO can open the batting, but a successful opening stand needs a partner at the other end.

Research by the Corporate Leadership Council (2005) revealed that through using leadership and change training programmes, companies can substantially increase the potential for change initiatives to be successful. The research also indicated that companies that initiate too much change too quickly actually negatively affect the motivation of employees and their performance.

Learnings from previous Six Sigma programmes suggest that management awareness has been a key factor in the successful application of Six Sigma in large organisations. Various methods have been followed including

1. Consultants' presentation to an offsite board meeting (e.g. General Electric)
2. Participation of senior managers in another organisation's leadership workshop (e.g. GSK and Raytheon)
3. Study visits by senior managers to an 'experienced' organisation (e.g. Noranda's visit to General Electric, DuPont and Alcoa)

Popular Six Sigma literature has advocated that Six Sigma cannot be implemented successfully without top management commitment (Pyzdek, 2003; Ladhar, 2007; Breyfogle, 2008). Moreover, Basu (2009) further argues that if the chief executive does not have a passion for quality and continuous improvement, and if that passion cannot be transmitted down through the organisation, then paradoxically the ongoing driving force will be from the bottom up. It can be argued that the apparent lack of total commitment but tacit support of management has empowered the middle management and acted as an incentive to demonstrate tangible results.

Small and medium-sized firms can learn from the experience of larger organisations, and indeed there can be mutual benefits for the larger organisation through an exchange of fact-finding missions. A service industry organisation could well benefit by exchanging these sorts of visits with successful Six Sigma companies in the finance sector such as American Express, Lloyds TSB and Egg Plc.

During the development of the management awareness phase, it is useful to produce a board report or 'white paper' summarising the findings and benefits. This account has to be well written and concise, but it should not be rushed. It is recommended that you allow between 4 and 12 weeks for fact finding, including visits, and the writing of the 'white paper'.

24.4.2 Step Two: Initial Assessment

Once the agreement in principle from the board is achieved, it is recommended that an initial 'health check' should be carried out at the organisation to develop a 'fitness for purpose' approach. There are many good reasons for conducting an initial assessment before formalising a Green Six Sigma programme. These include the following:

1. Having a destination in mind, and knowing which road to take, are not helpful until you find out where you are to start with.
2. You should get to know the organisation's needs through analysis and measurement of the initial size and shape of the business and its problems/concerns or threats. Once these are ascertained, then the techniques of Green Six Sigma can be tailored to meet these needs.
3. The initial assessment acts as a springboard by bringing together a cross-functional team and reinforces the 'buy-in' at the middle management level.
4. It is likely that most organisations will have pockets of excellence along with many areas where improvement is obviously needed. The initial assessment process highlights these at an early stage.
5. The health check must take into account the overall vision/mission and strategy of the organisation, so as to link Green Six Sigma to the key strategy of the board. Thus the health check will serve to reinforce or redefine the key strategy of the organisation.

There are two essential requirements leading to the success of the assessment (health check) process:

1. The criteria of assessment (check list) must be holistic covering all aspects of the business and specifically address the key objectives of the organisation.

2. The assessing team must be competent and 'trained' in the assessment process. (Whether they are internal or external is not a critical issue.)

It is sensible that the assessment team be trained and conversant with basic fact-finding methods, such as those used by industrial engineers. Some knowledge of the European Foundation for Quality Management (EFQM, 2003) would be most useful.

Once the health check assessment is completed a short report covering strengths and areas for improvement is required. It is emphasised that this report should be short (not the 75-page detailed account required for the EFQM). In writing the document the company might require the assistance of a Six Sigma consultant. The typical time needed for the health check is 2 to 6 weeks.

24.4.3 Step Three: Programme Brief and Organisation

This is the organisation phase of the programme requiring a clear project brief, the appointment of a project team and the development of a project plan. All elements are essential since 'major, panic driven changes can destroy a company; poorly planned change is worse than no change' (Basu and Wright, 1997).

The brief must clearly state the purpose, scope objectives, benefits, costs and risks associated with the programme. A Green Six Sigma syllabus is a combination of Total Quality Management, Lean Management, Six Sigma and culture change supervision. It is a huge undertaking and requires the disciplined approach of project management. 'Programme management is a portfolio of projects that change organisations to achieve benefits that are of strategic importance' (MSP, 2007).

One risk at this stage is that management might query the budget for the programme, and there might be some reluctance to proceed. If this is the case, then it is obvious that management has not fully understood the need for change. This is why it is stressed the importance of the first step, 'Management Awareness'. However, reinforcement could be needed during step three underpinned with informed assumptions and data including a cost/benefit/risk analysis. Unless management is fully committed there is little point in proceeding.

There is no rigid model for the configuration of the Green Six Sigma team. Basic elements of a project structure for a major change programme

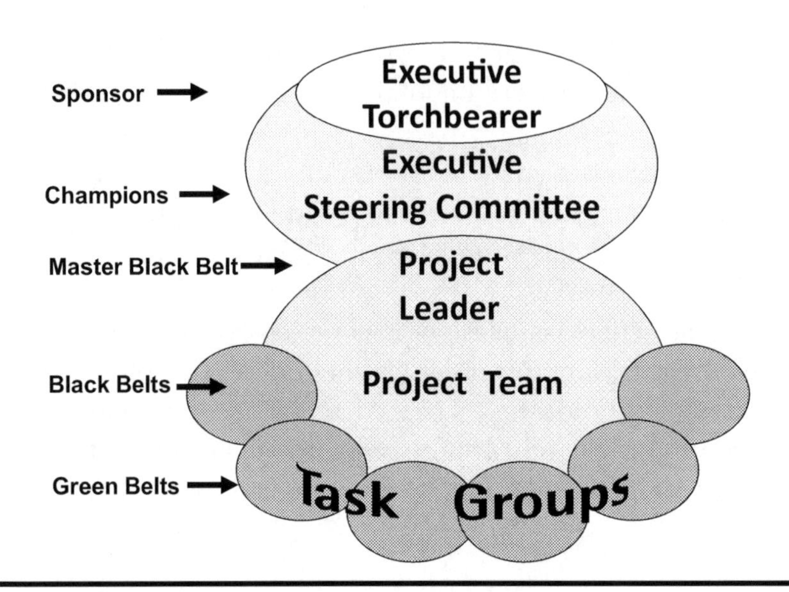

Figure 24.2 Green Six Sigma programme organisation.

can be found in Basu and Wright (1997). A tested Green Six Sigma model is shown in Figure 24.2.

24.4.3.1 Executive Torchbearer

Figure 24.2 shows an executive torchbearer, who ideally will be the CEO and will be the official sponsor for Green Six Sigma. There is a correlation between the higher up the organisation the torchbearer is and the greater the success of the programme. The role of the torchbearer is to be the top management focal point for the entire programme and to chair the meetings of the executive steering committee. Being a torchbearer may not be a time-consuming function, but it is certainly a very important role to give the programme a high focus, to expedite resources and to eliminate bottlenecks.

24.4.3.2 Executive Steering Committee

To ensure a high level of commitment and ownership to the project, the steering committee should be drawn from members of the board plus senior management. Their role is to provide support and resources, to define the scope of the programme consistent with corporate goals, to set priorities and consider and approve the programme team recommendations. In Six Sigma terminology they are the champions of processes and functional disciplines.

24.4.3.3 Programme Leader

The programme leader should be a person of high stature in the company—a senior manager with broad knowledge of all aspects of the business and possessing good communication skills. The leader is the focal point of the project and also the main communication link between the executive steering committee and the programme team. Often the programme leader will report direct to the torchbearer.

The programme leader's role can be likened to that of a consultant. The function of the leader is to a great extent similar to Hammer and Champy's 'czar' in *Reengineering the Corporation* (1993). In other words, the programme leader's task is to

- Provide necessary awareness and training for the project team, especially regarding multifunctional issues
- Facilitate the work of various project groups and help them develop and design changes
- Interface across functional departments

In addition to the careful selection of the programme leader, two other factors are important in forming the team. First, the membership size should be kept within manageable limits. Second, the members should bring with them not only analytical skills but also an in-depth knowledge of the total business covering marketing, finance, logistics, technical and human resources. The minimum number of team members should be three, with a maximum of seven. Any more than even this figure can lead to a series of practical difficulties such as arranging meetings, communicating and keeping to deadlines. The dynamics within a group of more than seven people allows a pecking order to develop and for sub-groups to emerge. The team should function as an action group, rather than as a committee that deliberates and makes decisions. Their role is to

- Provide objective input into the areas of their expertise during the health check stage
- Lead activities when changes are made

For the programme leader the stages of the project include

- Education of all the people in the company
- Gathering the data

- Analysis of the data
- Recommending changes
- Regular reporting to the executive steering committee and to the torchbearer

Obviously the programme leaders cannot do all the work themselves. A programme leader has to be the type of person who knows how to make things happen and who can motivate and galvanize other people to help achieve this aim.

24.4.3.4 Programme Team

The members of the programme team represent all functions across the organisation and they are the key agents for making changes. Members are carefully selected from both line management and a functional background. They will undergo extensive training to achieve Black Belt standards. Our experience suggests that a good mix of practical managers and enquiring 'high flyers' will make a successful project team. They are very often the process owners of the programme. Most of the members of the programme team are part time. As a rule of thumb no less than 1% of the total workforce should form the programme team. In smaller organisations the percentage will of necessity be higher so that each function or key process is represented.

24.4.3.5 Task Groups

Task groups are spin-off sets formed on an ad hoc basis to prevent the programme team getting bogged down in detail. A task group is typically created to address a specific issue. The topic could be relatively major such as the Balanced Scorecard, or comparatively minor such as the investigation of losses in a particular process. By nature, the task group members are employed directly on to the programme on a temporary basis. However, by supplying basic information for the programme, they gain experience and Green Belt training. Their individual improved understanding and 'ownership' of the solution provide a good foundation for sustaining future changes and ongoing improvements.

24.4.3.6 Time Frame

A preliminary time plan with dates for milestones is usually included in the programme brief.

24.4.3.7 The 'Do' Steps

Once the programme and project plan have been agreed by the executive steering team it should receive a formal launch. It is critical that all stakeholders, including managers, employees, unions, key suppliers and important customers are clearly identified. A high-profile programme launch targeted at stakeholders such as these is desirable.

24.4.4 Step Four: Leadership Workshop

All board members and senior managers of the company need to learn about the Green Six Sigma programme before they can be expected to give their full support and input into the scheme. Leadership training is a critical success factor. Leadership workshops can begin simultaneously with Step One, but should be completed before Step Five, see Figure 24.1. Workshops will last between 2 and 5 days and will cover the following issues:

1. What are Six Sigma and Green Six Sigma?
2. Why do we need Green Six Sigma?
3. What will it cost and what resources will be required?
4. What will it save, and what other benefits will accrue?
5. Will it interrupt the normal business?
6. What is the role of the programme leader and the executive committee?

24.4.5 Step Five: Training Deployment

The training programme, especially for the team members, is rigorous. One might question whether it is really necessary to train in order to achieve Black Belt certification. Indeed formal certification might not be essential. However, there is no doubt that without the in-depth training of key members of the programme, little value will be added in the short term, and certainly not over a longer period. The training/learning deployment creates a team of experts. It is presupposed that programme members will be experts in their own departments and processes as they are currently being run. It is expected that they will have the capability of appreciating how the business as a whole will be organised in the future. Green Six Sigma will equip them with the tools for the business overall in order to achieve world-class performance.

Apart from the rigorous education in techniques and tools, it is emphasised that the training will change how the members will look at things. Training is an enabler not only to understand the strategy and purpose of change but—as evidenced by the experience of American Express—it will help members to identify

Project replication opportunities
Leveraging the results of the programme
Identification and elimination of areas of rework
Drivers for customer satisfaction
Leverage of Green Six Sigma principles into new products and services

Smaller organisations are very often concerned about the cost of training, especially the money paid out to consultants and for courses. In a Green Six Sigma programme teaching costs can be minimised by the careful selection of specialist consultants and through the development of in-house training programmes.

24.4.6 Step Six: Project Selection and Delivery

The project selection process usually begins during the training deployment step.

Project selection, and subsequent delivery, is the visible aspect of the programme. A popular practice is to begin by having easy, and well-publicised, successes (known as 'harvesting low-hanging fruit'). We recommend that 'quick wins' should be aimed for (or 'just do it' projects).

In a similar fashion Ericsson AB applied a simplified 'Business Impact' model for larger schemes. They categorise ventures under three headings:

Cost Takeout
Productivity
Cost Avoidance

A variable weighting is allocated to each category as shown in Table 24.1.

Business impact = Cost takeout + 0.5 Productivity + 0.2 Cost avoidance – Implementation cost

For smaller, 'just do it' projects, it is a good practice to establish an 'ideas factory' to encourage task groups and all employees to contribute to savings

Table 24.1 Categories of Savings

Level	*Cost Takeout*	*Productivity and Growth*	*Cost Avoidance*
Definition	'Hard' savings - Recurring expense prior to Six Sigma - Direct costs	'Soft' savings Increase in process capacity so you can 'do more with less', 'do the same with less', 'do more with the same'	Avoidance of anticipated cost or investment which is not in today's budget
Example	- Less people to perform activity - Less money required for same item	- Less time required for an activity - Improved machine efficiency	- Avoiding purchase of additional equipment - Avoiding hiring contractors
Impact	Whole unit	Partial unit	Not in today's cost
Weighting	100%	50%	20%

and improvement. Very often small projects from the ideas factory require negligible funding.

24.4.6.1 Project Review and Feedback

One important point of the project selection and delivery step is to monitor the progress of each task and to control the effects of the changes so that expected benefits are achieved. The programme leader should maintain a progress register supported by a Gantt chart, defining the change, expected benefits, resources, time scale and expenditure (to budget), and showing the people who are responsible for each of the identified actions.

This phase of review and feedback involves a continuous need to sustain what has been achieved and to identify further opportunities for improvement. It is good practice to set fixed dates for review meetings as follows:

Milestone review (at least quarterly)
= Executive steering committee,
Torchbearer, and
programme leader
Programme review (monthly)
= programme leader and team,
with a short report to torchbearer

The problems/hold-ups experienced during projects are identified during the programme review with the aim of the project team taking action to resolve sticking points. If necessary, requests are made of the executive steering committee for additional resources.

24.4.7 Step Seven: Measurement of Success

The fundamental characteristic of a Six Sigma or Green Six Sigma programme that differentiates it from a traditional quality curriculum is that it is results orientated. Effective measurement is the key to understanding the operation of the process, and this forms the basis of all analysis and improvement work. In a construction project the milestones are both tangible and physically obvious, but in a change programme such as Green Six Sigma the modifications are not always apparent. It is essential to measure, display and celebrate the achievement of milestones in a Green Six Sigma programme, and in order to improve and sustain its results importance of performance management is strongly emphasised. The process and culture of measurement must start during the implementation of changes.

The components of measurement of success should include

Project tracking
Green Six Sigma metrics
Balanced Scorecard
Self-assessment review (e.g. EFQM or Baldridge)

There are useful software tools available such as Minitab (https://www.minitab.com) for carrying out detailed tracking of larger Six Sigma projects. However in most programmes, the progress of savings generated by each project can be monitored on an Excel spreadsheet. It is recommended that summaries of results are reported and displayed each month. Examples of forms of displays are shown in Figures 24.3 and 24.4.

24.4.7.1 Green Six Sigma Metrics

Green Six Sigma metrics are required to analyse the reduction in process variance and the reduction in the rate of defects resulting from the appropriate tools and methodology.

A word of caution: Black Belts can get caught up with the elegance of statistical methods and this preoccupation can lead to the development of a statistical cult. Extensive use of variance analysis is not recommended.

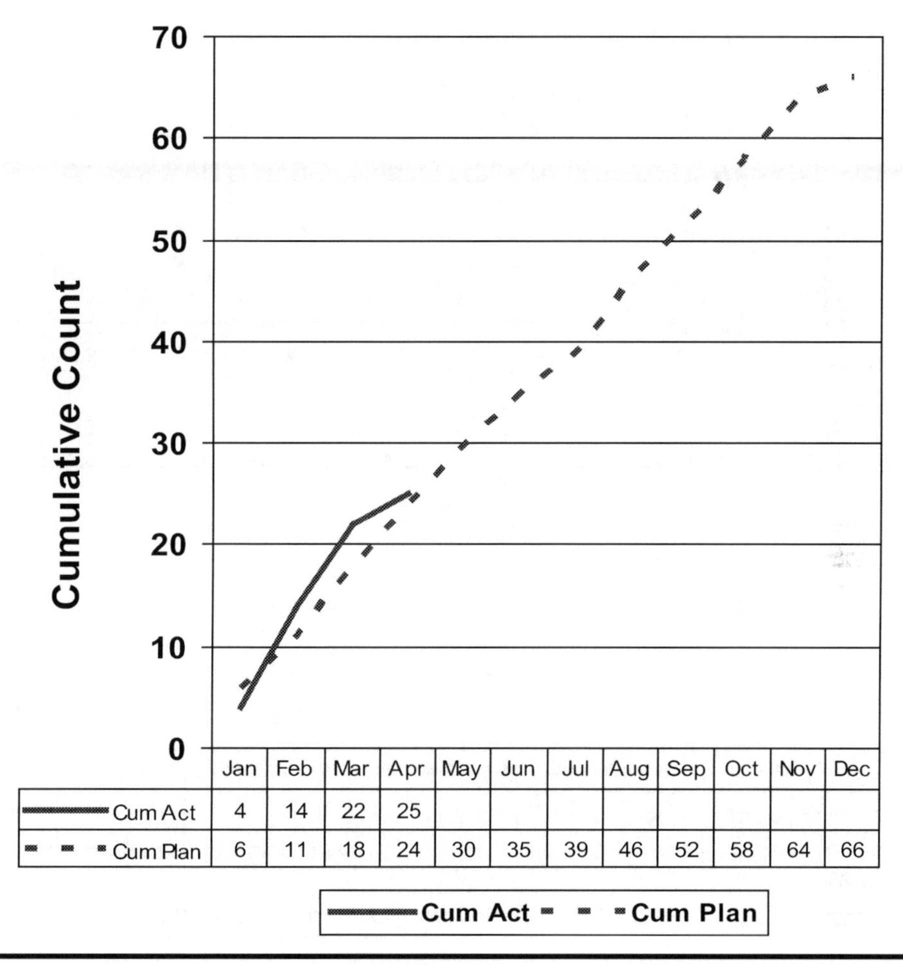

	Jan	Feb	Mar	Apr	May	Jun	Jul	Aug	Sep	Oct	Nov	Dec
Cum Act	4	14	22	25								
Cum Plan	6	11	18	24	30	35	39	46	52	58	64	66

Figure 24.3 Project planned and completed.

The following Green Six Sigma metrics are useful, easy to understand and easy to apply:

Cost of poor quality (COPQ) ratio
Defects per million opportunities (DPMO)
First-pass yield (FPY)
Carbon footprint

24.4.7.2 Cost of Poor Quality

$$\frac{\text{Internal failure \$ + External failure \$ + Appraisal and prevention \$ + Lost Opportunity \$}}{\text{Monthly Sales \$}}$$

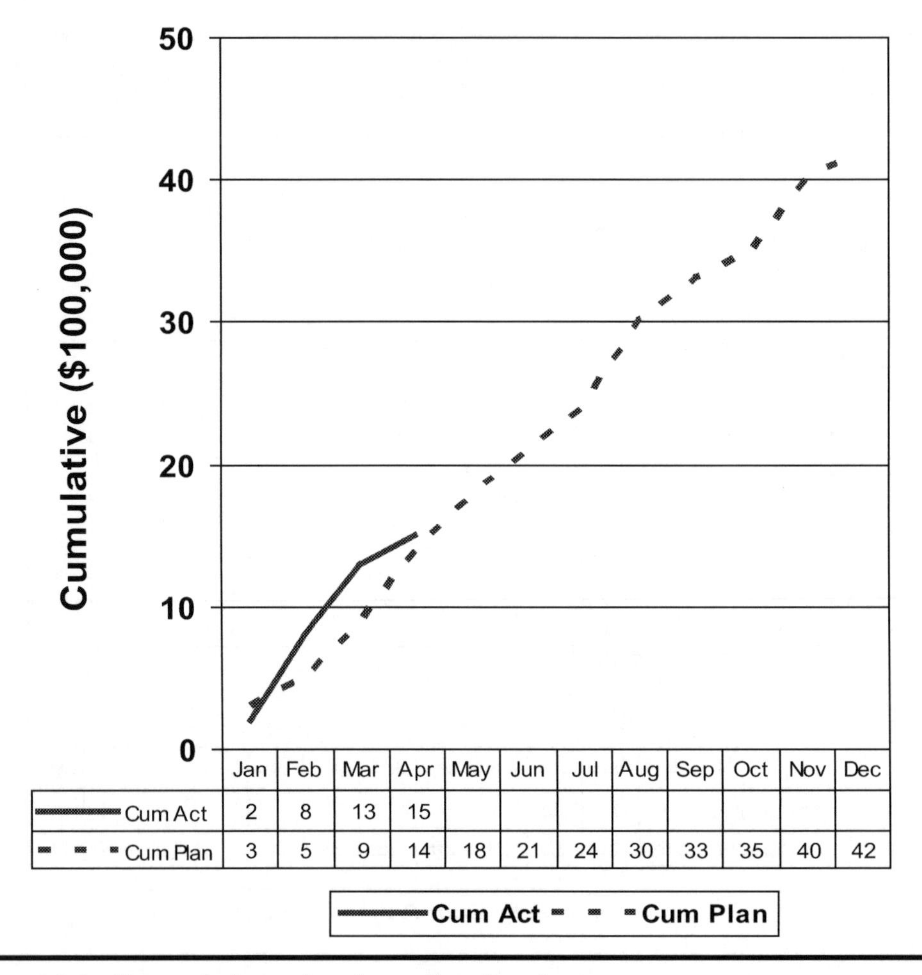

	Jan	Feb	Mar	Apr	May	Jun	Jul	Aug	Sep	Oct	Nov	Dec
Cum Act	2	8	13	15								
Cum Plan	3	5	9	14	18	21	24	30	33	35	40	42

Figure 24.4 Value of planned and completed projects.

24.4.7.3 Defects per Million Opportunities

$$\frac{\text{Total number of defects} \times 1{,}000{,}000}{\text{Total units and opportunities per unit}}$$

24.4.7.4 First-Pass Yield

$$\frac{\text{Number of units completed without defects and rework}}{\text{Number of units started}}$$

24.4.7.5 Carbon Footprint

A carbon footprint is the total amount of greenhouse gas emissions caused by an individual, event, organisation or process expressed as carbon dioxide equivalent. There are digital tools available (e.g. Emitwise, emitwise.com) for automatic calculations of carbon footprints.

By measuring and monitoring Green Six Sigma metrics each month, opportunities for further improvement will be identified.

As already emphasised in Chapter 5, a carefully designed Balanced Scorecard is essential for improving and sustaining business performance. It is generally agreed that the Balanced Scorecard is applicable for a stable process and thus should be appropriate after the completion of the Green Six Sigma programme. This may be so but unless the measures of the Balanced Scorecard are properly defined and designed for the purpose at an early stage, its effectiveness will be limited. Therefore, it is strongly recommended that during the Green Six Sigma programme the basics of the Balanced Scorecard should be established in order to manage the company-wide performance system.

The fourth component of measurement is the 'Self-Assessment and Review' process. There are two options to monitor the progress of the business resulting from the Green Six Sigma programme, and either of the following can be used:

1. A simple checklist to assess the overall progress of the programme
2. A proven self-assessment process such as the European Foundation of Quality Management (EFQM) or the American Malcolm Baldridge system

In the initial health check appraisal stage, using EFQM is advocated, and thus the methodology will already have been applied. Additionally it gives further experience in the self-assessment process, which will enable future sustainability. Finally, it will provide the foundation should the organisation wish at a later stage to apply for an EFQM or Baldridge award.

24.4.8 Step Eight: Culture Change and Sustainability

A culture change must *not* begin by replacing middle management by imported 'Black Belts'. Winning over as opposed to losing middle management is essential to the success of Green Six Sigma, or for that matter any quality initiative.

What is required is that the all-important middle management, and everyone else in the organisation, understands what Green Six Sigma is, and possesses the culture of quality.

The Green Six Sigma culture is shown in Table 24.2.

Table 24.2 Green Six Sigma Culture

1. Total vision and commitment of top management throughout the programme
2. Emphasis on measured results and the rigour of project management
3. Focus on training with short-term projects and results, and long-term people development
4. Use of simple and practical tools
5. Total approach across the whole organisation (holistic)
6. Leverage of results by sharing best practice with business partners (suppliers and customers)
7. Sustaining improvement by knowledge management, regular self-assessment and senior management reviews
8. Sustaining environmental requirements by monitoring the carbon footprints of solutions

Green Six Sigma requires a balanced culture comprising the key characteristics of the previously mentioned four categories. If an organisation is predominantly one type, then some cultural change will be required. Training Deployment, see Step Five of Figure 24.1, includes preparation for culture change. But education alone will not transform the mindset required for Green Six Sigma.

24.4.8.1 Communication

Finally, the key to sustaining a Green Six Sigma culture is the process of good communication. Methods of communication include

- A Green Six Sigma website, specifically developed, or clearly visible on the corporate website
- Specially produced videos
- A Green Six Sigma monthly newsletter
- Internal emails, voicemails, memos with updated key messages—*not* slogans such as 'work smarter not harder' and other tired clichés
- Milestone celebrations
- Staff get-togethers, such as special morning teas, a Friday afternoon social hour, or 'town hall'–type meetings
- An 'ideas factory' or 'think tank' to encourage suggestions and involvement from employees

24.4.9 Step Nine: Improve and Sustain

'Improve and sustain' is the cornerstone of a Green Six Sigma programme. This is similar to Tuckman's (1965) fifth stage of team dynamics for project teams ('Forming, Storming, Norming, Performing and Mourning'). During the Mourning step the project team disbands and members move onto other ventures or activities. They typically regret the end of the project and the breakup of the group, and the effectiveness or maintenance of the new method and results gradually diminish. In Chapter 8 to Chapter 13 we discussed in some detail that in order to achieve sustainability, four key processes must be in place:

1. Performance management
2. Senior management review
3. Self-assessment and certification
4. Knowledge management

The 'end game' scenario should be carefully developed long before the completion of the programme. There may not be a sharp cut-off point like a project handover and the success of the scenario lies in the making of a smooth transition without disruption to the ongoing operation of the business.

As part of the performance management, improvement targets should be gradually, and continuously, stretched and more advanced tools considered for introduction. For example the DFSS (design for Six Sigma) is resource hungry (Basu, 2009), and can be considered at a later stage in a Green Six Sigma programme. With Six Sigma, the aim is to satisfy customers with robust 'zero defect' manufactured products. In order to do so, DFSS is fully deployed covering all elements of manufacturing, design, marketing, finance, human resources, suppliers and key customers (including the supplier's suppliers and the customer's customers).

At an advanced stage of the programme, milestone review should be included in senior management operational review team meetings (such as the sales review meetings and operational planning meetings/committees). In other words, milestone review should occur not only within the Green Six Sigma executive steering committee.

It is recommended that a pure-play EFQM (or other form of self-assessment) should be incorporated as a six-monthly feature of the Green Six Sigma programme. Even if the company gains an accolade such as an EFQM or Baldridge award, the process must continue indefinitely.

Two specific features of knowledge management need to be emphasised. First, it is essential that the company seek leverage from Green Six Sigma results by rolling out the process to other business units and main suppliers. Second, it is equally important to ensure that career development and reward schemes are firmly in place to retain the highly trained and motivated Black Belts. The success of the sustainability of Green Six Sigma occurs when the culture becomes simply an undisputed case of 'this is the way we do things'.

24.4.9.1 Time Scale

The time scale of Green Six Sigma implementation will last several months and is, of course, variable. The duration not only depends upon the nature or size of the organisation but also on the business environment and the resources available. Four factors can favourably affect the time scale:

1. Full commitment of top management and the board
2. Sound financial position
3. Correct culture (workforce receptive to change)
4. A competitive niche in the marketplace

It is good practice to prepare a Gantt chart containing the key stages of the programme and to use it to monitor progress. Figure 24.1 also shows a typical time plan for a Green Six Sigma agenda in a single-site medium-sized company. The diagram shows an order of magnitude only, and the sequence could well vary. The timeline is not linear, stages overlap and frequent retrospection should occur in order to learn from past events and work towards future progress.

24.5 Green Six Sigma for 'Stalled' Six Sigma

Some organisations have already attempted to implement a Six Sigma (or a Total Quality Management) programme, but the process has stalled. Results are not being achieved and enthusiasm is waning; in some cases the programme has effectively been abandoned. The reasons for stalling are various but often progression has ground to a halt due to an economic downturn (such as that experienced in the telecommunications industry in 2001), a change in top management or a merger or takeover.

There are some organisations in this category that applied Lean Six Sigma for a specific project but not across the whole organisation, e.g. Network Rail in Case Example 20.3. These organisations will benefit by focusing on a net zero carbon strategy supported by an organisation-wide Green Six Sigma approach.

Green Six Sigma: Not a 'Quick Fix'

During restructuring, or if the company is in survival mode, the implementation of Green Six Sigma is not appropriate. Green Six Sigma is not simply a 'quick fix', a plaster hastily stuck over gaping wounds, and in such cases it is not sufficient. Instead the underlying, more serious issues, must be addressed. After applying the short-term cost-saving measures of a survival strategy, when the business has stabilised and a new management team is in place, then Six Sigma can be restarted. However this time it should be done correctly, and using the Green Six Sigma approach.

It is likely when restarting that many of the steps will not need to be repeated, including training/learning deployment. However, in a restart there is one big issue that makes life more difficult, and that is credibility. How do you convince all the necessary people that it will work the second time around, when things did not come together at the first attempt? This will put special pressure on Step Eight, Culture Change. The employees could well be tried of excessive statistics and complex Six Sigma tools. Thus the selection of appropriate tools is a strong feature of Green Six Sigma.

The Green Six Sigma programme for a re-starter will naturally vary according to the condition of the organisation, but the programme can be adapted within the framework shown in Figure 24.1. The guidelines for each step are as follows:

1. Management Awareness. If there is none, then you cannot re-start.
2. Initial Assessment. This has to be the re-start point. Where are we, where do we want to go?
3. Programme Brief and Organisation. The programme will need to be re-scoped and new teams formed.
4. Leadership Workshop. This will be essential, even if management has not changed.
5. Training/Learning Deployment. Appropriate tools should be selected. If past team members, or particular Black Belts, are not happy with the terminology of the old programme, then new expressions should be used. The title 'Black Belt' in itself is not sacrosanct and might be changed. If the 'old' experts are still in the organisation, then training

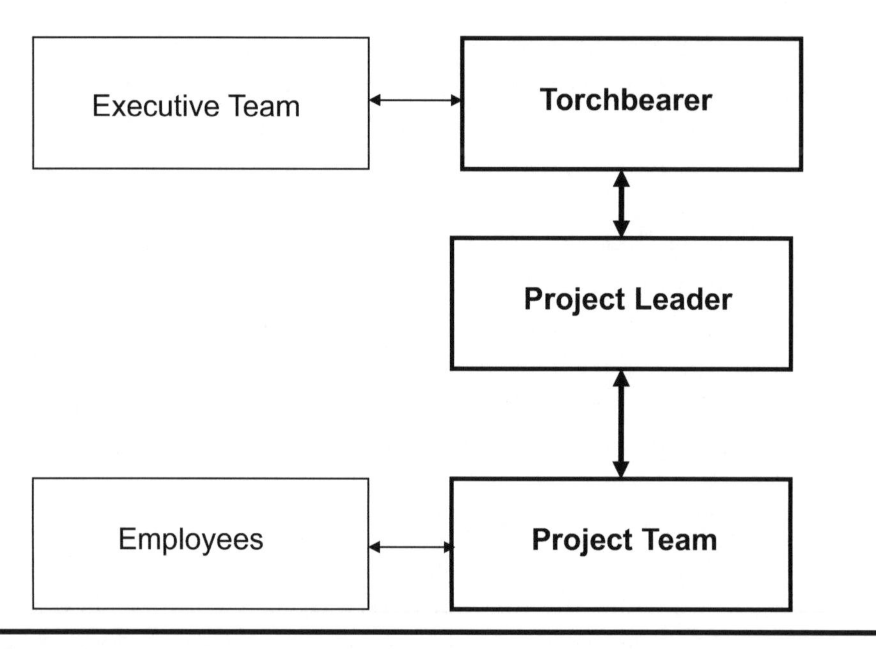

Figure 24.5 Green Six Sigma structure for SMEs.

time might be reduced; for example a workshop of only 1 week's duration might be sufficient.

6. Project Selection and Delivery. Same as the full Green Six Sigma programme, harvest the 'low-hanging fruit'.
7. Measurement of Success Review. Looking at the old measures, ascertain what worked, what did not, and follow the full Green Six Sigma programme.
8. Culture Change. This will be critical. Top management support must be extremely evident. Reward and appraisal systems will have to be aligned to Green Six Sigma.
9. Improve and Sustain. Same as for Green Six Sigma.

24.6 Green Six Sigma for Small and Medium-Sized Enterprises

There are many SMEs that are delivering products and services for climate change initiatives especially for retrofitting buildings. These are the companies engaging in the manufacture of heat pumps, home insulation and the circular economy.

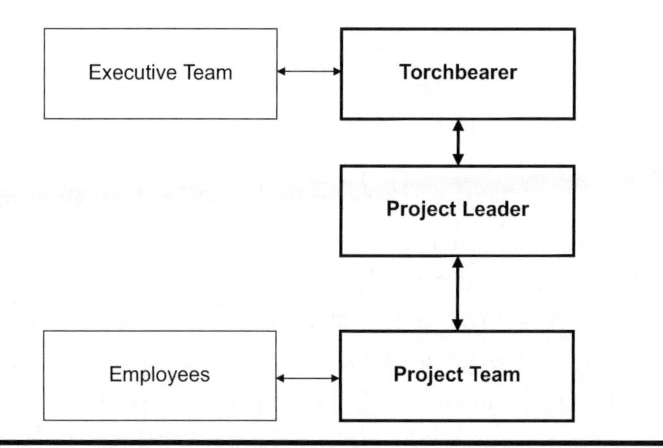

Figure 24.6 Green Six Sigma structure for SMEs.

The organisation structure of the programme will vary according to the nature and size of the organisation. For SMEs a typical structure is as shown in Figure 24.6. In small enterprises the programme leader might be part time. In all other cases the programme leader will be full time.

SMEs may follow the DMAIC Lite process (Basu, 2011). Similar to the full DMAIC (define, measure, analyse, improve, control) process, the DMAIC Lite procedure has three structured steps (viz. define, measure and analyse; improve and control). However, the boundaries between the Measure and Analyse phases and also between the Improve and Control phases are more flexible in the DMAIC Lite version. DMAIC Lite should be followed by the Sustainability tools of Green Six Sigma.

24.7 Green Six Sigma for Successful Organisations

There are some organisations that have been consciously engaged in successful Lean Six Sigma projects to achieve clean energy or green transports. The examples in this category include Ivanpah Solar Electric Generating System (Case Example 18.1) and Tesla electric vehicles (Case Example 20.2). The approach for these projects should be to focus on the sustainability tools of Green Six Sigma. This is also a case for 'we've completed Six Sigma, but where to now?'

William Stavropoulos, the CEO of Dow Chemical, is reported to have once said, 'The most difficult thing to do is to change a successful company'. It is true that employees of firms enjoying a high profit margin with some dominance in the market are likely to be complacent and to feel comfortable with the status quo. Perhaps it is even more difficult to remain at the top or to sustain success if the existing strategy and processes are not adaptable to change. Darwin famously observed, 'It is not the strongest species that survive, nor the most intelligent, but the ones most responsive to change'. It is possible for the management of some companies, after the completion of a highly successful Six Sigma programme, to find their attention diverted to another major initiative such as e-business or business-to-business alliances. Certainly new schemes must be pursued, but at the same time the long-term benefits that could be achieved from Six Sigma should not be lost. Green Six Sigma for sustainability—staying healthy—is the answer.

If a company has succeeded with Six Sigma, then the time is now right to move onto Green Six Sigma to achieve Step Nine: Improve and Sustain.

24.8 External Consultants

Many businesses, especially SMEs, are often concerned with the cost of consultants for a Six Sigma programme. Large consulting firms and academies for Six Sigma could well expect high front-end fees. However, with Green Six Sigma the recommended approach is to be selective in the use of outside consultants. We advocate that you should use outside consultants to train your own experts, and to supplement your own expertise and resources when necessary. No consultant will know your own company as well as your own people. For this reason the use of external consultants is not favoured in the role of the programme leader.

In a Green Six Sigma programme the best use of consultants is in the following steps.

Step Two: Initial Assessment. Here one would use a Six Sigma expert or an EFQM consultant to train and guide your team

Steps Four and Five: Leadership Workshop, and Training/Learning Deployment. Outside consultants will be needed to facilitate the leadership workshops, and to tutor your own 'Black Belts'. Once trained, your own 'Black Belts' will in turn train 'Green Belts' and develop new 'Black Belts'.

Step Eight: Culture Change. An outside consultant is best suited to develop a change management plan for change of culture.

24.9 Summary

This chapter provides practical guidelines for selecting appropriate tools and techniques and 'making it all happen' in a quality programme like Green Six Sigma. Many a Six Sigma exercise started with high expectations and looked good on paper. Many an organisation has been impressed by success stories of Six Sigma, but unsure of how to start. The implementation plan shown here will enable any organisation at any stage of a Green Six Sigma initiative to follow a proven path to success and to sustain benefits. The implementation plan has nine steps beginning with Management Awareness right through to the ongoing process of Improve and Sustain. There is no end to the process of striving for and measuring improvement.

In the spirit of Green Six Sigma, fit for purpose, this framework can be adjusted and customised to the specific needs of any organisation. Explicit comments have been included in this chapter for the implementation of Green Six Sigma in SMEs where resources are constrained (see Section 24.4.1 and Section 24.4.3). Instead of all nine steps in Figure 24.1, SMEs should focus primarily on steps 1, 3, 5, 6 and 9.

At all stages of the programme, it is essential not only that the executive steering committee and the torchbearer are kept informed (and in turn the torchbearer will keep the board up to date), but that there is open communication with all members of the organisation, so that everyone is aware of the aims, activities and successes of the programme.

As a final thought, with the 'clear and present danger' of climate change upon us, it is hard to be optimistic about the future. However, it is also possible that we have a fact-based worldview of climate change and when we all work together we can hope for and look to achieve a greener world for our future generations.

GREEN TIPS

- The United Nations, with the support of national governments, should sponsor the global rollout of Green Six Sigma as they did for Work Study in the 1960s.
- Universities should include courses to train and certify Green Six Sigma Black Belts as part of their undergraduate curriculums.

- Green Six Sigma should be appropriately applied to all climate change initiatives, whether they are new starters, completed projects or stalled projects, to accelerate the process and ensure sustainability of outcomes.
- Green Six Sigma can be tailored to apply to SMEs engaged in the supply and installation of home retrofitting projects.
- Green Six Sigma can be an effective catalyst to implement climate change initiatives.

Chapter 25

Afterword

COP26: The Glasgow Climate Pact

25.1 Introduction

As I write this 'Afterword' in December 2021, the UN COP26 summit in Glasgow has already taken place. The event was held during the first 2 weeks of November 2021 in order to thrash out a deal to curb global warming and accelerate the shift to cleaner economies. Nearly 40,000 people participated in the 2-week summit, which ended in overtime on Saturday 13 November with leaders signing up to the Glasgow Climate Pact after frantic last-minute talks.

Although my suggested outcomes of COP26 as outlined in Section 22.2 were not fully achieved, the spirit of change was there and clearly discernible. This was strengthened by the final outcome of the Glasgow Climate Pact document.

Key pledges of the Pact were as follows:

- A pledge was made by 140 countries to reach net zero emissions by 2050. This target includes the reduction of 90% of current global greenhouse gas emissions.
- More than 100 countries, including Brazil, vowed to reverse deforestation by 2030.
- Over 40 countries have undertaken to move away from fossil fuel subsidies and the use of coal. And 190 countries including China and India pledged to 'phase down' coal.

DOI: 10.4324/9781003268239-30

- India promised to draw half of its energy requirement from renewable sources by 2030 and to achieve net zero emissions by 2070.
- China committed to achieving net zero emissions by 2060.
- The governments of 24 developed countries and a group of major car manufacturers committed to 'work towards all sales of new cars and vans being zero emission globally by 2040'. However, it should be noted that major car manufacturing nations including the US, Germany, China, Japan and South Korea did not join this pledge.
- The pact includes a doubling of money for adaptation by 2025, by comparison with 2019 levels, but the $100 billion target is likely only to be met by 2023.

25.2 Side Deals

- The US and the European Union spearheaded a global methane cutting initiative in which around 100 countries have promised to reduce methane emissions by 30% from 2020 levels by 2030.
- The US and China, the world's two biggest carbon emitters, also announced a joint declaration to cooperate on climate change measures to accelerate their efforts to combat global warming.
- Companies and investors also made a slew of voluntary pledges that would phase out gasoline-powered cars, decarbonise air travel, protect forests and ensure more sustainable investment.

25.3 Analysis

The Glasgow Pact for the first time includes language that asks countries to reduce their reliance on coal and to roll back fossil fuel subsidies. This is crucial since fossil fuels are the primary drivers of man-made climate change. The wording was contentious, though, including the much-reported request from India and China that the deal call on countries to 'phase down', instead of 'phase out' unabated coal.

The agreement acknowledges that commitments made by nations so far to cut emissions of planet-heating greenhouse gases are nowhere near enough to prevent planetary warming from exceeding 1.5° above pre-industrial temperatures. Before COP26 the level of emission was 52.2 gigatonnes and after taking into account the various pledges we could get to 41.9

gigatonnes. However, we need to reach the key figure of 26.6 gigatonnes by 2030. We should conclude on a hopeful note though—arguably the goal of keeping the 1.5°C target is still alive, although only just.

Some tangible signs of progress are agreements by more than 100 countries to reverse deforestation by 2030, and to cut more than 30% of methane emissions by the same date. However, it was evident that COP26 lacked ambition on cutting global usage of fossil fuels.

Although there is some movement in the doubling of money for adaptation available to vulnerable countries, the delivery of the $100 billion per year by developed countries is disappointing.

Encouraged by the momentum to change, we should all—and not just our leaders—spend the key period preceding 2030 by focusing on technologies, policies, retrofitting, adaptation and habits that will put us all on a road map to eliminating at least half of greenhouse gases. I also hope that the tools and holistic approach of Green Six Sigma will help to accelerate this change.

Appendices

Appendix 1: Management Models

If a man does not know to what part he is steering
No wind is favourable to him.

—Seneca

A1.1 Introduction

Fortunately in the real world of business management it is still the ideas and solutions, not models, that matter. However, it occurred to us that a book dealing with the practical ideas of implementing tools and techniques towards achieving improved quality and operational excellence would be incomplete without the most popular management models. The models do not offer solutions to quality problems, but they help to assess a 'big picture' and often reduce the complexities involved.

As a conscious effort of differentiating the models from the problem-solving tools and techniques we included them in the Appendix. The abundance of models on offer is a source of bewilderment for many managers and consultants alike. We have drawn up a shortlist of 14 of the most frequently cited management models. The description of each model has the following structure:

- Description
- Application
- Final Thoughts

A1.2 Ansoff's Product/Market Matrix

A1.2.1 Description

The product/market matrix of Ansoff (1987) is a sound framework for identifying market growth opportunities. As shown in Figure A1.1, the *x*-axis shows the dimensions of the product, and the *y*-axis represents the current and the new market.

There are four generic growth strategies arising from Ansoff's grid:

1. Current Product/Current Market: The strategy for this combination is 'market penetration'. Growth will take place through the increase of market share for the current product/market mix.
2. Current Product/New Market: In this situation the strategy for growth is 'market development'. The pursuit will be for exploring new markets for current products.
3. New Product/Current Market: The strategy of 'product development' is followed to replace or to complement the existing products.
4. New Product/New Market: The strategy of 'product diversification' is pursued when both the product and market are new in the business.

Ansoff has also identified a number of specific strategies for the diversification quadrant depending on the different market and product combinations:

- Vertical integration: When the organisation decides to move into the suppliers' business.
- Horizontal diversification: When entirely new products are introduced in the existing market.

A1.2.2 Application

Ansoff's model is traditionally in the domain of marketing managers for projecting the direction of business growth. In addition to establishing the scope of the product and market mix, the model has been applied in other aspects of shaping the corporate strategy including business growth, competitive advantage, defensive/aggressive technology, synergy and make or buy decisions.

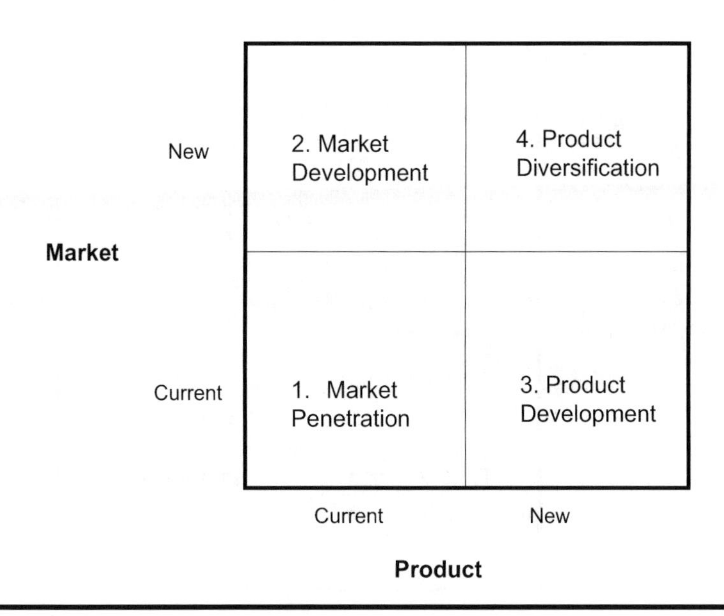

Figure A1.1 Ansoff's product/market matrix.

When used in conjunction with the corporate objectives of an organisation, the five aspects of the model can be applied in the development of the business strategy.

A1.2.3 Final Thoughts

The model on its own is ineffective to evaluate the best strategy, but it provides an excellent framework for exploring strategic discussions on products and markets. This accounts for the fact that it is still popular with marketing strategists and business school students.

A1.3 Basu's Outsourcing Matrix

A1.3.1 Description

The matrix of outsourcing strategy by Basu and Wright (1997) is a useful framework for deciding whether to 'make or buy' and selecting the supply partnership. As shown in Figure A1.2, the model uses the core strength (e.g.

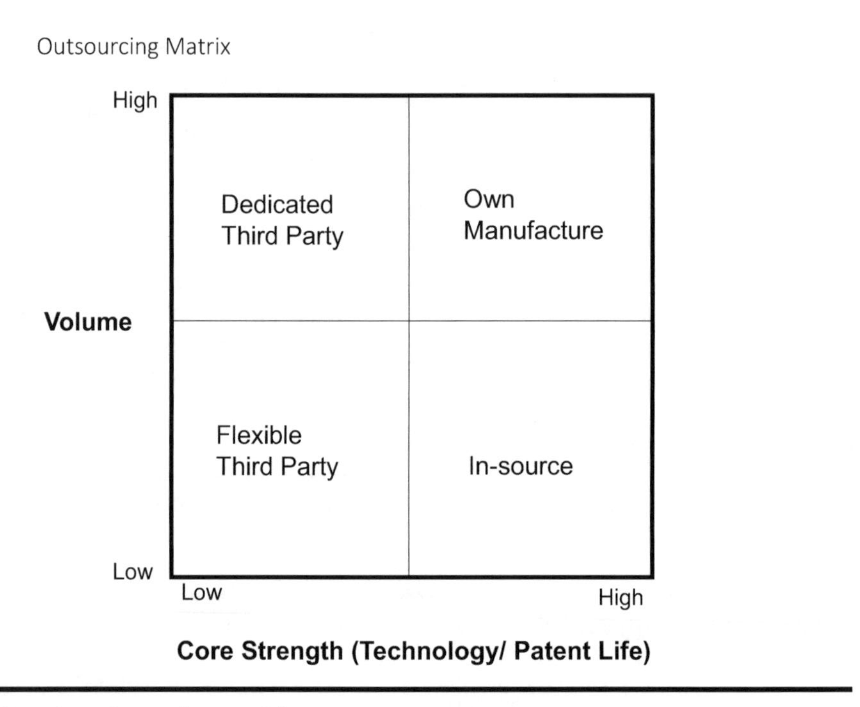

Figure A1.2 Basu's outsourcing matrix.

Technology or Patent Life) as the x-axis and the volume of the product as the y-axis.

The sourcing strategy of products is considered according to where they appear on the grid as follows:

1. High Technology/High Volume: The products in this quadrant are suitable for 'in-house' manufacture. It will be appropriate to invest to retain the core strength over a longer period.
2. High Technology/Low Volume: When the volume is low the strategy should be to 'in-source'. This means that either the global manufacture of the product is centralised in a single site or the capacities of high-technology resources are utilised by bringing in demand from other companies.
3. Low Technology/High Volume: After a period, the technological advantage of a product reduces and it becomes a mere commodity. If the volume is high then a supply partnership is agreed with a 'dedicated third party'.
4. Low Technology/Low Volume: If the volume of the commodity products is relatively low then the product is manufactured by more than one

third-party supplier and it does not require a longer-term partnership agreement.

A1.3.2 Application

The reduction of tariffs, free market agreements, improved logistics and e-commerce as well as cost-effective manufacturing capabilities in the so-called third-world countries have created an explosion of outsourcing operations. Few corporations are carrying out all manufacturing operations in-house. The trend is also similar in the service industry and public sectors.

This model has extensive application, especially in larger organisations, by varying the components of the core strength. We have shown only two components, viz. technology and patent life. The matrix can only provide a strategic pointer which should be further evaluated by external factors (e.g. PESTLE [political, economic, social, technical, legal, environmental] analysis) and internal factors (e.g. financial appraisal and risk analysis). Other dimensions such as quality, responsiveness, dependability, flexibility and innovation are also analysed to establish the outsourcing strategy.

A1.3.3 Final Thoughts

Basu's outsourcing matrix is a sound framework for identifying the options of outsourcing and categorising them for further analysis. Like other analytical grids it also suffers from the limitation of oversimplification.

A1.4 The BCG Matrix

A1.4.1 Description

The Boston Consulting Group (BCG) matrix is one of the early two-dimensional models for analysing and prioritising a product portfolio. A key assumption made by the BCG is that a company should have a portfolio of products that generates both high growth and low growth in a marketplace.

As shown in Figure A1.3, the matrix contains two dimensions, market share and market growth, and creates four categories of product in the portfolio, 'Star', 'Cash Cow', 'Wild Cat' and 'Dog'.

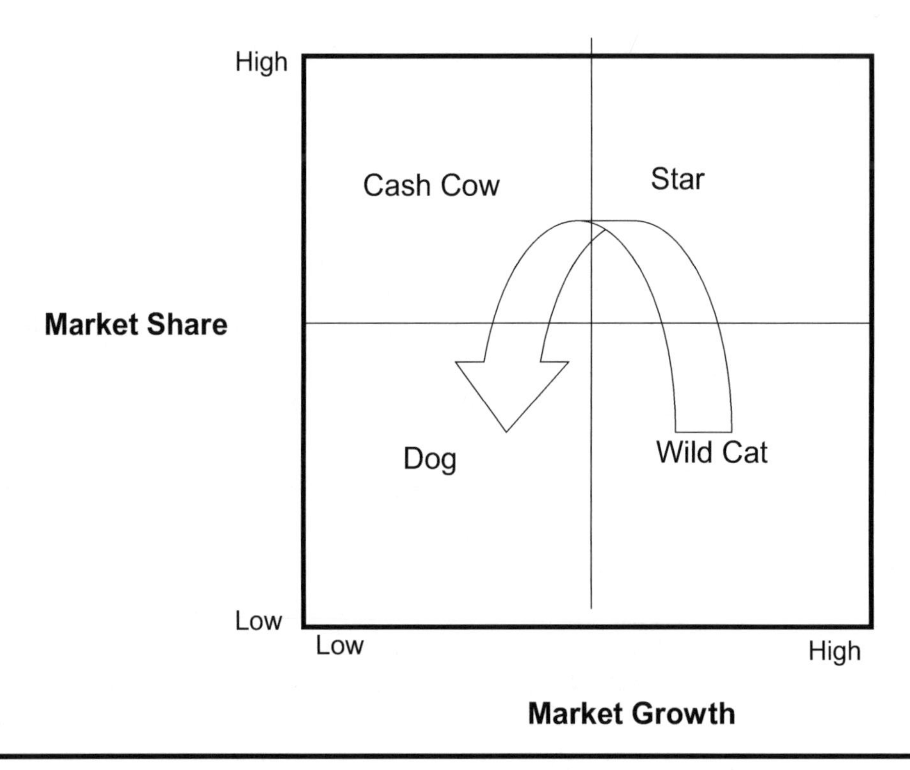

Figure A1.3 The BCG matrix.

1. High Growth/High Share: The products enjoy a relatively high market share in strongly growing markets. The 'Star' products are profitable and good candidates for further investment.
2. Low Growth/High Share: A product in this category is a 'Cash Cow' where the market is no longer growing but the market share is still high. The products are profitable but no investment is necessary to sustain the market share.
3. High Growth/Low Share: For a 'Wild Cat' the market share is low but the demand in the market is growing rapidly. The products are in this category and investment in the future may create big profits though this is not guaranteed.
4. Low Growth/Low Share: When both the market share and its growth are low the product becomes a 'Dog'. The products become unprofitable and are usually divested or discontinued.

A1.4.2 Application

Since its inception in the early 1970s, larger companies with a range of products have applied the BCG matrix as a broad framework for allocating resources among different business units. The matrix draws attention to the cash flow, investment characteristics and needs of an organisation's various divisions. The divisions or products groups evolve over time. Wild Cats become Stars, Stars become Cash Cows and Cash Cows become Dogs in an ongoing counterclockwise motion.

In spite of its popularity the model has its limitations. For example,

- Analysing every business as either Stars, Cash Cow, Wild Cats or Dogs is an oversimplification.
- Market growth rate and relative market share are only two factors of competitive advantage. The link between market share and profitability is also questionable.
- The framework assumes that each business division is independent of the others. In some businesses, a 'Dog' may be helping other units as a corporate strategy.

A1.4.3 Final Thoughts

The BCG matrix has been overused and is not free from flaws. However, it is a useful tool in shaping the direction of the product portfolio, but it should not be employed as the sole means of determining investment strategies for products.

A1.5 Belbin's Team Roles

A1.5.1 Description

Belbin (1985) defined a team role as 'a tendency to behave, contribute and interrelate with others in a particular way'. Based on his research of over 9 years on the behaviours of managers all over the world, Belbin defined nine team roles divided into three groups (see Figure A1.4).

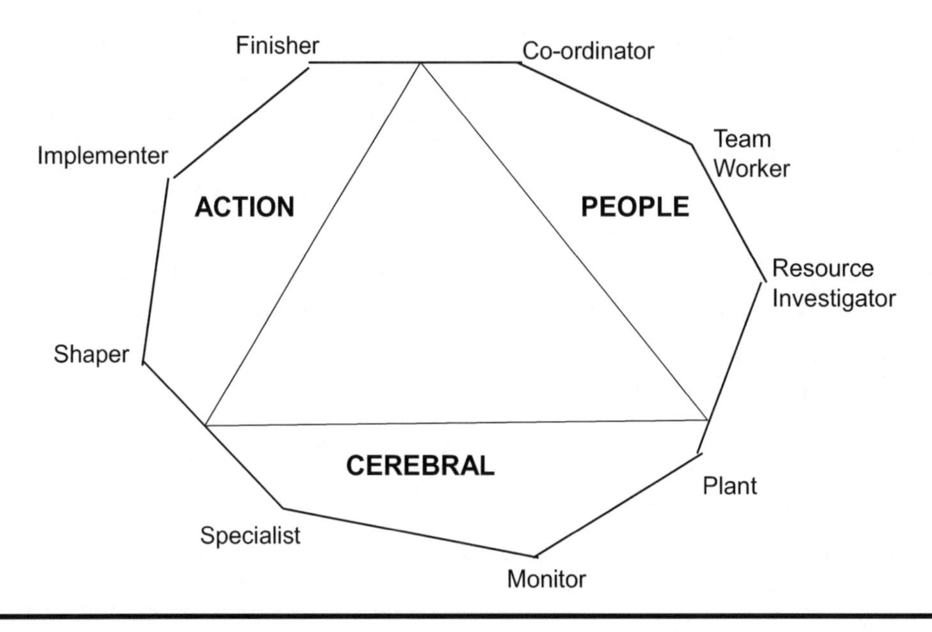

Figure A1.4 Belbin's team roles.

These are

Action-Oriented Roles: Shaper, Implementer and Finisher
People-Oriented Roles: Co-ordinator, Teamworker and Resource Investigator
Cerebral Roles: Plant, Monitor and Specialist

The characteristics of each role have been defined as follows.

The Shaper is challenging, dynamic and thrives on pressure. This person has the drive and courage to overcome obstacles. The Shaper is prone to provocation and is likely to offend people's feelings.

The Implementer is disciplined, reliable, conservative and efficient. This person turns ideas into practical actions. The Implementer is somewhat inflexible and slow to respond to change.

The Finisher or Completer is painstaking, conscientious and anxious. The Finisher delivers on time. This person is inclined to worry unduly and is reluctant to delegate.

The next category of worker is the Co-ordinator who is mature, confident and a good chairperson. This individual clarifies goals, promotes decision-making and delegates well. However, the Co-ordinator can often be seen as manipulative.

By contrast, the Teamworker is co-operative, perceptive and diplomatic. This person listens, balances and averts friction. However, this type of person is indecisive in crunch situations.

An animated, passionate 'people person', the Resource Investigator is extrovert, enthusiastic and communicative. This person explores opportunities and develops helpful contacts. Although over-optimistic, those belonging to this section tend to lose interest once initial enthusiasm has waned.

The Plant is also known as the Creator or the Inventor. This person is creative and imaginative and brilliant at times. The Plant ignores anything incidental and is too preoccupied to communicate effectively.

Sober, discerning and strategic, the Monitor evaluates options and judges accurately, but lacks drive and the ability to inspire others.

Finally, the Specialist is a single-minded, dedicated self-starter. This person provides rare knowledge, but contributes on a narrow front and dwells on technicalities.

A1.5.2 Application

With the support of appropriate assessment, Belbin's team roles have been applied in leadership, people development and project team building activities.

There are different forms of assessment:

- Self-assessment or psychometric profiling by answering multiple choice questions
- Assessment by mentor or line manager during performance appraisal
- Team assessment or 360° appraisal when members grade each other

The analysis of Belbin's team roles is particularly useful when an assignment requires the composition of a team or procurement of members of specific skills and combinations of roles. Belbin suggests that team members with complementary roles are more successful.

A1.5.3 Final Thoughts

The model is conceptually rational but in practice often encourages debates. The objective basis of assessment is questionable. A team, based on this model, may look good on paper but fail to function properly in practice. It also contains too many (e.g. nine) categories.

A1.6 Economic Value Added

A1.6.1 Description

Economic value added (EVA) is the estimate of the intrinsic worth of a company by looking at the present value of all future expected free cash flows in the organisation. This is simpler than it sounds. For example, if a company's capital is $100 million, including debt, shareholders' equity and the cost of using that capital (i.e. interest on debt) is $10 million a year, the company will add economic value for its shareholders only when its projects are more than $10 million a year.

EVA was developed by Stern Stewart & Company who compiles annual performance rankings of large publicly owned companies. EVA is also a measure of the surplus value created in an investment.

EVA accounts for the cost of doing business by deriving a capital charge. The calculation for EVA is simply

$$EVA = OPBT - T - (TCE \times WACC)$$

where
OPBT = Operation profit before tax
T = Tax
TCE = Total capital employed
WACC = Weighted average cost of capital

A positive EVA indicates that the company has created value. Often companies become so focused on earnings that they lose sight of the cost of generating those earnings. EVA has become a popular tool for being linked with the executive bonus. It is useful to note that EVA and return on investment (ROI) are closely related as

$$EVA = (ROI - WACC) \times TCE$$

Thus as long as ROI is more than WACC, the company will add value.

A1.6.2 Application

EVA is a popular financial performance measure in the current business environment. It has been applied in many areas of business purposes including

- Setting targets
- Determining bonuses for managers

- Communicating with investors
- Valuation during acquisition or merger

EVA takes into account the opportunity cost of working capital and relates it to the ROI. Thus it is more meaningful than the more traditional measures such as Earning Per Share.

As shown in Figure A1.5, Stern suggests four stages for successful EVA application.

First establish the rules to convert accounting data to form an EVA calculation. Then track EVA results unit by unit on a monthly basis to complete the requirements of measurement.

With regard to management, budgeting and planning techniques are adjusted to reconcile with EVA measures. By linking salary incentives with EVA, managers are motivated to increase the shareholder value. Thus individuals are encouraged to think and work in the interest of the shareholders. It has the potential to change the mindset and thus transform the company culture.

Market value added (MVA) is the complementary measure to EVA devised by Stern Stewart & Company. MVA is the difference between a company's fair market value (in the stock price) and the economic book value of capital employed.

A1.6.3 Final Thoughts

EVA is a powerful measure with potentials to motivate managers to act and think like they own or are directors of the business. However, its merits

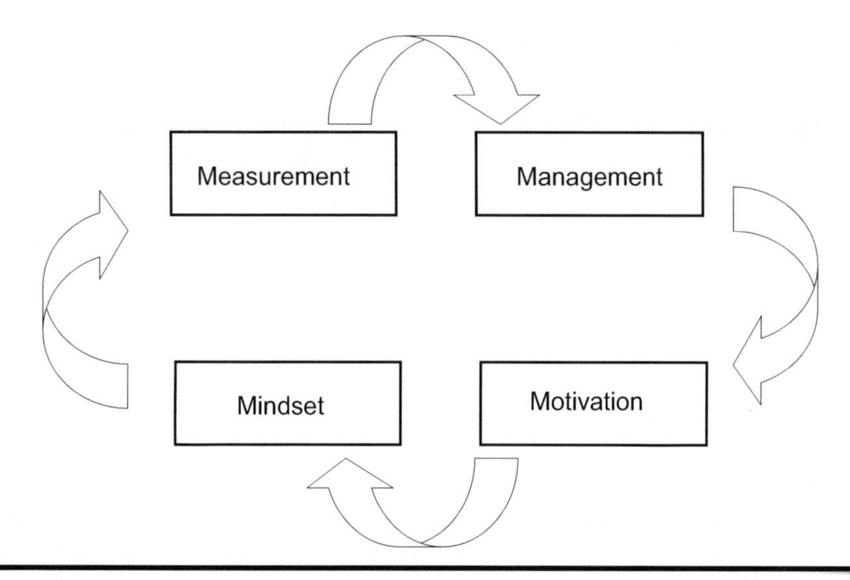

Figure A1.5 EVA application cycle.

should be weighed against two important drawbacks. First there are several anomalies in trying to calculate the true cost of working capital. Second, managers are often discouraged when the share price of the company falls even after showing a theoretical growth in EVA.

A1.7 The Fifth Discipline

A1.7.1 Description

Senge (1994) described five key disciplines that constitute the main elements of a learning organisation:

1. Personal Mastery
2. Mental Models
3. Shared Vision
4. Team Building
5. Systems Thinking

Senge calls Systems Thinking as the 'Fifth Discipline' because it makes other disciplines work together.

Personal Mastery relates to the ability to focus energy and personal vision and to see reality clearly and objectively.

Mental Models determine how we can see and interact with the world. This discipline helps to open up alternative perspectives and insight and balances enquiry and advocacy.

Shared Vision emerges from personal visions fostering the commitment of the group. The more people share the vision, the more likely it is that the goal becomes achievable.

Team Learning relates to the discipline of learning together. It distinguishes between dialogue (i.e. free exploration of complex issues) and discussion (i.e. the presentation of different opinions in search of the best solution).

The Fifth Discipline, Systems Thinking, provides a way of understanding practical issues to see the interrelationship between processes and cause and effect chains. Systems Thinking relies on 'feedback' which refers to how actions can cause or counteract each other. Senge also suggests that we speak of 'archetypes' when actions influence each other in a closed loop. System 'archetypes' are basic and understandable cycles that systems run through.

A1.7.2 Application

Senge's Fifth Discipline attracted both enthusiasts and critics. Its critics find the principles philosophical and theoretical. However, the enthusiasts are impressed by Senge's definition of a learning organisation underpinned by the five disciplines. It is worth noting that both critics and enthusiasts have requested tips for the daunting task of putting the Fifth Discipline into practice.

As shown in Figure A1.6, each discipline has different aspects of essence, principle and practice.

The Fifth Discipline brings the concept of 'learning organisations' where people continually expand their capacity to re-create the results they truly desire, where collective aspiration is set free and where people are continually finding out how to learn together.

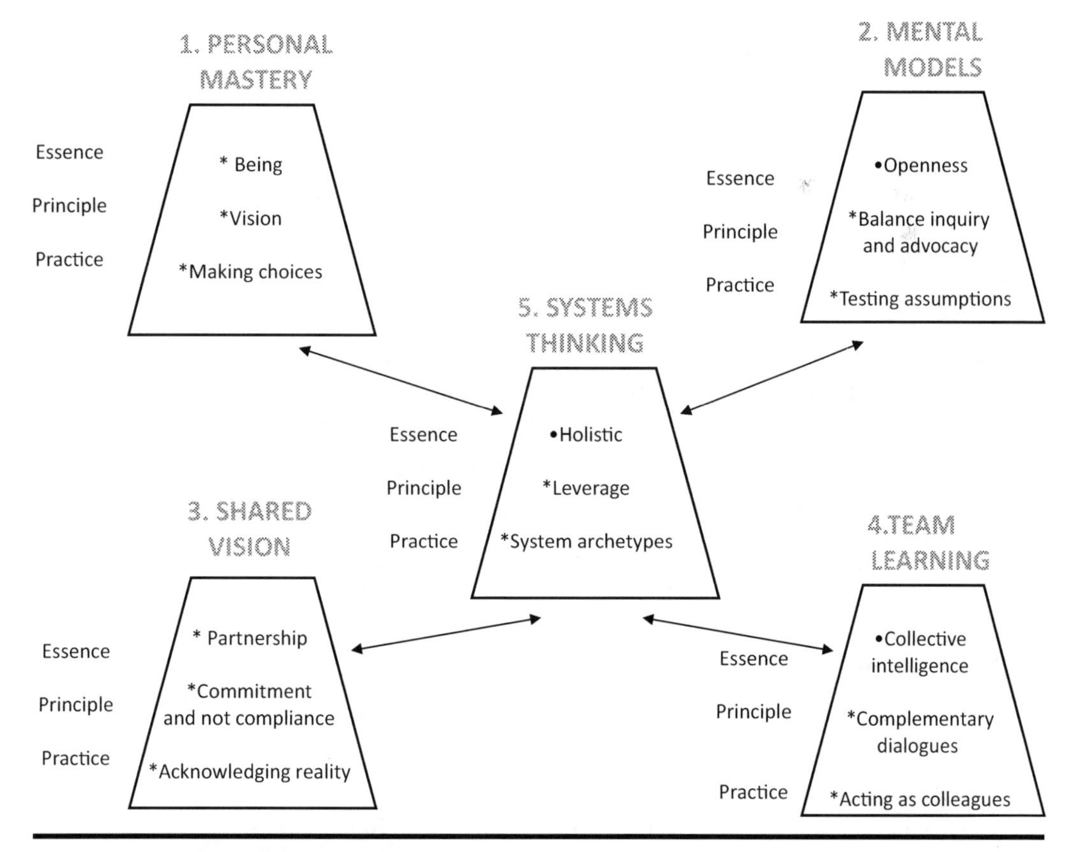

Figure A1.6 The Fifth Discipline.

The disciplines, at least in principle, can be applied beyond business organisations. For example, these are also applicable to family, community groups and to all society.

A1.7.3 Final Thoughts

The Fifth Principle is likely to be perceived as an 'ivory tower' approach to solving practical problems. It is like someone asking how to start a car and receiving an answer in terms of the laws of physics. However, the concept of 'learning organisations' is invaluable in the current environment of collaborative management and supplier partnership.

A1.8 The McKinsey 7-S Framework

A1.8.1 Description

The McKinsey 7-S framework was originally developed by Pascale (1990) as a way of thinking more effectively about organising a company strategy. McKinsey Management Consultants divided the seven organisational elements into what they called 'hard' and 'soft' elements as follows (see Figure A1.7).

The first three Ss are the 'Hard' or tangible elements, that is, Strategy, Schedule and Systems.

Strategy relates to the informed choices an organisation makes to achieve its business objectives, such as prioritising the investment projects.

Structure refers to the organisational hierarchy and division of functions and activities.

Systems concerns the primary and secondary processes that the organisation employs to get operations completed, such as manufacturing resource planning, supplier relationship management etc.

'Soft' elements comprise a further four concepts beginning with 'S': Shared Values, Style, Staff and Skills.

The core beliefs and expectations that employees have about their organisation can be termed its Shared Values.

Style relates to the unwritten but visible behaviour of the management in conducting business.

Staff is of course the people employed in the organisation.

The collective capability of an organisation is known as its Skills. These are independent of individuals but dependent on the other six S's.

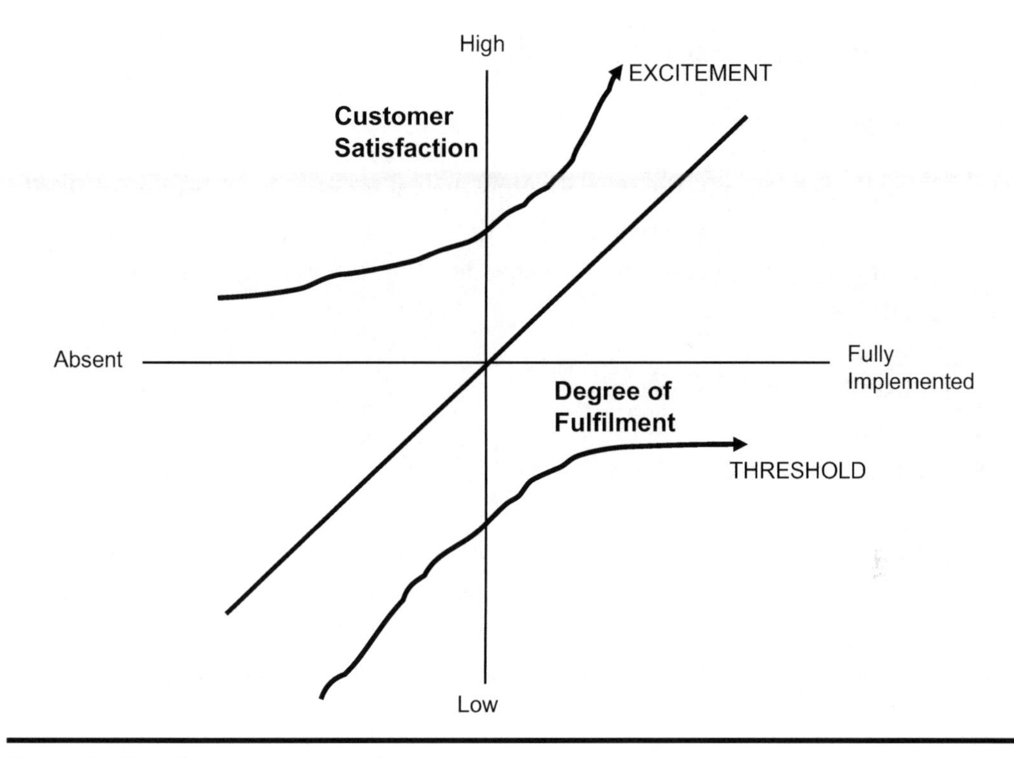

Figure A1.7 The 7-S Framework.

A1.8.2 Application

The 7-S framework has been used by consultants as a good checklist to analyse key elements of an organisation. At a more sophisticated level the framework can be used for assessing the viability of a strategic plan. In this case, the 7-Ss are considered as 'compasses' and each element is examined to assess whether or not they are all pointing in the same direction.

By constructing a matrix, the potential conflict between each element can be identified for further analysis. The decisions are made as to how to effect changes in the strategy in order to balance the conflicts. Pascale (1990) argues that smarter companies use conflicts to their advantage. They power the engine of root cause analysis in pursuit of excellence.

A1.8.3 Final Thoughts

The use of the 7-S Framework as a checklist is useful, but it suffers the risk of being a dressed-down application. The seven elements, especially those in the 'Soft' category, are broad based and lack a detailed a checklist to make them more specific.

A1.9 Kano's Satisfaction Model

A1.9.1 Description

The Kano model of customer satisfaction demonstrates that blindly fulfilling customer requirements has a risk if the supplier is not aware of different types of customer requirements. The model divides product attributes into three categories:

- Threshold or basic needs
- Performance needs
- Excitement needs

As shown in Figure A1.8 the model provides three customer requirements in two dimensions. The first dimension measures the degree to which the customer requirement is fulfilled. The second dimension (*y*-axis) is the customer's subjective response of satisfaction to the first dimension.

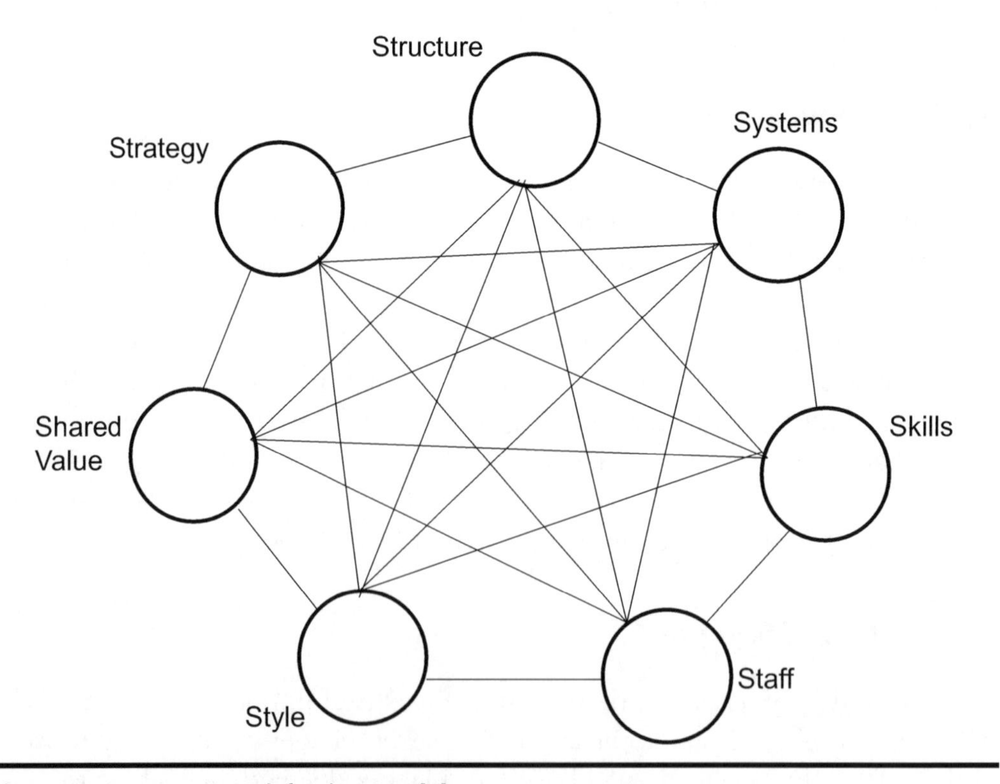

Figure A1.8 Kano's Satisfaction Model.

Threshold (or basic) attributes are the expected attributes of a product and do not provide an opportunity for product differentiation. Examples of this category are car brakes or a telephone dial tone.

Performance attributes are those for which more is better and are often referred to as 'fundamental' quality. Customers overtly state these needs. Examples are price, delivery and performance.

Excitement attributes are 'latent' needs but can result in high levels of customer satisfaction. Customers are unaware of these needs until they experience them. Examples are cupholders and Post-it notes.

A1.9.2 Application

The information obtained from the analysis of the Kano model especially related to performance and excitement attributes, provides valuable input to the quality function deployment (QFD) process.

The basic tool has two sided questions; one is asked in a positive case and the other is asked in a negative scenario. For example,

1. Rate your satisfaction if the product has the attribute.
2. Rate your satisfaction if the product did not have the attribute.

The customer's response should be one of the following:

a) Satisfied
b) Neutral
c) Dissatisfied
d) Don't care

Surveys are then tabulated. Basic attributes generally receive 'Neutral' to Question 1 and the 'Dissatisfied' response to Question 2. Consideration should be given to a 'Don't care' response as they will not increase customer satisfaction. Prioritisation of matrices can be useful in determining which excitement attributes would maximise customer satisfaction.

A1.9.3 Final Thoughts

The Kano model is useful in identifying customer needs and analysing competitive products. The model establishes that to be competitive products and services must execute flawlessly all three attributes of quality—basic, performance and excitement. Exceeding customer's performance expectations creates a competitive advantage.

A1.10 Mintzberg's Organisational Configuration

A1.10.1 Description

According to Mintzberg (1990) an organisation's structure is determined by the environmental variety which in turn is determined by both environmental complexity and the pace of change. He defines four basic types of organisational form (see Figure A1.9) which are associated with four combinations of complexity and change.

Mintzberg also defines five basic organisational subunits (see Figure A1.10) to help explain each of the four organisational forms.

The typical examples of the functional roles related to each subunit are shown in Table A1.1 in the context of a manufacturing company.

The four organisational configurations in Mintzberg's scheme depend on distinctive mechanisms for co-ordination. The power is divided throughout the organisation by different forms of decentralisation according to the configuration. The relevant characteristics of each configuration are summarised in Table A1.2.

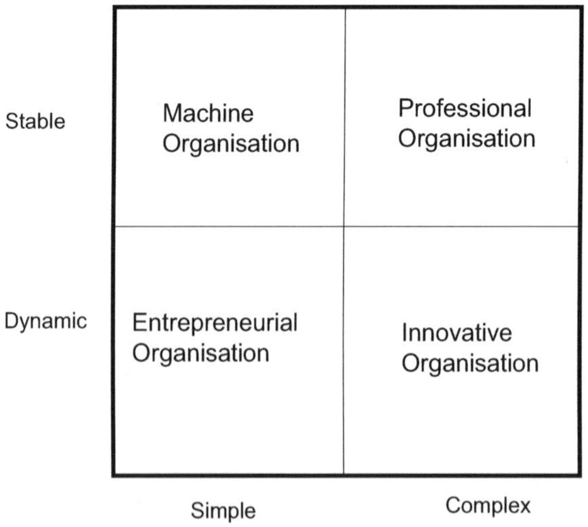

	Simple	Complex
Stable	Machine Organisation	Professional Organisation
Dynamic	Entrepreneurial Organisation	Innovative Organisation

Figure A1.9 Mintzberg's taxonomy.

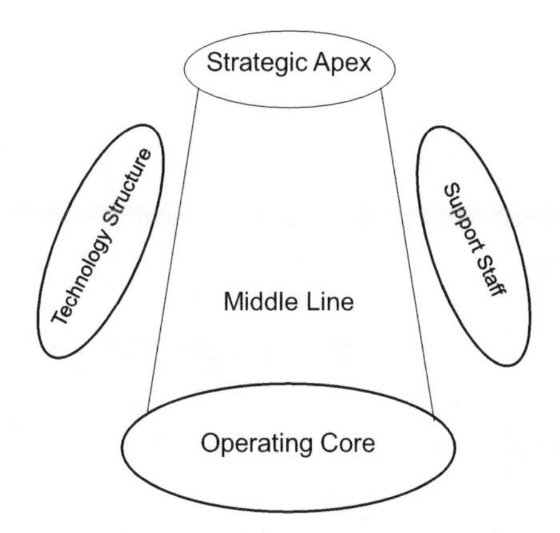

Figure A1.10 Organisational subunits.

Table A1.1

Subunit	*Roles*
Strategic apex	Board of directors
Technostructure	Strategic Planning, Engineering, Systems Analysis
Support staff	Human Resources, Payroll, Maintenance, Catering
Middle line	VP Operations, VP Marketing, Sales managers
Operating core	Machine operators, Assemblers, Warehouse operators, Salespersons

Table A1.2

Configuration	*Co-ordinating Mechanism*	*Key Subunit*	*Decentralisation*
Entrepreneurial Organisation	Direct Supervision	Strategic Apex	Vertical and Horizontal Centralisation
Machine Organisation	Standardisation of Work	Techno-Structure	Limited Horizontal Decentralisation
Professional Organisation	Standardisation of Skills	Operating Core	Horizontal Decentralisation
Innovative Organisation	Mutual Adjustment	Support Staff	Selected Decentralisation

In his revised writing, Mintzberg introduced other variants of configuration, such as Diversified Organisation, Missionary Organisation and Political Organisation.

A1.10.2 Application

Mintzberg's taxonomy offers a number of practical advantages:

- It focuses the way in which power is divided throughout the organisation.
- Having determined the category of an organisation, you can also determine what changes are needed to make it internally consistent.
- It enables the analysis of the co-ordinating mechanism most appropriate for the current configuration and the need for changes.

When an organisation lacks the driving subunit and a prime co-ordinating mechanism, it is more likely to become a political organisation. There is no defined co-ordinating mechanism or a predominant subunit in a political organisation, and thus subunits fight to fill the power vacuum.

In a practical environment, different configurations may form the whole organisation. For example, the university is a Professional Organisation but support units may be composed of other configurations. Support subunits that perform routine functions may have a Machine Organisation, but technocratic sub-units may be managed as an innovative Organisation or a Professional Organisation.

A1.10.3 Final Thoughts

Doubtless, Mintzberg's model helps us to understand the relationship between the nature of an organisation and its co-ordinating mechanism. With four basic criteria, it is difficult to 'fit' an organisation to the typical configuration. Most of the larger organisations are hybrids of multiple configurations.

A1.11 Porter's Competitive Advantage

A1.11.1 Description

The fundamental basis of above average profitability of a company within its industry determines its competitive advantage. According to Porter (1985),

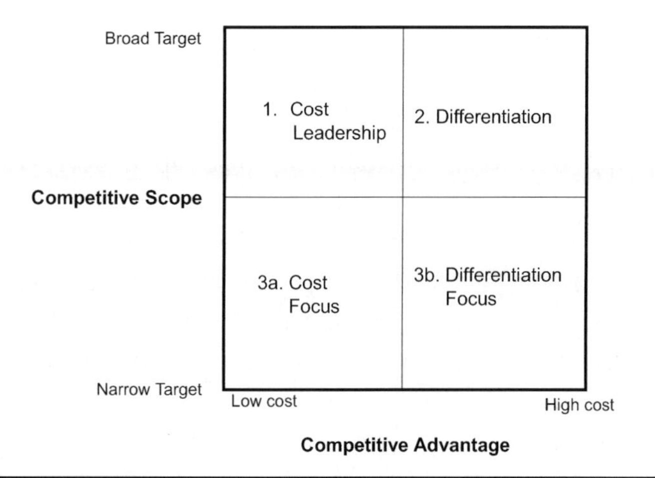

Figure A1.11 Porter's Competitive Advantage.

there are three generic strategies for achieving above average performance in industry:

- Cost leadership
- Differentiation
- Focus

As shown in Figure A1.11, the focus strategy has two variants, cost focus and differentiation focus.

Cost Leadership: A company in cost leadership sets out to become a low-cost producer in its industry. A low-cost producer must utilise all sources of cost advantage, such as economies of scale, asset utilisation, preferential procurement of raw materials and other factors to improve efficiency.

Differentiation: A company in a differentiation strategy aims to be unique in its industry among some features that are widely valued by customers regardless of cost. Differentiation requires higher investment in research, design and customer service.

Focus: The generic strategy of focus selects a segment in the industry and customises its strategy to satisfy the target segment. In cost focus a company aims towards a cost advantage in the target segment while in differentiation focus a company seeks differentiation in its target segment.

A1.11.2 Application

The model is so broad based that it is logical to assume that a company has no option but to choose one of the three generic strategies. There will always be a competitor in the industry that aims to be either cheaper or better differentiated than your company.

There is likely to be one cost leader in any industry, but it is possible to have multiple differentiators. Cost has only one dimension while differentiation can be achieved in several ways.

The initial step of applying this model is to determine the position of a company in the matrix. The next, and probably the most difficult, step is to map the main competitors in the grid. Having established the current relative position of the company appropriate steps can be taken either to maintain or improve it.

A1.11.3 Final Thoughts

Porter's generic strategy model reduces the guesswork in strategic planning but it only provides an initial pointer. Further analysis with the aid of other tools and techniques is required to formulate a practical and appropriate strategy for the company.

A1.12 Porter's Five Forces

A1.12.1 Description

Porter's (1985) model for competitive analysis relates to five different forces (see Figure A1.12):

1. New Entrants: Threat of entry from other organisations. Example: online banks challenging high street banks
2. Substitutes: Availability and competition from substitute products. Example: email as a threat to fax machines
3. Buyers: Bargaining power of buyers. Example: supermarkets' increasing power related to suppliers
4. Suppliers: Bargaining power of suppliers. Example: Trained specialist required by an industry
5. Existing competitors: Rivalry among existing competitors.
   Example: Vodafone and T-Mobile jockeying for position in the 3G market

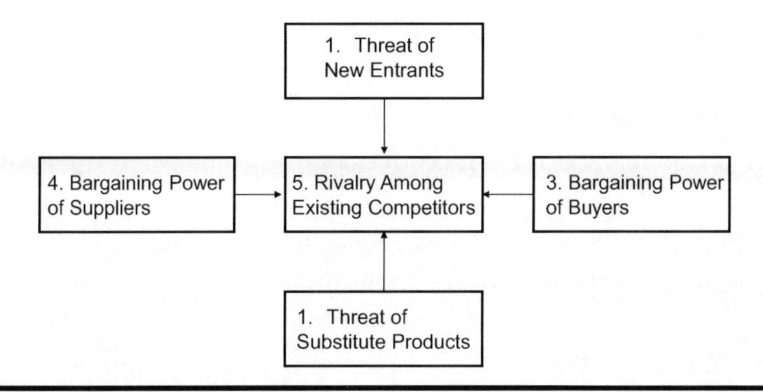

Figure A1.12 Porter's Five Forces.

A1.12.2 Application

The model emphasises the external forces of competition and how these can be countered by the company. It implies the danger inherent in focusing on your immediate competitors.

The elements involved with each force are shown in the following to prepare for competition.

New Entrants: Examine the entry barriers for new entrants including

- Economies of scale
- Brand identity
- Capital requirements
- Switching cost
- Access to distribution

Substitute: Analyse the determinants of substitution threats including

- Relative price performance
- Switching cost
- Buyer's inclination to substitute

Buyer: Examine to what extent buyers can bargain by considering

- Buyer volume
- Buyer information

- Decision maker's incentives
- Switching costs
- Differentiated products
- Impact on quality/performance

Supplier: Competitive forces from suppliers mirror those of buyers. Examine the determinants of supplier power including

- Differentiation of inputs
- Supplier volume
- Substitute inputs
- Switching cost
- Forward integration to your customers

Existing competitors: Analyse the rivalry determinants related to existing competitors. These factors could include

- Industry growth
- Diversity of competitors
- Fixed cost and asset bases
- Switching cost
- Brand identity
- Exit barriers

A1.12.3 Final Thoughts

Porter's model of the Five Forces has been most widely used in strategic analysis and business schools. However, the model is more useful for developing a reactive strategy. It is weak in developing a pro-active strategy building upon the core strengths of a company.

A1.13 Porter's Value Chain

A1.13.1 Description

According to Porter (1985), the competitive advantage of a company can be assessed only by seeing the company as a total system. This 'total system' comprises both primary and secondary activities (see Figure A1.13).

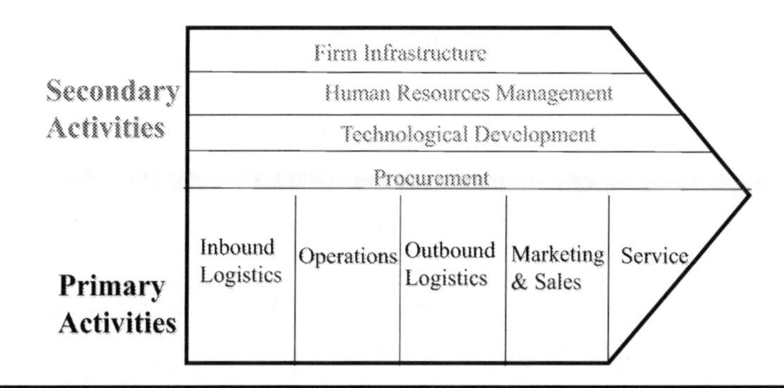

Figure A1.13 Porter's value chain.

The primary activities are made up of inputs, transformation processes and outputs. There are five primary activities, such as

- Inbound logistics
- Operations
- Outbound logistics
- Marketing and sales
- Services

The secondary activities are activities to support the primary activities. The four secondary activities are

- Infrastructure
- Human resources management
- Technological development
- Procurement

Inbound logistics involve relationships with suppliers and include all activities required to receive, store, list and group input materials (e.g. warehousing).

Operations are all activities to transform input into outputs (e.g. assembly).

Outbound logistics include all activities required to collect, store and distribute the output (e.g. distribution management).

Marketing and sales are all activities that convince customers to purchase company products (e.g. advertising).

Service includes all activities required to maintain products after sale (e.g. providing spare parts).

Infrastructure ties companies' various parts together (e.g. general management).

Human resources management includes the recruitment, education, compensating and, if necessary, dismissing of employees (e.g. training department).

Procurement is the acquisition of inputs or resources for the company (e.g. purchasing).

Technology development relates to the hardware, software, procedures and technical knowledge brought to transform inputs into outputs (e.g. R&D).

A1.13.2 Application

Value chain was originally developed to support general strategic analysis. According to Porter, an organisation may gain a competitive advantage by managing its value chain more effectively than its competitors.

As many of the value chain activities are independent, the value chain model can also be used to represent the linkages between activities. For example, a manufacturing organisation is likely to have important external linkages between its sales and marketing activities and the procurement activity of its customer.

The value chain model has been used as a versatile tool for top-level analysis, such as a visualisation of a company or a competitor, comparison of competitive strengths and as a potential match for mergers and acquisitions or strategic alliances.

With the emergence of e-business, the application of the value chain has been extended to the concept of a 'virtual' value chain.

A1.13.3 Final Thoughts

Like other examples of Porter's models, the value chain is an excellent tool for top-level concepts. For a detailed analysis leading to a solution, other appropriate tools and techniques are necessary. The value chain should not be confused with the detailed features of value stream mapping (see I6 in Chapter 11).

A1.14 Turner's Project Goals and Methods Matrix

A1.14.1 Description

The Goals and Methods Matrix was developed by Turner and Cochrane (1993). It identifies four types of projects according to the clarity of goals and

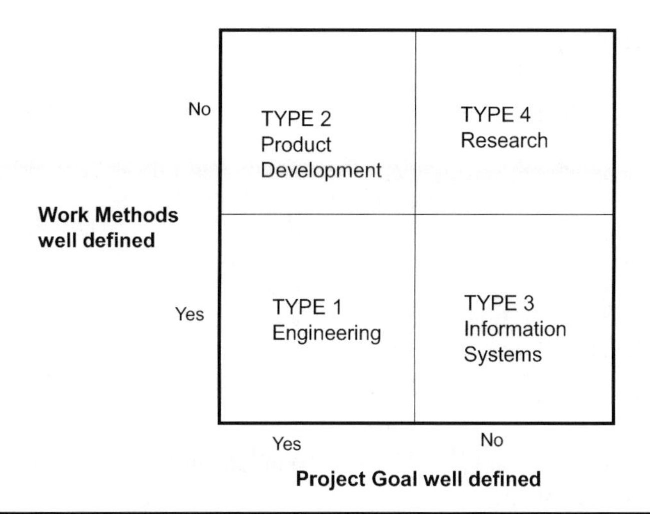

Figure A1.14 Method and Goal Matrix.

methods. As shown in Figure A1.14, the x-axis shows the Project Goals and the y-axis illustrates the Work Methods. The four types of projects are

Type 1: Projects with well-defined goals and well-defined methods of achieving them are in this category. Example: engineering projects
Type 2: Projects in this category are those with well-defined goals but poorly defined work methods. Example: product development projects
Type 3: Projects with poorly defined goals but well-defined methods are in Type 3. Example: information systems projects
Type 4: This type has both poorly defined goals and work methods. Example: research projects

A1.14.2 Application

It is useful to identify the type of project with the aid of this model. Type 1 projects with well-defined goals and methods have the best chance of success. On the other hand, Type 4 projects with only ill-defined goals and methods have the greatest likelihood of failure. The aim for all projects should be to define the goals first, followed by the work method. An organisational change project may start as a Type 4 project but can move gradually to a Type 1 project for a higher success rate.

Turner suggests that the identification of the project type can help to define the leadership requirement of a project manager. For example, the role of a project manager is as a conductor in Type 1, a coach in Type 2, a master craftsman in Type 3 and an eagle in Type 4.

A1.14.3 Final Thoughts

The Goals and Methods Matrix is a useful tool to identify the approach to a project start. However, it is important to emphasise that a clear definition of objectives and scope is essential for all types of projects.

A1.15 Wild's Taxonomy of Systems Structures

A1.15.1 Description

Ray Wild (2002) introduces the taxonomy of systems structures so that the high-level feasibility of an operating system can be examined with regard to its desirable objectives. Wild's taxonomy comprises seven structures (see Figure A1.15), four for manufacturing and supply operations and three for service operations.

Using simple systems terminology as shown in Figure A1.16, all operating systems may be seen to comprise input, process and outputs and either inventory or customer queue (except DOD).

Following Wild's labels of systems structures they are briefly described as follows:

SOS: Manufacture or supply from stock and the customer is served from an inventory of finished goods. Example: drugs manufacture

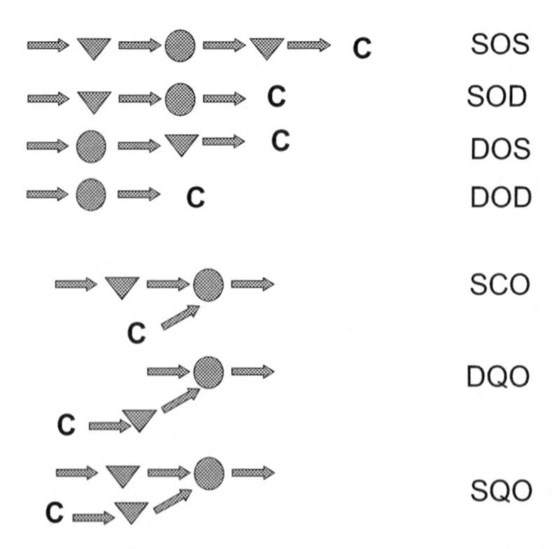

Figure A1.15 Wild's taxonomy.

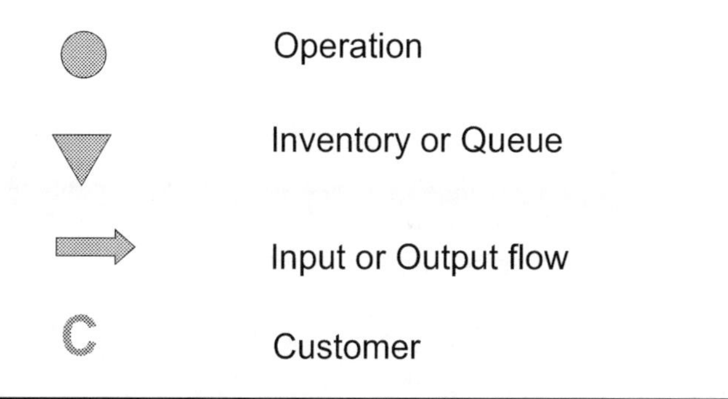

Figure A1.16 Symbols in Wild's taxonomy.

DOS: Manufacture or supply from source and the customer is served from a stock of finished goods. Example: off-shore oil production
SOD: Manufacture or supply from stock and goods are made to the customer's order. Example: retail shop
DOD: Manufacture and supply from source direct to customer and no input or output inventories are held. Example: builder
SCO: Service from stocked input resources but no customer queuing exists. Example: Fire service
DQO: Service from source without dedicated input resources but customer queuing exists. Example: Bus service
SQO: Service from stocked input resources and customer queuing exists. Example: checking in for Economy air travellers

A1.15.2 Application

These basic systems structures are useful to describe an operating system for both the total operation and its sub-systems. The system structure is then examined for the desired objectives of the operation. For example, if the objective is to deliver a high level of customer service, then it is more appropriate to have a structure with buffers offered by stocks (S) or queues (Q).

Customers exert some 'push' in the service models (SCO, DQO and SQO) while in manufacturing and supply models (SOS, DOS, SOD and DOD) customers 'pull' the system.

The choice of a system structure in both manufacturing and service will depend on its appropriateness, feasibility and desirability. If the structure is not desirable then it can be changed with the appropriate change in policy. For example, the original structure SQO for all customers can be changed to SCO for priority customers and SQO or DQO for non-priority customers.

A1.15.3 Final Thoughts

Wild's taxonomy works well for the overall operation. For a more complex operation it does not pinpoint the ineffectiveness of the system. Other tools, such as the flow diagram or process mapping, should be more appropriate to describe or analyse a multistage complex operation.

Appendix 2: Random Nominal Numbers

$\mu = 0,$				$\sigma = 1$			
	(1)	(2)	(3)	(4)	(5)	(6)	(7)
1	0.464	0.137	2.455	−0.323	−0.068	0.296	−0.288
2	0.060	−2.526	−0.531	−1.940	0.543	−1.558	0.187
3	1.486	−0.354	−0.634	0.697	0.926	1.375	0.785
4	1.022	−0.472	1.279	3.521	0.571	−1.851	0.194
5	1.394	−0.555	0.046	0.321	2.945	1.974	−0.258
6	0.906	−0.513	−0.525	0.595	0.881	−0.934	1.579
7	1.179	−1.055	0.007	0.769	0.971	0.712	1.090
8	−1.501	−0.488	−0.162	−0.136	1.033	0.203	0.448
9	−0.690	0.756	−1.618	−0.445	−0.511	−2.051	−0.457
10	1.372	0.225	0.378	0.761	0.181	−0.736	0.960
11	−0.482	1.677	−0.057	−1.229	−0.486	0.856	−0.491
12	−1.376	−0.150	1.356	−0.561	−0.256	0.212	0.219
13	−1.010	0.589	−0.918	1.598	0.065	0.415	−0.169
14	−0.005	−0.899	0.012	−0.725	1.147	−0.121	−0.096
15	1.393	−1.163	−0.911	1.231	−0.199	−0.246	1.239
16	−1.787	−0.261	1.237	1.046	−0.508	−1.630	−0.146

(*Continued*)

(Continued)

	$\mu = 0,$			$\sigma = 1$			
17	−0.105	−0.357	−1.384	0.360	−0.992	−0.116	−1.698
18	−1.339	1.827	−0.959	0.424	0.969	−1.141	−1.041
19	1.041	0.535	0.731	1.377	0.983	−1.330	1.620
20	0.279	−2.056	0.717	−0.873	−1.096	−1.396	1.047
21	−1.805	−2.008	−1.633	0.542	0.250	0.166	0.032
22	−1.186	1.180	1.114	0.882	1.265	−0.202	0.151
23	0.658	−1.141	1.151	−1.210	−0.927	0.425	0.290
24	−0.439	0.358	−1.939	0.891	−0.227	0.602	0.973
25	1.398	−0.230	0.385	−0.649	−0.577	0.237	−0.289
26	0.199	0.208	−1.083	−0.219	−0.291	1.221	1.119
27	0.159	0.272	−0.313	0.084	−2.828	−0.439	−0.792
28	2.273	0.606	0.606	−0.747	0.247	1.291	0.063
29	0.041	−0.307	0.121	0.790	−0.584	0.541	0.484
30	−1.132	−2.098	0.921	0.145	0.446	−2.661	1.045
31	0.768	0.079	−1.473	0.034	−2.127	0.665	0.084
32	0.375	−1.658	−0.851	0.234	−0.656	0.340	−0.086
33	−0.513	−0.344	0.210	−0.736	1.041	0.008	0.427
34	0.292	−0.521	1.266	−1.206	−0.899	0.110	−0.528
35	1.026	2.990	−0.574	−0.491	−1.114	1.297	−1.433
36	−1.334	1.278	−0.568	−0.109	−0.515	−0.566	2.923
37	−0.287	−0.144	−0.254	0.574	−0.451	−1.181	−1.190
38	0.161	−0.886	−0.921	−0.509	1.410	−0.518	0.192
39	−1.346	0.193	−1.202	0.394	−1.045	0.843	0.942
40	1.250	−0.199	−0.288	1.810	1.378	0.584	1.216

Appendix 3: Introduction to Basic Statistics

A3.1 Statistics

Statistics is the art and science of using numerical facts and figures. In Wikipedia statistics is defined as a mathematical science pertaining to the collection, analysis, interpretation or explanation, and presentation of data.

There are three kinds of statistics:

- - Descriptive statistics
- - Inferential statistics
- Causal modelling
- Descriptive statistics primarily deals with the description and interpretation of data by figures, graphs and charts.
- Inferential statistics is the science of making decisions in face of uncertainty by using techniques such as sampling and probability.
- Causal modelling is part of inferential statistics. It is aimed at advancing reasonable hypotheses about underlying causal relationships between the dependent and independent variables.
- In Six Sigma projects the most frequently used statistics is descriptive statistics; the most useful distribution is normal distribution.

A3.2 Descriptive Statistics

- Data distribution
- Normal distribution, other distributions
- Measures of central tendency

- Mean, median, mod
- Measures of dispersion
- Standard deviation, variance and range

A3.3 Normal Distribution

- Most data tend to follow the **normal distribution** or bell-shaped curve. One of the key properties of the normal distribution is the **relationship** between the **shape of the curve and the standard deviation.**
- **99.73% of the area of the normal distribution** is contained between −3 sigma and +3 sigma from the mean. Another way of expressing this is that 0.27% of the data falls outside 3 standard deviations for the mean.

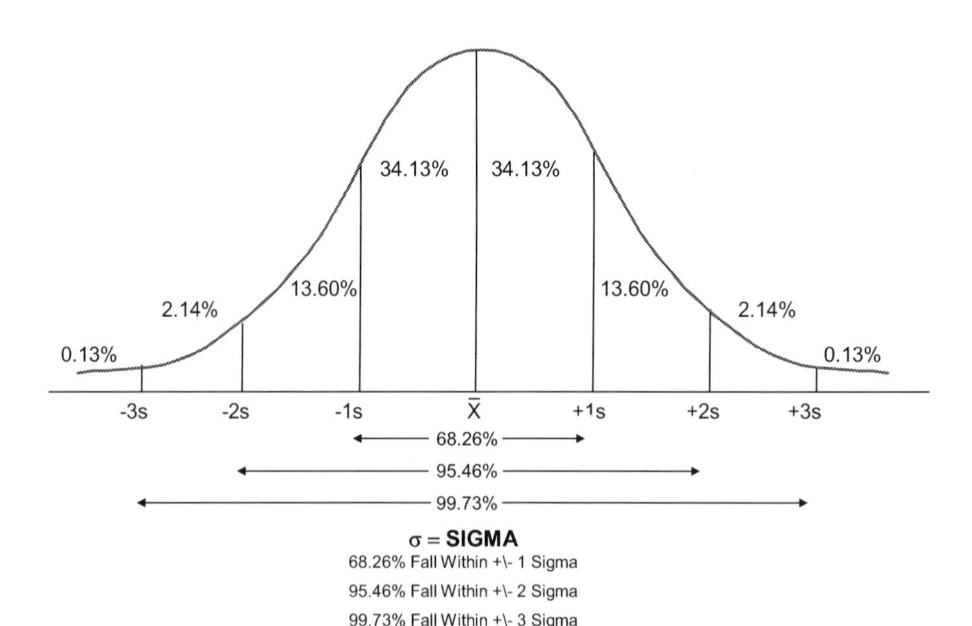

Population is the total number in the study (e.g. census), and a sample is a small subset of the population. Mean and standard deviation are expressed by Greek letters (μ, σ) in the population and by Roman letters (X, s)

A3.4 Measures of Central Tendency

Measures of central tendency are determined by mean, median and mode.

Mean is the arithmetic average of a set of data. It is a measure of central tendency, not a measure of variation. However, it is required to calculate some of the statistical measures of variation.

$$\text{Sample Mean } \overline{X} = \frac{\sum_{i=1}^{n} X_i}{n-1}$$

where
$\overline{x}$ = Sample mean
X_i = Data point i
n = Group sample size

Median is the central or middle data point, for example in the series 32, 33, 34, 34, 35, 37, 37, 39, 41 the median is 35

Mode is the highest data point, for example in the series 32, 33, 34, 34, 35, 37, 37, 39, 41 the mode is 41.

A3.5 Measure of Dispersion

The measure of dispersion is determined by Standard Deviation, Variance and Range and the shape of the distribution is measured by Skewness and Kurtosis.

Range is the simplest measure of dispersion. It defines the spread of the data and is the distance between the largest and the smallest values of a sample frequency distribution.

Variance is the average of the square values of the deviations from the mean.

Standard Deviation is defined as follows:

$$s = \sqrt{\sum_{i=1}^{n} \frac{(x_i - \overline{x})^2}{n-1}}$$

It is the square root of Variance and in a more intuitive definition, think of it as the 'average distance from each data point to the mean'.

A3.5.1 Skewness and Kurtosis

Skewness measures the departure from a symmetrical distribution. When the tail of the distribution stretches to the left (smaller values) it is negatively skewed and when it stretches to the right it is positively skewed.

Kurtosis is a measure of a distribution peakness or flatness. A curve is too peaked when the kurtosis exceeds +3 and is too flat when kurtosis is less than −3.

A3.6 Sources of Variation

There are two types or sources of variation as shown in the following table:

Type	Definition	Characteristics
Common cause	*No way to remove* *Influenced by one of the 6 Ms*	***Always present*** *Expected* *Normal* *Random*
Special cause	*Can be removed* *Influence of the 6Ms*	***Not always present*** *Unexpected* *Not normal* *Not random*

To distinguish between Common cause and Special causes variation, use display tools that study variation over time such as Run Charts and Control Charts.

A3.7 Process Capability

- Process Capability refers to the ability of a process to produce a product that meets given specification.
- It is measured by one or a combination of four indices:
  DPM Defects per million
  σ level Number of standard deviation from centre
  Cp Potential process index
  Cpk Process capability index

- σ level = minimum (USL – μ)/σ, (μ – LSL)/σ
  where USL = Upper specification level
  LSL = Lower specification level
  σ = Standard deviation
  μ = Mean
- Cpk = Process capability index
  = Minimum (USL – μ)/3σ, (μ – LSL)/3σ
  = σ level/3

$$C_p = \frac{|USL - LSL|}{6\sigma}$$

A3.8 Process Sigma

Process Sigma is an expression of process yield based on DPMO (defects per million opportunities) to define variation relative to customer specification. A higher-value Process Sigma is an indication of a lower variation relative to specification.

Historical data indicate a change of Sigma value by 1.5 when long- and short-term (actual) process sigma values are compared. The long-term values are higher by 1.5 (e.g. 6 Sigma is long-term Process Sigma for 3.4 DPMO).

A sample Process Sigma calculation is as follows:

Given 7 defects (D) for 100 units (N) processed with 2 defect opportunities (O) per unit.

Defects per opportunity, DPO = D/(N × O) = 7/(100 × 2) = 0.035
Yield = (1 – DPO) × 100 = 96.5%
From the yield conversion table (see Appendix 3),
Long-term Process Sigma = 3.3
Actual Process Sigma = 3.3 – 1.5

A3.9 Inferential Statistics

Inferential statistics helps us make judgments about the population from a sample. A sample is a small subset of the total number in a population. Sample statistics are summary values of sample and are calculated using all the values of the sample. Population parameters are summary values of the population but they are not known. That is why we use sample statistics to infer population parameters.

Sample size SS = $(DC \times V/DP)^2$

where
DC = Degree of confidence = Number of standard errors for the degree of confidence
V = Variability = Standard deviation of the population
DP = Desired precision = Acceptable difference between the sample estimate and the population value

A **null hypothesis** refers to a population parameter and not a sample statistic. Based on the sample data the researcher can accept the null hypothesis or accept the alternative hypothesis.

The ***t*-test** compares the actual difference between two means in relation to the variation in the data (expressed as the standard deviation of the difference between the means). It is applied when sample sizes are small enough (less than 30) and thus using an assumption of normality. This is a test of the null hypothesis that the means of two normally distributed populations are equal.

Analysis of variance (**ANOVA**) is used to assess the statistical differences between the means of two or more groups of population. ANOVA can also examine research problems that involve several independent variables.

The F-test asseses the differences between the group means when we use ANOVA.

F = Variance between groups/variance within groups. Larger F-ratios indicate significant differences between the groups and most likely the null hypothesis will be rejected.

A3.10 Causal Modeling

Many business objectives are concerned with the relationship between two or more variables. Variables are linked together if they exhibit **co-variation**, that is when one variable consistently changes relative to other variable. **Co-efficient of correlation** is used to assess this linkage. Large coefficients indicate high co-variation and a strong relationship and vice versa.

In causal modelling studies it is assumed that some variables cause or effect other variables, and the aim is to confirm or refute the hypothesized relationships and estimate their nature and robustness. The value of a model is to add clarity and understanding to an otherwise complex situation, to understand the relationship between variables and to enable a theory to be tested. The analysis of causal models is carried out by structural equation modelling (SEM) based techniques.

The advantage that SEM-based techniques have over the first-generation techniques (such as multiple regression analysis) is the greater flexibility that a researcher has for the interplay of theory and data. SEM enables the researcher to construct immeasurable variables measured by indictors (also called observed variables) as well as measurement errors. Hence it overcomes the limitations of first-generation techniques such as multiple regression.

A3.11 SPSS

The data analysis and statistical techniques presented in this appendix are all available in the popular software package SPSS (https://www.ibm.com/products/spss-statistics).

Appendix 4: Yield Conversion Table

These are estimates of log-term sigma values. Subtract 1.5 from these values to obtain actual sigma values.

Sigma	*DPMO*	*Yield%*	*Sigma*	*DPMO*	*Yield%*
6.0	3.4	99.99966	2.9	80757	91.9
5.9	5.4	99.99946	2.6	90801	90.3
5.8	8.5	99.99915	2.7	155070	88.5
5.7	13	99.99866	2.6	135666	86.4
5.6	21	99.9979	2.5	158655	84.1
5.5	32	99.9968	2.4	184060	81.6
5.4	48	99.9952	2.3	211855	78.8
5.3	72	99.9928	2.2	241964	75.8
5.2	108	99.9892	2.1	374253	72.6
5.2	159	99.984	2.0	308538	69.1
5.0	233	99.977	1.9	344578	65.5
4.9	337	99.966	1.8	382089	61.8
4.8	483	99.952	1.7	420740	57.9
4.7	687	99.931	1.6	460172	54.0
4.6	968	99.90	1.5	500000	50.0
4.5	1350	99.87	1.4	539828	46.0

(*Continued*)

(Continued)

Sigma	*DPMO*	*Yield%*	*Sigma*	*DPMO*	*Yield%*
4.4	1866	99.81	1.3	579260	42.1
4.3	2555	99.74	1.2	617911	38.2
4.2	3467	99.65	1.1	665422	34.5
4.1	661	99.53	1.0	691482	30.9
4.0	6210	99.38	0.9	725747	27.4
3.9	198	99.18	0.8	758036	24.2
3.8	10724	98.9	0.7	788145	21.2
3.7	13903	98.6	0.6	815940	18.4
3.6	17864	98.2	0.5	841345	15.9
3.5	22750	97.7	0.4	864334	13.6
3.4	28716	97.1	0.3	884930	11.5
3.3	35930	96.4	0.2	903199	9.7
3.2	44565	95.5	0.1	919243	8.1
3.1	54799	94.5			
3.0	66897	93.3			

Appendix 5: Carbon Footprint Fact Sheet

A5.1 Carbon Footprint

'A carbon footprint is the total greenhouse gas (GHG) emissions caused directly and indirectly by an individual, organisation, event or product'.[1] It is calculated by summing the emissions resulting from every stage of a product or service's lifetime (material production, manufacturing, use, and end-of-life). Throughout a product's lifetime, or life cycle, different GHGs may be emitted, such as carbon dioxide (CO_2), methane (CH_4) and nitrous oxide (N_2O), each with a greater or lesser ability to trap heat in the atmosphere. These differences are accounted for by calculating the global warming potential (GWP) of each gas in units of carbon dioxide equivalents (CO_2e), giving carbon footprints a single unit for easy comparison.

A5.2 Sources of Emissions

A5.2.1 Food

- Food accounts for 10–30% of a household's carbon footprint, typically a higher portion in lower-income households. Production accounts for 68% of food emissions, while transportation accounts for 5%.
- Food production emissions consist mainly of CO_2, N_2O and CH_4, which result primarily from agricultural practices.
- Meat products have larger carbon footprints per calorie than grain or vegetable products because of the inefficient transformation of plant

energy to animal energy, and due to the methane released from manure management and enteric fermentation in ruminants.[5]

- Ruminants such as cattle, sheep and goats produced 178 million metric tons CO_2e of enteric methane in the US in 2018.
- Eliminating the transport of food for 1 year could save the GHG equivalent of driving 1,000 miles, while shifting to a vegetarian meal 1 day a week could save the equivalent of driving 1,160 miles.
- A vegetarian diet greatly reduces an individual's carbon footprint, but switching to less carbon-intensive meats can have a major impact as well. For example, beef's GHG emissions per kilogram are 7.2 times greater than those of chicken.

Greenhouse Gases Contribution by Food Type in Average Diet Pounds of CO_2e per Serving

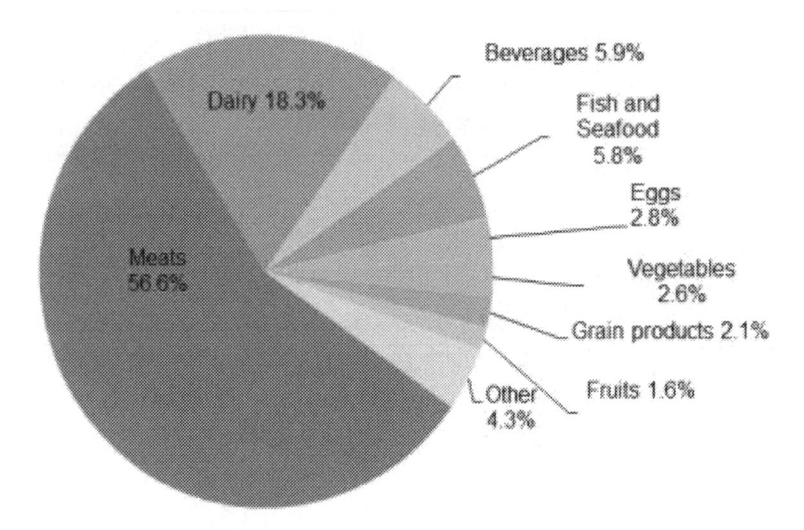

Source: University of Michigan (2021).

A5.3 Household Emissions

- For each kilowatt hour generated in the US, an average of 0.953 pounds of CO_2e is released at the power plant. Coal releases 2.2 pounds, petroleum releases 1.9 pounds, and natural gas releases 0.9 pounds. Nuclear, solar, wind and hydroelectric release no CO_2 when they produce electricity, but emissions are released during upstream production activities (e.g. solar cells, nuclear fuels, cement production).

- Residential electricity use in 2018 emitted 666.5 metric million tonnes CO_2e, 10% of the US total.
- Residential space heating and cooling are estimated to account for 44% of energy in US homes in 2020.
- Refrigerators are one of the largest users of household appliance energy; in 2015, an average of 720.5 lbs CO_2e per household was due to refrigeration.
- In the US each year, 26 metric million tonnes CO_2e are released from washing clothes. Switching to a cold water wash once per week, a household can reduce its GHG emissions by over 70 lbs annually.

A5.4 Personal Transportation

- US fuel economy (mpg) declined by 12% from 1987 to 2004, then improved by 30% from 2004 to 2018, reaching an average of 25.1 mpg in 2018. Annual per capita miles driven increased 9% since 1995 to 9,919 miles in 2018.
- Cars and light trucks emitted 1.1 billion metric tons CO_2e or 17% of the total US GHG emissions in 2018.
- Of the roughly 66,000 lbs CO_2e emitted over the lifetime of an internal combustion engine car (assuming 93,000 miles driven), 84% come from the use phase.
- Gasoline releases 19.6 pounds of CO_2 per gallon when burned, compared to 22.4 pounds per gallon for diesel. However, diesel has 11% more BTU per gallon, which improves its fuel economy.
- The average passenger car emits 0.78 pounds of CO_2 per mile driven.
- Automobile fuel economy can improve 7–14% by simply observing the speed limit. Every 5 mph increase in vehicle speed over 50 mph is equivalent to paying an extra $0.13 to $0.25 per gallon.
- Commercial aircraft GHG emissions vary according to aircraft type, trip length, occupancy rates, and passenger and cargo weight, and totalled 130.8 million metric tonnes CO_2e in 2018. In 2018, the average domestic commercial flight emitted 0.39 pounds of CO_2e per passenger-mile.
- Domestic air travel fuel efficiency (passenger miles/gallon) rose by 118% from 1990 to 2018, largely due to increased occupancy. Emissions

per domestic passenger-mile decreased 44% from 1990 to 2018, due to increased occupancy and fuel efficiency.

- In 2018, rail transportation emitted 42.9 million metric tonnes CO_2e, accounting for 2% of transportation emissions in the US.

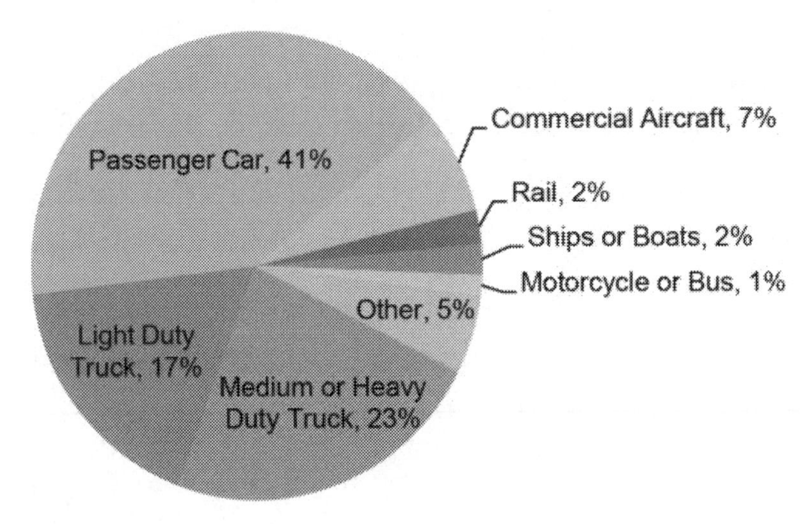

Source: University of Michigan (2021).

Glossary

ABC Analysis	It is based on a Pareto analysis grouping units usually according to the share of annual cost. Units having 80% annual cost are considered in the 'A' classification, units with the bottom 5% share are 'C' items and units with costs in between are in the 'B' category.
Activity Network Diagram	A network analysis technique to allow a team to find the most efficient path and realistic schedule of a project by graphically showing the completion time and sequence of each task.
Affinity Diagram	Used to generate a number of ideas by a team and then organise natural groupings among them to understand the essence of a problem.
AI	Artificial Intelligence demonstrated by technology as opposed to natural intelligence displayed by living creatures.
ASHP	Air source heat pump.
Bar Chart	Also known as Gantt Chart, a bar chart indicates scheduling activities. Horizontal bars show the various activities with the length of the bar proportional to the duration of a particular activity.
Benchmarking	Rating an organisation's products, processes and performances with other

organisations in the same or another business. The objective is to identify the gaps with competitors and the areas for improvement.

Best Practice This refers to any organisation that performs as well as or better than the competition in quality, timeliness, flexibility and innovation. Best practice should lead to world-class performance.

Big Data Analytics Big data analytics involve complex applications with predictive models and 'what if' analysis by analytics systems.

Black Belts Experts in Six Sigma methods and tools. Tools include statistical analysis. Black Belts are project leaders for Six Sigma initiatives; they also train other staff members in Six Sigma techniques.

Blockchain Type of database of information structured in table format to allow for easier searching for specific information.

Box Plot A box plot displays the five-number summary of a set of data. The five-number summary is the minimum, first quartile, median, third quartile and maximum. The spacings in each subsection of the box plot indicate the degree of dispersion and skewness of the data.

BPR Business Process Re-engineering has been described as a manifesto for revolution. The approach is similar to taking a clean piece of paper and starting all over by identifying what is really needed to make the mission of the organisation happen.

Brainstorming This is a free-wheeling group session for generating ideas. Typically a group meeting of about seven people will be presented with a problem. Each

	member will be encouraged to make suggestions without fear of criticism. One suggestion will lead to another. All suggestions, no matter how seemingly fanciful, are recorded and subsequently analysed.
Carbon Dioxide Equivalent	Also expressed as CO_2e, Carbon Dioxide Equivalent is a metric measure of greenhouse gases as carbon dioxide tonnes to compare the global warming potential from various greenhouse gases.
Carbon Footprint	A carbon footprint is the total set of greenhouse gases (GHG) emissions caused by an organisation, event or product. For simplicity of reporting, it is often expressed in terms of the amount of carbon dioxide or its equivalent of other GHGs emitted. A carbon footprint can be measured by undertaking a GHG emissions assessment of an individual, organisation or a country.
Cause and Effect Diagram	The cause and effect, fish-bone or Ishikawa diagram was developed by Kaoru Ishikawa. The premise is that generally when a problem occurs the effect is very obvious, and the temptation is to treat the effect. With the Ishikawa approach the causes of the effect are sought. Once the cause is known and eliminated the effect will not be seen again.
CCS	Carbon Capture and Storage is another version of Direct Air Capture.
CGIAR	Consultative Group for International Agricultural Research.
Check Sheet	Method of systematically recording data from historical sources or observations

	as they happen. The patterns and trends can be clearly detected and shown.
CIM	Computer-Integrated Manufacturing.
Circular Economy	A circular economy aims for a world without waste by recycling and reusing. It is also an important tenet of Green Six Sigma.
Cloud Computing	On-demand availability of computer system resources, especially data storage and computing power.
CO_2 PPM	Measurement of CO_2 concentration expressed as parts per million per unit volume of water.
Continuous Improvement	Always looking for ways to improve a process or a product, but not necessarily making radical step changes. If the basic idea is sound then building on it will improve quality. In Japan this is known as kaizen.
Control Chart	Tool in Statistical Process Control to monitor the number of defects found in a product or a process over time and study the variation and its source.
COP	Conference of Parties. These are organised every year to review the progress of climate change initiatives.
COPQ	Cost of Poor Quality. The cost of poor quality is made up of costs arising from internal failures, external failures, appraisal, prevention and lost opportunity costs. In other words all the costs that arise from non-conformance to a standard. Chapter 3 discusses COPQ in some detail.
CPFR	Collaborative Planning Forecasting and Replenishment. It aims to enhance supply chain integration of stakeholders through joint visibility and replenishment of products.

CRM Customer Relationship Management. It is the development of the database and strategies necessary to have the maximum client relationships in terms of quality, cost, reliability and responsiveness.

CSR Corporate Social Responsibility.

CTQ In Six Sigma, CTQ stands for Critical to Quality. This simply means the identification of factors that are critical for the achievement of a level of quality.

Cycle Time Elapsed time between two successive operations or the time required to complete an operation.

DAC Direct Air Capture is a device that absorbs CO_2 and the output is stored.

DFSS Design for Six Sigma. See Chapter 9 for detailed discussion. The steps are Define, Measure, Analyse, Design and Validate.

Digital Six Sigma (DSS) Reported to be a revitalization of Six Sigma methodology by Motorola through the following improvements:

- Leveraging new digital tools to drive project success
- Digitizing business processes to 'permanently' enforce optimal process compliance
- Tracking vital processes through digital databases
- 'Permanently' and proactively eliminating sources of variation that cause defects

DMADV Define, Measure, Analyse, Design and Verify. It is a modified version of DMAIC for Research and Development.

DMAIC Cycle Define, Measure, Analyse, Improve and Control. It is an integral part of a Six Sigma or Lean Six Sigma initiative.

DMAIC Lite This DMAIC process for SMEs.

DMAICS Cycle of DMAIC with an addition for Sustainability. It is an integral part of a Green Six Sigma initiative.

DOE Design of Experiments. The process of examining options in the design of a product or service. Controlled changes of input factors are made and the resulting changes to outputs noted. Losses from poor design include not only direct loss to the company from reworking, and scrap, but includes for the user downtime due to equipment failure, poor performance and unreliability.

DPMO Defects per million opportunities. This is the basic measure of Six Sigma. It is the number of defects per unit divided by the number of opportunities for defects multiplied by 1,000,000. This number can be converted into a Sigma value. For example Six Sigma = 3.4 per million opportunities.

E-business Electronic business is more than the transfer of information using information technology. E-business is the complex mix of processes, applications and organisational structures.

EDI Electronic Data Exchange.

EFQM The European Foundation for Quality Management. It is derived from the American Malcom Baldridge Quality award. It is an award for organisations that achieve world-class performance as judged by independent auditors against a checklist.

Electrofuel Electrofuel is net zero carbon-free fuel produced by combining the hydrogen in water and the carbon from CO_2.

EPA Environmental Protection Agency in US.

ERP Enterprise Resource Planning, is the extension of MRP II systems to the management of complete business functions including Finance and Human Resources.

EV Electric Vehicle.

Failure Mode and Effect Analysis FMEA was developed in the aerospace and defence industries. It is a systematic and analytical quality planning tool for identifying for new products or services, at the design stage, what could go wrong during manufactures, or when in use by the customer. It is also a tool for risk analysis.

First Pass Yield FPY, also known as RTY, is the ratio of the number of completely defects free without any kind of rework during the process units at the end of a process and the total number of units at the start of a process. The theoretical throughput rate is often regarded as the number of units at the start of the process. RTY/FPY is used as a key performance indicator to measure overall process effectiveness.

Fish-bone diagram The fish-bone, Ishikawa, or cause and effect diagram was developed by Kaoru Ishikawa. The premise is that generally when a problem occurs the effect is very obvious, and the temptation is to treat the effect. With the Ishikawa approach the causes of the effect are sought. Once the cause is known and eliminated the effect will not be seen again.

FIT SIGMA See FIT SIGMA incorporates all the advantages and tools of TQM, Six Sigma and Lean Sigma. The aim is to get an

	organisation healthy (fit) by using appropriate tools for the size and nature of the business (fitness for purpose) and to sustain a level of fitness; also see TQM, Six Sigma and Lean Sigma. FIT SIGMA is a holistic approach.
Flow Process Chart	A flow process chart sets out the sequence of the flow of a product or a procedure by recording all the activities in a process. The chart can be used to identify steps in the process, value-adding activities and non-value-adding activities.
Gantt Chart	See Bar Chart.
GCP	The Global Carbon Projects established 2001 as a global research project on carbon emissions.
GEF	The Global Environmental Faculty set up in 1992 at the Rio Earth Summit.
Green Belts	Staff trained to be Six Sigma project leaders; they work under the guidance of Black Belts. See Black Belts.
Greenhouse Gases	Polluting gases in the environment causing global warming. These are carbon dioxide, methane, nitrous oxide, water vapour and ozone.
Green Premium	Difference between the cost of a green solution and the existing option.
Green Six Sigma	Adaptation of Lean Six Sigma and FIT SIGMA for climate change initiatives.
GSHP	Ground source heat pump.
Histogram	A histogram is a descriptive and easy-to-understand chart of the frequency of occurrences. It is a vertical bar chart with the height of each bar representing the frequency of an occurrence.
ICAP	International Carbon Action Partnership.
IEA	International Energy Agency.
IoT	Internet of things to describe a network

of physical objects connected by sensors with Internet.

IPCC Intergovernmental Panel on Climate Change is the most effective body of international scientists to monitor climate change activities.

IPMVP International Performance Measurement and Verification Protocol.

IPO (Input-Process-Output) Diagram All operations or processes have inputs and outputs. The process is the conversion of inputs into outputs. Analysis of inputs should be made to determine factors that influence the process, for example input materials from suppliers meeting specification, delivery on time and so on.

Ishikawa The Ishikawa, or fish-bone, or cause and effect diagram was developed by Kaoru Ishikawa. The premise is that generally when a problem occurs the effect is very obvious, and the temptation is to treat the effect. With the Ishikawa approach the causes of the effect are sought. Once the cause is known and eliminated the effect will not be seen again.

ISO 9000 To gain ISO 9000 accreditation an organisation has to demonstrate to an accredited auditor that they have a well-documented standard and consistent process in place which achieves a defined level of quality or performance.

JIT Just in Time. It was initially a manufacturing approach where materials are ordered to arrive just when required in the process, no output or buffer stocks are held, and the finished product is delivered direct to the customer. Lean Six Sigma incorporates the principals of JIT.

Kaizen Kaizen is a Japanese word derived from a philosophy of gradual day-by-day betterment of life and spiritual enlightenment. This approach has been adopted in industry and means gradual and unending improvement in efficiency and/or customer satisfaction. The philosophy is doing little things better so as to achieve a long-term objective.

Kanban Kanban is the Japanese word for 'card'. The basic kanban system is to use cards to trigger movements of materials between operations in production so that a customer order flows through the system.

KPIs Key Performance Indicators include measurement of performance such as asset utilisation, customer satisfaction, cycle time from order to delivery, inventory turnover, operations costs, productivity, and financial results (return on assets and return on investment).

Kyoto Protocol The Kyoto Protocol was adopted in Kyoto in December 1997 by 192 countries. It extends the 1992 UNFCCC objective to reduce greenhouse gas emissions based on the scientific consensus on global warming.

Lean Six Sigma Lean Six Sigma was initially a manufacturing approach where materials are ordered to arrive just when required in the process, no output or buffer stocks are held, and the finished product is delivered direct to the customer. Lean Six Sigma incorporates the principals of Six Sigma and is related to the supply chain from supplier and supplier's supplier, through the process to the customer and the customer's customer; also see Just in Time (JIT).

MFA Material Flow Analysis is an analytical method to quantify and balance the flow of materials in a well-defined system.

Minitab Statistical package developed at the Pennsylvania State University and distributed by Minitab Inc. Today, Minitab is often used in conjunction with the implementation of Six Sigma and other statistics-based process improvement methods.

Mistake Proofing Refers to making each step of production mistake free. This is also known as poka-yoke. Poka-yoke was developed by Shingo (also see SMED) and has two main steps: (1) preventing the occurrence of a defect and (2) detecting the defect.

Monte Carlo Technique Simulation process. It uses random numbers as an approach to model the waiting times and queue lengths and also to examine the overall uncertainty in projects.

MRP (II) Manufacturing resource planning is an integrated computer-based procedure for dealing with all of the planning and scheduling activities for manufacturing, and includes procedures for stock re-order, purchasing, inventory records, cost accounting and plant maintenance.

Muda Muda is Japanese for waste or non-value-adding. The seven activities that are considered are Excess production, Waiting, Conveyance, Motion, Process, Inventory and Defects.

NDCs Nationally Determined Contributions are plans and targets for the participating countries of the Paris Climate Agreement.

NGO Non-governmental Organisation.

Nominal Group Technique Brainstorming tool for a team to come to a consensus of the relative importance of ideas by completing importance rankings into a team's final priorities.

Normal Distribution Graph of the frequency of occurrence of a random variable. The distribution is continuous, symmetrical, bell shaped and the two tails extend indefinitely; also known as Gaussian Distribution.

OEE Overall Equipment Effectiveness is the real output of a machine. It is given by the ratio of the good output and the maximum output of the machine for the time it is planned to operate.

Pareto Wilfredo Pareto was a 19th-century Italian economist who observed that 80% of the wealth was held by 20% of the population. The same phenomenon can often be found in quality problems. This is also known as the 80/20 rule.

Paris Climate Agreement The Paris Climate Agreement (also known as the Paris Accord) is an international treaty on the climate crisis. Its stated aim is to radically reduce global carbon emissions and restrict the rise in the earth's temperature to less than 2°C. This treaty was signed by all 189 participating countries in Paris in December 2015 at COP 21.

PDCA The Plan-Do-Check-Act cycle was developed by WE Deming. It refers to Planning the change and setting standards, Doing what is necessary to make the change happen, Checking that what is happening is what was intended (standards are being met), and Acting by taking action to correct back to the

	standard.
Performance Charts	Upper control and lower control limits (UCL/LCL) are used to show variations from specification. Within the control limits performance will be deemed to be acceptable. The aim should be over time to reduce the control limits.
PESTLE	Political, Economic, Social, Technical, Legal and Environmental, is an analytical tool for assessing the impact of external contexts on a project or a major operation and also the impact of a project on its external contexts. There are several possible contexts including - Political - Economic - Social - Technical - Environmental
PET	Polyethylene terephthalate. It is a form of polyester which is recyclable.
PEV	Plug-in Electric Vehicle.
PM2.5	Pollution for particulate matter.
Poka-Yoke	Refers to making each step of production mistake free. This is known as mistake proofing. Poka-yoke was developed by Shingo, also see SMED, and has two main steps: (1) preventing the occurrence of a defect and (2) detecting the defect.
Process Capability	Process Capability is the statistically measured inherent reproducibility of the output (product) turned out by a process that is within the design specifications of the product.
Process Mapping	Process Mapping is a tool to represent a process by a diagram containing a series of linked tasks or activities which produce an output.

Project A project is a unique item of work for which there is a financial budget and a defined schedule.

Project Charter A Project Charter is a working document for defining the terms of reference of each Six Sigma project. The charter can make a successful project by specifying necessary resources and boundaries that will in turn ensure success.

Project Management Planning, scheduling, budgeting and control of a project using an integrated team of workers and specialists.

PV System Photovoltaic System in solar energy.

QFD Quality Function Deployment is a systematic approach of determining customer needs and designing the product or service so that it meets the customer's needs the first time and every time.

Quality Circles Quality circles are teams of staff who are volunteers. The team selects issues or areas to investigate for improvement. To work properly teams have to be trained, first in how to work as a team (group dynamics) and second in problem-solving techniques.

Quality Project Teams Quality Project Teams are a top-down approach to solving a quality problem. Management determines a problem area and selects a team to solve the problem. The advantage over a Quality Circle is that this is a focused approach, but the disadvantage might be that members are conscripted rather than volunteers.

Regression Analysis Tool to establish the 'best-fit' linear relationship between two variables. The knowledge provided by the Scatter

	Diagram is enhanced with the use of regression.
Rolled Throughput Yield	RTY, aka First Pass Yield (FPY), is the ratio of the number of completely defects free without any kind of rework during the process units at the end of a process and the total number of units at the start of a process. The theoretical throughput rate is often regarded as the number of units at the start of the process. RTY/FPY is used as a key performance indicator to measure overall process effectiveness.
RU/CS	Resource Utilisation and Customer Service. Analysis is a simple tool to establish the relative importance of the key parameters of both Resource Utilisation and Customer Service and to identify their conflicts.
Run Chart	Graphical tool to study observed data for trends or patterns over a specific period of time.
5 S's	These represent a set of Japanese words for excellent housekeeping (Sein—Sort, Seiton—Set in place, Seiso—Shine, Seiketso—Standardise, and Sitsuke—Sustain).
SaaS	Software as a Service is a software licensing model in which software is licensed on a subscription or rental basis.
S&OP	Sales and operations planning is derived from MRP and includes new product planning. Demand planning, supply review, to provide weekly and daily manufacturing schedules, and financial information.
Scatter Diagram	These diagrams are used to examine the relationship between two variables.

	Changes are made to each, and the results of changes are plotted on a graph to determine cause and effect.
SCM	Supply Chain Management.
Sigma	Sigma is the sign used for standard deviation from the arithmetic mean. If a normal distribution curve exists, one sigma represents one standard deviation either side of the mean and accounts for 68.27% of the population. This is more fully explained in Chapter 3.
SIPOC	Supplier, input, process, output and customer: a high-level map of a process to view how a company goes about satisfying a particular customer requirement in the overall supply chain.
Six Sigma	Six Sigma is a quality system which in effect aims for zero defects. Six Sigma in statistical terms means six deviations from the arithmetic mean. This equates to 99.99966% of the total population, or 3.4 defects per million opportunities.
SME	Small and medium-sized enterprise with staff headcount less than 250.
SMED	Single Minute Exchange of Dies. This was developed for the Japanese automobile industry by Shigeo Shingo in the 1980s and involves the reduction of changeover of production by intensive work study to determine in-process and out-process activities and then systematically improving the planning, tooling and operations of the changeover process.

SPC	Statistical Process Control (SPC) uses statistical sampling to determine if the outputs of a stage or stages of a process are conforming to a standard. Upper and lower limits are set, and sampling is used to determine if the process is operating within the defined limits.
SPSS	Statistical Process for the Social Sciences is a software used for interactive statistical analysis.
SAF	Sustainable Aviation Fuel.
SUD	Sustainable Drainage System.
The Seven Wastes	Muda is the Japanese word for waste or non-value-adding. The seven activities that are considered are Excess production, Waiting, Conveyance, Motion, Process, Inventory and Defects; also see Muda.
SWOT	Strengths, Weaknesses, Opportunities and Threats: a tool for analysing an organisation's competitive position in relation to its competitors.
Tolerance charts	Upper control and lower control limits (UCL/LCL) are used to show variations from specification. Within the control limits performance will be deemed to be acceptable.
TPM	Total Productive Maintenance requires factory management to improve asset utilisation by the systematic study and elimination of major obstacles—known as the 'six big losses'—to efficiency. The 'six big losses' in manufacturing are breakdown, set up and adjustment, minor stoppages, reduced speed, quality defects, and start-up and shutdown.
TPS	Toyota Production System.

TQM — Total Quality Management is not a system it is a philosophy embracing the total culture of an organisation. TQM goes far beyond conformance to a standard, it requires a culture where every member of the organisation believes that not a single day should go by without the organisation in some way improving its efficiency and/or improving customer satisfaction.

UCL/LCL — Upper control and lower control limits are used to show variations from specification. Within the control limits, performance will be deemed to be acceptable. The aim should be over time to reduce the control limits.

UNEP — United Nations Environment Programme.

UNFCCC — United Nations Framework Convention of Climate Change.

Value Analysis — Very often a practice in purchasing, is the evaluation of the expected performance of a product relative to its price.

Value Chain — According to Michael Porter the competitive advantage of a company can be assessed only by seeing the company as a total system. This 'total system' comprises both primary and secondary activities; also known as Porter's Value Chain.

Value Stream Mapping — VSM is a visual illustration of all activities required to bring a product through the main flow, from raw material to the stage of reaching the customer.

World Class — World Class is the term used to describe any organisation that is making rapid and continuous improvement in performance and who is considered to be using 'best practice' to achieve world-class standards.

Zero defects

Philip Crosby made this term popular in the late 1970s. The approach is right thing, right time, right place and every time. The assumption is that it is cheaper to do things right the first time.

References

Altshuller, G.S. (1994), *Creativity as an Exact Science: The Theory of the Solution of Inventive Problems*, Translated by Williams, A., Gordon and Breach Science Publishers, Philadelphia.

Amar, K. and Davies, D. (2007), 'Evaluating Six Sigma in the Indonesian SME Context', *Proceedings of the 5th ANZAM and 1st Asian Pacific Symposium*, ANZAM, pp. 1–12.

Ansoff, I. (1987), *Corporate Strategy*, Penguin Books, London.

Apak, S., Atay, E. and Tuncer, G. (2017), 'Renewable Hydrogen Energy and Energy Efficiency in Turkey in the 21st Century', *International Journal of Hydrogen Energy*, Vol. 42, No. 4, pp. 2446–2452.

Asif, M., deBruijn, E.J., Douglas, A. and Fisscher, A.M. (2009a), 'Why Quality Management Programs Fail', *International Journal Quality and Reliability Management*, Vol. 26, No. 8, pp. 778–794.

Asif, M., de Bruijn, E.J., Fisscher, O.A.M., Searcy, C. and Steenhuis, H.J. (2009b), 'Process Embedded Design at Integrated Management Systems', *International Journal of Quality and Reliability Management*, Vol. 26, No. 3, pp. 261–282.

Barfod, N. (2007), 'Man behøver ikke være stor for at være lean (You Don't Have to be Big to be Lean)', *Børsen*, Vol. 5, No. 15, pp. 13–14.

Barth, H. and Melin, M. (2018), 'A Green Lean Approach to Global Competition and Climate Change in the Agricultural Sector—A Swedish Case Study', *Journal of Cleaner Production*, Vol. 204, pp. 183–192.

Basu, R. (2001), 'New Criteria of Performance Management', *Measuring Business Excellence*, Vol. 5, No. 4.

Basu, R. (2002), *Measuring e-Business in the Pharmaceutical Sector; A Strategic Assessment of New Business Opportunities*, Reuters, London.

Basu, R. (2003), *Measuring eBusiness in the Pharmaceutical Sector*, Reuters Business Insight, London.

Basu, R. (2004), 'Six Sigma in Operational Excellence': Role of Tools and Techniques', *International Journal of Six Sigma and Competitive Advantage*, Vol. 1, No. 1, pp. 44–64.

Basu, R. (2009), *Implementing Six Sigma and Lean*, Elsevier, Oxford.

Basu, R. (2011), *FIT SIGMA: A Lean Approach to Building Sustainable Quality Beyond Six Sigma*, John Wiley & Sons, Chichester, UK.

Basu, R. (2012), *Managing Quality in Projects*, Gower Publications, Routledge, Abingdon, UK.

Basu, R. and Wright, J.N. (1997), *Total Manufacturing Solutions*, Butterworth and Heinemann, Oxford, UK.

Basu, R. and Wright, J.N. (2003), *Quality Beyond Six Sigma*, Butterworth and Heinemann, Oxford, UK.

Basu, R. and Wright, J.N. (2017), *Managing Global Supply Chains*, Routledge, Abingdon, UK.

Belbin, R.M. (1985), *Management Teams: Why They Succeed or Fail*, Heinemann, London.

Belhadi, A., Kamble, S., Cherafi, Z.A. and Touriki, F.E. (2020), 'The Integrated Effect of Big Data Analytics, Lean Six Sigma and Green Manufacturing on the Environmental Performance of Manufacturing Companies: The Case of North Africa', *Journal of Cleaner Production*, Vol. 252.

Berr (2008), 'Annual Report and Accounts 2008/2009', The Department of Business, Enterprise and Regulatory Reform, Gov.UK.

Black, J.T. and Phillips, D.T. (2010), 'The Lean to Green Evolution', *Industrial Engineer*, Vol. 42, No. 7, pp. 46–51.

BP. (2020), *BP Statistical Review of World Energy*, BP Plc, St James's Square, London.

Brassard, M. and Ritter, D. (1994), *Memory Jogger*, Methuen Publishing, Methuen, MA.

Brassard, M., Finn, L., Ginn, D. and Ritter, D. (2002), *The Six Sigma Memory Jogger*, GOAL/QPC, New Hampshire.

Breyfogle III, F.W. (2008), 'Better Fostering Innovation: 9 Steps That Improve Lean Six Sigma', *Business Performance Management*, September 2008.

British Quality Foundations (BQF). (1999), *The European Foundation of Quality Management (EFQM) Model*, British Quality Foundation, London.

Bromwich, M. and Bhimani, A. (1989), *Management Accounting: Research Studies*, CIMA, London.

BSI. (2009), *BS EN ISO 9004:2009. Managing for Sustained Success of an Organisation. A Quality Management Approach*, British Standards Institution, London.

Buzan, T. (1995), *The Mind Map Book*, Penguin Publishing Group, London.

Camp, RC (1989), 'Benchmarking, the Search for Industry Best Practices the Lead to Superior Performance, ASQC Industry Press, Milwaukee.

Carlsen, S (2009), 'The role of management in the successful implementation of Lean Six Sigma in a Danish manufacturing company' Unpublished MBA dissertation, Henley Business School, UK.

Carnall, C. (1999), *Managing Change in Organisations*, Prentice Hall, London.

Carrington, D. (2021), 'A Billion Children at "Extreme Risk" from Climate Impacts—Unicef', *The Guardian*, August 20, 2021.

Chaurasia, B., Garg, D. and Agarwal, A. (2019), 'Lean Six Sigma Approach: A Strategy to Enhance Performance of First Through Time and Scrap Reduction in Automotive Industry', *International Journal Business Excellence*, Vol. 17, No. 1, pp. 42–57.

Cherrafi, A., Elfezazi, S. and Govindan, K. (2017), 'A Framework for the Integration of Green and Lean Six Sigma for Superior Sustainability Performance', *International Journal of Production Research*, Vol. 55, No. 15, pp. 4481–4515.

Choudhary, S. (2003), *Design for Six Sigma*, Prentice Hall, London.

Choudhary, S., Nayak, R., Dora, M., Mishra, N. and Ghadge, A. (2019), 'An Integrated Lean and Green Approach for Improving Sustainability Performance: A Case Study of a Packaging Manufacturing SME in the U.K', *Production Planning & Control*, Vol. 30, No. 5–6, pp. 353–368.

Churchman, C.W., Ackoff, R.L. and Arnoff, E.L. (1968), *Introduction to Operations Research*, John Wiley & Sons, New York.

Codling, S (1995), 'Best Practice Benchmarking', Gower Press, Hants.

Corporate Leadership Council. (2005), 'Leadership Competencies in a Lean Environment', *Corporate Executive Board*, August 2005, Catalog no. CLC13TKO7X, London.

Cranfield, D.J. and Taylor, J. (2008), 'Knowledge Management and Higher Education: A UK Case Study', *Electronic Journal of Knowledge Management*, Vol. 6, No. 2, pp. 85–100.

Crosby, P. (1979), 'Quality Without Tears', In Turner, J.R. (ed.), *The Handbook of Project Based Management: Improving the Processes for Achieving Strategic Objectives*, McGraw-Hill, London.

Dale, B.G. and Cooper, R. (1993), 'Improvement framework', *The TQM Magazine*, Vo. 5., No. 1, pp. 23–26.

Dale, B.G. (1999), *Managing Quality*, Blackwell Publishers, Oxford, Chapter 18, pp. 366–389.

Dale, B.G. (2000), *Managing Quality*, Blackwell Publishing, Malden, MA.

Dale, B.G. and McQuater, R. (1998), *Managing Business Improvement and Quality*, Blackwell Publishers, Oxford.

Davies, C.E. (2005), 'South West Region: Six Sigma for SMEs', *Management Services*, Summer 2005, pp. 6–7.

de Bono, E. (2016), *Six Thinking Hats*, Penguin Books, London.

De Carlo, N. (2007), *Lean Six Sigma*, Alpha Books, New York.

Deming, W.E. (1986), *Out of the Crisis*, MIT Centre for Advanced Research, Cambridge, Boston.

De Oliviera, J.A.P. (2009), 'The Implementation of Climate Change Related Policies at Subnational Level: An Analysis of Three Countries', *Habitat International*, Vol. 33, No. 3, pp. 253–259.

Duhon, B. (1998), 'It's All in Our Heads', *Inform*, Vol. 12, pp. 8–13.

Earth Systems Science Data. (2020), *A Comprehensive and Systematic Dataset for Global, Regional and National Greenhouse Gas Emissions by Sector* (Edited by Elger, K., Carlson, D., Kemp, J. and Peng, G.), Copernicus Publications, Gottingen, Germany.

Easton, G. and Jarrell, S. (1998), 'The Effects of Total Quality Management on Corporate Performance', *Journal of Business*, Vol. 71, No. 3, pp. 253–307.

Eckes, G. (2001), *Making Six Sigma Last*, John Wiley & Sons, Chichester, England.

The Economic Times. (2009), 'Vestas Saves $10 Million Through Lean Six Sigma Initiative', *The Economic Times*, Mumbai, August 18, 2009.

Elkington, J. (1994), 'Towards the Sustainable Corporation: Win-Win-Win Business Strategies for Sustainable Development', *California Management Review*, Vol. 36, No. 2, pp. 90–100.

Erwin, J. and Douglas, P.C. (2000), 'It Is Difficult to Change Company Culture', *Journal of Supervision*, Vol. 51, No. 11.

European Commission (2003), *SME Definition – User Guide*, European Commission, Brussels.

European Foundation of Quality Management. (2003), *The EFQM Excellence Model*, European Foundation of Quality Management, Brussels.

Feigenbaum, A.V. (1956), 'Total Quality Control', *Harvard Business Review*, Vol. 36, No. 6, November–December.

Feigenbaum, A.V. (1961), *Total Quality Control*, McGraw-Hill, New York.

Fiedler, T., Pitman, A.J., Mackenzie, K., et al. (2021), 'Business Risk and the Emergence of Climate Analytics', *Nature Climate Change*, Vol. 11, pp. 87–94.

Finkelstein, S. (2006), 'Why Smart Executives Fail: Four Case Histories of How People Learn the Wrong Lesson from History', *Business History*, Vol. 48, No. 2, pp. 153–170.

Gabor, A (2000), 'He made America think about quality', Fortune, Vol. 42, No. 10.

Gates, B. (2021), *How to Avoid a Climate Disaster*, Penguin Random House, New York.

Global Carbon Project. (2019), *Global Carbon Budget 2021* (Edited by Jittrapirom, P. and Canadell, P.), Global Carbon Project, Canberra.

George, G., Merrill, R.K. and Schillebeeckx, J.D. (2020), 'Digital Sustainability and Entrepreneurship: How Digital Innovations Are Helping Tackle Climate Change and Sustainable Development', *Sage Publications: Entrepreneurship Theory and Practice*, Vol. 45, No. 4, pp. 1–28.

George, M.L. (2002), *Lean Six Sigma: Combining Six Sigma Quality with Lean Production Speed*, McGraw-Hill, New York.

Gernaat, D.E.H.J., de Boer, H.S., Daioglou, V., et al. (2021), 'Climate Change Impacts on Renewable Energy Supply', *Nature Climate Change*, Vol. 11, pp. 119–125.

Gijo, E.V. and Sarkar, A. (2013), 'Application of Six Sigma to Improve the Quality of the Road for Wind Turbine Installation', *The TQM Journal*, Vol. 25, No. 3, pp. 244–258.

Global Renewables. (2020), *Global Renewable Outlook: Energy Transformation 2020*, IRENA, Abu Dhabi.

Goel, S. and Chen, V. (2008), 'Integrating the Global Enterprise Using Six Sigma: Business Process Reengineering at General Electric Wind Energy', *International Journal of Production Economics*, Vol. 113, No. 2, pp. 914–927.

Goldratt, E.M. (1999), *The Theory of Constraints*, North River Press, New York.

Gravin, D. (1984), 'What Does Product Quality Really Mean?', *Sloan Management Review*, Vol. 25, No. 2.

The Guardian. (2021), 'UK Weather: Met Office Issues Strong Wind Warnings', *The Guardian*, March 10, 2021.

Hammer, M. and Champy, J. (1993), *Re-Engineering the Corporation*, Nicholas Brealey Publishing, London.

Hartman, E. (1991), 'How to Install TPM in Your Plant', *8th International Maintenance Conference*, Dallas, November 12–14.

Hauser, J.R. and Clausing, D. (1988), 'The House of Quality', *Harvard Business Review*, May–June, 1988.

ICAP. (2021), *Emissions Trading Worldwide: Status Report 2021*, International Carbon Action Partnership, Berlin.

ILO. (1978), *Work Study (Third Edition)*, International Labour Organisation, Geneva.

Imai, M. (1986), *Kaizen: The Key to Japan's Competitive Success*, McGraw-Hill, New York.

IPCC. (2014a), *AR5 Climate Change: Mitigation of Climate Change, IPCC's Fifth Assessment Report*, Intergovernmental Panel of Climate Change, Geneva.

IPCC. (2014b), *Glossary*, Intergovernmental Panel of Climate Change, Geneva.

IPCC. (2021), *AR6 Climate Change 2021: The Physical Science Basis*, Intergovernmental Panel of Climate Change, Geneva.

Juran, J.M. (1951), *Quality Control Handbook*, McGraw-Hill, New York.

Juran, J.M. (1988), *Juran on Leadership for Quality: An Executive Handbook*, Free Press, New York.

Juran, J.M. (1989). *Juran on Leadership for Quality: An Executive Handbook*, Free Press. New York.

Juran, J.M. (1999), *Juran on Leadership for Quality: An Executive Handbook*, Free Press, New York.

Kalkar, P., Phule, D. and Chittanand, A. (2018), 'Dealing with the Sustainability Challenge with Lean Six Sigma Framework', *International Journal of Management*, Vol. 9, No. 3, pp. 21–31.

Kano, N. (ed.) (1996), *A Guide to TQM for Service Industries*, Asian Productivity Organisation, Tokyo.

Kaplan, R. and Johnson, T. (1987), *Relevance Cost: The Rise and Fall of Management Accounting*, Harvard Business School Press, Boston, USA.

Kaplan, R.S. and Norton, D.P. (1992), 'The Balanced Scorecard – Measures That Drive Performance', *Harvard Business Review*, Boston, USA.

Kaplan, R.S. and Norton, D.P. (1996), *The Balanced Scorecard*, Harvard Business School Press, Boston.

Kaplan, R.S. and Norton, D.P. (2004), 'Measuring the Strategic Readiness of Intangible Assets', *Harvard Business Review*, February 2004.

Karlof, B and Oslblom, S (1994), 'Benchmarking', John Wiley & Sons, Chichester, UK

Kaswan, M.S. and Rathi, R. (2020), 'Green Lean Six Sigma for Sustainable Development: Integration and Framework', *Environmental Impact and Assessment Review*, Vol. 83.

Kendon, M., et al. (2021), 'State of the UK Climate 2020', *International Journal of Climatology*, Vol. 41, No. S2, pp. 1–76.

Khadri, S. (2013), 'Six Sigma Methodology for the Environment Sustainable Development', *Mechanism Design for Sustainability*, Palgrave McMillan, New York, pp. 61–76.

Kharub, M., Sharma, G. and Sahoo, S.K. (2018), 'Investigating the Cause of Poor Efficiency in Thermal Power Plant—A Six Sigma Based Case Study', *CVR Journal of Science & Technology*, Vol. 14.

Kolarik, W.J. (1995), *Creating Quality*, McGraw-Hill, New York.

Kotabe, M. and Helsen, K. (2000), *Global Marketing Management*, John Wiley & Sons, New York.

Kumar, M., Antony, J., Singh, R.K., Tiwari, M.K. and Perry, D. (2006), 'Implementing the Lean Sigma Framework in an Indian SME: A Case Study', *Production Planning and Control*, Vol. 17, No. 4, pp. 407–423.

Ladhar, H. (2007), 'Effective Lean Six Sigma Deployment in a Global EMS Environment', *Circuits Assembly*, March 2007.

Ledolter, J. and Burnhill, C.W. (1999), *Achieving Quality Through Continual Improvement*, John Wiley & Sons, Chichester.

Lee, J., Maria, L. and Vaishnav, P. (2021), 'The Impact of Climate Change on the Recoverability of Airlines Network', *Transport and Environment*, Vol. 95, Elsevier.

Lemos, M., Kirchhoff, C. and Ramprasad, V. (2012), 'Narrowing the Climate Information Usability Gap', *Nature Climate Change*, Vol. 2, pp. 789–794.

Lewin, K. (1951), *Field Theory in Social Science: Selected Theoretical Papers*, Harper-Collins, New York.

Liker, J.K. (2004), *The Toyota Way*, McGraw-Hill, New York.

Ling, R.C. and Goddard, W.E. (1988), *Orchestrating Success*, John Wiley & Sons, Chichester.

Maskell, BH (1996). 'Making Numbers Count', Productivity Press, Oregon.

Matsushita, K. (2020), *Japan's Response to the Issue of Climate Change*, Sasakawa Peace Foundation, Tokyo.

McElroy, B. and Mills, C. (2000), 'Managing Stakeholders', In Turner, R. and Simister, S. (eds), *Gower Handbook of Project Management*, 3rd edition, Gower Publishing, Aldershot.

McLymont, R. and Zuckerman, A. (2001), 'Slipping into ISO 9000: 2000', Quality Digest, USA, August 2001.

McQuater, R.E., Wilcox, M., Dale, B.G., Boaden, R.J. (1995), 'Issues and Difficulties Associated with Quality Management Techniques and Tools: Implications for Education and Training', In Kochhar, A.K. (ed.), *Proceedings of the Thirty-First International Matador Conference*, Palgrave, London.

Mengel, T. (2008), 'Outcome-Based Project Management Education for Emerging Leaders—A Case Study of Teaching and Learning Project Management', *International Journal of Project Management*, Vol. 26, No. 3, pp. 275–285.

Mintzberg, H (1990), 'Mintzberg on Management: Inside our Strange World of Organisation', The Free Press, New York.

Mizuno, S. and Akao, Y. (eds) (1994), *Qfd: The Customer-Driven Approach to Quality Planning and Deployment*, Asian Productivity Organisation, Tokyo.

Moroney, M.J. (1973), *Facts from Figures*, Penguin Books, London.
Mowris, R.J., Jones, E. and Jones, A. (2006), 'Incorporating IPMVP and Six Sigma Strategies into Energy Efficiency Program Design, Implementation, and Evaluation', 2006 ACEEE Summer Study on Energy Efficiency in Buildings.
MSP. (2007), *Managing Successful Programme*, 3rd edition, The Office of Government Commerce, London.
Negroponte, N. (1995), *Being Digital*, Hodder and Stoughton, London.
Ninerola, A., Ferrer-Ruler, R. and Vidal-Sune, A. (2020), 'Climate Change Mitigation: Application of Management Production Philosophies for Energy Saving in Industrial Processes', *Lean Manufacturing, Operational Excellence and Sustainability (Special Edition)*, Universitat Rovira i Virgili, Spain.
Oakland, J.S. (2003), *TQM: Text with Cases*, Butterworth Heinemann, Oxford.
Ohno, T. (1988), *Toyota Production System*, Productivity Press, Cambridge, Boston.
Pallant, J. (2010), *SPSS Survival Manual*, Open University Press, Maidenhead, UK.
Parasuraman, A, Zeithamel, V. and Berry, L. (1988), 'SERVQUAL: A Multi-Item Scale to Measure Consumer Perceptions in Service Quality', *Journal of Retailing*, Vol. 64, No. 1, pp. 12–10.
Pascale, R.T. (1990), *Managing on the Edge: How Successful Companies Use Conflict to Stay Ahead*, Simon and Schuster, New York.
Pearce, I. (1992), 'Thinking Green', *Environmental Mananagement and Health*, Vol. 3, No. 2, pp. 6–12.
Pew Research Center. (2021), *European Electric Vehicle Industry, 2021–2024*, Pew Research Center, Washington, DC.
PMBoK. (2006), *A Guide to Project Management Body of Knowledge*, 3rd edition, Project Management Institute, Newton Square, PA.
Polman, P. and Winston, A. (2021), *Net Positive*, Harvard Business Review Press, Boston.
Porter, M.E. (1985), *Competitive Strategy*, Free Press, New York.
Prasad, A.G., Saravanan, S., Gijo, E.V., Dasari, S.M., Tatachar, R. and Suratkar, P. (2016), 'Six Sigma-Based Approach to Optimise the Diffusion Process of Crystalline Silicon Solar Cell Manufacturing', *International Journal of Sustainable Energy*, Vol. 35, No. 2, pp. 190–204.
Pyzdek, T. (2003), *The Six Sigma Handbook*, McGraw-Hill, New York.
Retro-First. (2021), 'The Big Zero Report', *Architects' Journal*, London.
RIBA. (2021), *RIBA 2030 Climate Challenge. Version 2*, Royal Institute of British Architects, London.
Ruggles, R. (1998), 'The State of Notions: Knowledge Management in Practice', *California Management Review*, Vol. 40, No. 3, pp. 80–89.
Saaty, T.L.(1959), *Mathematical Methods of Operations Research*, McGraw-Hill, New York.
Sayle, A.J. (1991). *Meeting ISO 9000 in a TQM World*, AJSL, Great Britain.
Schmidt, S.R., Kiemele, M.J. and Bardine, R.J. (1999), *Knowledge Based Management*, Academy Press, Colorado Springs.
Schweikhart, S.A. and Dembe, A. (2009), 'Applicability of Lean and Six Sigma Techniques to Clinical and Translational Research', *Journal of Investigative Medicine*, Vol. 57, No. 7, pp. 748–755.

Senge, P. (1990), *The Fifth Principle*, Random House, Sydney.
Senge, P. (1994), *The Fifth Discipline Fieldbook: Strategies and Tools for Building a Learning Organisation*, Currency, New York.
Seyam, S., Dincer, I. and Chaab, M.A. (2021), 'Novel Hybrid Aircraft Propulsion System Using Hydrogen, Methanol, Ethanol and Dimethyl Ether as Alternative Fuels', *Energy Conversion and Management*, Vol. 238, Elsevier.
Sharp, E.B., Dailey, D.M. and Lynch, M. (2011), 'Understanding Local Adoption and Implementation of Climate Change Integration Policy', *Sage Journals*, Vol. 47, No. 3, pp. 433–457.
Shettar, M., Hiremath, P., Nikhil, R. and Chouhan, V.R. (2015), 'Kaizen: A Case Study', *International Journal of Engineering Research and Applications*, Vol. 5, No. 5, pp. 101–103.
Shingo, S. (1985), *Non-Stock Production*, Productivity Press, Cambridge, MA.
Shingo, S. (1988), *Non-Stock Production*, Productivity Press, Cambridge, MA.
Shirose, K. (1992), *TPM for Workshop Learners*, Productivity Press, Cambridge, MA.
Skinner, S. (2001), 'Mastering Basic Tenets of Lean Manufacturing-Five Ss', *Manufacturing News*, Vol. 8, No. 11.
Slack, N., Barndon-Jone, A., Johnson, R. and Betts, A. (2012), *Opeartions and Process Management*, Pearson Education, Harlow, UK.
Stamatis, D.H. (1999), *Six Sigma and Beyond*, St Lucie Press, New York.
Statistica Research Department. (2021), *Share of Electrified Railroad Routes in European Countries* (Edited by Salas, E.B.), Statistica Research Department, Hamburg.
Sterman, J., Repenning, N.P. and Kofman, F. (1997), 'Unanticipated Side Effects of Successfull Quality Programmes: Exploring a Paradox of Organizational Improvement', *Management Science*, Vol. 43, No. 4, pp. 503–521.
Stern, N. (2006), 'Stern Review: The Economics of Climate Change', *HM Treasury*.
Sujova, A., Simanova, L. and Marcinekova, K. (2016), 'Sustainable Process Performance by Application of Six Sigma Concepts: The Research Study of Two Industrial Cases', *Sustainability*, Vol. 8, No. 3.
Treacy, M. and Wiersema, F. (1993), 'Customer Intimacy and Other Value Discipline', *Harvard Business Review*, January–February 1993, pp. 84–93.
Tuckman, B.W. (1965), 'Development Sequence in Small Groups', *Psychological Bulletin*, Vol. 63, pp. 384–399.
Tuckman, B.W. (2001), 'Development Sequence in Small Groups', *Psychological Bulletin*, Vol. 63, No. 6, pp. 384–399.
Turner, J.R. and Cochrane, R.A. (1993), 'The Goals and Methods Matrix', *International Journal of Project Management*, Vol. 11, No. 2.
Turner, J.R. and Simister, S.J. (2000), *Gower Handbook of Project Management*, Gower Publishing, Aldershot.
UK Carbon Trust. (2008), *Carbon Trust Standard 2008*, Carbon Trust, London.
UNEP. (2019), *Emissions Gap Report 2019*, United Nations Environment Programme, New York.
United Nations. (1987), *Report of the World Commission on Environment and Development*, UN General Assembly Resolution, December 11, 1987, New York.

United Nations Foundation. (2020), *Key Findings of the Sustainable Development Goals Report in 2020*, United Nations, New York.

University of Michigan. (2021), *2021 Factsheets Collection*, School of Environment & Sustainability, University of Michigan, Ann Arbor.

US EPA. (2019), *Global Greenhouse Emission Data*, United States Environment Protection Agency, Washington, DC.

Wallace, T.F. (1990), *MRPII: Making It Happen*, John Wiley & Sons, Chichester.

Waller, D.L. (2002), *Operations Management*, Thomson Learning, London, p. 157.

Warner, P. (2019), 'Norwegian Cruise Line Hurtigruten Trials Biodiesel on Cruise Ship', *Biofuels International*, London, October 31, 2019.

Watt, I. (2012), 'Lessons from Corporate Claims of Carbon Neutrality', *Forum for the Future*, November 7, 2012, *Major Projects Association*, London.

Wei, Y.M., Kang, J.N., Liu, L.C., et al. (2021), 'A Proposed Global Layout of Carbon Capture and Storage in Line with a 2°C Climate Target', *Nature Climate Change*, Vol. 11, pp. 112–118.

Weinstein, L.B., Castellano, J., Petrick, J. and Vokurka, R.J. (2008), 'Integrating Six Sigma Concepts in an MBA Quality Management Class', *Journal of Education for Business*, Vol. 83, No. 4, pp. 233–238.

Welch, J. and Byrne, J.A. (2001), *Jack: What I Have Learned Leading a Great Company and Great People*, Headline Book Publishing, London.

Wessel, G. and Butcher, P. (2004), 'Six Sigma for Small and Medium-Sized Enterprises', *TQM Magazine*, Vol. 16, No. 4, pp. 264–272.

Whitlock, R. (2020), 'What Place for Hydrogen? An Interview with Professor Armin Schnettler of Siemens', *Renewable Energy Magazine*, London, April 2020.

Wild, R. (2002), *Operations Management*, Continuum, London.

Womack, J.P. and Jones, D.T. (1998), *Lean Thinking*, Touchstone Books, London.

Womack, J.P., Jones, D.T. and Roos D. (1990), *The Machine that Changed the World*, Touchstone Books, London.

Yu, J., Williams, E., Ju, M. and Yang, Y. (2010), 'Forecasting Global Generation of Obsolete Personal Computers', *Environmental Science and Technology*, Vol. 44, No. 9, pp. 3232–3237.

Index

Page numbers in *italics* indicate a figure and page numbers in **bold** indicate a table on the corresponding page.

B

D

G

J

K

M

T

Printed in the United States
by Baker & Taylor Publisher Services